FRONTIERS IN GEOCHEMISTRY: GLOBAL INORGANIC GEOCHEMISTRY

FRONTIERS IN GEOCHEMISTRY: GLOBAL INORGANIC GEOCHEMISTRY

Konrad Krauskopf Volume 1

International Book Series, Vol. 5

W. G. Ernst, editor

Bellwether Publishing, Ltd. for the
Geological Society of America

Published by

Bellwether Publishing, Ltd.
8640 Guilford Road, Suite 200
Columbia, MD 21046-2655
http://www.bellpub.com
for the Geological Society of America

Exclusively marketed throughout the world by:

Geological Society of America
P. O. Box 9140
Boulder, CO 80301-9140
email: pubs@geosociety.org
or order from the GSA Bookstore on the World Wide Web at
http://www.geosociety.org/bookstore.

Library of Congress Control Number: 2002115508
ISBN 0-9665869-4-8

Printed by The Sheridan Press, Hanover, Pennsylvania.

FRONTIERS IN GEOCHEMISTRY: KONRAD KRAUSKOPF
VOLUME 1. GLOBAL INORGANIC GEOCHEMISTRY

W. G. Ernst, editor

Contents

Editor's Preface

A symposium entitled "Frontiers in Geochemistry" was held at Stanford University on December 11 and 12, 1999, in honor of Professor Konrad B. Krauskopf. The technical meeting and papers resulting from it, published in separate installments in Volumes 41–43 of *International Geology Review*, are an attempt to recognize Konnie's lifetime of extraordinary achievement in both geology and geochemistry. This remarkable scientist received his Ph.D. in Chemistry at the University of California, Berkeley in 1934, and a second doctorate in Geology at Stanford University in 1939. This unusually broad but doubly deep educational background, plus concurrent service as university instructor in the physical sciences, equipped Konnie for a scholarly career characterized by breathtaking interdisciplinary contributions. He has made numerous seminal contributions in all aspects of academic performance that we deem critically important—research, instruction, and public/professional service. And, he is still at it.

Konnie has published a diverse set international-quality investigations, broadly arching across the fields of hard-rock geology, petrology, geochemistry, and mineral deposits. Detailed studies include illuminating the parageneses of granitoids and basement terranes in the Pacific Northwest, the volcanic eruptions of Parícutin in the Transmexican Volcanic belt, and the regional petrologic evolution of coastal Norway. In parallel, he has generated both mineral deposit and general geologic maps for the California Division of Mines and the U. S. Geological Survey, chiefly in the Sierra Nevada and the White-Inyo ranges of eastern California.

He has quite literally written the book(s) on the application of the principles of physics and chemistry to the Earth, having provided geoscientists with pioneering, discipline defining texts in geochemistry and physical geology over a span of five decades. These forward-looking works have focused on and illuminated the fundamental chemical and physical foundations of the Earth sciences. Special emphases have included elucidation of aqueous solution-metal complex equilibria as well as thermodynamic applications to solid-melt-fluid partitioning, published at a time when most Earth scientists were mapping quadrangles (he did that too, having published seven quadrangle maps!). Acclaimed books include: *The Third Planet*; *Introduction to Geochemistry*; *Fundamentals of Physical Science*; *The Physical Universe*; *Radioactive Waste Disposal and Geology*; and *Introduction to Physics and Chemistry*. Most of these texts have run through several (up to eight) editions.

In terms of service to the nation beyond military geology in WW II, Krauskopf served for more than a decade as a member, then chair, of the U. S. National Research Council Board on Radioactive Waste Management, and was responsible for an outstanding, problem-defining and -quantifying National Academy report on the subject.

Konnie has received a great many honors. He is a member of the National Academy of Sciences, and the American Philosophical Society. He was president of the American Geological Institute in 1964 (and received its Ian Campbell Medal in 1984, and its Legendary Geoscientist Award in 2000), the Geological Society of America in 1967 (and was awarded its Arthur L. Day Medal in 1961), and the Geochemical Society in 1970 (and received its V. M. Goldschmidt Medal in 1982). He was honored with the Mineralogical Society of America Distinguished Public Service Award in 1994.

I can think of no other geochemist who has so faithfully served the Earth science profession in such far-ranging ways, being enormously impactful in all of them, as geologist, geochemist, and science/technology advisor to the Nation. This two-volume work, *Frontiers in Geochemistry*, is an insufficient tribute to the legendary scientific accomplishments of Konnie, but it is a start!

— W. G. Ernst

GLOBAL GEOCHEMISTRY

A Tale of Two Planets

DON L. ANDERSON

California Institute of Technology, Seismological Laboratory 252-21, 1200 E. California Boulevard, Pasadena, California 91125

Abstract

Most of Earth's volcanism occurs at plate boundaries—mid-ocean ridges and island arcs. Most of the ^{3}He and other noble gases also are expelled from the interior at these locations. A small amount, compared to the global budget, of magma (<10%) and ^{3}He (<<1%) escapes at mid-plate volcanoes, the so-called "hotspots." Locations of hotspots are controlled by stresses and cracks, not by deep, narrow mantle upwellings. Most, or all, of Earth's volatile inventory appears to have been brought in by a late veneer. The high ^{3}He content and high He/Ne and He/Ar ratios of mid-ocean ridge basalts, plus the excess ^{129}Xe and ^{136}Xe, suggest that the ridge basalt reservoir is relatively undegassed and that there is no primordial undegassed reservoir. The two prevailing views of the origin of Earth are summarized in the context of this information.

Introduction

It was the best of times, it was the worst of times,
it was the age of wisdom, it was the age of foolishness,
it was the epoch of belief, it was the epoch of incredulity,
it was the season of Light, it was the season of Darkness,
it was the spring of hope, it was the winter of despair.

* * * * * * *

I have known a vast quantity of nonsense…
Don't trust the conventional idea.

—Charles Dickens
A Tale of Two Cities

KARL POPPER advocated a critical discussion of alternative views in order to advance science. This was also common in ancient Greece. The principle of multiple working hypotheses has been common since the early days of geology. Popper thought that the scientific community ought to be an open society in which no theory, however dominant, is ever sacred. Thomas Kuhn viewed *scientific normalcy* as a closed society of closed minds; a strenuous and devoted attempt to force nature into the conceptual boxes supplied by textbooks and professional education. These two views of the philosophy of science are about as disparate as views on the origin and evolution of our planet.

There are two planets being discussed today by scholars of the Earth. One is the world of the isotope geochemists, and one is the world of the physicists. One formed in the "worst of times" (violent accretion), and one in the "best of times" (cool tranquil conditions).

Geochemistry has its roots in alchemy and magic and transmutation. The *Primordial Earth* and the *Philosopher's Stone* have much in common, as the secrets of the universe are contained in each. Paradoxes and secrets have always been a part of alchemy. Physics has its roots in natural philosophy, mathematics, astronomy, and logic. The ideal Earth that exists in the imagination is much more real than our perceived planet, for it must obey the laws of physics, and there are no paradoxes.

Two Views of Earth

The Earth of the isotope geochemist has a *Primordial* nucleus; the 650 km discontinuity of the seismologists (Anderson, 1967) is assumed by alchemists to separate the *Primordial Earth* from the *Degassed Upper Mantle* (DUM), the DUM-reservoir, and the source of mid-ocean ridge basalts (MORB) (Porcelli and Wasserburg, 1995). Ocean islands are pipelines to *Primordial Earth*. A high ^{3}He/^{4}He ratio, the *Magic Ratio*, of ocean-island basalts (OIB) implies excess ^{3}He, the primordial isotope; this implies a Primordial undegassed reservoir, which in turn implies a deep isolated reservoir (lower mantle). The crust formed from DUM, the mantle above the 650 km discontinuity, and the rest of the mantle is pristine (O'Nions and Tolstikhin, 1994). When rocks melt, the gases are immediately put into the atmosphere. There is no pristine heat (Kellogg and Wasserburg, 1990).

The Earth that is built by physics and cosmology has a solid-solid phase change, not a chemical change, at 650 km (Anderson, 1967), and the whole Earth has been through the melting, vaporizing, and drying process of accretion (Anderson, 1989; Gaffey, 1997). The 650 km discontinuity in the mantle occurs because the major minerals in the transition zone, $(Mg,Fe)_2SiO_4$ (spinel) and $(Mg,Fe)SiO_3$ (majorite), become unstable and collapse to a mixture of $(Mg,Fe)SiO_3$ (perovskite) and $(Mg,Fe)O$ (magnesiowüstite). These solid-solid phase changes typically have negative Clapeyron slopes that inhibit the transfer of material across the boundary. There is no evidence from seismology or mineral physics that this is a major chemical change. There may be a chemical change deeper in the mantle, nearer to 1000 to 1300 km (Wen and Anderson, 1997a, 1997b). The volatile elements including water and the noble gases were mainly added to the planet toward the end of the accretion process, the late veneer (Wänke et al., 1984; Anderson, 1993; Javoy, 1998). A late veneer of only 1% of the mass of Earth brings in all the volatiles that are known to exist and the excess siderophiles that are not in the core. The low totals of ^{3}He and non-atmospheric heavy noble gases in OIB are taken at face value. They imply that high *Magic Ratios* are the result of low ^{4}He, the daughter isotope of U and Th (*UTh*), and this implies a refractory and secondary non-fertile source, such as the residue of mantle after basalt extraction (Anderson, 1998a, 1998b). The low He/Ne and He/Ar ratios of OIB, including the ones with high *Magic Ratios*, show that melt-vapor equilibrium was involved at some stage in their genesis, not just crystal-melt separation. The atmospheric signature of the heavy noble gases in OIB suggests contamination by near-surface material (Farley and Neroda, 1998), as do the oxygen isotopes and oxidation state (Anderson, 1996).

There are many paradoxes associated with the *Standard Models* of whole-Earth chemistry (Anderson, 1998a, 1998b). These include the Lead paradox, the Helium heat-flow paradox, the ^{40}Ar paradox, the Xenon paradoxes, and the missing-^{3}He paradox. Many of these paradoxes can be resolved if: (1) there are regions of Earth that are no longer being sampled by igneous processes; (2) high *Magic Ratios* result from low ^{4}He, not high ^{3}He; (3) recycling into the shallow mantle is important; (4) most of Earth has been processed, degassed, and differentiated, mainly during accretion; and (5) gases degassed from magmas can be trapped in the shallow mantle.

The amount of ^{40}Ar in the atmosphere is about two-thirds of all the ^{40}Ar produced over the entire history of the planet, much greater than produced by DUM (Anderson, 1989). Most of Earth must be fairly efficiently degassed to explain this result and the abundance of other key elements in the crust. Since perfectly efficient degassing is unlikely, there is little room for Primordial mantle. In the DUM model, the ^{40}Ar of the degassed mantle, very efficiently degassed, is supplemented by gas leaking from the Primordial undegassed reservoir. This is an amendment, actually a reversal, of the *Standard Model*, which states that all the oceans, gas, and crust came out of the upper 30% of the mantle, based on old abundance tables of the crust and of primordial mantle. The lower mantle, the remaining 70%, must be primordial and undegassed. We now know that about 70% of the mantle must have been processed (Anderson, 1989). The tables have been turned, but the *Standard Model* has remained, with some immunizing auxiliary hypotheses.

Only 10% or so of the ^{4}He produced by *UTh* is accounted for (O'Nions and Oxburgh, 1983). The rest remains in Earth or the amount of *UTh* has been grossly overestimated. These are just a few of the paradoxes faced by modern mantle geochemistry.

Volcanoes

There are two current world views on "midplate" volcanoes, which are as different as they can be. In one, volcanoes are the result of thermal instabilities at the base of the planet—the pot-on-the-stove metaphor. "This must be so since these volcanoes emit 'excess ^{3}He', and this must be so because the ^{3}He/^{4}He ratio is high."

In the other view, when a crack forms, something must come up. Buoyant magma forces its way out when the lithosphere is put into tension. "High ^{3}He/^{4}He ratios imply low ^{4}He, and therefore low U and Th, and therefore a shallow refractory reservoir." The mantle is hot, and no excess heat is required to make a volcano.

The Noble Gases

Mid-ocean ridge basalts (MORB) have He isotope ratios ranging from 6 to 16 Ra, where Ra is the atmospheric ratio (not Radon). The distribution is strongly peaked at 8 Ra (Farley and Neroda, 1998). The general interpretation of this distribution is that the MORB mantle carries rather uniform ratios in the range from 7 to 9 Ra, but the MORB source may be modified as a result of input from other sources. This other source is usually assumed to be a hotspot or a plume or *Primordial Undegassed* lower mantle.

If one does not discriminate against certain types of spreading ridges because of some bias about their origin, then one obtains the whole range of terrestrial $^3He/^4He$ ratios by looking at all the spreading ridges on Earth, regardless of age, elevation, or perceived parentage. The North Atlantic, Iceland, Red Sea, Ethiopia, and East African Rift are new ridges, and have some high $^3He/^4He$ grains in their rocks. Backarc basins are cut off from the lower mantle by slabs, but are spreading ridges and have high R. The Easter, Galapagos, Azores, Shona, and Bouvet segments of the mid-ocean ridge system are complex, but they sample the shallow mantle, albeit in a different way than mature linear ridges.

He isotope ratios in ocean-island basalts vary from ~4 to ~32 Ra. Small seamounts and mantle xenoliths also exhibit a large range of values. The distinguishing characteristic of OIB samples, particularly those with high He ratios, is the very low absolute abundance of He and the other noble gases, and the presence of atmospheric contamination for the heavy noble gases (Kurz et al., 1998; Farley and Poreda, 1993).

It is usually assumed that the sources of MORB and OIB noble gases are quite different, with OIB being derived from a primitive, undegassed mantle source. This leaves as a series of paradoxes the low He content, the low He/Ar and He/Ne ratios of OIB (Anderson, 1998a, 1998b), and the excess ^{129}Xe and ^{136}Xe in MORB (Kurz et al., 1998), which are from ancient, short-lived isotopes. These properties suggest that it is the MORB source that is the more ancient and undegassed. OIB samples with very high Ra values have now been recovered from 5 km depth in the ocean, and these also have very few ^3He atoms compared to MORB samples (Valbracht et al., 1997). One can no longer argue that OIB glasses are more degassed than MORB because of the low pressure of emplacement.

Most basalts probably are mixtures or blends from a variety of depths. Seismic tomography suggests that ocean ridges are fed from at least 200 to 300 km depth (Anderson et al., 1992). The upper 100 to 200 km of the mantle is contaminated by dehydration of subducted material and the trapping of residual, small-degree melts (Anderson, 1996). The upper 100 km of our planet is composed of buoyant crust and buoyant depleted peridotite, products of melt extraction. Melts can pond and evolve beneath these buoyant and strong lids. When the outer shell of Earth is under horizontal compression, it is strong. However, when it is under extension, it is weak, and buoyant magma can rise to the surface, interacting with the various layers that it penetrates. Gases liberated by early cracking may be quite different from gases provided by the deeper, gas-rich MORB source. When the precursors to MORB reach degassing depths, the nature and quantity of gases sampled or expelled change drastically. Any magma blends will be dominated by the high-He MORB magmas.

Mid-ocean ridge basalts represent large-degree melts and large volumes of the mantle must be processed in the continuous process of seafloor generation. The sampling of mantle by a ridge can be compared to a giant steam shovel. In contrast, seamounts and islands ladle up their samples, whereas xenoliths are sampled by spoons and tweezers.

It is inevitable that these different kinds of samplings result in different distributions of trace elements and isotopes. A given stretch of ocean ridge may be the result of blending materials from hundreds of kilometers in both depth and distance. Seamount eruptions may sample only the top 10 km of the mantle. The first rift, prior to drift, may do the same.

Individual small samples exhibit a large range in values, and the variance among small samples is large. By averaging or blending many samples, the distribution becomes more peaked, and the variance is much smaller. The median value of samples from islands and other "hotspot" locations such as Ethiopia and Yellowstone is often very close to the MORB median value for R, suggesting that the samples may be derived from the same population, with different sampling strategies.

Many of the world's high $^3He/^4He$ ratio "hotspots" are in the Pacific. The South Pacific "hotspots" are anomalous compared to predictions of the hotspot and plume theories with respect to heat flow, swells, and age progression (Okal and

Batiza, 1987). There is also a tectonic context for each of these islands that suggests that high R may involve new cracks in the outer shell (McNutt et al., 1997; Favela and Anderson, 1999).

For example: (1) Samoa is in an area of high bending stresses caused by change in dip and strike of the Pacific plate boundary; (2) the Lau Basin is an unstable ridge system; (3) French Polynesia is in a region where volcanism is controlled by stress of the lithosphere (McNutt et al., 1997); (4) Easter Island is in a region of microplates and ridge propagators, new cracks in the plate; (5) the Juan Fernandez and San Felix Island chains are offshore from the volcano gap in the Andes, and may be tears associated with flat subduction; (6) the Galapagos Islands are on a young plate being torn apart by diverging slab pulls; and (7) the Loihi seamount is being built on a loading crack on the flank of Mauna Loa.

Mature volcanoes approach MORB-like values for their helium ratios. Thus, there appears to be an element of lithospheric or shallow mantle control on the evolution of the $^3He/^4He$ ratio in oceanic islands. Ridges sample deeper and broader, and later, than initial cracks. Early ridges tend to have high $^3He/^4He$ ratios. Inactive or dying islands and ridges tend to have low R.

The Two Planets

There is a stark contrast in the ways different disciplines view a planet. Sometimes one learns best from metaphors or parables. The stilted prose, obscurity, and subtlety of scientific writing sometimes hides important conflicts. Jonathan Swift and Miguel de Cervantes recognized the need for satire in exposing the foibles of man. They effectively made their points. The *Looking Glass Worlds* of modern deep Earth science are best contrasted in short stories. The two worlds are as different as Oz and Wonderland, and we are Alice in Oz.

The Pristine Planet

The story begins today. One can best understand this planet by looking at it as it is. After all, the adult reflects the integrated effects of an entire life and the traumas of birth and childhood are in the distant past. Alchemists are best qualified for this chore, since their isotopes do this integration and tell us the whole history of the world. The wizards have long been amused by the coincidence of the world's heat flow and that generated by *Primordial Matter*, the building blocks of worlds. From the sky, there falls yet today some remnants of the early heavens, wealthy in carbon and water and other vapors and the elements named after the gods Uranus and Thor, and their many daughter elements, some named after Helios and Plumbos. The Uranium and Thorium and Potassium of the world fuel the internal furnaces, and the heat escapes through the surface. About a part in three of the world's tenuous and vaporous elements are in the fluid envelopes and crust (according to ancient parchments), proving that two parts in three of the planet are pristine. The heat flow coincidence, although not perfect, proves a steady situation—generation equals liberation. This world must have started cold, since there is no need for primordial heat from a cooling crystallizing orb, the world of Kelvin the Lord. Geophysics and Dante tell us that the inner world is composed of spheres, separated by barriers, each one different. At least, that is the interpretation provided by alchemy.

These nether regions, however, are not sins of mankind but spheres of different essence. The outer third of this world has yielded up its essence to make the atmo-sphere, the hydro-sphere, and crusto-sphere, the three main spheres of outer Earth. These were stripped out of the para-sphere, the all-around nearby region under the world we stand on. Alchemists call this the *Depleted*, or *Degassed* shell, a pale remnant of its former self, a mere cinder. This *Uppersphere*, or *Parasphere* (see the glossary presented in the Appendix), fuels the furnaces under the oceans where new crust exudes from the interior, surfacing the bottoms of the watery parts of the planet.

The deeper world retains the oceans and vapors and is *Pristine* in all respects. It is called *Primordial World*. It is undepleted, undegassed, a sample of the way things were. It is isolated from the upper world, except for narrow tubes that rise up to islands in the middle of the seas.

The stuff of islands is not like ordinary stuff, the stuff of ocean floors, beneath the loose debris. It is as if the molten rocks of the seafloor have been sprinkled with bits of crust and seafloor debris and seawater, and other things made in the outermost reaches. The molten rocks of islands also have bubbles of the rarest of gases. The heavier of these gases are much like those in the atmo-sphere or those in seawater. All of these materials must have come up the narrow tubes from *Primordial World*, the nether depths, to the remote islands. The islands are built on the crust and sediments and debris, in the middle of the sea, but it is known that the molten rocks come from Primordial matter because of the most ephemeral of the rare gases, the most noble of them all. It is a curious thing that although islands are full of the matter surrounding them—the air, the sea, their foundation—it is the deepest world that provides these constituents. *Alchemy* thrives on riddles, puzzles, mysteries, and paradoxes.

Island rocks must come from nether depths, through narrow tubes, because the ratio of tri-helios (^{3}He) to quatra-helios (^{4}He), that is the *Magic Ratio* (^{3}He/^{4}He), is greater than in the vapors of the atmosphere. Rocks from the seafloor, the *Morbish* rocks, also have *Magic Ratios* higher than air, but it is known that these rocks are from the upper degassed degenerate part of the world.

^{3}He is *Primordial Matter*. It must have been in the world from the beginning of time. ^{4}He is *Common Matter*; it is made from *Uranium* and *Thorium* (*UTh*), just like much of the heavy *Lead* of this world. These elements are not Primordial, they are daughters and granddaughters of the godly elements, progeny of Thor and Uranus.

Since the *Magic Ratio* of some island rocks is great, they *must* have excessive *Primordial ^{3}He*. This is the mantra of alchemy. Since island rocks have less ^{3}He than *Morbish* rocks, they must have lost it prior to eruption; otherwise it is a Paradox, since it is known that they come from a *Pristine Undegassed* world, deep inside the planet.

Since it is known that high *Magic Ratios* mean excess of the Primordial isotope of Helios, despite low actual quantities, the island rocks *must* come from *Undegassed Pristine* depths. Therefore, narrow tubes must extend to below the barrier between upper degenerate and lower pristine worlds.

The Logic is clear. High ^{3}He/^{4}He means high ^{3}He means rocks from undegassed depths means *Primordial Matter* means deepest depths, the pristine remnant of the cold beginnings of the world. The always low ^{3}He, when excess ^{3}He is expected, means that these rocks always lose their gases before they are seen.

The *Primordial World* has an abundance of *UTh* and therefore abundance of the daughter ^{4}He. If there is abundance of ^{4}He and if the *Magic Ratio* is nevertheless high, there must be abundance of ^{3}He and other vapors even less fragile than the Helios elements. When one does the sums, the *Primordial World* at the end of the straw must look like the most pristine of the stones from the sky. This world is a gas-rich world, with only a fraction of the airs and oceans liberated. These airs and oceans have been in the world since the beginning and have stayed in because of the easy childhood and the balmy beginnings. Other worlds may have had traumatic births and steamy youths and are cinders in space, but not this one. Other worlds may have been boiling and obtained their vapors at the end from icy rocks and stars with tails, but not this one. This one is the *Primordial World*, its essence is locked in the depths; the watery envelopes are just a taste of things to come. The *Prima facies* of this world can yield multiple oceans through its forty straws. The floods of legend are nothing compared to the potential of this watery world locked in the depths.

All the deep heat of this world comes from degeneracy of the godly elements and lowly potash. No Primordial heat survives. A caloric of heat in the depths becomes instantly a caloric of heat at the top. In all things, there is a balance.

The Pummeled Planet

Oedipus asked, "How can I not be glad to know my birth?" History can also be written, as Horace says, "ab ovo." This planet can best be understood by examining its birth traumas and its youth, for these early experiences control its life ever after. Philosophers of Physiks are best qualified for this task since they always start by "considering a spherical whatever," which is a good place to start for ovos and orbs. This planet formed in a hot place, near Helios, and grew hotter as it was pummeled, for it is known that gravity and speed turn into heat. This planet grew from cinders and burned as it grew. It is a cinder in space. As Brothers Dante (1555) and Burnet (1691) knew, the primordial nether spheres were sorted by gravity, with iron at the bottom and dross at the top. Gases from this celestial furnace rose and escaped, for the planet was too small and too hot to retain the vapors. Only the heaviest elements remained in this world—earthen and ferrous and sulfurous. Even rocks boiled. Oceans of magma covered the land. Aeons of cooling before the first rock, and stable dross. Iron dregs were already in the netherest depths. The shoalest third of the siliceous part of this planet cooled to crystalline peridot. Below, crushed by pressure, a new crystalline form, the same but denser—a new phase. Not a boundary between two kinds of different matter, or between Primordial stuff and non-, or gas and no-gas, but a permeable wall separating burnt and dry processed planet. A dry, hot world, already turned inside out by birth and afterbirth pains and constant pummeling, resetting its plutonic clock.

An almost finished planet, a cinder, a vapor-free planet; sweeping up debris. Snowballs and watery rocks perturbed into this world by the Jovial Giants, the gas balls, helped cool the world. Life-giving juices only at the end and only on top. Hydrous, carbonaceous, and noble vapors, finally welcomed by this cooling world. The gases came from space, not the nether depths. Regassing, not degassing, is the rule of this world. Ingress from above watered the shallow depths. The patina of the world, the late veneer, surfaced the planet—the icing on the cake, a layered cake built on a metal platter.

On a hot growing world, light stuff rises. On a cold cooling world, the vapors condense and soak into rocks and sink back down, but mostly not. As Earth cools it holds more of its recycled water (and carbon) from above.

This world's supply of the noblest gases rises at ridges, part of the late veneer acquired when the planet was cold and accepted the gentle rain from heaven. The Magic Ratio of Morbish rocks is high, eight times air, with enormous amounts of helium, neon, and other nobles. The amount of ^{4}He, and the nature of xenon, indicate ancient sources.

The remote islands leak a little ^{3}He and little other ancient gases, but lots of air-like heavy nobles. The high Magic Ratios of remote island rocks indicate little of the parent U and Th in the island sources—sources that lost their *UTh* when magmas left and buoyant residues with bubbles remaining, floated. Almost no nobles are exhaled at islands—only that from ancient bubbles, trapped in the upper shells of the world, sullied by air and ocean, the *Perisphere* (see glossary in Appendix). This is a pummeled, dry, separated world, its adult state forever reflecting a turbulent, refractory youth, but blessed by its soothing veneer of recently acquired fluids. The nether depths are forever isolated from the top, their secrets locked in crystalline dense phases, long since dejuiced, and sealed from the top by gravity. The watery envelopes of this world are most of what there is or ever was. This is not a gas-rich world. This is not the *Primordial World* of myth and legend.

In this cooling planet, there is no Primordial virgin unprocessed matter, but there is Primordial heat. Fully half of the heat flowing out of the top is leftover pristine heat. Only half is daughter heat from the godly elements. Therefore, there is much less *UTh* in this world and much fewer gaseous daughter atoms.

REFERENCES

Anderson, D. L., 1967, Phase changes in the upper mantle: Science, v. 157, p. 1165–1173.

______, 1989, Theory of the Earth: Boston, Blackwell Sci. Publ., 366 p.

______, 1993, ^{3}He from the mantle; primordial signal or cosmic dust?: Science, v. 261, p. 170–176.

______, 1996, Enriched asthenosphere and depleted plumes: INT. GEOL. REV., v. 38, p. 1–21.

______, 1998a, The Helium Paradoxes: Proc. National Acad. Sci. USA, v. 95, p. 4822–4827.

______, 1998b, A model to explain the various paradoxes associated with mantle noble gas geochemistry: Proc. National Acad. Sci. USA, v. 95, p. 9087–9092.

Anderson, D. L., Tanimoto, T., and Zhang, Y.-S., 1992, Plate tectonics and hotspots: The third dimension: Science, v. 256, p. 1645–1651.

Burnet, T., 1691, Sacred Theory of the Earth (R. Norton, ed.): London.

Dante Alighieri, 1555, La divina comedia (transl. by M. B. Anderson): New York, Heritage Press, 492 p.

Farley, K., and Neroda, E., 1998, Noble gases in the Earth's mantle: Ann. Rev. Earth Planet. Sci., v. 26, p. 189–218.

Farley, K., and Poreda, R. J., 1993, Mantle neon and atmospheric contamination: Earth Planet. Sci. Lett., v. 114, p. 325–339.

Favela, J., and Anderson, D. L., 1999, Extensional tectonics and global volcanism: Submitted, November 1999.

Gaffey, M. J., 1997, The early solar system: Origins of Life, v. 27, p. 185–203.

Javoy, M., 1998, The birth of the Earth's atmosphere: Chem. Geol., v. 147, p. 11–25.

Kellogg, L. H., and Wasserburg, G. J., 1990, The role of plumes in mantle helium fluxes: Earth Planet Sci. Lett., v. 99, p. 276–289.

Kurz, J., Staudacher, T., and Allègre, C., 1998, Plutonium-fission xenon found in Earth's mantle: Science, v. 280, p. 877–880.

McNutt, M. K., Caress, D. W., Reynolds, J., Jordahl, K. A., and Duncan, R. A., 1997, Failure of plume theory to explain midplate volcanism in the Southern Austral Islands: Nature, v. 389, p. 479–483.

Okal, E. A., and Batiza, R., 1987, Hotspots: The first 25 years, *in* Keating, B. H., et al., eds., Seamounts, islands, and atolls: Washington, DC, Amer. Geophys. Union, Geophys. Monog., v. 43, p. 1–11.

O'Nions, R., and Tolstikhin, I., 1994, Behavior and residence times of lithophile and rare gas tracers in the upper mantle: Earth Planet. Sci. Lett., v. 124, p. 131–138.

O'Nions, R. K., and Oxburgh, E. R., 1983, Heat and helium in the Earth: Nature, v. 306, p. 429–431.

Porcelli, D., and Wasserburg, G. J., 1995, Mass transfer of helium, neon, argon, and xenon through a steady-state upper mantle: Geochim. et Cosmochim. Acta, v. 59, p. 4921–4937.

Valbracht, P., Staudacher, T., Malahoff, A., Moreira, M., and Allègre, C., 1997, Noble gas systematics of deep rift zone glasses from Loihi Seamount, Hawaii: Earth Planet. Sci. Lett., v. 150, p. 399–411.

Wänke, H., Dreibus, G., and Jagoutz, E., 1984, Mantle chemistry and accretion history of the Earth, *in* Kroener, A., Hanson, G. N., and Goodwin, A. M., eds., Archean geochemistry: The origin and evolution of the Archean continental crust: Berlin, Springler-Verlag, p. 1–24.

Wen, L., and Anderson, D. L., 1997a, Slabs, hotspots, cratons, and mantle convection revealed from residual seismic tomography in the upper mantle: Phys. Earth Planet. Inter., v. 99, p. 131–143.

Wen, L., and Anderson, D. L., 1997b, Layered mantle convection: A model for geoid and topography: Earth Planet. Sci. Lett., v. 146, p. 367–377.

APPENDIX. Glossary

para-: Alongside, as in *paradigm*; beyond, as in *paradox*; wrongly, harmfully, as in *paralysis*.

peri-: About, around, encircling, close at hand, adjacent, as in *periapt*—an amulet worn as a protection against mischief, or *perimeter*. Also reminds one of *peridot*, the most common upper mantle mineral.

perisphere: A shallow region of the mantle defined on the basis of location, history, and chemistry, in contrast to *lithosphere*, *asthenosphere*, and *upper mantle*. Not to be confused with *parasphere*.

The shallowest mantle appears to be *enriched*, not *depleted*. This is probably a result of *metasomatism*, *slab dehydration*, and *trapped melts*.

The upper layers of Earth in geochemical models are referred to as "asthenosphere," "lithosphere," "upper mantle," "convecting mantle," and so on. These are physical subdivisions, and the chemistry is unknown. Subdivisions needed in geochemistry are "reservoirs" or "components," and these need bear no relation to seismic or physical subdivisions. The mid-ocean ridge basalt (MORB) reservoir is often referred to as Depleted Upper Mantle (DUM). It is *depleted* in incompatible trace elements but is *fertile* and gas rich. Barren peridotites, buoyant because of basalt (garnet) extraction, are *depleted* and *infertile*. In all geochemical models, the MORB reservoir is thought to be shallow *(alongside)* since MORB seems to be widely available (ridges). Since *asthenosphere* means "weak" and *lithosphere* means "strong," and upper mantle refers to a depth interval, not a reservoir, the term *parasphere* may be appropriate for the inferred characteristics of this reservoir. It should be noted that MORB is only common at mature spreading centers, not at newly opening cracks, so there may be a shallow *enriched* reservoir that can contaminate MORB in the initial stages of continental break-up and seafloor spreading.

The Solar System's Earliest Chemistry: Systematics of Refractory Inclusions

Trevor R. Ireland[1]

Department of Geological and Environmental Sciences, Stanford University, California 94305-2115

and Bruce Fegley, Jr.

Planetary Chemistry Laboratory, Deptartment of Earth and Planetary Sciences, Washington University, St. Louis, Missouri 63130

I was going to church with Graeme, he was only a little fellow then, and I heard a rumbling noise. I looked up and saw smoke and wondered what the dickens it was.

— Mr. Reg Orr, Murchison, Australia

Abstract

Refractory inclusions, or CAIs (calcium-aluminium–rich inclusions) are a unique ingredient in chondritic meteorites. As the name suggests, they are enriched in refractory elements, essentially reflecting a condensation sequence of phases from a cooling gas of solar composition. However, the widespread preservation of diverse isotopic anomalies is not compatible with the inclusions having been in a gaseous form. Rather, the CAIs appear to represent mixtures of condensate and refractory residue materials. The condensates formed from cooling solar gas and fractionation of that gas produced variations in the abundances of refractory elements according to volatility. Solar condensate has isotopically normal Ca and Ti isotopic compositions and has $^{26}Al/^{27}Al$ of the canonical value for the solar system at 5×10^{-5}. Residues of material falling in toward the Sun are probably aluminous oxides such as corundum and hibonite, and preserve diverse Ca and Ti isotopic anomalies. Meteoritic inclusions from the Murchison meteorite show the best polarization of these components. Spinel-hibonite-perovskite inclusions (SHIBs) predominantly have normal Ca and Ti isotopes, $^{26}Al/^{27}Al$ at 5×10^{-5}, and ultrarefractory fractionated REE patterns. Single hibonite crystal fragments (PLACs) have diverse Ca and Ti isotopic compositions and low $^{26}Al/^{27}Al$ because of the initially high proportion of ^{27}Al in the residue. REE patterns in PLACs are variable in terms of the ultrarefractory fractionation of their REE patterns, as indicated by Tm/Tm*, but are dominated by depletion in the less refractory REE Eu and Yb. Both PLACs and SHIBs homogenized with ^{16}O-rich gas, enriched relative to terrestrial O by up to 7%, thus removing any isotopic heterogeneity from the PLAC precursors. CAIs formed close to the Sun where condensation and re-evaporation of REE was possible, and were then ejected back to planetary radii where they were eventually accreted onto planetesimals.

Introduction

The study of isotopic anomalies in meteorites has forged ahead in recent years with the discovery of interstellar grains that probably formed as direct condensates around other stars (Zinner, 1998). Almost lost in the rapidly exploding data set are refractory inclusions or CAIs (calcium-aluminium–rich inclusions), which have played a pivotal role in concepts of solar system formation. CAIs are rich in refractory elements and attest to high-temperature processes in the solar nebula (Grossman, 1980). They are viewed as the most primitive materials produced in the solar system because they have the oldest Pb/Pb ages of any objects at ~4.566 Ga (Allègre et al., 1995) and they have excess ^{26}Mg, from decay of the short-lived radionuclide ^{26}Al with an initial $^{26}Al/^{27}Al$ ratio of 5×10^{-5} (Lee et al., 1977), which is far higher than any other type of solar-system object. Furthermore, CAIs contain ubiquitous isotopic anomalies in elements such as O, Ca, and Ti; their preservation suggests formation prior to

[1]Current address: Research School of Earth Sciences, The Australian National University, Canberra, ACT 0200, Australia.

homogenization of source materials in the solar system. While CAIs are not direct stellar condensates, the size of some of the CAI isotopic anomalies rivals those in interstellar grains, suggesting minimal processing between stellar condensation and crystallization into their present form in the solar system.

CAIs are a common component of carbonaceous chondrites but it was not until 1969 that researchers were presented with an abundance of these objects (Marvin et al., 1970). Two large carbonaceous chondrites fell in that year: in Mexico, the Allende CV3 chondrite fell with a recovered weight of approximately 2000 kg (Clark et al., 1970); and in Australia, the Murchison CM2 chondrite fell with a recovered weight of approximately 100 kg (Fuchs et al., 1973). Inclusions from both of these meteorites were set upon with state-of-the-art analytical equipment designed for analysis of the lunar return samples. However, despite extensive study over the past 30 years, their origins remain enigmatic. Their original interpretation as condensates from the solar nebula (Marvin et al., 1970; Grossman, 1972) was based on the compatibility of CAI mineralogy with the sequence of phases predicted to condense from a cooling gas of solar composition, and mineralogical textures such as fluffy aggregates of microscopic crystals that appear to be direct gas-solid condensates. Furthermore, Allende CAIs are typically enriched in refractory trace elements by about a factor of 20 with respect to chondritic abundances, suggesting that they could represent the first 5% of condensed rock-forming elements. Such an origin was also consistent with the non-detection (at that time) of isotopic anomalies in the lithophile elements because in the gas phase, isotopic anomalies are quickly homogenized leading to the average solar isotopic abundances.

The condensation model was so simple and appealing that even with the discovery of ubiquitous (and diverse) isotopic anomalies in elements such as O and Ti, the concept of local gas reservoirs with distinct compositions was advanced. In this case, local heterogeneities are maintained in the solar system in the gas phase, and the precipitation of distinct CAIs reflected the local gas composition. However, in detail, the systematics of these inclusions suggest that their formation is a more complicated process. In part, the systematics have been determined by high-temperature processing, but it is likely that relict solid material was involved as well, and the inclusions may have been through more than one heating cycle after their initial formation.

This paper reviews the characteristics of CAIs and attempts to place them in a framework for their formation in the solar system. The main source of CAIs has been the Allende and Murchison meteorites. Other meteorites have some distinctive characteristics in their CAIs, but in some ways, every CAI is unique, and there is no point in examining the unique aspects of every inclusion if an overall picture of CAI formation is to be obtained. The CAIs from Allende and Murchison basically contain the full range of CAIs known, and so discussion will be limited to the major characteristics of CAIs from each of these meteorites.

The literature is rich in chemical and isotopic data from many CAIs (e.g., Clayton and Davis, 1988), but there has been only limited success in the interpretation of what these characteristics are telling us. Thus a major goal of this paper is to place a limited number of chemical and isotopic characteristics of CAIs in an internally consistent framework. The characteristics chosen are the major isotopic and chemical features that set CAIs apart from other inclusions in chondrites. The features include the refractory nature of CAIs, the volatility fractionations in the REE patterns, and isotopic abundances of magnesium, titanium, and oxygen.

Refractory Inclusions

CAIs from the Allende CV3 chondrite

Grossman (1975) petrographically divided Allende CAIs into two main types (A and B), each of which had two subgroups—"fluffy" and "compact." Type A CAIs were composed predominantly of melilite with less spinel, whereas Type B CAIs have pyroxene as the dominant mineral with melilite, spinel, and plagioclase. However, the compositional limits proposed by Grossman (1975) appear to be rather restrictive, as noted by Mason and Taylor (1982) who analyzed a suite of inclusions for trace elements. Essentially Types A and B might be more appropriately considered as endmembers of the compositional range. Fluffy CAIs appear to have the characteristics of condensates, being friable assemblages of small crystals, whereas the compact CAIs are more consistent with a molten origin. However, such a distinction cannot be uniquely determined, and the petrographic distinctions between condensate, melt, and xenocrystic material cannot be made with great confidence.

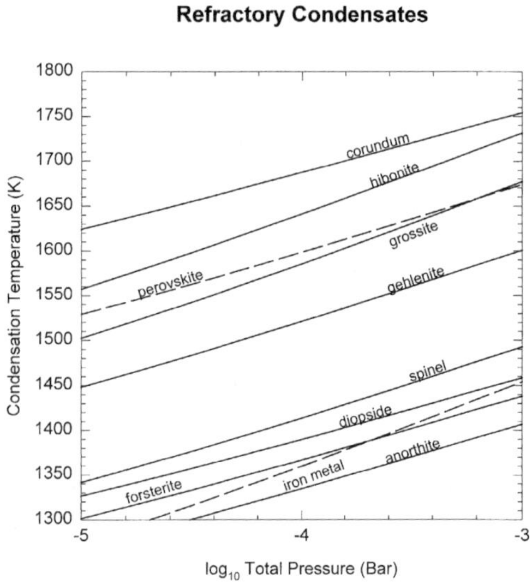

FIG. 1. Condensation temperatures as a function of total nebular pressure for refractory minerals found in CAIs and several lower-temperature phases. Modified from Lodders and Fegley (1993) and Lauretta and Lodders (1997). The different phases shown are stable on and below the labeled lines.

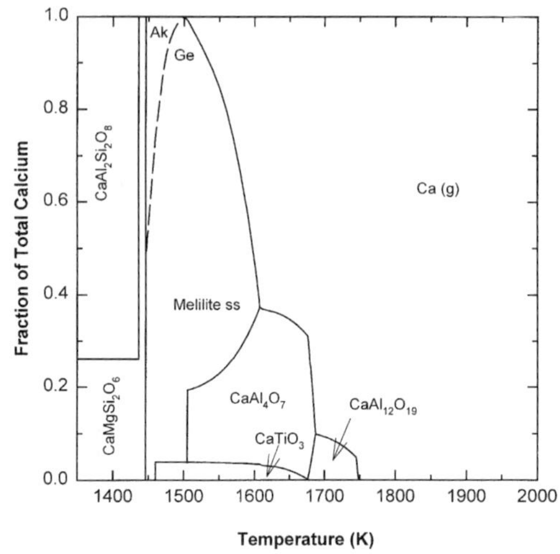

FIG. 2. Equilibrium chemistry of calcium in the solar nebula at 10^{-3} bar total pressure. The dominant gas is monatomic Ca, which condenses into different Ca-bearing minerals with decreasing temperature. The first condensate is hibonite ($CaA_{12}O_{19}$), followed by grossite ($CaAl_4O_7$) and perovskite ($CaTiO_3$). Most of the Ca gas is eventually consumed by formation of melilite solid solution. The dashed line shows the amounts of gehlenite and åkermanite in melilite. Calcium is 50% condensed at 1615 K and is distributed between Ca-bearing clinopyroxene (represented by diopside) and plagioclase (represented by anorthite) at low temperatures. Modified from Lauretta and Lodders (1997).

What can be readily seen is that the mineralogy of the Allende CAIs is consistent with the phases predicted to condense from a cooling gas of solar composition at pressures of 10^{-5} to 10^{-3} bar (Grossman, 1972). As illustrated in Figure 1, the first phases to condense from a solar-composition gas should be refractory oxides such as corundum (Al_2O_3), calcium aluminates (hibonite [$CaAl_{12}O_{19}$] and grossite [$CaAl_4O_7$]), and perovskite ($CaTiO_3$). With decreasing temperature, these phases back-react with nebular gas to form other less refractory oxides and aluminosilicates. For example, Figures 2 and 3 show that Ca aluminates react with nebular gas to form melilite. Initially the melilite is almost pure gehlenite ($Ca_2Al_2SiO_7$), but as temperatures continue to decrease, increasing proportions of åkermanite ($Ca_2MgSi_2O_7$) become incorporated into the melilite. Further decreases in temperature lead to formation of spinel ($MgAl_2O_4$), clinopyroxenes (represented in Figure 1 by diopside [$CaMgSi_2O_6$]), plagioclase feldspar (represented in Figure 1 by anorthite [$CaAl_2Si_2O_8$]), forsterite (Mg_2SiO_4), and Fe metal alloy. The mineral sequence is somewhat dependent upon the total nebular pressure (e.g., see the intersecting grossite and perovskite lines or the forsterite and iron metal lines in Figure 1), and the details of the gas-solid reactions as a function of temperature are somewhat complex (Figs. 2 and 3).

However, the overall picture that emerges is that the Allende CAI mineralogy represents the middle to lower range of temperatures, because both the highest-temperature phases (e.g., corundum, hibonite, and grossite) and the lower-temperature phases (anorthite, forsterite, and Fe metal alloy) are generally absent. It appears that the most refractory oxides have back-reacted with nebular gas to form less refractory phases such as melilite, pyroxene, spinel, and plagioclase, and that most Allende CAIs were chemically isolated from nebular gas prior to condensation of Fe alloy and significant amounts of anorthite and forsterite.

The high-temperature origin of the mineralogy is also consistent with the trace-element systematics of CAIs. A common attribute for CAIs is a light-rare-earth-element (LREE) abundance of ~10–20 times chondritic levels. If the CAIs were produced during the condensation of a gas of solar composition, then this enrichment can be interpreted as all

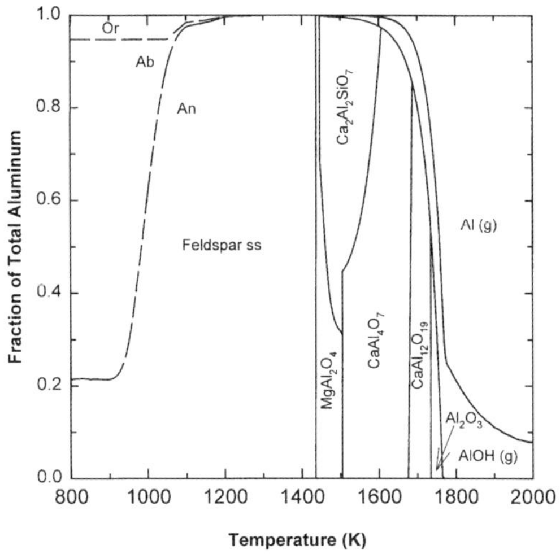

FIG. 3. Equilibrium chemistry of aluminum in the solar nebula at 10^{-3} bar total pressure. The dominant gases are monatomic Al and AlOH, which initially condense into corundum. Hibonite and other Al-bearing condensates form with decreasing temperature. Aluminum is 50% condensed at 1735 K and is distributed between orthoclase (or), albite (ab), and anorthite (an) at lower temperatures. Modified from Lauretta and Lodders (1997).

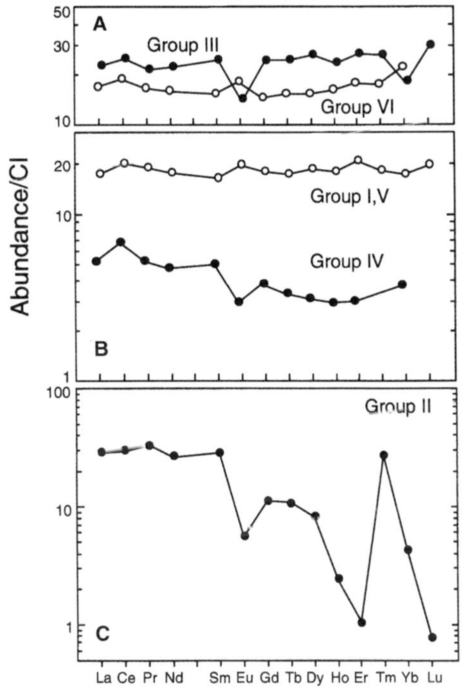

FIG. 4. REE abundance patterns most commonly found in Allende refractory inclusions. Mason and Taylor (1982) divided the patterns into six groups. The CAIs typically have LREE abundances of ~10–20 x chondritic levels. Group I has a relatively unfractionated pattern with only a small positive Eu anomaly, Group II has a highly fractionated pattern with much lower abundances of the heavy REE Gd-Er and Lu, Group III has negative anomalies of Eu and Yb, Group IV are olivine-rich inclusions that have a relatively unfractionated pattern at 2–4 x chondrite levels, Group V is unfractionated, and Group VI has small positive anomalies at Eu and Yb.

of the REE being condensed into only a few phases, and to get a 20-fold enrichment, only 5% of the rocky material (e.g., Mg, Si, Fe, Ca, Al, Ti, etc.) would have condensed. Mason and Taylor (1982) described six discrete groups of patterns that represent some subtle, and some not so subtle, variations from flat chondritic abundances (Fig. 4). Group I has a relatively unfractionated pattern with only a small positive Eu anomaly, Group II has a highly fractionated pattern with much lower abundances of the heavy-rare-earth-element (HREE) Gd-Er and Lu, Group III has negative anomalies of Eu and Yb, Group IV has a relatively unfractionated pattern at 2–4 times chondrite levels, Group V is unfractionated, and Group VI has small positive anomalies at Eu and Yb (Figs. 4A–4C).

Only a loose correspondence with texture was noted: Groups I, V, and VI are predominantly melilite-rich chondrules, Groups II and III are mainly fine-grained aggregates, and Group IV is composed of olivine-rich aggregates and chondrules. Fegley and Ireland (1991) noted that there is a significant potential bias in the relative abundances of these patterns in Allende CAIs. The easiest CAIs to obtain from Allende are the largest inclusions, which range in size up to 2 cm. These inclusions have predominantly Group I, V patterns. Smaller CAIs have Groups II and III patterns more commonly.

REE patterns in CAIs can be described as the result of three different processes, two of which are related to volatility, and the other resulting from magmatic partitioning. The volatility effects are illustrated in Figure 5 for Allende CAIs from three main groups, with the elements reordered into a relative volatility sequence. For the Group I, V pattern in Figure 5A, both the REE pattern and volatility pattern show near-flat chondrite-normalized patterns at approximately 20 times CI. The most volatile element V does appear to show a slight relative depletion, being only ~10 times CI. In Figure 5B,

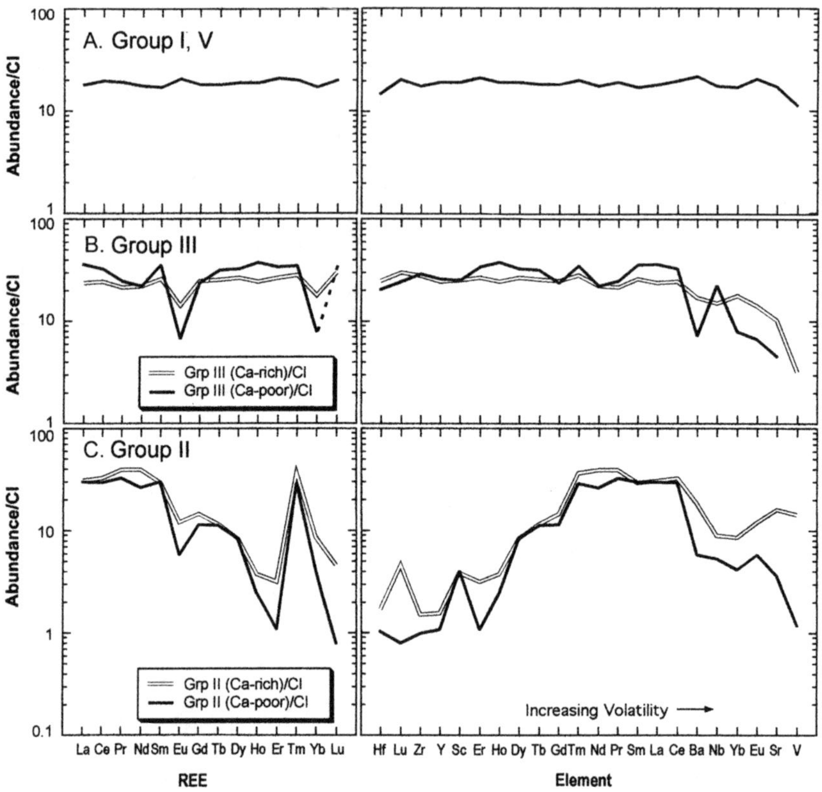

FIG. 5. The most common REE patterns are shown on the left, and trace-element patterns in order of increasing volatility (or decreasing condensation temperature) are shown on the right. The Group I,V pattern is relatively unfractionated in REE as well as the other refractory lithophile elements shown. Group III patterns show Eu and Yb anomalies, which are the most volatile of the REE, and this is consistent with the trace-element pattern, which shows a progressive depletion in the more volatile elements with unfractionated more-refractory elements. Group II inclusions show depletions in the most volatile and the most refractory elements. Data from Kornacki and Fegley (1986) subdivided into Ca-rich and Ca-poor inclusions for Groups II and III.

Group III patterns are shown for Ca-rich and Ca-poor CAI compositions after Kornacki and Fegley (1986). Eu and Yb are the most volatile REE, and these elements, along with the other relatively volatile elements Ba, Nb, Sr, and V are depleted. The Ca-rich CAIs exhibit a very smooth progressive depletion, whereas the Ca-poor inclusions are somewhat more scattered. In Figure 5C, the Group II pattern is seen to be depleted in the most refractory, as well as the most volatile elements, while the light REE and Tm are present in similar abundances to Group I, V and III patterns.

The depletions in the most volatile elements in the Group III pattern can be produced either by condensation, in which case the condensate must have been isolated from the gas prior to condensation of the more volatile elements, or by distillation, whereby a CAI with a chondritic pattern, albeit elevated in overall abundance, is raised in temperature until the more volatile elements are evaporated. The Group II pattern, which is depleted in the most refractory and the most volatile REE, is apparently consistent only with a condensation origin. Obtaining that pattern requires removal of the most ultrarefractory elements, either in an earlier condensate that is then isolated from the remaining gas, or left behind as a residue (Boynton, 1975). CAIs with ultrarefractory-enriched patterns have not been found in Allende, but were first found in Murchison (Boynton et al., 1980), and will be described below.

Thus the different REE patterns observed in CAIs are predominantly the result of volatility-con-

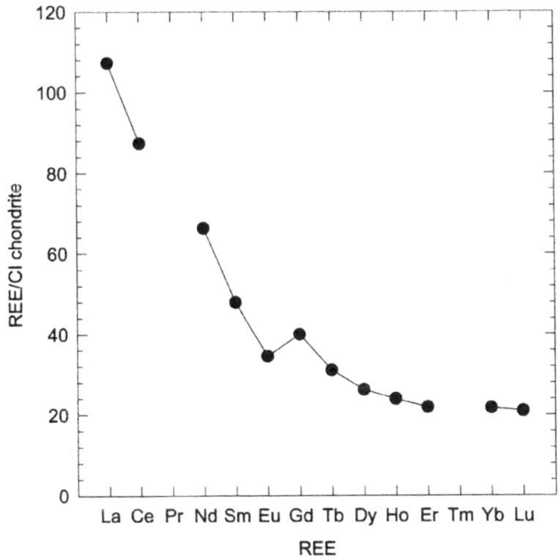

FIG. 6. REE abundance pattern in a terrestrial basalt (BCR-1) from Haskin et al. (1971). Analytical values for Pr or Tm do not exist because these elements are difficult to measure in terrestrial samples by neutron activation analysis.

trolled fractionations rather than igneous partitioning. The latter pattern, shown for a terrestrial basalt in Figure 6, has a characteristic Eu depletion caused by the incorporation of Eu^{2+} into Ca-bearing minerals, and a smooth attenuation due to ionic size.

Volatility-controlled fractionations result from the different volatilities and speciation of the REE in nebular gas. A schematic illustration of these effects is shown in Figure 7. With the exception of Yb, present mainly as the monatomic vapor, and Ce, present mainly as CeO_2 gas with a minor fraction as CeO gas, the major REE gases in the solar nebula are the monoxides. Temperature-dependent partitioning of the REE between nebular gas and refractory minerals is controlled by condensation into solid solution in a host phase such as perovskite (or hibonite) via net thermochemical reactions exemplified by

$$\mathrm{LaO\ (g) + 0.5H_2O\ (g) =}$$
$$\mathrm{LaO_{1.5}\ (in\ CaTiO_3) + 0.5H_2\ (g)}$$

for REE present dominantly as monoxide vapors,

$$\mathrm{Yb\ (g) + 1.5H_2O\ (g) =}$$
$$\mathrm{YbO_{1.5}\ (in\ CaTiO_3) + 1.5H_2\ (g)}$$

for monatomic Yb, and

$$\mathrm{CeO_2\ (g) + 0.5H_2\ (g) =}$$
$$\mathrm{CeO_{1.5}\ (in\ CaTiO_3) + 0.5H_2O\ (g)}$$

for most of Ce that is present as CeO_2 (g).

The Gibbs energies for the different condensation reactions do not vary smoothly across the lanthanide series. As a result, the partitioning of REE between gas and grains is irregular, with the LREE generally being more volatile than the HREE. Thermodynamic calculations predict that Eu, Ce, and Yb are the most volatile REE, and that Lu, Er, and Ho are the most refractory REE in nebular gas (e.g., Kornacki and Fegley, 1986). Because of the importance of CeO_2 gas, Ce partitioning between gas and grains is sensitive to the oxidation state of the surrounding environment, with more oxidizing conditions leading to less Ce being incorporated in condensed phases at a given temperature.

CAIs have played a major role in the search for isotopic vestiges of presolar material. Of paramount importance in the search for isotopic anomalies was the discovery of widespread anomalies in O isotopes (Clayton et al., 1973), likely caused by excess ^{16}O, at a level of up to 5% enrichment in Allende anhydrous materials. The characteristic feature on the oxygen three-isotope diagram is a slope 1 mixing line, as opposed to the slope 1/2 terrestrial mass fractionation line, between different mineral species from individual CAIs (Fig. 8A). Spinel and fassaitic pyroxene are the most enriched minerals, whereas melilite and anorthite are close to normal in composition. This behavior has been modeled as: (1) production of an ^{16}O-rich reservoir; (2) crystallization of CAIs in equilibrium with that reservoir; followed by (3) back-reaction in the solar nebula with re-equilibration dependent upon the individual minerals. A small subset of inclusions (FUN inclusions, see below) have an additional step in that they require the production of a mass-fractionated composition prior to back-reaction with isotopically normal O (Fig. 8B). Thus the third most abundant element in the solar system has an extremely large isotopic anomaly, requiring extreme heterogeneity in the solar nebula. The concept of a simple, completely vaporized, homogeneous cloud of solar gas is untenable. At the very least, a progressive change in the isotopic composition of the material being processed in the solar nebula is required.

Following the discovery of oxygen isotope effects, a major effort was instigated in solid-source mass spectrometry. Quantities of material required for analysis on thermal ionization mass spectrometers fell from micrograms to nanograms, and precision capability improved from percent to per mil and then to ε unit (10-4). These advances made possible isotopic analysis of Mg (Gray and Compston, 1974;

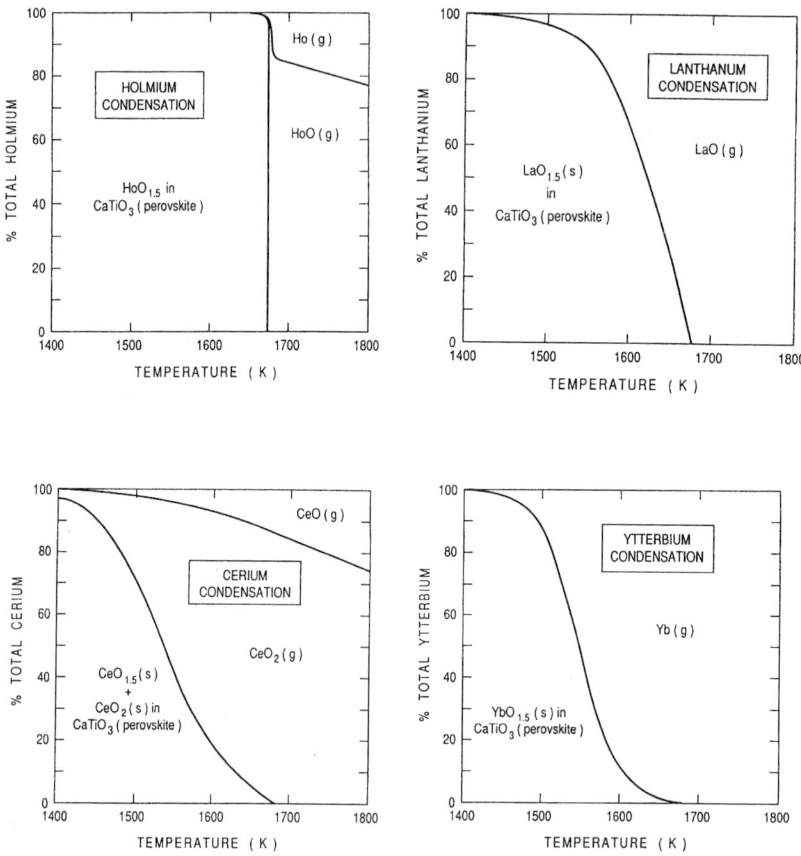

FIG. 7. Equilibrium chemistry of Ho, La, Ce, and Yb in the solar nebula at 10^{-3} bar total pressure. These four plots exemplify condensation of ultrarefractory (Ho), refractory (La), oxidizable (Ce), and volatile (Yb) REE. Methodology for the condensation calculations is described by Kornacki and Fegley (1986).

Lee and Papanastassiou, 1974), which led to the discovery of the extinct radionuclide ^{26}Al in Allende CAIs (Lee et al., 1977). Then excesses in ^{50}Ti were quantified (Heydegger et al., 1979; Niederer et al., 1980; Niemeyer and Lugmair, 1981) and thenceforth excesses in heavy isotopes of other elements of the Fe abundance peak, viz. Ca, Cr, Fe, Ni, and Zn (Birck and Allègre, 1984; Jungck et al., 1984; Birck and Lugmair, 1988; Loss and Lugmair, 1990; Völkening and Papanastassiou, 1989, 1990; Fig. 9).

The excesses in the n-rich isotopes of these elements have been interpreted as representing the addition of neutron-rich material from near the mass cut of a supernova into the solar system. The multiple-zone mixing model of Hartmann et al. (1985) is particularly successful in predicting the abundances of ^{48}Ca, ^{50}Ti, and ^{54}Cr in terms of material that has experienced large neutron excesses in a Type II supernova. However, Meyer et al. (1996) have indicated that the extreme entropy of the Type II supernova would likely bypass the lighter elements, and they postulate that most ^{48}Ca is produced in rare Type I supernova events.

In most Allende CAIs, the isotopic anomalies for the Fe abundance peak elements are typically of the order of 1‰. However, in a small subset of inclusions, larger isotopic anomalies are found. For example δ^{50}Ti in inclusion C1 is –4‰, whereas in EK1-4-1 it is +4‰ (Niederer et al. 1985), while EK1-4-1 has a ^{58}Fe excess of +29‰ (Völkening and Papanastassiou, 1989). These isotopic anomalies are associated with mass fractionation in the elements (i.e., the isotopes are also systematically fractionated according to their

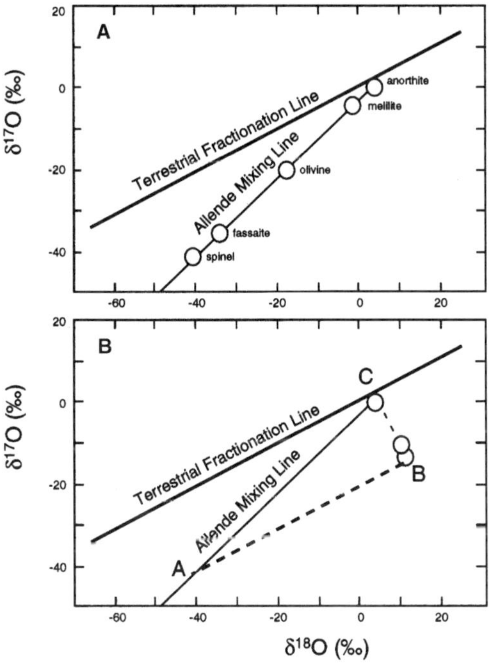

FIG. 8. A. Oxygen isotopic systematics for "normal" Allende CAIs. Minerals from an individual CAI, as well as bulk CAI measurements, define a mixing line with slope close to unity, in contrast to the slope 0.5 line defined by kinetic processing of oxygen isotopes on Earth (terrestrial fractionation line). These data suggest that CAIs formed from a reservoir enriched in ^{16}O by at least 50‰ relative to terrestrial abundances. Heterogeneity between minerals is probably related to reequilibration of melilite and anorthite with isotopically normal O, whereas spinel and fassaite (pyroxene) largely retain their initial composition (Clayton et al., 1977). B. Oxygen isotopic systematics for FUN Allende CAIs. These CAIs indicate an additional mass fractionation event that changes the initial composition from A to B followed by back-reaction with normal O at C. Data for the FUN inclusion HAL is shown (Lee et al., 1980).

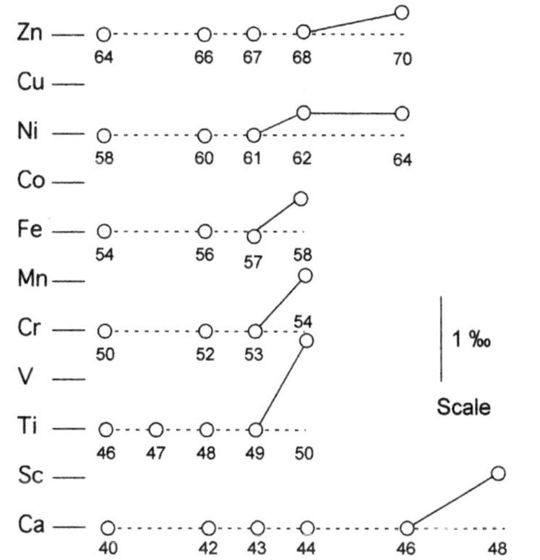

FIG. 9. Isotopic anomalies of the Fe group elements in "normal" Allende CAIs, showing enrichments of the heaviest isotopes of each element.

masses). The observation of similar effects in Mg, Si, and O led to these inclusions being termed FUN inclusions for their fractionated and unknown nuclear components. For these elements, an unequivocal partitioning between the F and the UN effects may not be possible because the normalizing isotopes used for the mass fractionation determination could also be affected by the UN component. There are no distinguishing mineralogical characteristics for these inclusions to distinguish them from other "normal" Allende CAIs.

The isotopic compositions of other elements such as Sr, Zr, Ba, Nd, and Sm have also been measured from a limited number of Allende CAIs (see Lee, 1988). These elements too have a story to tell, but there are too few analyses for them to be used in a systematic way to address the formation of CAIs. Specific reference will be made to isotopic compositions of some of these elements in evaluating formation models.

Systematics of Allende CAIs

A common interpretation from Allende CAIs is that there is no systematic relationship between petrography, presence of $^{26}Al/^{27}Al$ at 5×10^{-5}, Ti isotopic composition, or ^{16}O enrichment (MacPherson et al., 1988). Nor are there any fundamentally distinctive characteristics of FUN inclusions. The FUN inclusions, with their large isotopic effects, are perhaps the most cited of any of the CAIs. Although these inclusions were extremely important in terms of initially documenting isotopic anomalies, their isotopic anomalies are now orders of magnitude less than those found in Murchison CAIs, at least in terms of the elements Ca and Ti. The coexistence of mass fractionation and nuclear anomalies in these inclusions is an interesting aspect, but so-called "normal" CAIs are still isotopically anomalous without the fractionation effects. Thus while the FUN inclusions are remarkable, their properties are not so unique as to require specific accommodations rather than just being within the continuum of CAI variations.

CAIs from the Murchison CM2 chondrite

Murchison is the most plentiful and accessible CM2 chondrite, so most analyses of CM2 CAIs are from this meteorite. There does not appear to be any marked distinction in the inclusions from Murchison compared to those from other CM2 meteorites. A feature of CM2 refractory inclusions is the widespread presence of hibonite ($CaAl_{12}O_{19}$), and rarely, the occurrence of corundum (Bar-Matthews et al., 1982; MacPherson et al., 1984, 1988; Hinton et al., 1988). Thus, these CAIs appear to contain assemblages that formed and were isolated from the nebular gas at an overall higher condensation temperature than Allende CAIs (e.g., see Fig. 1).

Murchison CAIs are typically of the order of 500 μm maximum dimension; thus they are difficult to analyze using the same methodologies as those originally employed in Allende CAI research. Sufficient material is available in an Allende CAI for neutron activation measurement of trace elements, and also for crushing, dissolution, and separation of the desired chemical species for mass-spectrometric isotopic analysis. The advent of ion microprobe analysis opened up Murchison CAIs to routine analysis for trace-element concentration as well as isotope measurements (see Ireland, 1995). The advantage of ion microprobe analysis is that analyses can be carried out *in situ*, allowing petrographic context to be utilized in selecting areas, even of a given mineral, for analysis. Thus for example, different generations of a single mineral can be analyzed. Ion microprobe analysis has sometimes been referred to as "virtually nondestructive," but this is demonstrably not the case for relatively small extraterrestrial objects such as these. The drawbacks of ion microprobe analysis are the requirements for high mass resolution to separate molecular isobars, and the relatively lower precision for ion microprobe analysis compared to some other mass spectrometric techniques. However, the panacea for ion microprobe analysis is that large isotopic variations can be found in CAIs, for which ultimate precision is not an issue.

A feature of Murchison CAIs is the common occurrence of blue hibonite. Hibonite is predicted to be the second major condensate phase after corundum from a cooling gas of solar composition. Perhaps the most distinctive type of Murchison CAIs are single grains of essentially pure hibonite that are transparent but display a blue pleochroism through thick sections. These have been referred to as DJ (blue chip) fragments (Hinton et al., 1988) or PLAC (PLAty Crystals; Ireland, 1988). However, the most common type of hibonite-bearing inclusions are those within spinel. These hibonites appear bright blue even as relatively small (typically 10×30 μm) inclusions. This type has been referred to as SHIB (Spinel-HIBonite; Ireland 1988). Other types of inclusion described include hibonite-glass inclusions, where the hibonite occurs as crystals within glass or devitrified glass of pyroxene composition (Ireland et al., 1991; Simon et al., 1998), and the HAL-type hibonite inclusions named after the Allende FUN inclusion (Ireland et al., 1992; Russell et al., 1998).

The chemistry of hibonite allows the substitution of Mg^{2+} and Ti^{4+} for $2Al^{3+}$ (Allen et al., 1978). In the PLAC hibonite, and in hibonite from glass inclusions, TiO_2 substitution typically reaches a maximum of 2 wt% TiO_2, but in SHIB hibonite, concentrations as high as 10 wt% are found (Ireland, 1990). The blue color of hibonite is related to the presence of a small amount of Ti^{3+} (direct substitution for Al^{3+}; Ihinger and Stolper, 1986). Thus the SHIB hibonites are the brightest blue because of the overall higher Ti concentration, and thereby higher Ti^{3+} concentration. The HAL-type inclusions are noteworthy in having relatively low Ti concentrations (< 1 wt % as TiO_2) and essentially no Mg.

The trace-element chemistry of Murchison CAIs like that in Allende is dominated by volatility-controlled fractionation, but is also distinctive in the petrographic hibonite groups. PLAC hibonites typically have a Group III-type pattern with depleted abundances of Eu and Yb relative to the other REE (Ireland et al., 1988; Fig. 10A). SHIB hibonites have patterns that are quite similar to the Group II pattern, with depleted abundances of the ultrarefractory REE compared to the light REE abundances (Fig. 10B), and a few have ultrarefractory-enriched compositions (Fig. 10C). In the Murchison inclusions, however, the more volatile REE Eu and Yb are more variable and, in some cases, even elevated beyond the light REE concentrations. The glass spherules have hibonite that is somewhat variable in composition, and the glass can have trace element abundances associated with equilibrium hibonite-melt distribution coefficients (Beckett and Stolper, 1994), but not always (Ireland et al., 1991).

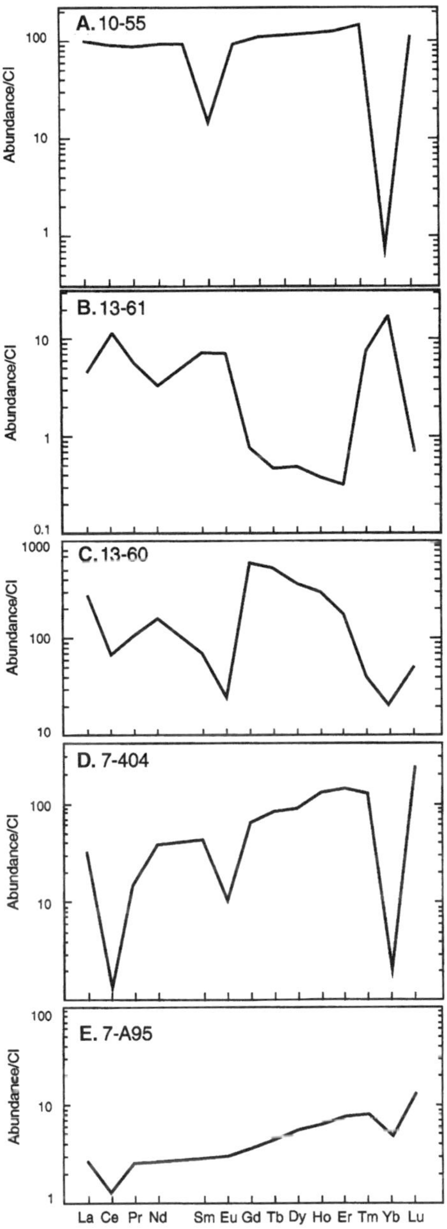

FIG. 10. REE abundance patterns from Murchison CAIs show anomalies in the abundances of the relatively volatile REE Ce, Eu, and Yb, as well as fractionations in the ultrarefractory REE ranging from depletions to enrichments.

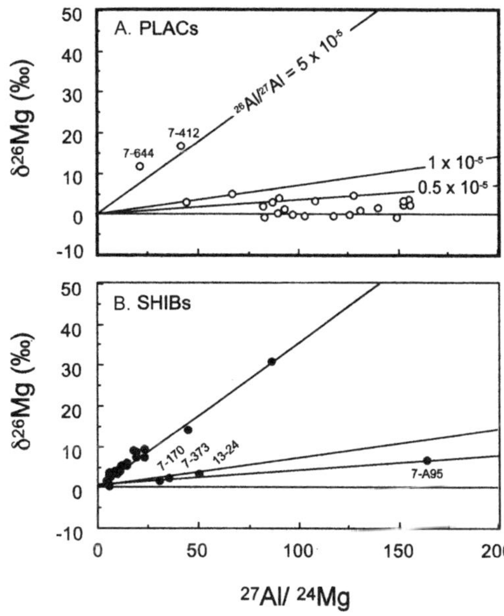

FIG. 11. Al-Mg isochron diagrams for Murchison CAIs. In general, PLAC hibonites show little or no excess ^{26}Mg due to the decay of ^{26}Al, whereas SHIBs show $^{26}Al/^{27}Al$ at the canonical solar system abundance of 5×10^{-5}.

The HAL-type inclusions have strong depletions in Ce, which is a characteristic feature of distillation in an oxidizing environment (Ireland et al., 1992; Fig. 10D). Other inclusions can have relatively unfractionated patterns with only minor anomalies in the more volatile REE (Fig. 10E).

Magnesium isotopic compositions most commonly show enrichments in ^{26}Mg due to ^{26}Al decay with a bimodal distribution between the canonical solar system $^{26}Al/^{27}Al$ of 5×10^{-5} for SHIB inclusions to $< 1 \times 10^{-5}$ for PLAC inclusions (Fig. 11). The glass spherules are typically nonradiogenic (i.e., low apparent initial $^{26}Al/^{27}Al$), and some inclusions have initial $^{26}Mg/^{24}Mg$ ratios that are lower than the terrestrial ratio (Ireland et al., 1991). The HAL-type inclusions have very low Mg concentrations and therefore have high Al/Mg ratios, which facilitates measurements of initial $^{26}Al/^{27}Al$. The HAL-type inclusions show variable inferred $^{26}Al/^{27}Al$, and also mass-fractionated Mg isotopes consistent with evaporation of Mg (Ireland et al., 1992).

Calcium and titanium isotopic compositions are the most anomalous in PLAC inclusions (Ireland, 1990), whereas most SHIB inclusions are within error of normal. The pattern is similar to that for Allende CAIs—that is, the anomalies are in the heaviest, most neutron rich isotopes. However, the size and range of the anomalies are much larger. Enrichments in ^{50}Ti and ^{48}Ca range up to 270‰ and

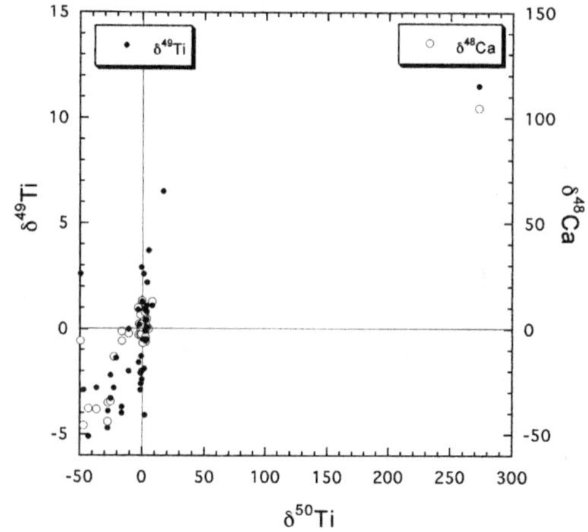

FIG. 12. Ca-Ti isotopic compositions in Murchison hibonites show extreme compositions compared to the Allende inclusions. Data for δ^{49}Ti and δ^{48}Ca are plotted against δ^{50}Ti. Both ^{49}Ti and ^{48}Ca anomalies are well correlated against ^{50}Ti. Selected data from Ireland (1990) and Zinner et al. (1986).

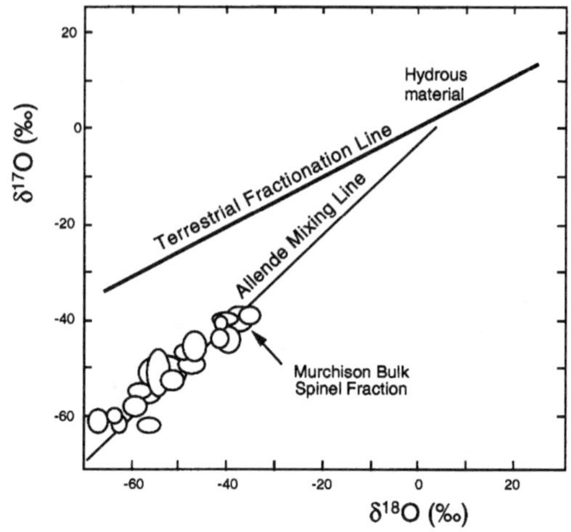

FIG. 13. O isotopic compositions in hibonite-bearing inclusions from Murchison show larger ^{16}O excesses than Allende CAIs extending down the Allende mixing line to ~70‰ excess levels.

100‰, respectively, while depletions also are common, with the largest being –70‰ and –60‰ for ^{50}Ti and ^{48}Ca, respectively (Zinner et al., 1986; Hinton et al., 1987; Ireland, 1990). Anomalies in ^{49}Ti are smaller but are well correlated with ^{50}Ti anomalies (Fig. 12). The glass inclusions have moderately variable Ca and Ti isotopic compositions, whereas the HAL-type inclusions typically have mass-fractionated Ca and Ti isotopes but relatively small "non-linear" effects after removal of the mass fractionation component.

Oxygen isotopic compositions (Fahey et al., 1987; Ireland et al., 1992) are not distinctive between the various groups and all show the large apparent ^{16}O effect seen in Allende CAIs (Fig. 13), albeit at a higher level of ^{16}O excesses, maximizing at around 70‰ as opposed to Allende CAIs at 50‰. Murchison CAIs are also affected by mixing, with a more normal isotopic component yielding variable compositions close to the Allende mixing line. HAL-type inclusions show isotopic fractionation away from the Allende mixing line, followed by mixing with isotopically normal material (i.e., the characteristic FUN inclusion systematics).

Systematics of Murchison CAIs

Murchison CAIs show a very strong polarization of morphological, chemical, and isotopic features between PLACs and SHIBs. PLACs have Group III–dominated REE patterns, while SHIBs are commonly Group II or ultrarefractory enriched. PLACs are the dominant source of Ca and Ti isotopic anomalies; only one SHIB is highly anomalous. Hutcheon et al. (1983) first noted the exclusivity of ^{26}Al and Ti isotopic anomalies in hibonite inclusions. This is largely a manifestation of PLAC hibonites, which have most of the large Ti isotopic anomalies, and do not contain excess ^{26}Mg. There is a propensity for inclusions with Group II–type patterns to have $(^{26}\text{Al}/^{27}\text{Al})_0$ of 5×10^{-5}, but this does not hold as strongly as the likelihood of SHIB inclusions to have Group II REE patterns and ^{26}Al/^{27}Al at 5×10^{-5}, and PLAC hibonites to lack ^{26}Al. This is particularly noted for SHIB 7-170, which has a Group II pattern, but also has a large ^{50}Ti deficit and therefore no apparent ^{26}Al.

The correlation of ^{26}Al with Group II REE patterns is largely a manifestation of SHIB inclusions. However, this is not entirely the case, and there are several PLACs with Group II patterns and excess ^{26}Mg at $5 \times 10^{-5} \times {}^{27}$Al. Similarly, some SHIB hibonites possess rather flat REE patterns and do not have canonical radiogenic Mg.

Even within the types of hibonite, two morphologically very similar inclusions can exhibit quite different isotopic systematics. For example, 7-734 and 7-170 (from Ireland, 1990) are morphologically the same, with hibonite-spinel cores surrounded by a rim sequence. However, 7-734 is a typical SHIB with Group II REE pattern, $^{26}Al/^{27}Al$ of 5×10^{-5}, and normal Ti isotopic composition. In contrast, 7-170 is the most unusual SHIB with a Group II REE pattern but no radiogenic ^{26}Mg excess, and a ^{50}Ti deficit of ~50‰. Thus the morphological and petrographic distinctions of PLACs and SHIBs largely reflect variations (pressure, temperature, gas composition) between volumes of the solar nebula in which the inclusions formed. SHIB 7-170 probably has included a proto-CAI that has more similarities to PLACs than to the typical Ti isotopes and Al-Mg systematics of SHIBs. But with thermal processing in the SHIB region, the PLAC-type precursor has been absorbed into the new CAI. The question then becomes, which of the components really are linked. Is the SHIB region dominated by Group II REE and $^{26}Al/^{27}Al$ of 5×10^{-5} and normal Ti with only a few seed inclusions that bring in inherited components such as the Ti and Al-Mg in 7-170? An interesting test would be to measure the Mg associated with 7-170 spinel to see if its composition is consistent with other SHIB Al-Mg at $^{26}Al/^{27}Al$ of 5×10^{-5}.

Even though there is some consistency between the groups of hibonites, there is a wide range of chemistry and isotope compositions that needs to be evaluated against one another. Although isotope systematics can be represented as a numerical value of an isotopic anomaly, REE have more generally been referred to in terms of the canonical REE groups. But the assignment to a group is rather subjective, and some inclusions bear attributes that are not found in any of the Allende REE patterns. As such we have developed a numerical scheme for quantifying REE characteristics of CAIs.

As previously noted, REE patterns in CAIs are the result of at least three different fractionation events. The volatility-controlled fractionations affecting the most refractory and the most volatile elements of the refractory lithophile sequence are of most use for classification purposes, whereas of minimal use is the magmatic fractionation between different phases. Fractionation in the most refractory elements is responsible for Group II patterns and the attribute that is commonly used to describe the Group II pattern is the presence of a Tm anomaly. This can be numerically defined as Tm/Tm*, where Tm* refers to the interpolated value between the Er and Lu CI-normalized abundances.

The more volatile lithophile elements could be represented by either Eu or Yb. We prefer to use Yb because of the possible fractionation of Eu^{2+} between Ca-bearing phases. However, calculation of

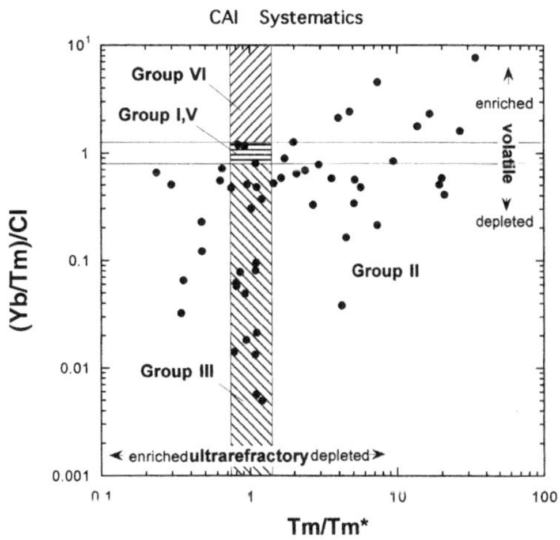

FIG. 14. REE classification of CAIs based on Tm and Yb anomalies to quantify ultrarefractory REE fractionation and volatile REE fractionation, respectively. Yb is used rather than Eu because of the propensity of Eu to exist in the Eu^{2+} oxidation state and be partitioned during magmatic processes into Ca-bearing phases. Tm is used as the reference element because it behaves as a LREE and provides a better estimate of volatile fractionation in the HREE. Tm* is calculated from the interpolated value between Er and Lu. Yb/Tm is normalized to the CI value such that (Yb/Tm) CI is the solar value. Shown on the plot are the regions where the Allende CAI patterns (Fig. 4) occur.

Yb/Yb* is not straightforward because of the possible superposition of the ultrarefractory fractionation. In terms of the volatile-element fractionation, a more useful reference point is the light REE. But to use the LREE, a substantial extrapolation is required, and it might also be compromised by variability in the LREE abundances such as in Murchison ultrarefractory-fractionated CAIs. However, as found in Allende Group II patterns, Tm is a good proxy for the LREE. Thus we use the CI-normalized Yb/Tm ratio to express the volatile-REE fractionation.

Data from approximately 50 Murchison CAIs, from Ireland et al. (1988) and Ireland (1990), are presented in this classification scheme in Figure 14. Allende CAIs of Groups I, III, V, and VI have no Tm anomalies and show variation only in Eu and Yb abundances. The positions these groups plot are shown in the hachured regions. Group II inclusions from Allende show depletions in both ultrarefractory and volatile REE, and hence plot in the lower right

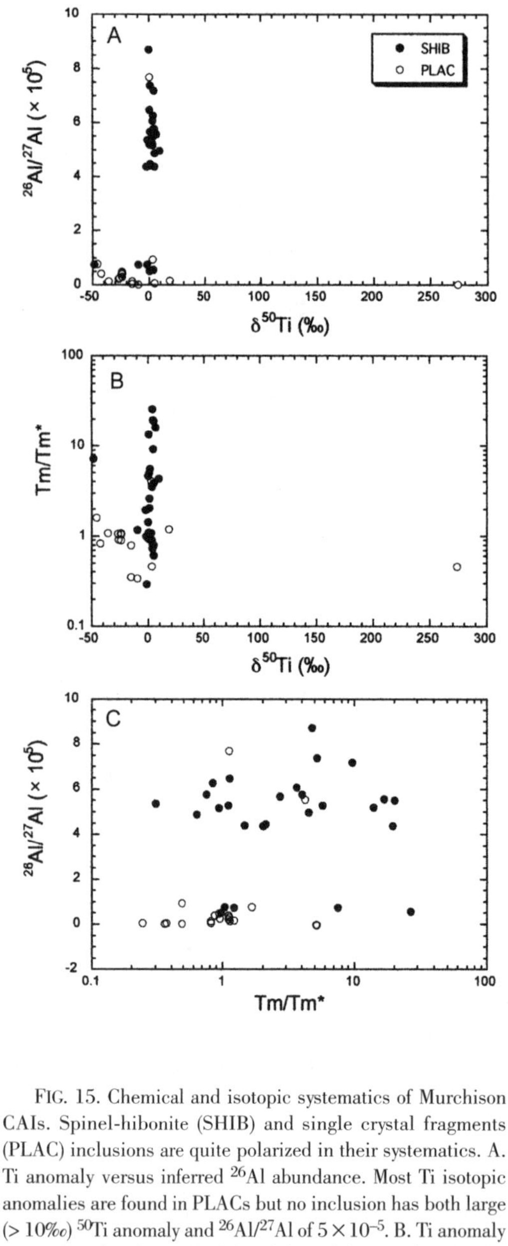

FIG. 15. Chemical and isotopic systematics of Murchison CAIs. Spinel-hibonite (SHIB) and single crystal fragments (PLAC) inclusions are quite polarized in their systematics. A. Ti anomaly versus inferred ^{26}Al abundance. Most Ti isotopic anomalies are found in PLACs but no inclusion has both large (> 10‰) ^{50}Ti anomaly and $^{26}Al/^{27}Al$ of 5×10^{-5}. B. Ti anomaly versus Tm abundance. PLAC and SHIB hibonites largely plot on orthogonal trends because of the propensity for SHIB to show ultrarefractory fractionations but no Ti anomalies, whereas PLAC have variable Ti isotopic compositions and more commonly volatile REE fractionation (Group III). C. ^{26}Al abundance vs. Tm anomaly. The bimodality of inferred initial ^{26}Al is apparent, but there is no distinct correlation between REE fractionation and ^{26}Al abundance despite SHIB hibonites dominating the ^{26}Al-rich inclusions. Ultrarefractory-element fractionations are also present in PLAC with low ^{26}Al abundances.

region of the diagram. Murchison CAIs cover a large range of this diagram, ranging from ultrarefractory enriched to depleted and volatile enriched and depleted. However, most inclusions are volatile-REE depleted and the only inclusions strongly enriched in these elements also show the strongest Tm anomalies. No inclusion with an ultrarefractory enriched pattern also has enriched volatile elements.

With a numerical quantification of REE patterns, isotopic compositions can be compared to the trace element systematics. The exclusivity of Ti isotopic anomalies and inferred ^{26}Al is shown in Figure 15A. In Figure 15B, Ti isotopic anomalies are plotted against Tm anomalies. The ultrarefractory-fractionated SHIB patterns produce a vertical pattern around the normal Ti isotopic composition, whereas the PLACs with diverse Ti isotopic compositions have more normal Tm abundances. In Figure 15C, bimodality of inferred ^{26}Al abundance is clearly apparent, but no correlation exists between Tm abundance and inferred ^{26}Al at a level of $5 \times 10^{-5} \times {}^{27}Al$.

The Early Solar System

High-temperature processes

The formation of CAIs in a hot, cooling solar nebula is consistent with the models for solar nebula formation proposed by Cameron (1962). Basically, gravitational in-fall produced sufficient kinetic energy to vaporize all solid materials. The widespread absence of isotopic anomalies in solar system materials, as measured at that time, was entirely consistent with such a view (Suess, 1965). Up until the discovery of ubiquitous isotopic anomalies in oxygen, isotopic variations had only been described from volatile elements such as H, Ne, and Xe (Reynolds, 1967) and those anomalies could be ascribed to late-stage in-fall of some exotic material that had escaped homogenization. However, the high-precision measurement of O isotopic compositions in Allende CAIs substantially changed this concept. At the very least, two major isotopic reservoirs were required to accommodate O isotopic compositions. Further isotopic variability in elements such as titanium require preservation of the isotopic integrity of much smaller amounts of material. While high temperatures are required to form the condensation-fractionated REE patterns, not all material could have been vaporized because isotopic heterogeneity would have been quickly removed.

Preservation of isotopic anomalies

In essence, the concept of an isotopic anomaly is rather misleading. The isotopic abundances of the elements in the solar system are sourced from many different nucleosynthetic sites, and those sites have specific physical and chemical conditions that drive nuclear reactions in specific ways. Thus our reference frame for examining isotopic compositions is the average of all the material from all sites that has come into the source region for Earth. As such, there is nothing especially significant about the exact mix that is present on Earth. There is even a distinct likelihood that radial variations exist in the solar nebula for certain isotope abundances, particularly those for elements that exist in volatile components. Determining isotopic anomalies is an attempt at unmixing some of the isotopic components that have not been completely homogenized during the mixing phases of the solar nebula. We use the average terrestrial isotopic compositions of the elements simply as a useful reference point to begin.

The concept of isotopic anomalies has changed over the years from the attempts to measure isotope abundances that differ from the terrestrial abundances by only small amounts, to one of characterizing distinct nucleosynthetic compositions in grains that have likely crystallized around another star. In this case, there is nothing particularly anomalous in the composition of the grain in its local neighborhood of formation, although its composition may differ greatly from that of terrestrial isotopic compositions. The isotopic composition of the grain simply reflects the mix of components that have contributed to the star, possibly augmented by nucleosynthesis in the star. As such, referencing the isotopic composition to terrestrial abundances may give a misleading inference as to the anomalous isotope component(s) present. Where anomalies are large, a specific comparison with terrestrial abundances can be dropped, and the ratios alone can be sufficiently diagnostic to allow association with a specific source. When the isotopic deviations are smaller, it may be more difficult to assign a specific component that characterizes the anomalous material.

Further complicating the assignment of anomalous components is the presence of deficits of certain isotopes relative to the terrestrial composition. A case in point is the presence of apparent deficits in ^{50}Ti in hibonite, which is complementary to large ^{50}Ti excesses. The interpretation of ^{50}Ti as the anomalous isotope is beyond question. The abundances of the light isotopes of Ti are typically within the range of normal terrestrial Ti, but more importantly, there is no difference in these abundances for compositions with large ^{50}Ti excesses compared to those with ^{50}Ti deficits. But, while a ^{50}Ti excess can simply be expressed as a number of atoms of ^{50}Ti in excess of that observed for isotopically normal Ti, the deficit cannot be represented as a negative number of atoms because such a number has no physical basis.

The interpretation of such isotopic deficits places greater constraints on the interpretation of isotopic anomalies than the interpretation of excesses. In the case of ^{50}Ti again, Hinton et al. (1987) proposed that the inclusion with the greatest deficit, BB-5 at –70‰, actually represented the original Ti-isotopic base composition for the solar nebula, and the present terrestrial ratio was augmented by the addition of a spike of isotopically anomalous Ti as represented in the inclusions with large enrichments in ^{50}Ti. This concept is part of the Late Supernova Injection (SNI) model originally proposed to account for excesses in ^{16}O and ^{26}Al in the early solar system, as produced by a supernova in close proximity to the solar nebula (Cameron and Truran, 1977). Thus, isotopic components were added to the solar system, which was originally depleted in these isotopes relative to the present. The base composition for any element is represented by the most depleted composition for the given isotope of that element. However, Fahey et al. (1987) dismissed the case for the SNI model for the deficit in ^{50}Ti because BB-5 has an ^{16}O excess that is even larger than the most ^{50}Ti-enriched hibonites. The base composition for ^{50}Ti is associated with an enriched ^{16}O composition, not the base composition of oxygen which would be expected for the SNI model. This behavior indicates uncorrelated behavior of O and Ti isotope systems.

An alternative interpretation of the Ti isotope depletions is that the inclusions contain isotopic components inherited from the site of nucleosynthesis, and have been preserved through the high-temperature processing in the solar system. This chemical memory model (Clayton, 1982) is relevant to all isotopic variations, and simply states that material can survive the extreme mixing and homogenization processes in the early solar system. In this case, refractory elements such as Ca, Ti, and Mg are more likely to preserve isotopic effects because they are less likely to be processed into the gas phase where isotopic variations may be easily homogenized.

Oxygen-16

Enrichments in ^{16}O in CAIs are a ubiquitous feature. In fact, the ubiquity of ^{16}O enrichments at basically the same level in all CAIs is rather puzzling. Despite wide variability in other isotopic systems such as Ti, oxygen is rather uniform. This is despite highly anomalous oxygen isotope systematics of putative presolar grains (Nittler et al., 1997; Choi et al., 1999), indicating that exotic O isotopes were present in the solar system mix originally.

Clayton et al. (1973) found that the extreme enrichment in Allende material was approximately 50‰ (relative to the intersection of the Allende mixing line and the terrestrial mass fractionation line which is at +10‰). Clayton and Mayeda (1984) found similar enrichments in Murchison refractory separates, although more extreme compositions with enrichments up to 70‰ have been found in some Murchison inclusions (Fahey et al., 1987; Virag et al., 1991; Ireland et al., 1992). The extreme compositions show evidence for back-mixing with a more normal oxygen isotopic composition, and the composition of a mineral is apparently related to its susceptibility to resist O equilibration by diffusion (Blander and Fuchs, 1975; Ryerson and McKeegan, 1994). Melilite and anorthite typically have more normal oxygen isotopic compositions than pyroxene, while spinel and hibonite typically show extreme enrichments. However, the nature of these isotope effects is not easily explained in terms of diffusion because diffusion gradients that should be observed are not there (Ryerson and McKeegan, 1994; McKeegan et al., 1998). FUN inclusions also show ^{16}O enrichments, but they are complicated by an apparent mass-dependent fractionation from the ^{16}O-enriched reservoir along a line of slope 1/2 (i.e., mass-dependent fractionation), followed by back-reaction with a near-normal component.

Excluding FUN CAIs, variable replacement of oxygen thus results in the CAI mixing line being located between the exotic ^{16}O-enriched component and the normal compositions. Best-fit slopes of these lines typically have been approximately 0.95, although microanalysis of CAIs suggests a slope of 1.00 (Young and Russell, 1998). The distinction between slope 0.95 and 1.00 is important for the two main competing mechanisms for production of an ^{16}O-enriched reservoir. The original interpretation of the ^{16}O-rich reservoir is that it represents a nucleosynthetic enrichment in the early solar system, possibly related to the supernova responsible for ^{26}Al production (Clayton et al., 1973, 1977).

An alternative explanation proffered by Thiemens and Heidenreich (1983) is that oxygen isotopes can undergo mass-independent fractionation because of differing molecular stability of nonsymmetric (e.g., $^{16}O-C-{}^{17}O$) versus symmetric (e.g. $^{16}O-C-{}^{16}O$) molecules (e.g., Thiemens, 1996). This or another mechanism occurs in the Earth's atmosphere, where mass-independent O isotope fractionations are observed in O_3, CO_2, and other O-bearing molecules. Chemical fractionation is thus a highly attractive model for explanation of the ^{16}O enrichment. However, an explicit illustration of the applicability of non–mass-dependent fractionation in the solar nebula has not been demonstrated. In particular, CAIs form at high temperature, where these mechanisms may not be viable, and gas-solid fractionations are important as opposed to gas-gas fractionation. Nevertheless, the viability of this mechanism is implicit upon a fractionation slope of exactly 1.00. If the slope of 0.95 were real, the exotic composition would be required to have an anomalous $^{17}O/^{16}O$ (or $^{18}O/^{16}O$) to explain the exotic reservoir, which cannot be explained by the mass-independent fractionation model. However, the presence of a slope 1.00 mixing line does not preclude the presence of a pure ^{16}O-rich nucleosynthetic reservoir.

Aluminium-26

The nature of ^{26}Al abundances in solar system materials has been reviewed recently by MacPherson et al. (1995). The main feature of the Al-Mg systematics of refractory inclusions is the rather well defined upper cutoff of 5×10^{-5} for $(^{26}Al/^{27}Al)_0$. This upper limit is very important and has several fundamental implications for the incorporation of ^{26}Al into the solar system.

The source of the ^{26}Al is still a matter of conjecture. Supernovae produce ^{26}Al at a level of $10^{-3} \times {}^{27}Al$ (Clayton and Leising, 1987) and a supernova source would suggest that there is a decay interval of approximately 3 Ma while the $^{26}Al/^{27}Al$ ratio decayed from 10^{-3} to 5×10^{-5}. However, the gamma line for ^{26}Al decay has been detected coming from the galactic center at an abundance that cannot be produced entirely by supernovae. Presolar SiC grains contain inferred ^{26}Al at levels higher than produced in supernovae, with mainstream SiC having ratios up to 10^{-3} or so, and grains that may have a supernova origin having ratios up to 0.2 (Amari et al., 1992). The rare presolar oxide grains, corundum and hibonite, also show large excesses in Mg, corresponding to a similar range as found in carbide grains (Virag et al., 1991; Choi et al., 1999). The presolar grains certainly indicate that there are suitable nucleosynthetic sources. However, in order to achieve a uniform abundance, the source would have to have a fixed production ratio and there would have to be a fixed decay interval for travel from the point of nucleosynthesis to the solar system and then crystallization in solid objects. A scenario incorporating a variable production ratio compensated by different decay intervals is too abstruse to be a viable mechanism. The maximum production ratio indicated from the presolar grains (0.2) indicates a maximum decay interval of 12 half-lives, or approximately 9 Ma, to achieve a solar system $^{26}Al/^{27}Al$ of 5×10^{-5}.

Additional temporal constraints come from ^{41}Ca. The half-life of ^{41}Ca is only 0.1 m.y., and its abundance in the early solar system is estimated to be $1.5 \times 10^{-8} \times {}^{40}Ca$ (Srinivasan et al., 1994, 1996). The production ratio of ^{41}Ca has been estimated at $0.01 \times {}^{40}Ca$ (Wasserburg et al., 1995), and a value very similar to this has been measured directly from presolar graphite grains (Amari et al., 1992). This production ratio suggests a decay interval of the order of 19 half-lives, or 1.9 m.y., to achieve the 1.5×10^{-8} solar system value. Wasserburg et al. (1995) evaluated a common source for ^{41}Ca and ^{26}Al from an AGB star and used additional constraints from other radionuclides such as ^{60}Fe and ^{107}Pd. They conclude that the decay interval can only be of the order of 0.5 m.y. for internal consistency of these chronometers in CAIs. At this time, the source would have had a $^{26}Al/^{27}Al$ of 8×10^{-5} and a $^{41}Ca/^{40}Ca$ of 5×10^{-7}, based on the apparent measured values in CAIs.

The possibility of producing ^{26}Al *in situ* in the early solar system through T-Tauri radiation or the like has been examined, but does not produce the different radionuclides (^{26}Al, ^{41}Ca, ^{53}Mn, ^{60}Fe, etc.) in the predicted abundances (Wasserburg, 1985; Wasserburg et al., 1995; Shu et al., 1996; Lee et al., 1998). This may reflect uncertainties in the modeling of the radiation processes or that different nuclides have different relative contributions from *in situ* production and fresh nucleosynthetic injection. A case in point is the recent discovery of $^{11}B/^{10}B$ anomalies, which indicate the presence of ^{10}Be in CAIs (McKeegan et al., 2000) that can only have a spallogenic source, because ^{10}Be is consumed in stars, not produced. Nevertheless, the presence of spallogenic Be still does not preclude the presence of nucleosynthetic contributions in other elements.

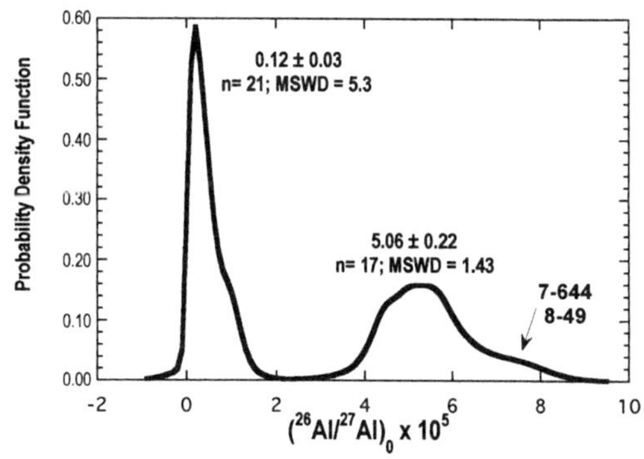

FIG. 16. Probability density function for ^{26}Al abundance in Murchison CAIs. Data are included only for CAIs with $^{27}Al/^{24}Mg > 10$. Two inclusions have marginally high $^{26}Al/^{27}Al$ values with respect to the main peak at $\sim 5 \times 10^{-5}$, but 17 inclusions are consistent with a single value at 5.06 ± 0.22 ($2\sigma_m$). Inclusions with low apparent initial ^{26}Al show a larger degree of scatter.

Another difficulty for modeling *in situ* production of an isotope is that the target nuclei are generally other elements, and so different chemical compositions can result in different apparent initial ratios.

Even with a uniform production ratio of the radiogenic nuclide, a uniform distribution in the proto-CAIs is not guaranteed, nor even expected. The preservation of isotopic anomalies suggests that packets of dust were processed in the solar nebula at high temperatures to produce melting and thence crystallization. Any old Al (i.e., pure ^{27}Al) present in that package will produce a lower apparent initial $^{26}Al/^{27}Al$. Thus the expectation for the distribution of ^{26}Al is that the maximum will only be attained when the CAI consists predominantly of freshly injected material. Thus the distributions shown by MacPherson et al. (1995) are unusual, in that there is a clear peak at 5×10^{-5} and another peak closer to zero, with some intermediate values. However, it is not clear whether these intermediate values are the result of disturbance, stochastic variability associated with low Al/Mg ratios, or temporal decay. Of great importance in assessing this behavior will be the abundance of ^{41}Ca in these inclusions because the shorter half life of ^{41}Ca will effect an exponential change in the abundance, whereas dilution or disturbance of ^{26}Al will produce a linear change of the same magnitude in ^{41}Ca. The main problem is that the abundance of ^{41}Ca is already difficult to measure at the 10^{-8} level, and the relative quantification between the predicted exponential decay and the linear dilution may be beyond present analytical capabilities.

Thus the maximum $^{26}Al/^{27}Al$ is the salient point of CAIs because any old ^{27}Al can simply dilute the solar $^{26}Al/^{27}Al$ to a lower value. CAIs can also be disturbed in terms of their Al-Mg systematics through redistribution of Mg in subsequent thermal processing. It is in fact surprising that so many CAIs do preserve a common $^{26}Al/^{27}Al$. This property is exemplified in Figure 16, which shows the probability density function for $^{26}Al/^{27}Al$ values of 40 Murchison hibonites analyzed by Ireland (1988, 1990). The pattern is essentially bimodal, consisting of CAIs with low $^{26}Al/^{27}Al$ ($< 10^{-5}$) and those around 5×10^{-5}. The populations are somewhat biased because the probability density function is binned in intervals of 1×10^{-6}, and hence the low values are contained in a fewer number of bins, which increases the height of the function.

Furthermore, the low $^{26}Al/^{27}Al$ values come from hibonites with high Al/Mg and are often free of spinel (PLACs), thus producing narrow gaussian profiles. SHIBs which contribute most of the 5×10^{-5} peak are characterized by having small hibonite crystals with lower Al/Mg, sitting in a spinel matrix and thus the measurement errors for SHIBs can be relatively high in terms of $^{26}Al/^{27}Al$. The peak with elevated $^{26}Al/^{27}Al$ has a mean of 5.06 ($\pm 0.22\ 2\sigma_m$) $\times 10^{-5}$ with an MSWD of 1.43. Two inclusions, 7-644 and 8-49 with apparently higher $^{26}Al/^{27}Al$, have been removed from the calculation. Nevertheless, the low MSWD for the remainder indicates that the analytical errors are largely consistent with there being only one component present. Even the standard deviation about this peak is only 1×10^{-5}.

Thus, CAI formation must occur over a very short interval of time, much less than one half-life of ^{26}Al (= 0.7 m.y.). It appears that the CAI data set we have comes from a temporally distinct period in the solar system. CAIs with low $^{26}Al/^{27}Al$ probably do not have temporal significance because of the possibility of adding old ^{27}Al associated with residues (as noted above). What is surprising is that there is no indication of more continuous mixing between these populations.

Trace-element fractionations

As noted above, the mineralogy of CAIs is consistent with high-temperature formation. But the chemical and isotopic systematics tell us more. The Group II pattern in particular is construed to be a condensate pattern and requires the separation of the most refractory elements in the first condensates and their removal from the locale prior to condensation of the Group II REE, which in turn must be isolated prior to condensation of the most volatile REE, Eu, and Yb (Boynton, 1975). The complementary ultrarefractory enriched pattern is found in some Murchison inclusions as well as in CAIs from other meteorites, although notably not Allende. The ultrarefractory pattern could be an early condensate removed from the gas phase, or a residue of material that was not quite completely evaporated. Isotopic-mass-fractionation measurements of hibonite-bearing inclusions show that on average, those with ultrarefractory-enriched patterns tend to have isotopically heavier Ti isotopes than those with the Group II (ultrarefractory-depleted patterns) (Ireland, 1990). This alone does not exclusively distinguish between a residue and condensate origin for the ultrarefractory-enriched material. The relative degrees of isotopic fractionation for either condensate or residue are a function of the composition of the starting material. Thus a residue will always be isotopically heavier than the starting material and will get heavier as evaporation continues. But the condensate evaporating from a residue will also become progressively heavier as the residue becomes heavier. Thus a late-stage condensate can ultimately become heavier than the original material.

The volatility fractionation of the REE and the other refractory lithophile elements has never been demonstrated experimentally and the ordering of these elements is largely based on a theoretically calculated volatility sequence modified using REE abundances from approximately 100 Allende CAIs (Kornacki and Fegley, 1986). That is, the ordering can be changed slightly so as to achieve the smoothest pattern of abundance versus condensation temperature. In detail, the apparent ordering can be compromised if significant fractionation of elements between two different phases takes place. This partitioning is a function of ionic radius; hence elemental abundances can change accordingly, and the resultant volatility sequence could change depending on the type of analysis that is carried out (for example, bulk analysis via INAA, or selected mineral analysis by ion microprobe). Trace-element partitioning appears as a smooth variation on the REE abundance pattern—for example, as a depletion of the REE with increasing ionic size in both hibonite and perovskite. This type of behavior is noticeable in hibonites with Group III patterns where the trend

TABLE 1. REE Classification of 396 CAIs

Meteorite	I, V	II	III	IV	VI	Ultra	UC	HAL	Total
Acfer 182[1,2]		17	1			3			21
Acfer 059 - El Djouf 001[1,2]		4							4
Allende[3]	54	36	12	3	6			1	112
Arch	1								1
Cold Bok.			2						2
Dhajala								1	1
Efremovka[4]	5	1					1		7
Essebi						1			1
Felix			1						1
Grosnaja	2								2
Kaba		1	7	12			2		22
Lancé	1	1							2
Leoville[4, 5]	8	4			1				13
Mighei[6]	10	23	6				28		67
Mokoia		5							5
Murchison[7]	5	33	35			7		2	82
Murray	1	5	2			1			9
Ornans		1				3			4
Semarkona		1							1
Vigarano[4,5]	12	5	2				1		20
ALH85085[8]	2	12	2		1		2		19
Total	101	149	70	15	8	15	34	4	396
Percent	25.5	37.6	17.7	3.8	2.0	3.8	8.6	1.0	

[1]Weber et al., 1995.
[2]Weber and Bischoff, 1994.
[3]Simon et al., 1998.
[4]Sylvester et al., 1993.
[5]Sylvester et al., 1992.
[6]MacPherson and Davis, 1994.
[7]Simon et al., 1996.
[8]Kimura et al., 1993.

of the heavy REE is easily seen. The corresponding phase containing the heavy REE is typically not present with the PLACs, but the glass in hibonite-glass spherules can show the complementary HREE-enriched pattern (Ireland et al., 1991).

Such fractionation is more difficult to identify in ultrarefractory-depleted patterns because this pattern is already depleted in the HREE. It can, however, be very noticeable on the ultrarefractory-enriched pattern where there is commonly a distinct roll-off in the abundances of the heaviest REE, which are the most refractory and should have the highest normalized concentrations. Four inclusions analyzed by Ireland (1990) all have roll-offs in the abundances of the HREE, indicating the role of REE partitioning during melting. There is no phase containing the remaining REE in these inclusions, although occasional metal-rich nuggets and other "hot spots" can contain high concentrations of ultrarefractory REE (Palme et al., 1982; Hinton et al., 1988) and a small proportion of these objects can dramatically affect the REE systematics. CAIs with their full complement of ultrarefractory REE are relatively uncommon (e.g., MH-115, Boynton et al., 1980; ALH85085 CAI-143, Kimura et al., 1993); both show flat patterns in the HREE, which are elevated above LREE abundances.

The production of the Group II pattern requires removal of an ultrarefractory-REE host phase (e.g., perovskite) from the nebular gas within a few degrees of its condensation temperature (Davis and Grossman, 1979), and so any fractionation scheme based on volatility requires some mechanism for holding the gas at exactly the right temperature to produce the fractionation. In a sense, the fractionation of these elements is all but miraculous by this mechanism, but the Group II pattern is extremely common in CAIs. Fegley and Ireland (1991) compiled REE data from 280 CAIs from 19 chondrites and found that the Group II pattern was present in one-third of all CAIs (Table 1). Moreover, the Group II pattern is evident even in bulk Allende (Fegley and Kornacki, 1984), reflecting both the high abundance of refractory inclusions in this meteorite and the high proportion of Group II CAIs. This is despite the possible selection bias of larger Allende CAIs having Group I or V patterns. In Murchison, for which a more random selection of CAIs is taken, Group II and III patterns are subequal, and only a few Group I, V patterns are evident.

Despite the common occurrence of Group II patterns in Allende, the complementary ultrarefractory-enriched pattern has never been found in that meteorite. Almost all patterns showing ultrarefractory-enriched patterns come from hibonite- and perovskite-dominated CAIs rather than the silicate-bearing Allende CAIs, although ultrarefractory-enriched patterns are found in the Ornans CO3 chondrite. In Murchison CAIs, the Group II–type patterns show distinct differences from Allende Group II in that they contain substantially higher Eu and Yb, and also typically show variability in the light REE abundances.

The complementarity of the ultrarefractory-depleted and ultrarefractory-enriched patterns from Murchison CAIs can be seen through the addition of the normalized abundances of two CAIs, 13-60 and 13-61 (Fig. 17). The simple addition of these two inclusions shows no major anomalies and results in a smoothly fractionated pattern. Particularly noteworthy is the removal of variability in the LREE, which is also seen to be related to volatility. As such, the differential partitioning between the Group II (condensate) and ultrarefractory-enriched pattern allows an estimation of the relative volatility of each element. However, the addition of the logarithms of the abundances belies the real mixing proportions

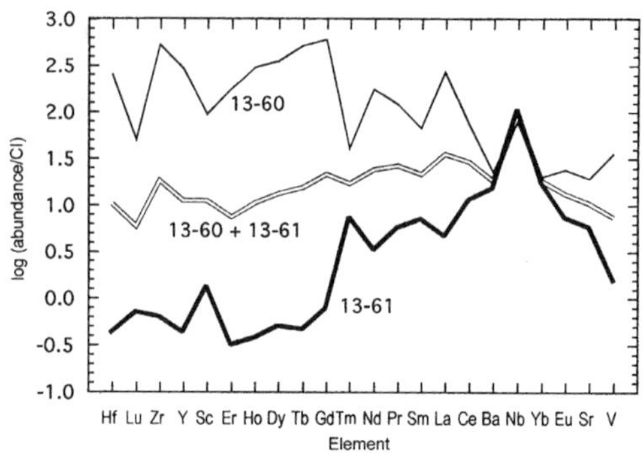

FIG. 17. Trace-element abundance plots for CAIs 13-60 and 13-61 ordered according to volatility. The hibonite patterns are more structured than Allende inclusions, probably because of magmatic fractionation of the REE during CAI crystallization. Nevertheless, summing the two patterns together reveals a relatively unfractionated pattern that indicates complementary volatile fractionations between these inclusions.

between these two components. The ultrarefractory-enriched patterns typically have much higher concentrations than do the ultrarefractory-depleted patterns. Perhaps, then, it is not surprising that the ultrarefractory-enriched pattern is so numerically underrepresented in CAIs.

The refractory/lithophile-element fractionations require that different chemical components of the inventory be isolated from each other at different temperatures. For example, the ultrarefractory-enriched Murchison patterns reflect a higher-temperature component than the ultrarefractory-depleted pattern. Despite these distinct temperature fractionations, the mineralogy of the individual CAIs is decoupled from the trace-element fractionation pattern. As such there is no distinction mineralogically between an ultrarefractory-enriched SHIB and an ultrarefractory-depleted SHIB. Furthermore, the mineralogical associations as governed by major-element abundances in CAIs can be construed to cover a range of condensation temperatures. For example, Allende CAIs require condensation of the REE through to partial condensation of silicon (as melilite, pyroxene, and feldspar) to achieve the silicate-dominated mineralogy, whereas Murchison CAIs are commonly regarded as higher-temperature objects because of the lack of silicates and the presence of hibonite and perovskite. However, Group II–type patterns are found in both Allende and Murchison CAIs. Thus, the mineralogy of CAIs appears to be a regional function of the solar nebula and is not necessarily related to a single-stage condensation event that covers all elements. It is likely that the original REE-bearing condensates were incorporated and dissolved into the proto-CAIs at a later stage.

Igneous fractionations are evident in CAIs through trace-element partitioning as well as major-element partitioning. This melting reflects secondary processing in the solar system. Most Murchison CAIs appear to have been molten, but there are fine-grained fluffy CAIs that have the physical characteristics of condensates. But real differences in trace-element chemistry between the textural condensates and the re-melted condensates do not exist. Evidently the condensation event imparts its own chemical memory on the material that constitutes the CAI. Re-melting the inclusion will simply repartition elements between solid and liquid. Yet at the temperatures required to melt these objects, it is still expected that some of the more volatile elements should be lost.

Such behavior is seen in HAL-type hibonite inclusions. HAL-type inclusions are characterized by isotopically heavy Ca and Ti (in terms of mass fractionation) as well as large Ce depletions (Ireland et al., 1992). These characteristics are consistent with thermal processing in an oxygen-rich environment, where Ce^{3+} is oxidized to the less refractory Ce^{4+}, which then evaporates from the object; schematically this process is represented by the net thermodynamic reaction

$$Ce_2O_3 \text{ (in solid solution)} + 0.5O_2 = 2CeO_2 \text{ (gas)}.$$

Similarly, the light isotopes of Ca and Ti are preferentially evaporated. Such behavior can be shown in experimental charges where less refractory materials are heated sufficiently to obtain hibonite-bearing residues (Ireland et al., 1992; Floss et al., 1996, 1998). However, other inclusions can show isotopic fractionation but do not show Ce depletions. This probably indicates that the effective oxygen fugacity was controlled by a combination of the H_2-rich nebular gas as well as by the more O_2-rich vapor evolved from the vaporizing CAIs. For instance, most CAIs contain reduced species as indicated by Ti^{3+}. However, the local environment may not be in equilibrium with the object being evaporated, and it is possible that sufficient time for equilibration might not exist if the time scale for melting is rapid. Furthermore, if inclusions become altered before re-melting, the nature of the evaporation event could be quite different.

It might be expected that FUN inclusions, being isotopically heavy in O, would have Ce depletions as well. Of the FUN inclusions, only HAL has a large Ce depletion, and it is more closely related to the group of hibonite inclusions. Other FUN inclusions apparently evaporated in a more reducing environment. But what is so special about FUN inclusions? Why is there a combination of a (relatively) large isotopic anomaly with mass fractionation? This association is not evident in Murchison CAIs with their much larger isotopic anomalies (in Ca and Ti at least). There is not a general progression in CAIs to compositions that are fractionated having larger non-linear anomalies. Furthermore PLACs and SHIBs were not necessarily the seeds for the larger CAIs from Allende. Very different sources for different types of inclusions probably are necessary.

Formation of CAIs

Time and place

CAIs are found commonly in carbonaceous chondrites, but also are found in many types of chondrites, including enstatite chondrites. The distinct characteristics of CAIs suggest that they have a common formation mechanism and perhaps the differences and commonalties among different types can tell us something about the actual mode of formation. However, there are innumerable problems in trying to ascertain what CAIs are really telling us about the early solar system. Were the inclusions formed during an early epoch when temperatures in the region were only then sufficiently high? Or, were CAIs formed continuously in a region through the early solar nebula, but could not escape later into the sampling region for chondritic parent bodies? We have a set of objects for which we have few constraints in terms of their original formation location and we have poor temporal resolution for timing of formation.

The entire population of CAIs is probably extremely biased, not only because of the dominance of a limited number of meteorites providing samples, but also because only those CAIs from a specific location and time period may have been entrained into the asteroidal parent bodies from which the source meteorites came. Other types of CAIs might have been formed at different stages, but these either fell into the Sun, or were swept up in the formation of the planets. Only the CAIs that ended up in the asteroid region have been preserved. In this case we must make a distinction between the CAI source region (the place where CAIs formed) and the CAI sampling region (that region of the source region from where the CAIs we see have been derived). The sampling region may sample material from all over the source region, but it need not. For instance, might another CAI sampling region closer to the Sun have picked up a larger proportion of ultrarefractory-enriched CAIs?

CAIs contain phases such as perovskite with high U and low initial Pb concentrations, and thus U/Pb ages with high precision can be obtained. On an absolute time scale, CAIs provide the oldest U-Pb ages in the solar system at 4,566 Ma (Allègre et al., 1995). While CAIs provide a useful U-Pb chronometry, most other objects in chondrites (and especially chondrules) have low U and relatively high initial Pb, making precise measurements difficult. The next major time marker that is known to high precision in the U-Pb system is the age of angrites, achondrite meteorites formed during differentiation and melting of small planetary bodies. The angrite Angra dos Reis has been precisely dated at 4,557.8 ± 0.7 Ma (Lugmair and Galer, 1992). Thus the total time required for U-Pb closure of CAIs through to crystallization and closure of differentiated magmas on parent bodies is on the order of 8 Ma. But despite the high analytical precision available in the U-Pb system, it is still not sufficient to decipher possible age differences in different CAIs. And no other absolute chronometer (K[Ar]-Ar, Rb-Sr, Sm-Nd, Re-Os, etc) can provide the precision that is available from U-Pb.

Relative age differences can be established through the use of short-lived radionuclides such as ^{26}Al, ^{53}Mn, ^{107}Pd, ^{129}I, etc. (Wasserburg, 1985; Podosek and Swindle, 1988), but these have proved somewhat problematic in terms of obtaining agreement among these systems. While any system can give an internally consistent picture of the chronology, intercomparison shows discrepent time scales (Podosek and Swindle, 1988).

A major problem in interpreting early solar system chronology is the difficulty of obtaining age constraints on the formation of chondrules. The difficulty with chondrules is that they contain high abundances of Mg and Cr, limiting the usefulness of ^{26}Al and ^{53}Mn decay schemes, and other schemes such as K-Ar, I-Xe, etc. are obfuscated by low-temperature processes that remobilize the parent elements and disturb the systems. Recent work in the ^{26}Al system for phases having high Al/Mg (especially anorthite) suggests that chondrules may have formed within a million years or so of the CAIs (e.g., Hutcheon et al., 2000). Chondrules also require extremely high temperatures to melt, and so the extension of the CAIs forming epoch through chondrule formation is possibly a natural extension. All indications are that CAI formation took place over a short period of time and was quickly followed by chondrule formation.

But even though CAI formation took place over an apparently very short interval of time, the association of a number of diametrically opposed features cannot be accommodated in a simple single-stage event. REE fractionations apparently require a high-temperature condensation event for at least half of the CAIs having a Group II/ultrarefractory-depleted or ultrarefractory-enriched pattern, and yet there are no apparent mineralogical differences, especially in the Murchison inclusions. The chemistry of Allende inclusions suggests diverse oxidation states—in many cases, REE are not oxidized (no Ce anomaly), but siderophiles such as Mo and W have apparently been oxidized at some stage (Fegley and Palme, 1985), while the presence of Ti^{3+} indicates reducing conditions. These apparently contradictory indications suggest different processing for different components of the CAIs before final assembly of all the packages into proto-CAIs. Even then, the petrography of Allende (in particular) CAIs suggests that CAIs were melted and re-melted in the early solar system. What then are the minimum components required in the early solar system?

Condensates and residues

Condensation is required to form the Group II pattern. But the preservation of diverse isotopic anomalies is probably only consistent with the survival of refractory residues. We have previously suggested that the Group II pattern (and by extension other ultrarefractory-fractionated patterns) may be a result of condensation around another star and that this came into the solar system as a chemical memory (Fegley and Kornacki, 1984; Ireland, 1990). Although condensation around other stars is certainly occurring, it is probably unlikely that such a large fraction of solar system CAIs should have this pattern. We thus explore the consequences of having condensates and residues present as two end-members in the early solar system.

The specific characteristic of condensation is the ultrarefractory-REE fractionated pattern. Notwithstanding some notable exceptions, this pattern is typically accompanied by normal Ca and Ti isotopic compositions and $^{26}Al/^{27}Al$ of 5×10^{-5}, as is particularly evident in the Murchison SHIB inclusions. We shall take these systematics as the attributes of the solar condensate.

Conversely, the presence of Ca and Ti isotopic anomalies is indicative of residual material that survived homogenization. In terms of Murchison inclusions, the PLACs are the carriers of these anomalies. Typically, these inclusions have Group III–type REE patterns with little or no excess ^{26}Mg associated with ^{26}Al decay.

It is unlikely that PLACs, SHIBs, or any other type of CAIs are pristine representatives of these two processes. However, these inclusions probably represent the best polarization of these components.

A longstanding issue with the PLACs is the absence of ^{26}Al effects, when the preservation of isotopic anomalies tends to suggest an early formation time. The prevalence of a maximum $^{26}Al/^{27}Al$ at 5×10^{-5} in the early solar system has been used as an argument that there was a uniform distribution of ^{26}Al and thence $^{26}Al/^{27}Al$ can be used as a chronometer (MacPherson et al., 1995). However, true uniformity of distribution would only be realized if there were no solid residues left in the early solar system—that is, all material in the solar system would need to be in the gas phase. As has been noted many times before, the preservation of diverse isotopic anomalies does not support this scenario. Any surviving refractory material with ^{27}Al cannot have a $^{26}Al/^{27}Al$ of 5×10^{-5}, and these objects cannot conform to a chronometric scheme based on ^{26}Al abundance.

The corollary to the presence of old ^{27}Al is that CAIs with $^{26}Al/^{27}Al$ of 5×10^{-5} must comprise Al that is almost entirely direct condensate of solar material. The canonical solar system value can only be obtained provided the CAIs formed early and the CAIs incorporated no old ^{27}Al. If CAIs formed over a short period of time (compared to the half-life), $^{26}Al/^{27}Al$ can be used as a tracer of the degree of ^{27}Al mixing. The PLAC hibonites incorporate residual material with diverse Ca and Ti isotopic compositions, but these objects are also likely to be rich in old ^{27}Al. It is thus the association of Ca and Ti isotopic anomalies with old ^{27}Al that yields the low apparent initial $^{26}Al/^{27}Al$ values, rather than the $^{26}Al/^{27}Al$ being a temporal indicator suggesting that PLAC hibonites are young.

Allende CAIs with near canonical $^{26}Al/^{27}Al$ should therefore contain little residue material and this is supported by their overall smaller Ti and Ca isotopic anomalies. FUN inclusions, with low $^{26}Al/^{27}Al$, have larger Ca and Ti anomalies, indicating a larger component of residue material. Thus the possible size of Ca and Ti isotopic anomalies is a natural consequence of the amount of residual material mixed in with a CAI. Still though, there is no particular indication as to why FUN inclusions should show large mass fractionation effects. In any case, the Allende CAIs are probably a poor set of inclusions to model early solar system behavior because of their size and the likely mixture of components present within them. The Ca-Ti compositions of Allende CAIs can be explained by a small amount of residue being incorporated into the proto-CAIs, which will not unduly upset the canonical $^{26}Al/^{27}Al$; a 20% addition of residue will only lower the $^{26}Al/^{27}Al$ from 5 to 4×10^{-5}, assuming the residue is dominated by ^{27}Al.

The association of Ca and Ti isotopic anomalies with low $^{26}Al/^{27}Al$ suggests that the presolar carriers responsible for these characteristics were aluminous and possibly even hibonite itself. There is no *a priori* expectation of a distinct trace-element signature for the residues. That is, any REE abundance might be possible depending on the r- to s-process abundance ratio as well as mixtures in the precursors. Isotopic compositions of REE would be illuminating in this respect, but we can already note that Zr isotopes in hibonite 13-13, which has a 27% excess of ^{50}Ti, are normal (Ireland, 1994), and therefore a local origin for the REE cannot be ruled out. Indeed the Group III pattern is the most common for PLACs. This pattern indicates the more refractory REE are in their solar abundances, and they may therefore be unassociated with the carriers of the Ca and Ti isotopic anomalies. If the PLAC precursor was mixed with material that had solar relative abundances of the REE, a Group III pattern would be produced simply by the melting of the proto CAIs and thence evaporation of Eu and Yb. These observations suggest that the residues themselves have little in the way of REE and that the PLAC REE systematics were established in the solar nebula.

The problem with oxygen

Large oxygen isotopic anomalies are a characteristic feature of CAIs, yet despite widely different Ca and Ti isotopic compositions, initial ^{26}Al abundance, and REE patterns, ^{16}O excesses in CAIs of several percent are the norm. In comparison, presolar oxides (including corundum and hibonite) have diverse oxygen isotopic compositions with highly anomalous $^{17}O/^{16}O$ and $^{18}O/^{16}O$ ratios (Hutcheon et al., 1994; Nittler et al., 1997; Choi et al., 1999). Furthermore, both PLACs (dominantly residues) and SHIBs (condensate), as well as Allende CAIs have the same oxygen isotope compositions, typically with an enrichment in ^{16}O of 50‰ relative to terrestrial oxygen. Even the most anomalous hibonite, Murchison inclusion BB-5, only shows an enrichment of 70‰ (Fahey et al. 1987).

The lack of any diversity in oxygen isotopic compositions is not consistent with the retention of original oxygen in the CAI-precursor residues, if the compositions of the presolar oxides are indicative of the possible range in compositions that existed. Clayton and Mayeda (1984) argued that the ^{16}O enrichment is a signature of solids in the solar nebula while normal O was in the gas phase, but this reflects possibly only the final stage in CAI formation. The lack of any diversity in O isotopic compositions in CAIs suggests that CAI precursors must have equilibrated with a ^{16}O-rich gas in the solar nebula, thereby erasing their original compositions. In order to homogenize the precursors, the ^{16}O-enriched gas must have been pervasive throughout the CAI-forming region, at least in the initial stages. One potential source is the evaporation of a ^{16}O-enriched phase with a low vaporization temperature, although there is no indication for the preservation of this phase in meteorites. Alternatively, a ^{16}O-enriched reservoir may have been produced by chemical processing of material in the accretion disk (Thiemens, 1996). Nevertheless, the oxygen isotopic systematics tend to suggest a progressive change from the extreme ^{16}O enrichment of ~70‰ found in Murchison CAIs, through 50‰ for Allende CAIs, and to 20‰ and less for chondrules (Leshin et al., 1997, 2000). The formation of planetary bodies evidently took place with O-isotope heterogeneity, as indicated by differing compositions of meteorite parent bodies, Earth, and Moon, but with compositions closer to what we would regard as normal. CAI O-isotope systematics do not reflect the protoliths or precursors in the solar nebula and so polarization of systematics is not be expected between condensates and residues.

Processes and mechanisms

The existence of CAIs attests to high temperatures being attained in the solar nebula. But the CAI systematics described above cannot be placed into a single-stage model of formation that is consistent with the condensation sequence and the preservation of isotopic anomalies. Both condensates and residues are required and both experienced high temperatures. In the simplest scenario, partial condensation can be invoked to produce the Group II REE fractionation, whereas isotopic anomalies are preserved in refractory residues such as Murchison PLAC CAIs. But the systematics are more complicated than this.

It appears that condensation of the refractory trace elements must be decoupled from the major-element components Ca, Al, Ti, Si, and Mg. The evidence for this is the lack of correlation between REE and lithophile patterns and bulk composition in Allende CAIs, the lack of correlation between siderophile patterns and bulk composition in Allende CAIs, and the petrographic similarity of ultrarefractory-enriched and ultrarefractory-depleted CAIs in Murchison. In Allende CAIs, it is possible that all the major elements condensed with the Group II pattern because of the possible continuity of condensation from light REE through to Si and Mg. However, in Murchison CAIs, it would not be expected that an ultrarefractory residue could have the same major-element composition as the condensate (Group II). At the very least, elemental volatility-controlled fractionation of Ca, Ti, and Al relative to Mg and Si would occur with the former elements present in the ultrarefractory-enriched CAIs, and the latter with the Group II condensates. This is clearly not the case.

Allende CAIs can simply form from a mix of condensates to produce the range of trace element patterns observed. The ultrarefractory trace elements may have mixed in with Group II condensate to reform a flat abundance pattern for the refractory REE (which is the Group III characteristic), or with fully chondritic relative abundances (Group I). However, the ultrarefractory component was not preserved in the Allende CAI formation area. This further suggests a lack of association of the major elements Al, Ca, and Ti, which themselves are highly refractory, with the REE condensation sequence, at least insofar as the required chemical inventory for Allende CAIs is concerned. In the Murchison SHIB CAI-forming zone, pockets of ultrarefractory-enriched and ultrarefractory-depleted REE are preserved. Eventually the REE components mixed with very similar proportions of the major elements to form SHIB CAIs with similar bulk compositions of major elements (i.e., dominated by spinel-hibonite).

Allende CAIs are dominated by melilite and pyroxene (both silicates), whereas the dominant mineralogy in SHIBs is oxide: spinel-hibonite-perovskite. There is no Si-bearing phase in SHIBs despite Si condensing in melilite at a higher temperature than spinel (Fig. 1). Basically, these inclusions represent a similar problem as Group II patterns do for REE, only in reverse. SHIBs preserve the high-temperature (REE) component in a lower-temperature matrix (spinel), and the intermediary silicate is missing. In the context of condensation, this would require that the melilite failed to condense before Mg came out as spinel, and the CAI was isolated from the gas prior to forsterite/diopside condensation. Alternatively, condensation of spinel may have been achieved in a Si-depleted region of the solar nebula that could have formed by vaporization of chondritic material, separation of the vapor and residue, and reaction of the Mg-rich vapor with the Al-rich residue to form the bulk compositions observed. Conversely, the evaporation experiments of Hashimoto (1983) indicate that Mg was more volatile than Si, and so evaporation and recondensation of spinel is possible.

CAIs experienced multiple reheating episodes, as indicated by textural relationships. However, most subsequent events had to be at temperatures sufficient to melt the CAIs, but not enough to cause complete evaporation of Mg. However, even at these temperatures Eu and Yb will be evaporated and recondensed, allowing a variety of REE patterns (such as Group I, III, V, and VI). The FUN inclusions may represent an early phase of CAIs where isotopic heterogeneities were more widespread and heating to higher temperatures after formation was possible. Even then, however, all elements cannot experience Rayleigh mass fractionation at the same time. For instance, Mg will be totally evaporated from the residue before Ti begins to evaporate.

Thus CAIs represent the culmination of many processes that provide inconsistent indications to the formation environment in the solar nebula. The lack of systematics in Allende CAIs is therefore not surprising. But the systematic behavior of Murchison CAIs in general allows the extraction of some details of the individual processes.

The solar nebula

From our present knowledge of the solar system, and inferences drawn from observations of other young stellar systems, we can set up some boundary conditions and see what the implications are for formation of CAIs. The Sun is now sufficiently hot to hold all elements in a gaseous form. It is likely that once ignited, the early sun could do the same. At the outer edges of the solar system, comets and the survival of presolar materials in meteorites indicate that not all presolar material was held in the gaseous phase (Fegley, 1999). In between these extremes, there is partial to complete volatilization, depending on the volatility of individual chemical species. This is also the condensation line for vaporized material in a cooling nebula. The condensation line is not a constant for the solar nebula, but depends on a number of parameters such as in-fall rate, opacity, etc.

A description of the formation of CAIs requires an understanding of the evolution of the solar nebula, which comprises the transformation of a molecular cloud into a proto-solar system. With the direct observation of proto-planetary systems with the Hubble Space Telescope, such discussion can be placed into a context that has not previously been possible. Cameron (1995) divided the evolution of the solar system into four stages, as summarized below:

Stage 1. Molecular cloud collapse. In-falling material produces the nebular disk from the molecular cloud core over a time period of a few times 10^5 years. Initially the disk mass is greater than the mass of the proto-Sun, but ultimately most matter is captured by the Sun.

Stage 2. Disk dissipation. The Sun forms at a rate of 2×10^{-5} solar masses per year—i.e., a total time of 50,000 years. Matter falling onto the accretion disk is transported to the proto-Sun—i.e., matter is transported inward, but angular momentum is transported outward through the geometrically thin disk. The major transport mechanisms are spiral density waves, disk-driven bipolar outflows, and the Balbus-Hawley magnetic instability.

Stage 3. Terminal accumulation of the Sun. The final accumulation lasts for 1 to 2 million years, with an accumulation rate decreasing from 10^{-7} to 10^{-8} solar masses per year. The proto-Sun becomes a classical T-Tauri in this phase. Planetary accretion occurs.

Stage 4. Loss of nebular gas. The Sun becomes a weak line T-Tauri star in a time period covering 3–30 million years. Material is no longer being accreted from the disk, and the T-Tauri wind removes gas from the inner nebula.

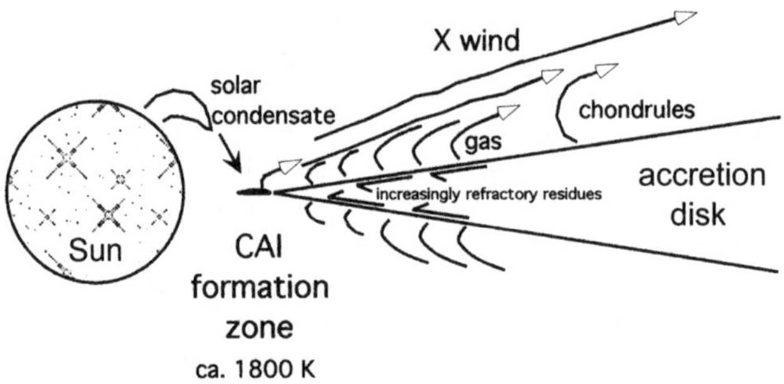

FIG. 18. Schematic view of the CAI formation process in the early solar system. CAIs are formed from a mixture of solar condensate and refractory residue falling in through the accretion disk. The solar condensate is responsible for the refractory lithophile fractionations typically observed in CAIs. The residues carry isotopic anomalies in refractory elements. CAIs are melted and transported to the planetary reaches of the solar system by protostellar winds such as the X-wind postulated by Shu et al. (1996).

Clearly, the highest temperatures in the solar nebula will be obtained with the highest in-fall rates and these occur in the earliest stages (predominantly stage 2) of nebula evolution. The entire span of Stages 1 and 2 is only on the order of 150,000 years, which is entirely consistent with the $^{26}Al/^{27}Al$ systematics of the SHIB inclusions. However, chondrule formation appears to occur at least one half-life after CAI formation, which is of the order of 750,000 years, and so the quandary of the mechanism of chondrule formation in the solar system remains.

The most likely location for the high temperatures required for CAI formation is close to the Sun and close to the mid-plane of the accretion disk of the solar nebula. Woolum and Cassen (1999) have examined T-Tauri stars and found temperatures at the mid-plane only exceed 1500 K at distances typically less than 0.5 a.u. Temperatures at earlier times (1 m.y. earlier) could have been substantially higher, allowing volatilization of silicates. But CAI temperatures require volatilization of REE, and this is likely to occur only in close proximity to the Sun.

In fact, it is likely that the only place where REE can exist entirely in the gas phase is within the Sun. Then condensates must be samples of the Sun that have been ejected to the CAI formation region. Such outflow could be provided by winds, flares, or jets of material from the early Sun. Progressive condensation of the gas to smokes or small grains (Nuth and Dodd, 1983) will occur as the outflow progresses, and so a mechanism for elemental fractionation exists in this scenario. Thus, the ultrarefractory carriers condense out closest to the Sun and this can be most readily re-accreted back to the Sun. In this case, a Group II pattern is to be expected as more common than the ultrarefractory-enriched pattern. These REE components are mixed in the CAI-forming region to greater or lesser degrees. Aluminium condensing from the Sun has the canonical $^{26}Al/^{27}Al$ ratio, and the oxygen isotopic composition in the gas of the condensation region is ^{16}O enriched. Furthermore, this oxygen largely equilibrates with infalling refractory residues, wiping out original O isotopic compositions. It remains an unknown as to whether this O composition reflects the nebula, the Sun, or the chemical processing in the region (Wiens et al., 1999). Condensate material can be accreted onto residues and form CAIs such as PLACs, which retain isotopic diversity in the refractory elements Ca and Ti, and retain old ^{27}Al.

The formation of CAIs must occur close to the Sun simply to attain the temperatures necessary to melt CAIs (~1800–2000 K). Once formed, these CAIs must be ejected out of the inner region to the asteroid zone, where they were eventually collected in chondritic materials. However, CAIs also could experience multiple heating events if the ejection mechanism is not very efficient and some CAIs can fall out and remain close to the CAI-forming region while others are transported away to the asteroid belt.

Such a scenario is very much in accord with the X-wind models of Shu and colleagues (Shu et al., 1996; Lee et al., 1998; Shang et al., 2000). In their model, CAIs form close to the Sun at ~0.06 AU, and are then thrown out to planetary distances from a bipolar outflow. The CAIs (and chondrules) are picked up by the aerodynamic drag of a magneto-centrifugally driven wind. The source of material for CAIs and chondrules is the infalling accretion disk.

Proto-CAIs and -chondrules experience flash heating as they are taken out of the shielded disk and into the sunlight. The general model is illustrated in Figure 18.

There remain several unsatisfactory corollaries to this model. If volatiles are driven off the rapidly melting objects, substantial isotopic mass fractionation should be present in the transformation from an aggregate that is near-chondritic to refractory. The exposure to energetic particles can produce short-lived radionuclides, although not in direct proportion to observations.

However, the presence of bipolar jets in young stellar objects is now an essential part of the evolution of a star on to the main sequence. This provides the source of energy for transport of CAIs from close to the Sun to the outer reaches of the solar system where they can be entrained in asteroidal bodies. In view of the chemical and isotopic systematics reviewed here, close proximity to the Sun is essential for the mixing of solar condensate with refractory residues that survive transport into the solar system.

Synthesis

CAIs represent the earliest formed objects in the solar system. They preserve chemical and isotopic features not found in any other objects and are a window into high-temperature processes in the solar system.

Trace-element patterns with ultrarefractory depletions and enrichments require a condensation event to fractionate these elements. The only place compatible with such extreme temperatures is the Sun itself. Condensates from solar gas eruptions or outflows are then mixed with proto-CAIs.

The ^{16}O enriched signature in CAIs might be a response to the extreme chemical processing taking place in the leading edge of the accretion disk. There is no correlation between O isotopic anomalies and anomalies in other elements, suggesting that any original nucleosynthetic anomalies have been wiped out.

Isotopic anomalies in refractory elements (especially Ca and Ti) are preserved in refractory residues.

Short-lived radionuclides may be produced by high-energy particle reactions, but the abundance of ^{26}Al and its correlation with other chemical and isotopic characteristics suggests that it is indeed a nucleosynthetic input. The canonical ^{26}Al/^{27}Al of the early solar system can only be obtained for condensates largely free of aluminous (i.e., old ^{27}Al) residues.

The final chemical and isotopic systematics cannot be described in terms of any single process, but require the superposition of condensation of solar gas, the preservation of isotopic anomalies in refractory residues, and the melting of mixtures of these components into the objects we now see. CAIs remain enigmatic, but the observed correlations in chemistry and isotopics, and better insights into the astrophysical environment, offer us the possibility of a better understanding of the chemistry of the early solar system.

Acknowledgments

This work was supported by NASA Cosmochemistry Program (Grant NAG5-7609 to TRI) and the NASA Origins of Solar System Program (Grant NAG5-4323 to BF). Kevin McKeegan and Sasha Krot kindly provided reviews that were of great use in the preparation of this manuscript.

REFERENCES

Allègre, C. J., Gèrard, M., and Göpel, C., 1995, The age of the Earth: Geochim. et Cosmochim. Acta, v. 59, p. 1445–1456.

Allen, J. M., Grossman, L., Davis, A. M., and Hutcheon, I. D., 1978, Mineralogy, textures, and mode of formation of a hibonite-bearing Allende inclusion: Proc. Lunar Planet. Sci. Conf., v. 9, p. 1209–1233.

Amari, S., Hoppe, P., Zinner, E., and Lewis, R. S., 1992, Interstellar SiC with unusual SiC compositions: Grains from a supernova?: Astrophys. Jour. (Lett.), v. 394, p. L43–L46.

Bar-Matthews, M., Hutcheon, I. D., MacPherson, G. J., and Grossman, L., 1982, A corundum-rich inclusion in the Murchison carbonaceous chondrite: Geochim. et Cosmochim. Acta. v. 46, p. 31–41.

Beckett, J. R., and Stolper, E., 1994, The stability of hibonite, melilite and other aluminous phases in silicate melts: Implications for the origin of hibonite bearing inclusions from carbonaceous chondrites: Meteoritics, v. 29, p. 41–65.

Birck, J. L., and Allègre, C. J., 1984, Chromium isotopic anomalies in Allende refractory inclusions: Geophys Res. Lett., v. 11, p. 943–946.

Birck, J. L., and Lugmair, G. W., 1988, Nickel and chromium isotopes in Allende inclusions: Earth Planet. Sci. Lett., v. 90, p. 131–143.

Blander, M., and Fuchs, L. H., 1975, Calcium-aluminum-rich inclusions in the Allende meteorite: Evidence for a liquid origin: Geochim. et Cosmochim. Acta, v. 39, p. 1605–1619.

Boynton, W. V., 1975, Fractionation in the solar nebula: Condensation of yttrium and the rare earth elements: Geochim. et Cosmochim. Acta, v. 39, p. 569–584.

Boynton, W. V., Frazier, R. M., and Macdougall, J. D., 1980, Identification of an ultrarefractory component in the Murchison meteorite: Lunar Planet. Sci., v. XI, p. 103–105.

Cameron, A. G. W., 1962, The formation of the Sun and planets: Icarus, v. 1, p. 13–69.

______, 1995, The first ten million years in the solar nebula: Meteoritics, v. 30, p. 133–161.

Cameron, A. G. W., and Truran, J. W., 1977, The supernova trigger for formation of the solar system: Icarus, v. 30, p. 447–461.

Choi, B.-G., Wasserburg, G. J., and Huss, G. R., 1999, Circumstellar hibonite and corundum and nucleosynthesis in asymptotic giant branch stars: Astrophys. Jour. (Lett.), v. 522, p. L133–L136.

Clark, R. S., Jr., Jarosewich, E., Mason, B., Nelen, J., Gomez, M., and Hyde, J. R., 1970, The Allende, Mexico, meteorite shower: Smithson. Contrib. Earth Sci., v. 5, p. 1–53.

Clayton, D. D., 1982, Cosmic chemical memeory: A new astronomy: Jour. Roy. Astron. Soc., v. 23, p. 174–212.

Clayton, D. D., and Leising, M., 1987, ^{26}Al in the interstellar medium: Phys. Rept., v. 144, p. 1–50.

Clayton, R. N., Grossman, L., and Mayeda, T. K., 1973, A component of primitive nuclear composition in carbonceous chondrites: Science, v. 182, p. 485–488.

Clayton, R. N., Hinton, R. W., and Davis, A. M., 1988, Isotopic variations in the rock-forming elements in meteorites: Phil. Trans. Roy. Soc. Lond., v. A325, p. 483–501.

Clayton, R. N., and Mayeda, T. K., 1977, Correlated oxygen and magnesium isotope anomalies in Allende inclusions, I: Geophys. Res. Lett., v. 4, p. 295–298.

______, 1984, The oxygen isotope record in Murchison and other carbonaceous chondrites: Earth Planet. Sci. Lett., v. 67, p. 151–161.

Clayton, R. N., Onuma, N., Grossman, L., and Mayeda, T. K., 1977, Distribution of the presolar component in Allende and other carbonaceous chondrites: Earth Planet. Sci. Lett., v. 34, p. 209–224.

Davis, A. M., and Grossman, L., 1979, Condensation and fractionation of rare earths in the solar nebula: Geochim. et Cosmochim. Acta, v. 43, p. 1611–1632.

Fahey, A. J., Goswami, J. D., McKeegan, K. D., and Zinner, E., 1987, ^{16}O excesses in Murchison and Murray hibonites: A case against a late supernova injection origin of isotopic anomalies in O, Mg, Ca, and Ti: Astrophys. Jour.(Lett.), v. 323, p. L91–L95.

Fegley, B., Jr., 1999, Chemical and physical processing of presolar materials in the solar nebula and the implications for preservation of presolar material in comets: Space Sci. Rev., v. 90, p. 239–252.

Fegley, B., Jr., and Ireland, T. R., 1991, Chemistry of the rare earth elements in the solar nebula: Eur. Jour. Solid State Inorg. Chem., v. 28, p. 335–346.

Fegley, B., Jr., and Kornacki, A. S., 1984, The geochemical behavior of refractory noble metals and lithophile trace elements in refractory inclusions in carbonaceous chondrites: Earth Planet. Sci. Lett., v. 68, p. 181–197.

Fegley, B., Jr., and Palme, H., 1985, Evidence for oxidizing conditions in the solar nebula from Mo and W depletions in refractory inclusions in carbonaceous chondrites: Earth Planet. Sci. Lett., v. 72, p. 311–326.

Floss, C., El Goresy, E., Zinner, E., Kransel, G., Rammansee, W., and Palme, H., 1996, Elemental and isotopic fractionations produced through evaporation of the Allende CV3 chondrite: Implications for the origin of HAL-type hibonite inclusions: Geochim. et Cosmochim. Acta, v. 60, p. 1975–1997.

Floss, C., El Goresy, E., Zinner, E., Palme, H., Weckwerth, G., and Rammansee, W., 1998, Corundum-bearing residues produced through the evaporation of natural and synthetic hibonite: Meteoritics& Planet. Sci., v. 33, p. 191–206.

Fuchs, L.H., Olsen, E., and Jensen, K. J., 1973, Mineralogy, mineral chemistry, and composition of the Murchison (C2) meteorite: Smithson. Contrib. Earth Sci., v. 10, p. 1–39.

Gray, C. M., and Compston, W., 1974, Excess ^{26}Mg in the Allende meteorite: Nature, v. 251, p. 495–497.

Grossman, L., 1972, Condensation in the primitive solar nebula: Geochim. et Cosmochim. Acta, v. 36, p. 597–619.

Grossman, L., 1975, Petrography and mineral chemistry of Ca-rich inclusions in the Allende meteorite: Geochim. et Cosmochim. Acta, v. 39, p. 433–454.

______, 1980, Refractory inclusions in the Allende meteorite: Ann. Rev. Earth Planet. Sci., v. 8, p. 559–608.

Hartmann, D., Woosley, S. E., and El Eid, M. F., 1985, Nucleosynthesis in neutron-rich supernova ejecta: Astrophys. Jour. (Lett.), v. 297, p. 837–845.

Hashimoto, A., 1983, Evaporation metamorphism in the early solar nebula—evaporation experiments on the melt FeO-MgO-SiO_2-CaO-Al_2O_3 and chemical fractionations of primitive materials: Geochem. Jour., v. 17, p. 111–145.

Haskin, L. A., Helmke, P. A., Paster, T. P., and Allen, R. O., 1971, Rare earths in meteoritic, terrestrial, and lunar matter, *in* Activation anal. geochem. cosmochem.: Oslo, Universitetsforlaget, Proc. NATO Advanced Study Institute, p. 201–218.

Heydegger, H. R., Foster, J. J., and Compston, W., 1979, Evidence of a new isotopic anomaly from titanium isotopic ratios in meteoric materials: Nature, v. 278, p. 704–707.

Hinton, R. W., Davis, A. M., and Scatena-Wachel, D. E., 1987, Large negative ^{50}Ti anomalies in refractory inclusions from the Murchison carbonaceous chondrite—Evidence for incomplete mixing of neutron-rich supernova ejecta into the solar system: Astrophys. Jour. (Lett.), v. 313, p. 400–428.

Hinton, R. W., Davis A. M., Scatena-Wachel, D. E., Grossman, L., and Draus, R. J., 1988, A chemical and isotopic study of hibonite-rich refractory inclusions in primitive meteorites: Geochim. et Cosmochim. Acta, v. 52, p. 2573–2598.

Hutcheon, I. D., Huss, G. R., Fahey, A. J., and Wasserburg, G. J., 1994, Extreme ^{26}Mg and ^{17}O enrichments in an Orgueil corundum: Identification of a presolar oxide grain: Astrophys. Jour. (Lett.), v. 425, p. L97–L100.

Hutcheon, I. D., Krot, A. N., and Ulyanov, A. A., 2000, ^{26}Al in anorthite-rich chondrules in primitive carbonaceous chondrites: Evidence chondrules post-date CAIs [abs.]: Lunar Planet Sci., v. XXXI, abs. no. 1869.

Hutcheon, I. D., Steele, I. M., Wachel, D. E. S., Macdougall, J. D., and Phinney, D., 1983, Extreme Mg fractionation and evidence of Ti isotopic variations in Murchison refractory inclusions: Lunar Planet. Sci., v. XIV, p. 339–340.

Ihinger, P. D., and Stolper, E., 1986, The color of meteoritic hibonite: an indicator of oxygen fugacity. Earth Planet. Sci. Lett., v. 78, p. 67–79.

Ireland, T. R., 1988, Correlated morphological, chemical and isotopic characteristics of hibonites from the Murchison carbonaceous chondrite: Geochim. et Cosmochim. Acta, v. 52, p. 2827–2839.

______, 1990, Presolar isotopic and chemical signatures in hibonite-bearing refractory inclusions from the Murchison carbonaceous chondrite: Geochim. et Cosmochim. Acta, v. 54, p. 3219–3237.

______, 1994, Isotopically normal zirconium in Murchison hibonite 13-13: Implications for a link between e and r processes: Astrophys. Jour. (Lett.), v. 434, p. L79–L81.

______, 1995, Ion microprobe mass spectrometry: Techniques and applications in cosmochemistry, geochemistry, and geochronology, *in* Hyman, M., and Rowe, M., Advances in analytical geochemistry, v. 2: Greenwich, UK, JAI Press, p. 1–118.

Ireland, T. R., Fahey, A. J., and Zinner, E. K., 1988, Trace-element abundances in hibonites from the Murchison carbonaceous chondrite: Constraints on high-temperature processes in the solar nebula: Geochim. et Cosmochim. Acta, v. 52, p. 2841–2854.

______, 1991, Hibonite-bearing microspherules: A new type of refractory inclusion with large isotopic anomalies: Geochim. et Cosmochim. Acta, v. 55, p. 367–379.

Ireland, T. R., Zinner, E. K., Fahey, A. J., and Esat, T. M., 1992, Evidence for distillation in the formation of HAL and related hibonite inclusions: Geochim. et Cosmochim. Acta, v. 56, p. 2503–2520.

Jungck, M. H. A., Shimamura, T., and Lugmair, G. W., 1984, Ca isotope variations in Allende: Geochim. et Cosmochim. Acta, v. 48, p. 2651–2658.

Kimura, M., El Goresy, A., Palme, H., and Zinner, E., 1993, Ca-, Al-rich inclusions in the unique chondrite ALH85085: Petrology, chemistry, and isotopic compositions: Geochim. et Cosmochim. Acta, v. 57, p. 2329–2359.

Kornacki, A. S., and Fegley, B., Jr., 1986, The abundance and relative volatility of refractory trace elements in Allende Ca, Al-rich inclusions: Implications for chemical and physical processing in the solar nebula: Earth Planet. Sci. Lett., v. 79, p. 217–234.

Lauretta, D. S., and Lodders, K., 1997, The cosmochemical behavior of beryllium and boron: Earth Planet. Sci. Lett., v. 146, p. 315–327.

Lee, T., 1988, Isotopic anomalies, *in* Kerrdige, J. F., and Matthews, M. S., eds., Meteorites and the early solar system: Tucson, Univ. of Arizona Press, p. 1063–1089.

Lee, T., Mayeda, T. K., and Clayton, R. N., 1980, Oxygen isotopic anomalies in Allende inclusion HAL: Geophys. Res. Lett., v. 7, p. 493–496.

Lee, T., and Papanastassiou, D. A., 1974, Mg isotopic anomalies in the Allende meteorite and correlation with O and Sr effects: Geophys. Res. Lett., v. 1, p. 225–228.

Lee, T., Papanastassiou, D. A., and Wasserburg, G. J., 1977, Aluminium-26 in the early solar system: Fossil or fuel?: Astrophys. Jour.(Lett.), v. 211, p. L107–L110.

Lee, T., Shu, F. H., Shang, H., Glassgold, A., and Rehm, K. E., 1998, Protostellar cosmic rays and extinct radioactivities in meteorites: Astrophys. Jour. (Lett.), v. 506, p. 898–912.

Leshin, L. A., Rubin, A. E., and McKeegan, K. D., 1997, The oxygen isotopic composition of olivine and pyroxene from CI chondrites: Geochim. et Cosmochim. Acta, v. 61, p. 835–845.

Leshin, L. A., McKeegan, K. D., and Benedix, G. K., 2000, Oxygen isotope geochemistry of olivine from carbonaceous chondrites [abs.]: Lunar Planet Sci., v. XXXI, abstract No. 1918.

Lodders, K., and Fegley, B., Jr., 1993, Lanthanide and actinide chemistry at high C/O ratios in the solar nebula: Earth Planet. Sci. Lett., v. 117, p. 125–145.

Loss, R. D., and Lugmair, G. W., 1990, Zinc isotope anomalies in the Allende meteorite: Astrophys. Jour. (Lett.), v. 228, p. L93–L98.

Lugmair, G. W., and Galer, S. J. G., 1992, Age and isotopic relationship among the angrites Lewis Cliff 86010 and Angra Dos Reis: Geochim. et Cosmochim. Acta, v. 56, p. 1673–1694.

MacPherson, G. J., and Davis, A. M., 1994, Refractory inclusions in the prototypical CM chondrite, Mighei: Geochim. et Cosmochim. Acta, v. 58, p. 5599–5625.

MacPherson, G. J., Davis, A. M., and Zinner, E. K., 1995, The distribution of aluminum-26 in the early solar system—a reappraisal: Meteoritics, v. 30, p. 365–386.

MacPherson, G. J., Grossman, L., Hashimoto, A., Bar-Matthews, M., and Tanaka, T., 1984, Petrographic studies of refractory inclusions from the Murchison meteorite: Jour. Geophys. Res. B., v. 89 (suppl.), p. C299–C312.

MacPherson, G. J., Wark, D. A., and Armstrong J. T., 1988, Primitive material surviving in chondrites: Refractory inclusions, *in* Kerrdige, J. F., and Matthews, M. S., eds., Meteorites and the early solar system: Tucson, Univ. of Arizona Press, p. 746–807.

Marvin, U. B., Wood, J. A., and Dickey, J. S., Jr., 1970, Ca-Al-rich phases in the Allende meteorite: Earth Planet. Sci. Lett., v. 7, p. 346–350.

Mason, B., and Taylor, S. R., 1982, Inclusions in the Allende meteorite: Smithson. Contrib. Earth Sci., v. 25, p. 1–39.

McKeegan, K. D., Chaussidon, M., and Robert, F., 2000, Incorporation of short-lived ^{10}Be in a calcium-aluminum-rich inclusion from the Allende Meteorite: Science, v. 289, p. 1334–1337.

McKeegan, K. D., Leshin, L., Russell, S. S., and MacPherson, G. J., 1998, Oxygen isotopic abundances in calcium-aluminum-rich inclusions from ordinary chondrites: Implications for nebular heterogeneity: Science, v. 280, p. 414–418.

Meyer, B. S., Krishnan, T. D., and Clayton, D. D., 1996, ^{48}Ca production in matter expanding from high temperature and density: Astrophys. Jour. (Lett.), v. 462, p. 825–838.

Niederer, F. R., Papanastassiou, D., and Wasserburg, G. J., 1980, Endemic isotopic anomalies in titanium: Astrophys. Jour. (Lett.), v. 240, p. L73–L77.

______, 1985, Absolute isotopic abundances of Ti in meteorites: Geochim. et Cosmochim. Acta, v. 49, p. 835–851.

Niemeyer, S., and Lugmair, G. W., 1981, Ubiquitous isotopic anomalies in Ti from normal Allende inclusions: Earth Planet. Sci. Lett., v. 53, p. 211–225.

Nittler, L. R., Alexander, C. M. O'D., Gao, X., Walker, R. M., and Zinner, E., 1997, Stellar sapphires: The properties and origins of presolar Al_2O_3 in meteorites: Astrophys. Jour. (Lett.), v. 483, p. 475–495.

Nuth, J., and Donn, B., 1983, Nucleation theory is not applicable to the condensation of refractory grains in the primitive solar nebula: Lunar Planet. Sci., v. XIV, p. 570–571.

Palme, H., Wlotzka, F., Nagel, K., and El Goresy, A., 1982, An ultra-refractory inclusion from the Ornans carbonaceous chondrite: Earth Planet. Sci. Lett., v. 61, p. 1–12.

Podosek, F. A., and Swindle, T. D., 1988, Extinct radionuclides, *in* Kerrdige, J. F., and Matthews, M. S., eds., Meteorites and the early solar system: Tucson, Univ. of Arizona Press, p. 1093–1113.

Reynolds, J. H., 1967, Isotopic abundance anomalies in the solar system: Ann. Rev. Nucl. Phys., v. 17, p. 253–316.

Russell, S. S., Huss, G. R., Fahey, A. J., Greenwood, R. C., Hutchison, R., and Wasserburg, G. J., 1998, An isotopic and petrologic study of calcium-aluminum-rich inclusions from CO3 meteorites: Geochim. et Cosmochim. Acta, v. 62, p. 689–714.

Ryerson, F. J., and McKeegan, K. D., 1994, Determination of oxygen self-diffusion in åkermanite, anorthite, diopside, and spinel: Implications for oxygen isotopic anomalies and the thermal histories of Ca-Al-rich inclusions: Geochim. et Cosmochim. Acta, v. 58, p. 3713–3734.

Shang, H., Shu, F. H., Lee, T., and Glassgold, A. E., 2000, Protostellar winds and chondritic meteorites: Space Sci. Rev., v. 92, p. 153–176.

Shu, F., Shang, H., and Lee, T., 1996, Toward an astrophysical theory of chondrites: Science, v. 271, p. 1545–1552.

Simon, S. B., Davis, A. M., and Grossman, L., 1996, A unique ultrarefractory inclusion from the Murchison meteorite: Meteorit. Planet. Sci., v. 31, p. 106–115.

______, 1998, Formation of an unusual compact Type A refractory inclusion from Allende: Meteorit. Planet. Sci., v. 33, p. 115–126.

Simon, S. B., Davis, A. M., Grossman, L., and Zinner, E. K., 1998, Origin of hibonite-pyroxene spherules found in carbonaceous chondrites: Meteorit. & Planet. Sci., v. 33, p. 411–424.

Srinivasan, G., Sahijpal, S., Ulyanov, A. A., and Goswami, J. N., 1996, Ion microprobe studies of Efremovka CAIs: II. Potassium isotope composition and ^{41}K in the early solar system: Geochim. et Cosmochim. Acta, v. 60, p. 1823–1835.

Srinivasan, G., Ulyanov, A. A., Hutcheon, I. D., and Goswami, J. N., 1994, Excess ^{41}K in the early solar system: Astrophys. Jour. (Lett.), v. 431, p. L67–L70.

Suess, H. E., 1965, Chemical evidence bearing on the origin of the solar system: Ann. Rev. Astron. Astrophys., v. 3, p. 217–234.

Sylvester, P. J., Grossman, L., and MacPherson, G. J., 1992, Refractory inclusions with unusual chemical compositions from the Vigarano carbonaceous chondrite: Geochim. et Cosmochim. Acta, v. 56, p. 1343–1363.

Sylvester, P. J., Simon, S. B., and Grossman, L., 1993, Refractory inclusions from the Leoville, Efremovka, and Vigarano C3V chondrites: Major element differences between Types A and B, and extraordinary refractory siderophile element compositions: Geochim. et Cosmochim. Acta, v. 57, p. 3763–3784.

Thiemens, M. H., 1996, Mass-independent isotopic effects in chondrites: The role of chemical processes, *in* Hewins, R., Jones, R., and Scott, E., eds., Chondrules and the protoplanetary disk: Cambridge, UK, Cambridge Univ. Press, p. 107–118.

Thiemens, M. H., and Hedenreich, J. E., 1983, The mass independent fractionation of oxygen: A novel isotope effect and its possible cosmochemical implications: Science, v. 219, p. 1073–1075.

Virag, A., Zinner, E., Amari, S., and Anders, E., 1991, An ion microprobe study of corundum in the Murchison meteorite: Implications for ^{26}Al and ^{16}O in the early solar system: Geochim. et Cosmochim. Acta, v. 55, p. 2045–2062.

Völkening, J., and Papanastassiou, D. A. P., 1989, Iron isotope anomalies: Astrophys. Jour. (Lett.), v. 347, p. L43–L46.

______, 1990, Zinc isotope anomalies: Astrophys. Jour. (Lett.), v. 358, p. L29–L32.

Wasserburg, G. J., 1985, Short-lived nuclei in the early solar system, *in* Black, D. C., and Matthews, M. S., eds., Protostars & planets II: Tucson, Univ. of Arizona Press, p. 703–737.

Wasserburg, G. J., Gallino, R., Busso, M., Goswami, J. N., and Raiteri, C. M., 1995, Injection of freshly synthesized ^{41}Ca in the early solar nebula by an AGB star: Astrophys. Jour. (Lett.), v. 440, p. L101–L104.

Weber, D., and Bischoff, A., 1994, The occurrence of grossite ($CaAl_4O_7$) in chondrites: Geochim. et Cosmochim. Acta, v. 58, p. 3855–3877.

Weber, D., Zinner, E., and Bischoff, A., 1995, Trace element abundances and magnesium, calcium, and titanium isotopic compositions of grossite-containing inclusions from the carbonaceous chondrite Acfer 182: Geochim. et Cosmochim. Acta, v. 59, p. 803–823.

Wiens, R. C., Huss, G. R., and Burnett, D. S., 1999, The solar oxygen-isotopic composition: Predictions and implications for solar nebula processes: Meteorit. Planet. Sci., v. 34, p. 99–107.

Woolum, D. S., and Cassen, P., 1999, Astronomical constraints on nebular temperatures: Implications for planetesimal formation: Meteorit. Planet. Sci., v. 34, p. 897–907.

Young, E. D., and Russell, S. S., 1998, Oxygen reservoirs in the early solar nebula inferred from an Allende CAI: Science, v. 282, p. 452–455.

Zinner, E., 1998, Stellar nucleosynthesis and the isotopic composition of presolar grains from primitive meteorites: Ann. Rev. Earth Planet. Sci., v. 26, p. 147–188.

Zinner, E., Fahey, A., Goswami, J. N., Ireland, T. R., and McKeegan, K. D., 1986, Large ^{48}Ca anomalies are associated with ^{50}Ti anomalies in Murchison and Murray hibonites: Astrophys. Jour. (Lett.), v. 311, p. L103–L107.

In Situ Studies of Iron under Pressure: New Windows on the Earth's Core

RUSSELL J. HEMLEY AND HO-KWANG MAO

Geophysical Laboratory and Center for High Pressure Research, Carnegie Institution of Washington, 5251 Broad Branch Road, N.W., Washington, DC 20015

Abstract

Experiments on iron with recently developed high-pressure techniques are providing new insights into the nature of the Earth's core. *In situ* high P-T synchrotron x-ray diffraction studies to above 160 GPa and 3000 K with double-sided, laser-heated diamond-anvil cells demonstrate that the hexagonal closed packed phase (ε-Fe) has a wide stability field extending from deep mantle to core conditions. Direct measurements of the melting line of ε-Fe to above 100 GPa with these techniques provide constraints on the temperature and density at the inner core boundary. Radial x-ray diffraction measurements carried out at room temperature to 220 GPa have constrained the elasticity, rheology, and acoustic velocities of ε-Fe at core pressures. Vibrational properties providing information on thermodynamic and elastic parameters have been measured to above 150 GPa using both synchrotron x-ray and laser Raman spectroscopies. X-ray diffraction studies of liquid iron reveal changes in short-range structure as functions of pressure and temperature. Insight into the problem of the identity and abundance of other (e.g., light) elements in the core has been obtained from direct studies of the Fe-O, Fe-S, and Fe-H binary systems as well as observations for several other key elements. Progress using various new techniques is reviewed, and new directions and questions are examined. We also compare recent experimental findings with a variety of theoretical predictions.

Introduction

AS THE MOST INACCESSIBLE REGION of the planet, the core has been the focus of considerable attention during the past several years. Despite this intensive study, numerous fundamental problems still remain for the Earth's enigmatic center (Fig. 1). Although the core comprises approximately one-third of the planet's mass, its composition is not known in detail. This question is directly coupled to the broader issue of the composition, formation, and evolution of the Earth as a whole, including the exchange of material between the mantle and core through time (e.g., McDonough, 2001). The temperature at the Earth's center is still uncertain, a problem intimately related to the compositional issue and one with profound implications for the power available for generating both the magnetic field as well as solid-state convection of the mantle. In fact, it is still not known what drives the dynamo (compositional or thermal forces) and the origin of its variations through time (Buffett, 2000). The solid inner core is reported to be elastically anisotropic (Morelli et al., 1986; Song and Helmberger, 1993; Vinnik et al., 1994; Glatzmaier and Roberts, 1996) and exhibits layered substructure (Song and Helmberger, 1998). It has also been proposed that the core superrotates relative to the planet as a whole (Song and Richards, 1996; Su et al., 1996), although the geophysical evidence is still controversial (e.g., Bréger et al., 1999). Not only is the origin of these phenomena unclear (Buffett, 2000), but the principal phase of the inner core (along with its composition and thermal state) have been matters of dispute (Stixrude and Brown, 1998). Moreover, understanding the process of core formation requires information on the fate of volatiles and interactions within a possible magma ocean (see Canup and Righter, 2000). When the inner core formed may be closely correlated with the appearance of life, since the existence of the inner core appears necessary for stabilizing the Earth's magnetic field (Buffett, 2000), which is crucial for deflecting harmful cosmic rays.

Solving these problems requires establishing the basic physics and chemistry of core-forming iron alloys. As the above problems are outstanding, two broad conclusions remain robust. First, as Birch (1964) pointed out, the core is dominantly iron, the most abundant element in the planet by weight. Hence, the high-pressure behavior of iron is the

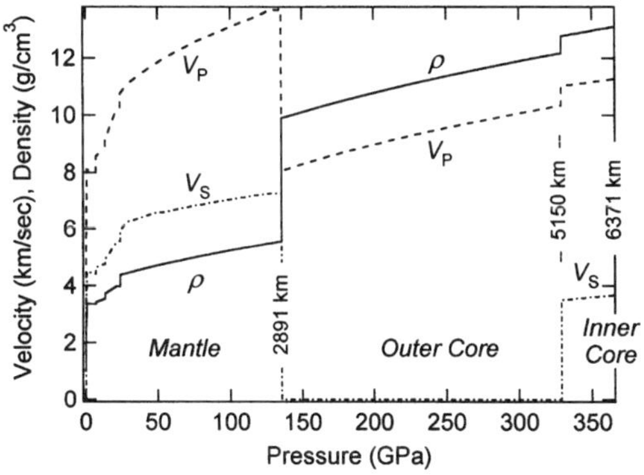

FIG. 1. The density ρ, shear-wave velocity V_S, and compressional wave velocity V_P profiles for the Earth's interior determined from the radially averaged seismological preliminary-reference Earth model (PREM) (Dziewonski and Anderson, 1981). The depths of various discontinuities are indicated.

starting point for understanding the nature of the core, which is clearly the key to a number of current geological, geochemical, geophysical, and even geobiological questions. Second, the physical properties of iron are dramatically altered by the extreme conditions of pressure and temperature that prevail deep within the planet. These changes are also manifested in the chemistry of iron; for example, elements that are chemically incompatible with iron at ambient pressure may form alloys or even new stoichiometric compounds at elevated pressure. These effects include high-pressure reactions of iron with volatiles, or atmophiles (Krauskopf and Bird, 1994), an observation with important implications for the planet's atmosphere early in its history. Because of these large changes in the physical and chemical properties of iron under deep Earth conditions, *in situ* measurements are crucial for the characterization of iron and iron-rich materials needed for understanding many enigmatic aspects of the core.

For years, information on candidate materials of the core under the relevant P-T conditions was available only from approximate extrapolation of ambient (or near ambient) pressure data, or in some cases from shock-wave experiments. However, static high-pressure methods not only can now reach the required conditions but also provide a wealth of information on the properties of these materials in these extreme thermodynamic states. In fact, many static compression techniques have been developed and/or improved in just the last two years. The phase relations of iron have been controversial in recent years, in part because of the lack of *in situ* phase determination at very high P-T conditions. Progress on this problem has been made with the advent of new laser heating methods coupled with simultaneous analysis of samples by synchrotron x-ray diffraction. A variety of additional synchrotron-based methods have been developed, including diffraction techniques that allow crucial studies of elasticity, strength, and texture of samples, and detailed measurements of the state of samples at core pressures. Knowledge of the elasticity and texture of iron is crucial for understanding recent (and controversial) seismological observations of the inner core, such as the high Q, the low shear velocity and the elastic anisotropy. Additional x-ray methods allow determination of electronic, magnetic, and vibrational properties, and optical spectroscopy (e.g., Raman scattering) has provided additional insight. These techniques are complemented by very recently developed methods for direct magnetic measurements. Moreover, theoretical techniques have improved in reliability and accuracy, providing quantitative predictions of high P-T behavior and fundamental understanding of observed phenomena that complement recent experimental results.

This paper provides an overview of selected applications of new high-pressure techniques to Fe at extreme P-T conditions in relation to current problems in understanding the nature of the Earth's core. Anderson (1995) and Stixrude and Brown (1998) provided earlier reviews of experimental studies of iron and the implications for the core. More recently, high-pressure experiments and the phase diagrams of core (and lower-mantle) materials from the standpoint of single-sided laser heating have been reviewed by Boehler (2000). Many additional and still more recent studies have been published in a variety of journals (including the physics literature) and may not be widely known in the geoscience community. The new data examined here allow us to resolve several outstanding controversies regarding iron at high pressures and temperatures. In addressing these problems, we also briefly examine some of the technical issues associated with the experimental work that have not been discussed previously in the literature.

Phase Diagram

Overview

Under ambient conditions, iron crystallizes in the body-centered cubic (bcc) α-Fe structure, and with increasing temperature transforms to face-centered cubic (fcc) γ-Fe, and then δ-Fe (also bcc) before melting. Evidence for polymorphism at high pressure was first presented by Bancroft et al. (1956), which was later confirmed to signal a transition to the hexagonal close packed (hcp) ε-Fe phase (Clendenen and Drickamer, 1964; Takahashi and Bassett, 1964). Since then, the ε-Fe phase has been studied with a large number of techniques (both static and dynamic compression) and over an expanding range of pressure and temperature. The phase diagram at high pressures has been investigated with diamond-cell internal resistive-heating methods, laser heating with direct visual observation, laser heating combined with synchrotron x-ray diffraction, large-volume x-ray diffraction methods, and shock-wave techniques. New phases have been proposed on the basis of the experimental data, and still others have been predicted theoretically.

Many of these results are mutually inconsistent. One of the most difficult and controversial problems has been the melting line at very high pressures and temperatures. This property provides a first-order constraint on the temperature of the core, and thus key information on the internal heat of the planet, and heat flow to the surface. Another major problem has been the numerous subsolidus phase transitions, starting with the suggestion of a possible new phase at high temperatures, roughly tracking the slope of the melting line at lower temperatures (Boehler, 1993). The putative phases have proposed structures that are close to dhcp and have been referred to as β and ε' by Saxena et al. (1995) and Yoo et.al. (1995). The latter identified the ε' phase as a metastable structure that exists due to incomplete transformation between hcp and fcc phases (Yoo et al., 1997). An orthorhombic phase of iron was later reported by Andrault et al. (1997, 2000) on the basis of x-ray measurements carried out at 30 to 100 GPa. *In situ* x-ray diffraction measurements with an internally heated, large-volume device to 32 GPa (and 1000 K) show no evidence for any additional phases (Funamori et al., 1996). The existence of another bcc phase has been proposed for very high P-T (i.e., core) conditions on the basis of theoretical calculations (Ross et al., 1990); however, subsequent theoretical work suggested that bcc iron is mechanically unstable at very high pressures due to the loss of magnetism with pressure (Soderlind et al., 1996; Stixrude et al., 1997), although distortions of the bcc structure could not be ruled out (e.g., Soderlind et al., 1996).

Resolution of these questions requires consideration of five crucial components:

1. *Generation of extreme pressures and temperatures.* Studies of Fe in diamond cells with laser heating have been conducted mostly below 50–100 GPa and 2500 K. Only a handful of laboratories have actually exceeded P-T of the Earth's core-mantle boundary (>3000 K and >135 GPa). The maximum simultaneous P-T conditions in static experiments are still far from that of the inner-outer core boundary (>330 GPa and ~6000 K; see below).

2. *Accurate measurements of temperatures at high pressures.* Whereas measurement of black-body temperatures of a large, uniformly heated sample is straightforward, the existence of a microscopic heating spot with steep temperature gradients, both across the laser beam spot (radial gradient) and through the iron sample (axial gradient) can amplify the problems of astigmatism and dispersion of the optical elements (including the diamond anvils) and introduce large errors. Recent implementations of the double-sided laser heating techniques show that the axial gradient can be eliminated. In addition, two near-infrared laser beams can be combined to generate a homogeneous flat-top power distribution within the area of interest (>30 μm); this is sufficiently larger than the sampling spot and completely avoids temperature errors introduced by the radial gradient outside the flat-top region.

3. *Phase identification under extreme P-T conditions.* Recent applications of synchrotron x-ray diffraction provide a positive identification of the solid phases. Prior to 1995, high-pressure phase transitions and melting of iron were detected on the basis of sudden changes in visual appearance or power-temperature relations. Such observations reflect changes of laser-sample coupling that can be caused by phase transitions but also other reasons. The reported subsolidus phase line subparallel to the melting curve was based on such observations (Boehler, 1993; Saxena et al., 1993). Between 30 and 80 GPa, we have re-examined the proposed line by x-ray diffraction, and found that above the line crystals grew and changed orientation rapidly, but the same ε-Fe phase was preserved, indicating a recrystallization (i.e., kinetic) boundary, not a phase line.

4. *Coordination of the analytical probe and heating spot on the sample.* Unambiguous phase identification requires the x-ray probe (5–10 μm in diameter), laser beams (20–40 μm), and sample in the gasket hole to be centered to the same point. This presents a great challenge, particularly

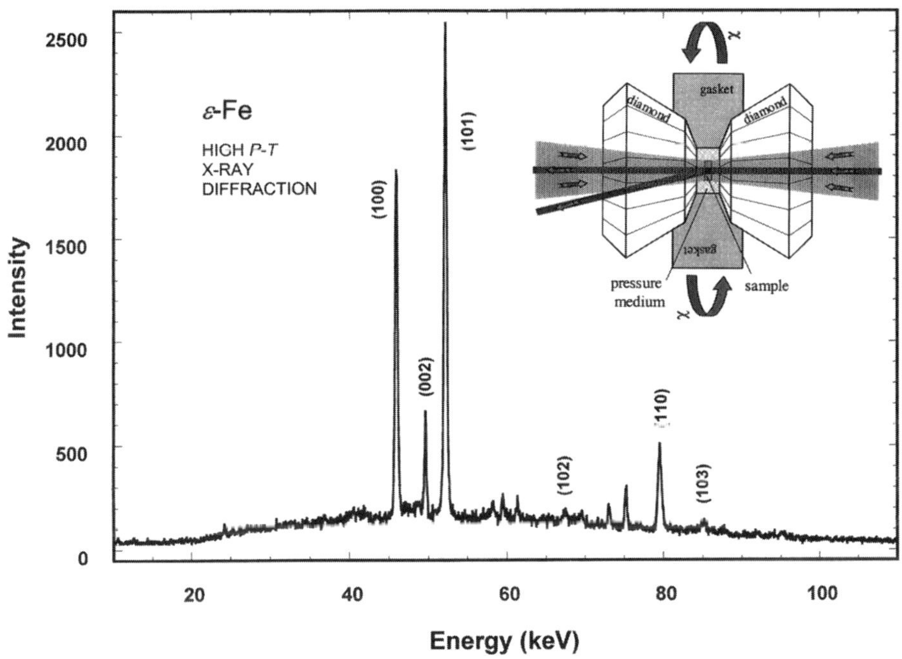

FIG. 2. X-ray diffraction of iron at 161 (±0.2) GPa and 2450 (±100) K (Ma et al., submitted). The half-widths of the most intense peaks (101) and (100) are close to the resolution limit of the detector. Five peaks from a hcp phase are observed, with refined lattice parameters a = 2.238(1), c = 3.594 (±0.004) Å, c/a = 1.605 (±0.007) and V = 15.595 (±0.018) Å3. The c/a ratio is similar to that measured at lower pressures and temperatures (Mao et al., 1990). Weak peaks are fluorescence or diffraction lines from the medium (Al_2O_3). The inset shows the schematic of the sample in the laser-heated diamond cell. The figure shows the configuration of *in situ* laser heating with x-ray diffraction in a diamond cell; the cell is rotated coaxially with the x-ray beam and heating laser (χ-rotation). The center of rotation, heating laser beam, temperature measurement optics, incoming x-ray beam, and collimation of the diffracted beam are aligned to within 1–2 mm. Two systems have been used; in the first, two Nd:YLF single-mode laser beams (TEM00 and TEM01 modes) are combined to generate a homogeneous flat-top power distribution (with radial gradient less than 3%) at the center region (installed at Sector 13 of the Advanced Photon Source). In the second system, split beams from a multimode Nd:YAG laser were used with the same system configuration (installed at beamline X17B of the National Synchrotron Light Source).

because under intense heating, the relative positions of samples, diamond cells, or the entire system may shift during the x-ray analysis. Positioning mismatch or diffuse wings of the x-ray beam can pick up additional diffraction peaks from materials outside the area of interest, including the Fe sample at low temperatures and different stress conditions and even the gasket, and may account for inconsistent results.

5. *Heating path and sample history.* True phase boundaries must be determined from reversals between thermodynamically stable phases. Some metastable phases are difficult to eliminate; for instance, the γ-Fe can be quenched to ambient temperatures at high pressures. The proposed dhcp phase can often be produced from γ-Fe in its metastable region or from ε-Fe in the presence of a steep temperature gradient, but not from ε-Fe under uniform and stable heating conditions.

Extended phase equilibrium studies of Fe that satisfy the above five components have been carried out to pressures of 161 GPa and temperatures above 3000 K. An example of an *in situ* x-ray diffraction measurement carried out under extreme P-T conditions is shown in Figure 2. Temperatures were measured with the thermal radiation techniques described previously, and the diffraction measured with a focused or collimated synchrotron beam and energy-dispersive techniques (Mao, 1998a). The advantage of higher resolution possible with the

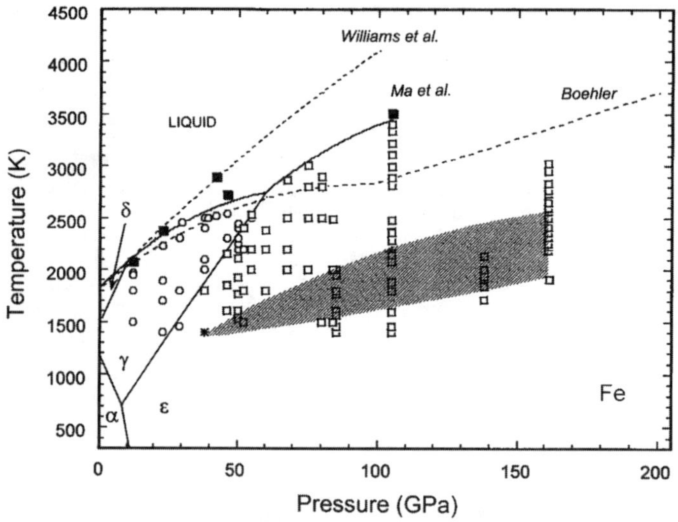

FIG. 3. Phase diagram of iron determined by synchrotron x-ray diffraction/double-sided laser heating experiments. The solid lines are from double-sided laser heating studies (Ma et al., submitted); see also Shen et al. (1998) and Shen and Heinz (1998). The dashed lines are selected previous static pressure results from Williams et al. (1987), and Boehler (1993); see Boehler (2000) for additional determinations. The points correspond to regions where the indicated phases were observed. The shaded area shows the region of observed recrystallization and texturing of the sample.

alternative angular dispersive x-ray diffraction technique is not useful because splittings of the iron diffraction peaks measured with both methods under these conditions are significantly larger than the instrument resolution; virtually all of the new peaks reported in angular dispersive studies can be equally well resolved with either technique. The energy dispersive technique has the advantage of allowing rapid measurements, permitting an investigation of real-time processes such as phase changes including crystallization within a sample. As melting is approached at high temperatures, measurement of the powder diffraction pattern is complicated by sample recrystallization and coarsening; this problem can be overcome by continuous rotation of the diamond cell (Ma et al., 2001).

These measurements allow direct and unambiguous identification of the fcc (γ-Fe) and hcp (ε-Fe) structures, as well as liquid, and thereby a determination of the ε-γ-*l* triple point at 60 (±5) GPa and 2800 (±200) K (Fig. 3). The triple point is close to that determined in earlier studies using similar techniques (Shen et al., 1998), and not far from the estimate presented by Boehler (1986) based on the internal resistive ("wire") heating technique. The results also are close to the very recent thermodynamic analysis of Anderson et al. (submitted). Ahrens et al. (in press) reported shock melting data for γ-Fe, concluding that the phase melts at 71 (±2) K and 2775 (±160) K (i.e., forcing the γ-ε-*l* triple point to be above 71 GPa). They also reported that the melting is 300 K lower than the reported lower bound by Shen et al. (1998). In reconciling these results, it should be pointed out that in the shock experiment, superheating effects cannot be ruled out, and the temperatures were calculated.

Question of additional subsolidus phases

As mentioned above, there have been reports of the existence of other phases (e.g., dhcp [Saxena et al., 1995; Saxena and Dubrovinsky, 2000] and an orthorhombic form [Andrault et al., 1997, 2000]) within the stability field identified above for ε-Fe. We have found that the additional diffraction lines used to identify the proposed new structures can be reproduced when the transformation from the ε to γ phases is incomplete (e.g., incomplete heating of the sample). This results in either a metastable polytype structure or a mixed phase sample (i.e., ε and γ-Fe) or sample containing a single phase (e.g., ε-Fe) at distinct pressures induced by localized heating (i.e., phase heterogeneity associated with large pressure gradients of >10 GPa over a few μm). When samples are thoroughly heated and transformations driven to completion, a simple pattern emerges (only ε-Fe is observed over the indicated P-T range above 60 GPa). An incomplete transformation can give rise to a metastable closed-packed structure that is intermediate between fcc and hcp (i.e., dhcp), or more likely, several such polytypes (or even a complex distribution of polytypes), which could also be viewed as extensive stacking faults. The form of the diffraction from a random stacking of close-packed layers in such systems was worked out long ago (Wilson, 1946). Mixed polytype structures of close-packed materials may be common in (martensitic) fcc-hcp transitions that are not driven to completion (when there is insufficient heating), e.g., Xe (Jephcoat et al., 1987). Notably, first-principles-based calculations find the dhcp structure to be unstable under these conditions (Price et al., 1999; Vocaldo et al., 1999; Laio et al., 2000).

The effects of temperature gradients and the high shear strength of iron at high pressure are critical factors in these laser-heating experiments. Simultaneous measurements from both sides of the sample show that temperature differences can be as high as 1000 K when the sample is heated from only one side (Ma et al., submitted). Because the x-ray beam produces signal from the heated and unheated (or poorly heated) zones, single-sided heating would give different results relative to those obtained with double-sided heating. For example, when the sample is heated from one side (or the heating spot is either too small or misaligned with respect to the x-ray beam), splittings of diffraction peaks can be measured for the ε- and α- phases, corresponding to stress differences within the samples of ~10 GPa across tens of μm. The magnitude of these stress differences is consistent with observations of the effect of pressure on the strength of Fe (Hemley et al., 1997), which is discussed in further detail below. The issue of strength effects seems not to have been fully considered in discussions of this controversy (Andrault et al., 1998, 2000; Dubrovinsky et al., 1998; Saxena and Dubrovinsky, 2000). Although the proposed transitions have been identified as reversible (Dubrovinsky, et al., 2000b), the presence of a mixed-phase sample as well as texture in a single phase can give rise to heterogeneity such that very slight and reversible movements of the sample induced by heating (e.g., by just a few μm) can give rise to the appearance of a "reversible" transition when in fact this is simply reversible movement of the sample. This effect has been observed in both laser-heating or internal resistive-heating experiments.

Boehler (1993) reported visible changes in Fe samples on heating within the ε-phase, and Saxena et al. (1993) found changes in thermal emission. The double-sided laser heating experiments with χ rotation reveal only ε-Fe between the triple point and to 161 GPa and ~3000 K. However, changes in the intensity of the hcp diffraction pattern can be observed in the region of the proposed transition (Ma et al., submitted). These systematic changes, which roughly track the melting line of Fe, can be attributed to a reduction of the strength of the iron. Such effects are observed in other materials (strength depends on the homologous temperature scaled to the melting point). At pressures higher than 60 GPa, only the ε-phase and the liquid state can be identified. These results established the wide stability field of ε-Fe, again consistent with our earlier results (Shen et al., 1998). We also note that this interpretation is compatible with thermodynamic analyses, which provide no evidence for additional phases (Anderson and Isaak, 2000). This result is consistent with recent theoretical calculations (Price et al., 1999; Vocaldo et al., 1999; Laio et al., 2000) as well as recent shock-wave data (Nguyen and Holmes, 2000). In the absence of other data, we are forced to conclude that the hcp phase of iron continues to be the most useful model for iron in the core.

Melting line

One of the principal goals of the high P-T study of iron is the determination of the melting temperature at the pressure of the outer core/inner core boundary (ICB, 330 GPa). Previous analyses of high-pressure melting data gave temperatures at this critical boundary of 7600 (± 500) K, according to Williams et al. (1987), and 4850 (± 200) K according to Boehler (1993). Figure 4 shows an extended phase diagram of iron, including the comparison with earlier static and shock-compression data, the results of thermodynamic calculations, and first-principles theory. The melting line has been measured to 105 GPa with double-sided laser heating methods (Ma et al., submitted). The result is in good agreement with thermodynamic analyses based on the recalculated shock temperatures (Chen and Ahrens, 1996), and it is also close to the results of a recent thermodynamic treatment of Anderson and Isaak (2000), which we now explicitly describe.

Anderson and Isaak (2000) fit the 1990 equation of state (EOS) data (Mao et al., 1990) to a Vinet form and used a Lindemann melting law with the Grü-

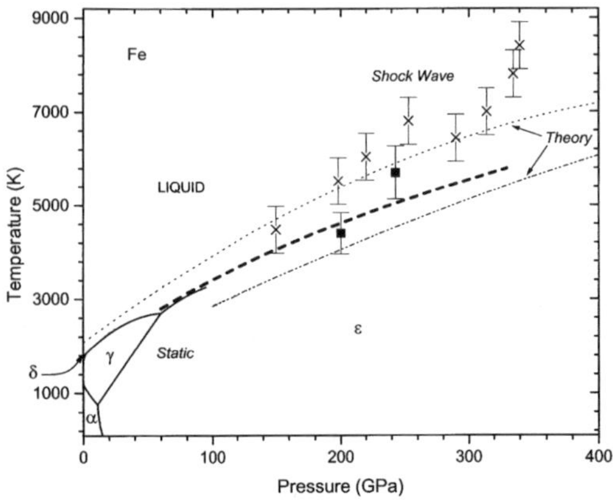

FIG. 4. Extended phase diagram of iron. The solid lines are from synchrotron x-ray/double-sided laser heating studies (Shen et al., 1998; Shen and Heinz, 1998; Ma et al., submitted). The thick dashed line is the Lindemann law extrapolation using a high-temperature volume-dependent Grüneisen parameter ($\gamma_0 = 1.9$ and $q = 1.0$) fixed by a thermal equation of state model (Anderson and Isaak, 2000). The shock-wave data correspond to reported Hugoniot results that cross the melting lines: solid squares are selected points from Brown and McQueen (1986); the crosses are from Yoo et al. (1993). The light dotted line shows the predicted melting curve from the first-principles calculations of Alfè et al. (1999); the light dash-dotted line is theoretical result of Laio et al. (2000), using a first-principles-based potential.

neisen parameter γ determined from a thermal equation of state. The Grüneisen parameter γ is defined as $\gamma = \alpha K_T V / C_V$, where α is the thermal expansion coefficient, K_T is the isothermal bulk modulus, and C_V is the constant volume heat capacity, all at a given V. As discussed further below, the volume dependence of γ can be written as:

$$\frac{\gamma V}{\gamma V_0} = \left(\frac{V}{V_0}\right)^q. \quad (1)$$

According to the Lindemann model,

$$\frac{dT_m}{dP} = \frac{2T_m}{K_T}\left(\gamma - \frac{1}{3}\right). \quad (2)$$

Combining these and integrating gives,

$$\frac{T_m}{T_{m_0}} = \left(\frac{V}{V_o}\right)^{2/3} \exp\left\{\frac{2\gamma_0}{q}\right\}\left[1 - \left(\frac{V}{V_0}\right)^q\right], \quad (3)$$

where T_{m0} is the melting temperature at the reference volume V_0. Anderson and Isaak (2000) showed that the EOS data from Mao et al. (1990) (Fig. 5A) are compatible with both the Hugoniot (Brown and McQueen, 1986) and thermal expansion data (Duffy and Ahrens, 1993). T_{m0} was fixed at 1600 K, which gives a temperature of the triple point well within the error of the melting data for ε-Fe determined by double-sided laser heating. The resulting curve shown in Figure 4 gives T_m = 5800 K (with an estimated error of ± 200 K) for the temperature at 330 GPa, which is an upper bound for the temperature at the ICB. A related earlier treatment gave 6100 K at the ICB (Anderson, 1998). The new results agree with the more extended thermodynamic analysis presented by Anderson et al. (submitted) and the approximate calculations of Brown and McQueen (1986).

The melting line determined from first-principles calculations (Alfè et al., 1999; Laio et al., 2000) brackets the experimental results; the Clapeyron slopes of the curves above 100 GPa are broadly similar to, but slightly higher than, the experimentally based extrapolation (Anderson and Isaak, 2000), despite the differences in approach (i.e., no melting law assumed in the theory). Indeed, it is likely that the theoretical calculation of the Clapeyron slope is more accurately determined than the predicted value of T_m. The thermodynamic estimate is close to the melting line of Stixrude et al. (1997) who estimated T_m = 6150 K at the ICB pressure, also based on a Lindemann law in which the parameters of the model (e.g., γ) were determined theoretically. The results also are in fair agreement with the first-principles calculations of Hausleitner and Hafner (1989) (for the fcc phase, instead of hcp, which was not examined). These simulations provide additional support for the use of a Lindemann model for iron melting to megabar pressures (the relative root-mean square atomic displacement $(\delta r/r)^{1/2}$ along the melting curve remains constant).

Brown and McQueen (1986) reported shock-melting data that showed evidence for a solid-solid transition at inner core pressures and temperature prior to melting (see also Brown et al., 2000). More recent shock-compression experiments apparently show no evidence for a transition in this pressure range (Nguyen and Holmes, 2000). These authors carried out sound velocity measurements to shock pressures of 400 GPa and found a single discontinuity identified as melting and no secondary transition. The temperatures along the shock Hugoniot are

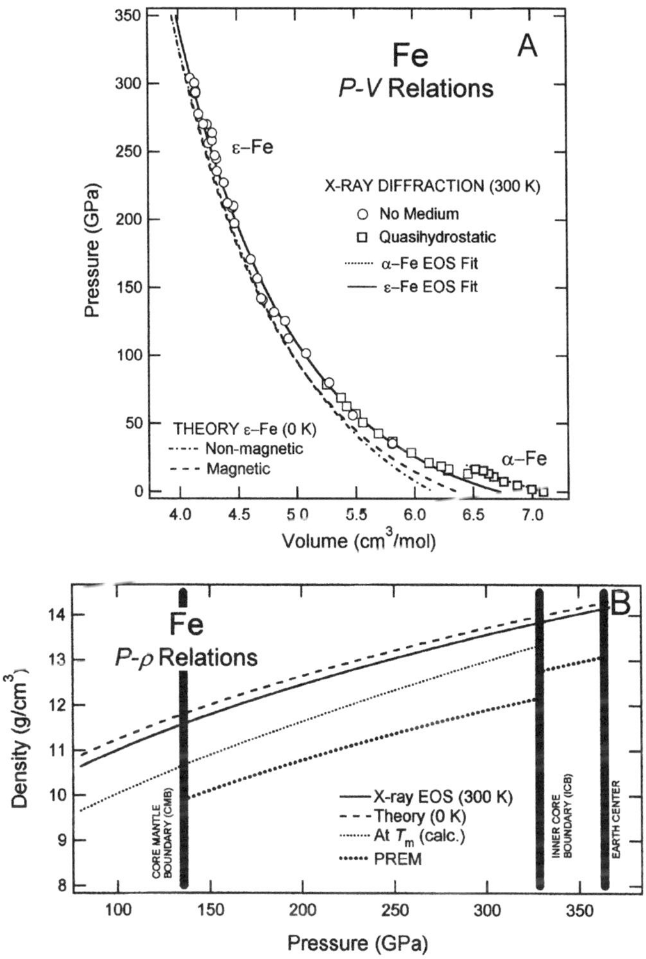

FIG. 5. A. Pressure-volume equation of state (EOS) for iron. The data points correspond to 300 K x-ray diffraction measurements with Ar and Ne pressure-transmitting media to 78 GPa (Jephcoat et al., 1986) and without a medium to 304 GPa (Mao et al., 1990). The solid line for ε-Fe is the Vinet EOS fit (see text); the thin dotted line for ε-Fe is the EOS fit from Mao et al. (1967); $V_0 = 6.73$ cm^3/mol, K_{T0}= 160.2 (±2.1) GPa, and K'_{T0}= 5.82 (±0.08). The dot-dashed line is the non-magnetic generalized gradient approximation (GGA) result (V_0 = 6.16 cm^3/mol, K_0 = 292 GPa, and K'_0= 4.4); the thick dashed line shows results of spin-polarized GGA calculations for an antiferromagnetically ordered structure consisting of layers of opposite spins normal to [100] (V_0 = 6.35 cm^3/mol, K_0 = 209 GPa, and K'_0 = 5.2) (Steinle-Neumann et al., 1999). The theoretical calculations were performed for a static lattice (effectively 0 K); inclusion of thermal pressure is estimated to shift the curve by a few percent. B. Pressure-density relations at core pressures, showing the above curves along with the calculated density along the melting line (Anderson and Isaak, 2000) and the PREM profile for the core (Dziewonski and Anderson, 1981).

either calculated or measured. Although direct measurements may be considered the most reliable, they have been problematic and the subject of much discussion in the literature. Boehler (2000) interpreted the shock data as indicating the onset of melting at 200 GPa and 4000 K (with a liquid-solid mixed phase region between 200 and 250 GPa). The higher temperature of the most recent melting curve discussed here requires a less steep Clapeyron slope to be compatible with shock-wave data. In fact, the indicated slope is close to those predicted theoretically (Laio et al., 2000).

Elasticity and Equation of State

P-V-T relations

Since the work of Birch (1952), a consideration of the density of iron at high pressure has been used to ascertain the composition of the core. The equation of state (EOS, i.e., density and bulk modulus) of ε-Fe has been previously measured to core pressures by static and dynamic methods (Brown and McQueen, 1986; Mao et al., 1990; Yoo et al., 1993; Nguyen and Holmes, 1998). The x-ray diffraction results to 304 GPa (Mao et al., 1990) are shown in Figure 5A. The pressure-density relation was found to be indistinguishable from that of the $Fe_{0.8}$-$Ni_{0.2}$ alloy within experimental uncertainty, yielding no constraint on the Ni abundance in the core from density and bulk modulus comparisons (see below). The bulk modulus was obtained from a fit to both the Birch-Murnaghan (Birch, 1978) and Vinet et al. (1987) equation-of-state forms. Because of the versatility of the Vinet et al. form (see Cohen et al., 2000) and to allow comparisons with other recent analysis using this EOS, we focus on the parameters obtained from this equation-of-state. To second order, the function is given as

$$P(x) = 3K_{T0}(1-x)^{-2} \times \exp\left[\frac{3}{2}(K_{T0} - 1)(1-x)\right], \tag{4}$$

where $x = (V/V_0)^{1/3}$, and best-fit parameters V_0 = 6.73 cm³/mol (reference volume), K_{T0} = 160.2 (±2.1) GPa (isothermal bulk modulus at V_0), and K'_{T0} = 5.82 (±0.08) (pressure derivative of K_{T0}) were obtained (Mao et al., 1990). The results are shown in Figure 5A, along with curves calculated at two levels of first-principles theory. Neglect of magnetic interactions in these calculations produces a curve with a significantly higher bulk modulus. Inclusion of antiferromagnetic ordering stabilizes the structure and has a significant effect at low pressure, particularly below 50 GPa (Steinle-Neumann et al., 1999). Inclusion of magnetic effects tends to improve the agreement with experiment, although the density is still overestimated relative to experiment. The question of magnetic interactions in ε-Fe is discussed further below.

Estimating the composition of the inner core based on density as well as bulk modulus or bulk sound velocity requires determination of the thermal expansivity α up to core pressures. The thermal contribution on the equation of state can also be characterized by the Grüneisen parameter γ, as described above. The thermal expansivity at core pressures now appears relatively well constrained. Dubrovinsky et al. (2000b) recently measured the thermal expansivity to 300 GPa and 1300 K, and these results are consistent with the earlier measurements of Funamori et al. (1996) at lower pressures (to 32 GPa and 1000 K). The average value near 200 GPa is 9.1×10^{-6} K^{-1}, consistent with shock-wave data (9.1×10^{-6} K^{-1} [Duffy and Ahrens, 1993]), α-ρ systematics (9.9×10^{-6} K^{-1} [Boehler, 2000]) and theory (1.0×10^{-6} K^{-1} [Wasserman et al., 1996; Stixrude et al., 1997]). These results appear to be in reasonable accord with (or are included in) explicit thermal pressure models in P-V-T equations of state (e.g., Anderson, 1995; Anderson and Isaak, 2000). The Grüneisen parameter γ gives the temperature dependence of the thermal pressure $dP_{th}/dT = \rho\gamma$, where $P_{th} = \int \alpha K_T dT$. The results for γ from various studies are listed in Table 1.

This broad agreement among results from various techniques provides a basis for determining the density of iron at the ICB (Fig. 5B). The PREM value at the ICB is 12.76 g/cm³ (Dziewonski and Anderson, 1981). The thermal model of Anderson and Isaak (2000) gives ρ = 13.30 g/cm³ at 330 GPa (at 5800 K). In contrast, the thermal expansion model of Anderson (1998) using the same EOS data give a slightly higher value of 13.38 g/cm³ (at 6100 K) at the ICB. Overall, the results imply a 4–5% density deficit for the inner core. At this level of precision, the effects of nonhydrostaticity (Hemley et al., 1997) and uncertainties in pressure calibration, i.e., Pt equation of state standard (Mao et al., 1990), rather than the temperature correction become the dominant sources of error in determining the density of solid iron at inner core pressures. There is, however, excellent consistency between these static compression results and independent shock-wave Hugoniot pressure-density data (Jephcoat et al., 1986; Stixrude et al., 1997). The present determination for the high P-T density of iron together with the lack of evidence for an additional high P-T phase of lower density and the analysis of the velocities described below thus provide a robust constraint on the abundance of light elements in the inner core.

The hcp *c/a* ratio also is obtained from x-ray diffraction. The ratio is observed to be only weakly dependent of pressure at 300 K, with c/a = 1.602 (±0.002) – 4.9 (±1.4) × 10^{-5} P determined in the

TABLE 1. Grüneisen Parameters of ε-Fe Determined by Various Techniques

γ_0	q_0	Technique	P-T range	Reference
1.5 (±0.2)	0	NRIXS[1]	20–42 GPa 300 K	Lübbers et al., 2000
1.68 (±0.17) 300 K	0.7 (±0.5)	Raman	15–150 GPa 300 K	Merkel et al., 2000 Goncharov et al., in prep.
1.78 (±0.06)	0.69 (±0.10)	X-ray diffraction	15–300 GPa 300–1300 K	Dubrovinsky et al., 2000a, 2000b
1.7	0.7	Thermodynamic analysis	0–240 GPa 300 K	Anderson, 1998
1.9	1.0	Thermodynamic analysis	0–240 GPa Along T_m	Anderson and Isaak, 2000
2.2 (±0.5)	1.62 (±0.37)	Shock-wave	~15–150 GPa to >4000	Jeanloz, 1979 McQueen et al., 1970
2.75	1.5	Theory	0–300 GPa to 3000 K	Wasserman et al., 1996

[1]Nuclear resonant inelastic x-ray scattering (see text).

measurements to 304 GPa; a larger (but still weak) pressure dependence was found in more precise quasihydrostatic experiments to 78 GPa [c/a = 1.606 (±0.002) – 1.2 (±0.3) × 10^{-4} P]. Low P-T x-ray diffraction studies indicated that the ratio remains essentially constant to 573 K (<24 GPa), although a slight increase (to 1.63 (±0.02)] was observed at 723 K (Huang et al., 1987). An apparent increase observed by Funamori et al. (1996) was not reversible. In contrast, the recent high P-T data to above 2000 K and 100 GPa show no major changes relative to ambient temperature. Theoretical calculations give results that are broadly consistent with, but lower than, these values—i.e., 1.58 increasing to 1.595 at 320 GPa, both at 0 K (Steinle-Neumann et al., 1999). However, a significant increase in c/a is predicted with increasing temperature (e.g., at constant pressure) to above 1.7 at 6000 K at core pressures (Wasserman et al., 1996; Stixrude et al., 2000); this trend has not been observed experimentally (up to the melting curve at 105 GPa and to ~3000 K at 161 GPa).

Acoustic velocities

Recent experimental and theoretical developments have provided methods to extract the single-crystal second elastic moduli in addition to the bulk modulus (Singh et al., 1998a, 1998b) (Fig. 6A). This is done by radial (as opposed to the conventional near axial) diffraction of a polycrystalline sample under pressure in an imposed non-hydrostatic stress field. The principle and basic methodology (but not applied to high-pressure data) has been developed independently by several groups (Bollenrath et al., 1967; Bittorf et al., 1998; Gnäupel-Herold et al., 1998; Howard and Kisi, 1999). Measurements of the d-spacings as a function of angle relative to the load axis reveals the elastic anisotropy of the material, which is controlled by the elastic moduli, or stiffness coefficients, at a given pressure. This measured angle dependence of the d-spacings, together with an independent determination of K and p can be used to constrain the additional elasticity parameters. If the uniaxial as well as the hydrostatic stress is measured on the sample, then the full tensor can be determined (see below). The technique has been applied to a variety of materials to date, including α-Fe and ε-Fe (Singh et al., 1998b), Re and Au (Duffy et al., 1998), FeO (Singh et al., 1998a, 1998b), FeS_2 (Merkel et al., submitted), and MgO (Merkel et al., in prep.), all at ambient temperature. We focus on recent applications to ε-Fe (Mao et al., 1998b; Singh et al., 1998b).

The dependence of d-spacings as a function of ψ and hkl, together with additional constraints provided by hydrostatic x-ray diffraction data are used. The stress state of the specimen compressed between two anvils is a superposition of the hydrostatic pressure, $\sigma_P = (\sigma_3 + 2\sigma_1)/3$, and the differential stress components, $t = \sigma_3 - \sigma_1$, where σ_3 and σ_1 are axial and radial stresses, respectively. The experimentally measured *d(hkl)* as a function of ψ can be derived from anisotropic linear elasticity theory (Singh and Balasingh, 1994; Uchida et al., 1996), and has the following form:

$$d(hkl) = d_P(hkl)\,[1 - 3\cos^2\psi]Q(hkl), \tag{5}$$

where $d_P(hkl)$ denotes the *d*-spacing under hydrostatic pressure σ_P. The values of *Q(hkl)* and $d_P(hkl)$(slope and intercept of this equation) are determined by linear regression. For elastically isotropic materials, the slope *Q(hkl)* is identical for all *hkl*s, the data obtained for ε-Fe at 220 GPa. For elastically anisotropic systems, *Q(hkl)* varies with *hkl*, as illustrated in Figure 6B. The shear modulus can be approximated as:

$$G = (t/6\bar{Q}) = (\sigma_3 - \sigma_P)/4\bar{Q}, \tag{6}$$

where $\bar{Q}$ denotes the average value of *Q(hkl)* for all observed *hkl* reflections (Singh et al., 1998b). The hydrostatic pressure component, σ_P, can be calculated from $d_P(hkl)$ and the hydrostatic P-V equation of state.

Two different types of experiments were performed, with a stress calibrant (Au) to 39 GPa and without a stress calibrant to 220 GPa, assuming a constant ratio of the bulk modulus to shear modulus, *K/G*, where the pressure was determined from Au or ruby (Mao, 1998b; Singh et al., 1998b). In the former a two-layer sample configuration is used to determine the axial stress, σ_3 (Fig. 6A). In this configuration, σ_3 is constant across the interface between the sample (*A*) and calibrant (*B*), but the radial stresses (σ_{1A} and σ_{1B}) and average pressures (σ_{PA} and σ_{PB}) are different in *A* and *B*. Values of σ_P and $\bar{Q}$ for the sample and calibrant are measured simultaneously by radial x-ray diffraction. The axial stress, σ_3, is determined from the above equation with the known value of *G* of the calibrant, which in general can be determined by high-pressure Brillouin scattering or extrapolated from lower-pressure ultrasonic measurements. Because of the extremely low shear strength of Au relative to Re; both $\bar{Q}$ and *t* of Au are a factor of ten smaller than those of Re. At 37 GPa, $t_{Au} = 0.52$ GPa and $t_{Re} = 5.8$ GPa. The shear modulus of Re determined with the gold calibrant compares well with ultrasonic measurements at low pressures (see Mao, 1998b).

The aggregate sound velocities are related to the (adiabatic) K_S and G_S by

$$V_P = \sqrt{\left(K_S + \frac{4}{3}G_S\right)/\rho}, \tag{7}$$

$$V_S = \sqrt{G_S/\rho}. \tag{8}$$

The analysis uses K_S [where $K_S = K_T(1 + \alpha\gamma T)$] and ρ determined by x-ray diffraction to 300 GPa discussed above (Jephcoat et al., 1986; Mao et al., 1990). The pressure dependence of V_P and V_S determined from this analysis is shown in Figure 6C. The results are in good agreement with ultrasonic measurements of V_P and V_S at 16.5 GPa, which were performed as a test of the method (Mao, 1998b). Agreement with theory is fair, and tends to improve at higher pressures. The velocities and moduli calculated from theory are significantly higher than the multi-anvil ultrasonic data at 16.5 GPa. As described above for the bulk modulus, these discrepancies can be attributed to remaining errors in the generalized gradient approximation (GGA) applied to iron and/or the stabilization of more complex spin arrangements than those considered theoretically (e.g., antiferromagnetic models described below) (Stixrude et al., 1997; Steinle-Neumann et al., 1999). The pressure dependence of V_S recently estimated from the pressure dependence of the Debye temperature obtained from x-ray diffraction intensities (Anderson et al., 2001) is significantly lower than the radial diffraction and theoretical results.

The radial diffraction data were obtained at room temperature; thus, additional analysis is required to estimate the effect at expected core temperatures and pressures. The shock-wave data can be used to estimate the effect of temperature on velocities at these temperatures. Assuming core temperatures of 6000 K, with $dV_P/d_T = -2.9 \times 10^{-4}$ km s^{-1} K^{-1} (Brown and McQueen, 1986), velocities of iron at core conditions are somewhat higher than those determined by seismological studies of the core (Mao et al., 1998b). Plots of V_P and V_S as functions of density provide additional and complementary compositional constraints, i.e., Birch's law (Birch, 1963). The radial diffraction results for V_P are shown in Figure 6D along with other determinations, including very recent inelastic x-ray scattering (IXS) measurements to 110 GPa (Fiquet et al.,

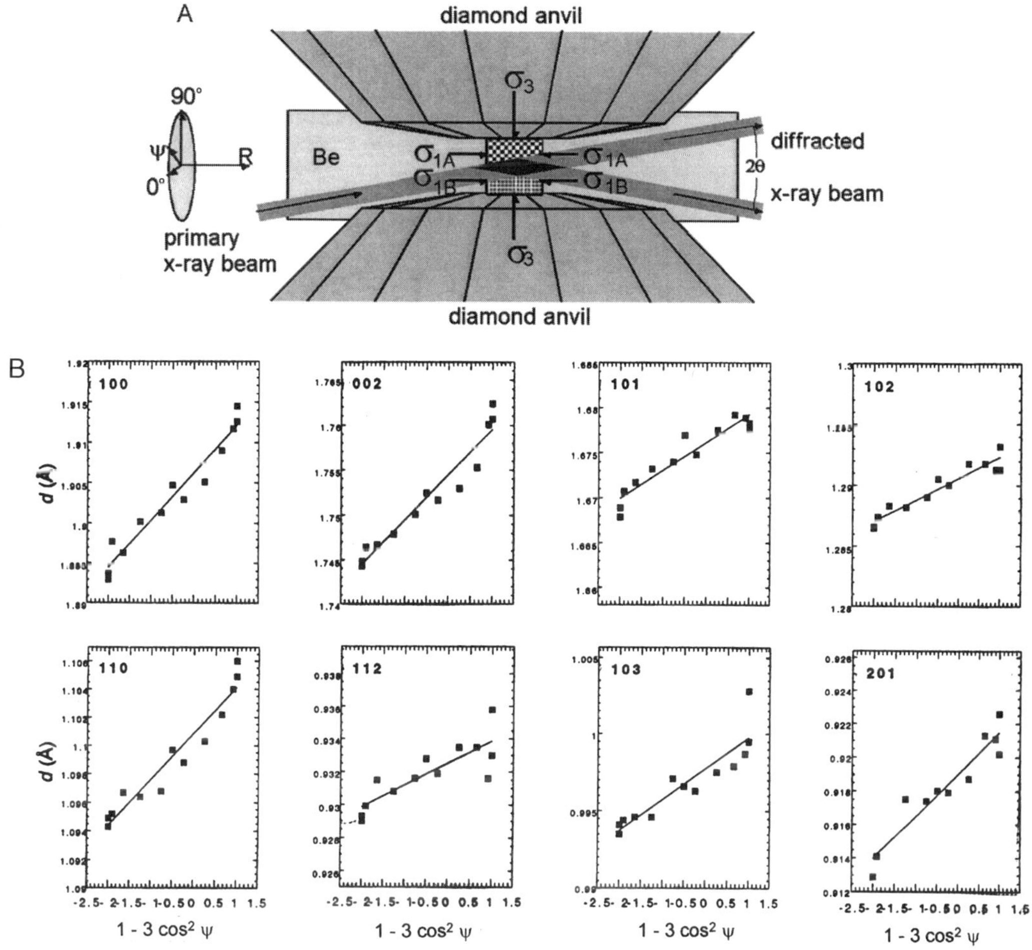

FIG. 6. A. Schematic diagram of diamond cell used for radial diffraction measurements. B. d-spacings of iron measured by radial x-ray diffraction at 220 GPa and 300 K. (*continues on facing page*).

2001). Extrapolation of V_P-density relation from the radial diffraction data shows 2% lower density than the inner core, suggesting possible heavier elements (e.g., Ni), in the inner core. However, the difference is within the experimental uncertainty for pure Fe, and improved accuracy is needed to determine accurate relative abundances in the inner core.

Elastic anisotropy

The magnitude (and even existence) of elastic anisotropy for the inner core has been the subject of much recent controversy. However, a body of evidence indicates that compressional waves travel ~2% faster along the polar relative to equatorial directions (see Dziewonski, 2000). A starting point for understanding this phenomenon from a materials point of view requires a determination of the elastic anisotropy of pure iron. As mentioned above, the anisotropy of ε-Fe at 220 GPa and 300 K is immediately apparent from the variation in slope of the d-spacings measured as a function of ψ at a given pressure (Fig. 6B). However, quantitative determination of the elastic anisotropy and the full second-order elastic moduli (stiffness coefficients C_{ij}s) from such data are the subjects of ongoing studies. The radial diffraction technique and anisotropic elasticity theory used to obtain the C_{ij}s has been well tested for cubic systems. In addition the V_P, V_S, and G appear to be close for hexagonal systems where the data can be cross checked against other

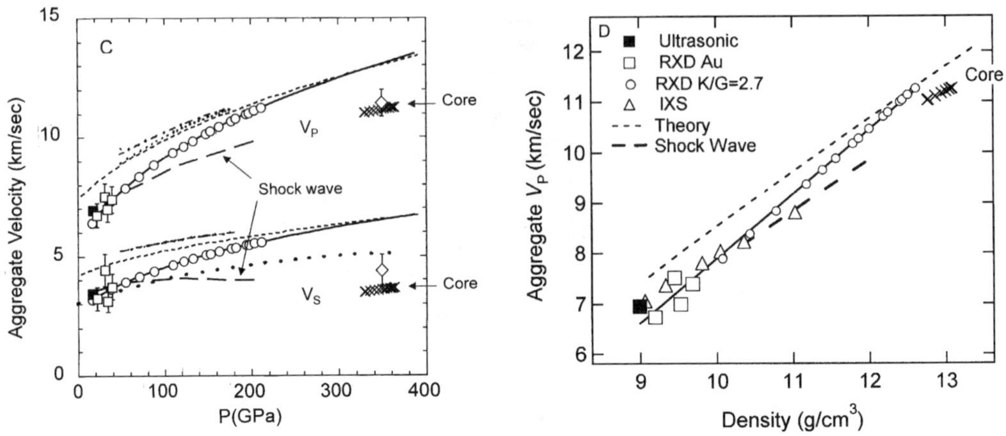

FIG. 6. (*continued*) C. Aggregate shear and compressional wave velocities for iron. The solid square is the ultrasonic measurement at 16.5 GPa. The open symbols are the lattice radial x-ray diffraction results; squares are measurements with Au stress calibration; the circles were obtained assuming a constant *K*/*G* (Mao et al., 1998b). The short dashed line is the static lattice tight-binding (non-magnetic) calculation from the study of Stixrude and Cohen (1995); the dash-dotted line is the later non-magnetic GGA result (Steinle-Neumann et al., 1999); the thin dotted line is from antiferromagnetic GGA calculations (Steinle-Neumann, pers. commun.). The long dashed line shows the shock-wave results from Brown and McQueen (1986). The thick dotted line is the recent estimate for V_S from analysis of x-ray diffraction intensities (Anderson et al., 2001). The crosses are the PREM model results for the inner core (Dziewonski and Anderson, 1981), and the open diamonds show the calculated velocities at core conditions obtained from the static compression data (Mao et al., 1998b). D. Aggregate acoustic velocity (V_P) as a function of density. The symbols are the same as in Figure 6C; the solid line is a fit to both sets of radial diffraction data. The triangles are the recent inelastic x-ray scattering (IXS) results (Fiquet et al., 2001).

methods. For hexagonal systems (ε-Fe and Re), the values of key moduli, including C_{44}, differ from first-principles theory (Steinle-Neumann et al., 1999), although there appear to be differences among theoretical results that depend strongly on approximations made in the calculations (cf., Soderlind et al., 1996; Laio et al., 2000).

The single-crystal acoustic velocity anisotropy is 24% at 300 K, assuming no effects of texture and anisotropy in the plastic deformation on the inversion for the single-crystal elastic moduli. As discussed by Mao et al. (1998b), different results are obtained if there are stress variations due to anisotropy of the shear strength (i.e., preferred slip systems) giving rise to development of lattice-preferred orientation. In fact, such texture is observed (Hemley et al., 1997; Wenk et al., 2000b), as discussed below. An extended theoretical method has been developed that self-consistently takes account of these effects in the calculation of the elastic moduli (Matthies et al., in prep.). We note that, in contrast to these experimental results, the 0 K calculations give an anisotropy of 4% at core pressures (Stixrude and Cohen, 1995). On the other hand, significant differences are found when the effects of temperature at high pressure are included (Stixrude et al., 2000). This is coupled to the large changes in *c*/*a* predicted in these calculations (Cohen, pers. commun.), findings that remain to be observed experimentally.

Vibrational Properties

Phonon density of states

Determination of the phonon densities of states (DOS) for various phases at high pressures and temperatures provides both a microscopic basis for, as well as a means of calculating, thermodynamic and elastic parameters essential for characterizing iron at depth and for interpreting seismological observations. Integration of the phonon density of states provides information on thermodynamic properties such as the Debye temperature (Θ_D), vibrational entropy (S_{vib}), and vibrational heat capacity (C_{vib}), as well as other parameters such as the vibrational kinetic energy (E_k) and zero-point energy (E_Z). For metals, there is an additional electronic contribution to the thermodynamic properties at Earth-interior temperatures, which has been examined at high pressure only theoretically (e.g., Boness et al., 1986; Boness and Brown, 1990; Stixrude et al., 1997). Measurements of the pure vibrational component provide fundamental constraints for theory, and are typically done with neutron scattering techniques, but this requires very large samples, and therefore is not yet possible at very high pressures (e.g., core conditions). Such information serves as useful input for thermodynamic models (e.g., Anderson and Isaak, 2000). Other estimates of the macroscopic thermodynamic properties have been obtained by shock-wave measurements (Brown and McQueen, 1986; Yoo et al., 1997) and theoretical calculations (Stixrude et al., 1997; Vocadlo et al., 2000) at core pressures and temperatures.

A new method based on ^{57}Fe nuclear resonant inelastic x-ray scattering (NRIXS) has been developed for determination of phonon densities of states (Figs. 7A and 7B) (Sturhahn et al., 1995). The technique is particularly suitable for pure Fe, in which all phonons naturally result from vibrations of Fe atoms, as demonstrated by the excellent agreement for the density of states of bcc Fe measured by inelastic neutron scattering (Minkiewicz et al., 1967) and NRIXS at ambient pressure (Hu et al., 1999). NRIXS measurements of ε-Fe were first performed to 42 GPa and the phonon density states determined (Lübbers et al., 2000). Subsequent measurements carried out to 153 GPa and with higher resolution have provided a more accurate and complete evaluation of the phonon properties of iron to core pressures (Mao et al., in press). The DOS at selected pressures and the phonon contribution to thermodynamic functions obtained from the analysis are shown in Figures 7C and 7D. Differences between the two NXIRS studies arise from the lower resolution of the earlier measurements (Wortmann, pers. commun.). For Θ_D there is good agreement between the newer results and the recent determination from x-ray diffraction intensities (Anderson et al., 2001); despite the scatter in the NRIXS results at lower pressure, the similar pressure shift indicates that the effective Grüneisen parameters obtained from the two studies are close (γ = 1.72 calculated by Anderson et al. [2001]).

The form of the phonon DOS as $E \rightarrow 0$ gives another determination of the aggregate sound velocity. Specifically, parabolic fitting of the low-energy portion of the phonon DOS provides a measure of the so-called Debye velocity V_D (Simmons and Wang, 1971), which is defined as

$$3V_D^{-3} = V_P^{-3} + 2V_S^{-3}. \tag{9}$$

Combined with accurate determinations of the bulk and shear moduli as a function of pressure, the measurement allows an independent determination of V_P and V_S. For ε-Fe, there is broad agreement between the acoustic velocities determined by NRIXS, various first-principles techniques, and zero-pressure ultrasonic measurements (Guinan and Beshers, 1968). For ε-Fe, there is good agreement between the NXIRS results and the analysis of the radial diffraction data described above, particularly at higher pressures (Mao et al., in press). The DOS curves calculated with first-principles calculations using soft pseudo-potentials are compatible with the NRIXS results, although there are significant differences at higher frequencies (Mao et al., in press).

Very recently, non-resonant x-ray inelastic scattering (IXS) phonon measurements have been reported for single-crystal α-Fe to 10 GPa (Klotz and Braden, 2000). Single-crystal measurements provide additional information on the phonons, including dispersion, single-crystal elastic tensor, and anisotropy, but have not yet been possible for ε-Fe, which becomes polycrystalline when it is converted from α-Fe. The IXS measurements of polycrystalline ε-Fe to 110 GPa at ambient temperature mentioned above have been limited to V_P (Fiquet et al., 2001); the results agree (within error bars) with the radial x-ray diffraction and NRIXS values, but do not provide the phonon DOS, V_S, and thermodynamic properties, as do the NRIXS measurements. IXS determination of the full phonon properties will require the growth of single-crystal ε-Fe by techniques such as laser-heating annealing. Both NRIXS and IXS measurements at high pressures may also be extended to very high temperatures (e.g., several thousand Kelvin) by laser heating.

Optical modes

Additional information has been obtained from optical light scattering measurements. For many years, it was thought that vibrational spectroscopy of metals under pressure would not be possible because their high reflectivity tends to preclude or significantly attenuate the measurement of vibrational mode structure by infrared spectroscopy and Raman scattering. Pioneering Raman scattering measurements of transition metals have been carried out by Olijnyk and co-workers (see Olijnyk and Jephcoat, 2000), although measurements on Fe proved very difficult. With the development of new

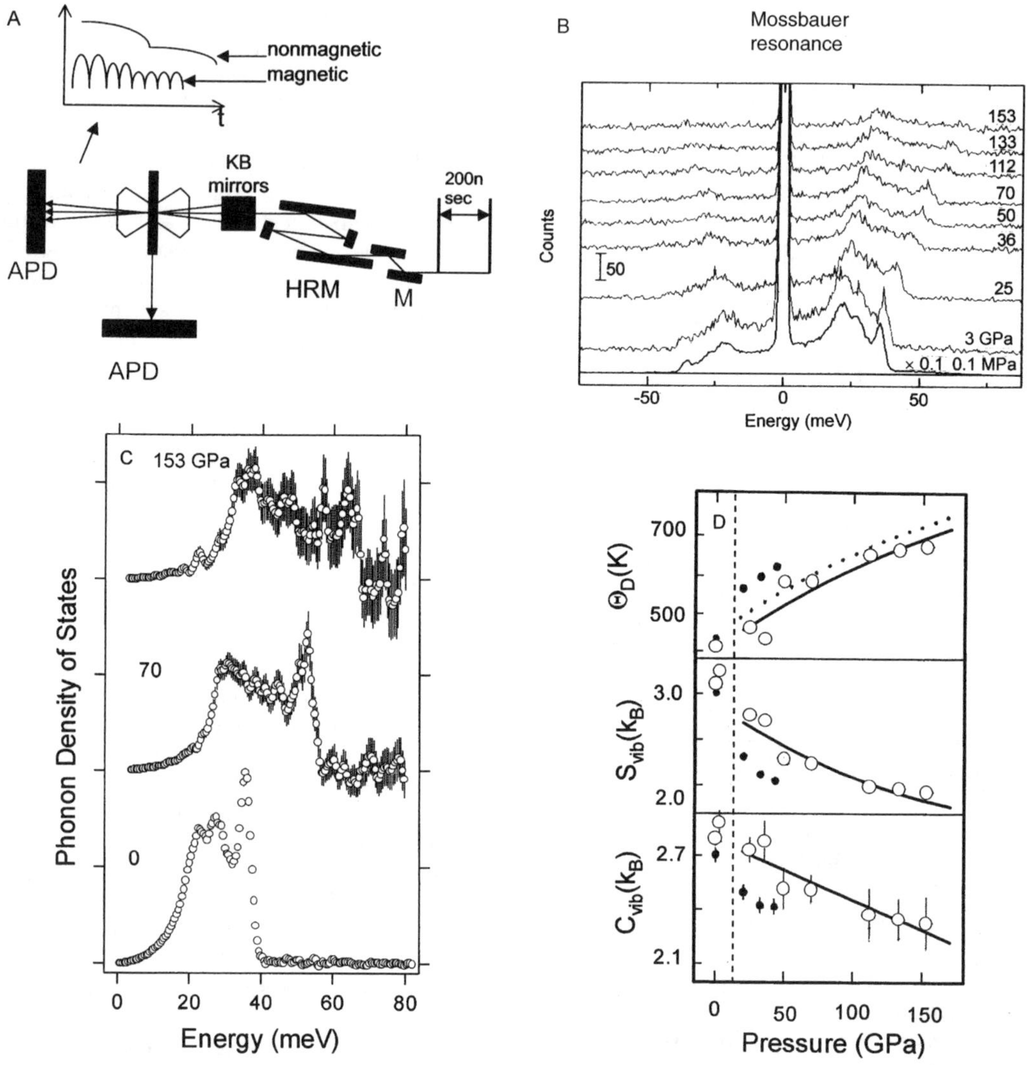

FIG. 7. Nuclear resonant inelastic x-ray scattering (NRIXS) measurements at 300 K (Mao et al., in press). A. Schematic of the experimental layout. The APDs are avalanche photodiode detectors; that on the direct beam is used for Mössbauer measurements (coherent nuclear forward scattering), whereas the side detectors are used for the fluorescence density of states measurements. The beam passes through a monochromator (M), high-resolution monochromator (HRM), and KB (Kirkpatrick-Baez) mirrors before impinging on the sample (Sector 3, Advanced Photon Source). B. Spectra measured showing the side-band structure on the 14.413 keV ^{57}Fe Mössbauer resonance corresponding to the phonon density of states. C. Phonon density of states at selected pressures extracted from the spectra. The circles (with error bars) show the experimental results obtained from NRIXS. D. Debye temperature (Θ_D), vibrational entropy (S_{vib}), and the vibrational contribution to the constant volume heat capacity (C_{vib}), all at 300 K (k_B is the Boltzmann constant); the open circles show the NRIXS results from Mao et al. (in press); the solid line is a guide to the eye; solid circles are the lower pressure NRIXS data from Lübbers et al. (2000); the dotted line is the recent Θ_D determination from x-ray diffraction intensities (Anderson et al., 2001). The vertical dashed line shows the α-ε transition.

methods, such as high-throughput spectrographs, sensitive charge-coupled device detectors, and ultrapure synthetic diamond anvils (having essentially no background luminescence), such weak Raman spectra became accessible. The application of these new techniques to ε-Fe showed that Raman

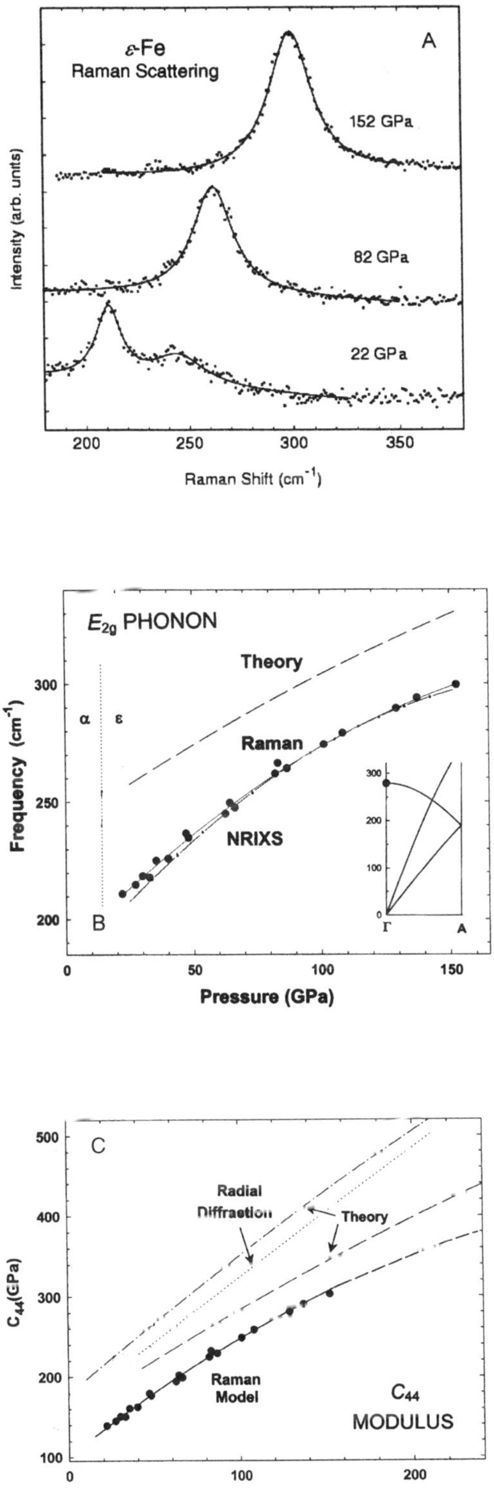

FIG. 8. A. Raman spectra of ε-Fe showing the E_{2g} vibrational mode (Merkel et al., 2000). B. The pressure dependence of the E_{2g} Raman mode (both compression and decompression). The solid line is a fit through the data, and nearly coincident dashed-dotted line is the corresponding peak in the phonon density of states obtained from the NRIXS measurements (Fig. 7). The upper dashed curve shows the results of pseudo-potential calculations (L. Vocadlo, pers. commun.; Goncharov et al., in prep.). The inset shows the qualitative zone-folding of the acoustic mode to produce the Raman-active E_{2g} mode, i.e., frequencies as a function of wave vector from Γ (zone center) to the A point in the Brillouin zone. C. C_{44} calculated from Raman data and compared with various determinations. The solid line is a fit to the experimental data (dashed line is an extrapolation). The radial diffraction (lattice strain) results are given by the dotted line (Mao et al., 1998b). The dot-dashed line is the LDA calculation of Soderlind et al. (1996); the middle dashed line is a simple fit to the LDA results of Steinle-Neumann et al. (1999) and Laio et al. (2000).

spectra could be measured to core pressures and room temperature (Fig. 8A); in fact, predicted Raman-active E_{2g} phonon was observed to 152 GPa (Merkel et al., 2000). The results also provide the mode-Grüneisen parameter for the E_{2g} band; the result is close to the bulk (thermodynamic) γ obtained from other measurements.

The E_{2g} phonon is of particular interest because this mode correlates with a zone-edge acoustic mode (Fig. 8B, inset). As such, its frequency ν can be compared with the first peak in the phonon density of states; excellent agreement is found. Moreover, its frequency can be related to the value of C_{44} and therefore the shear-wave velocity as a function of pressure (Fig. 8C). Assuming both two- and three-body interactions, the following relation can be derived (Upadhyaya et al., 1994):

$$C_{44} = 2\pi^2 M[(\sqrt{3}c)/(6a^2)]\nu^2 , \qquad (10)$$

where M is the atomic mass of the metal.

Jephcoat and Olinyk (pers. commun.) have examined the correlation between ν and the measured values for C_{44} for a large number of hcp metals and have found that the values agree to within <20%, with significantly better agreement found for materials in which the hcp phase has a wide stability field. The resulting pressure dependence of C_{44} estimated from the Raman mode frequency is lower, and has a smaller pressure derivative, than that obtained from the radial diffraction analysis. A simple ad hoc adjustment of the C_{44} gives closer agreement with the 0 K theoretical results (Merkel et al., 2000); however, because of tradeoffs in the best-fit solution, a final resolution of the problem awaits full optimization, including the texture and strength effects in the analysis of the radial diffraction.

Electronic and Magnetic Properties

The electronic and magnetic structure of iron at or near ambient pressure is generally well understood but its behavior at high pressure is less clear: variations in electronic structure as a function of compression control its high-pressure behavior, including polymorphism as well as numerous physical and chemical properties. New insight has come from both experiment and theory. For example, it is well established that the bcc phase is magnetic, and theoretical calculations indicate that the magnetism stabilizes the bcc structure. An additional recent advance has been the measurement of the pressure dependence of magnetic susceptibility of ferromagnetic materials, including polycrystalline iron, in diamond cells (Timofeev et al., 1999). The amplitudes of the peaks of iron samples (linear dimensions ~50 μm) remain essentially unchanged with increasing pressure up to 11 GPa (the maximum reached in this study), indicating that iron retains its ferromagnetism in this phase, consistent with high-pressure Mössbauer measurements (Williamson et al., 1972). The technique is now ready to be extended to higher pressures.

The electronic and magnetic properties of iron change upon transforming to the ε-phase. Both the reflectivity and conductivity of iron decreases markedly across the α-ε transition (e.g., Mao et al., 1987); measurements to 300 GPa showed that the low reflectivity continues to much higher pressure (Reichlin, Hemley, and Mao, unpubl.). Early Mössbauer studies to 17 GPa found no evidence for magnetic ordering in ε-Fe (Williamson et al., 1972). However, recent theoretical considerations suggest the existence of magnetic interactions in ε-Fe at low pressure: first-principles calculations that include antiferromagnetic ordering give a lower energy and show improved agreement for the bulk modulus when compared to the non-magnetic solution (Steinle-Neumann et al., 1999). Two different models for the antiferromagnetic ordering have been examined—atomic layers of opposite spins perpendicular to *c*, and layers of opposite spins normal to [100]. Because such ordering would be frustrated in the hcp structure, there is the further possibility of an incommensurate structure and a distinct magnetic collapse transition within the ε-phase, as in transition metal oxides and sulfides (Cohen et al., 1997; Hemley et al., 1998). These predictions suggest the need to examine in detail the pressure (and temperature) dependence of the electronic and magnetic properties of ε-Fe with new magnetic, synchrotron x-ray (e.g., coherent nuclear forward scattering) and neutron techniques. Preliminary studies indeed reveal measurable effects of pressure on the valence band densities of states and apparent local moment of Fe (Struzhkin et al., in prep.). These experiments utilize high-resolution emission measurements of the K_β multiplet fine structure recently applied to FeO and FeS (Badro et al., 1999; Rueff et al., 1999). Quantitative analyses of these new data will require calculations that go beyond now-standard band-structure approximations, a proper treatment of local magnetic interactions, and the renormalization of properties by temperature.

Rheology

The deformation of iron in the core has been the subject of numerous theoretical studies. Early on, Jeanloz and Wenk (1988) proposed that crystal alignment attained during deformation might be responsible for the seismic anisotropy of the inner core, and that this arises from convection in the inner core. The phenomenon has been examined in more detail in recent convection-model simulations that show that outer-core convection can give rise to elastic anisotropy, but do not support the hypothesis of inner-core convection (Wenk et al., 2000a). A recent theoretical study by Poirier and Price (1999) has predicted that at the microscopic level basal slip may be the dominant mode of deformation. The radial x-ray diffraction measurements described above provide direct information on the rheological properties of iron at core pressures. In fact, consideration of the plastic deformation in combination with the development of texture is essential for inverting the x-ray data to obtain the individual elastic tensor elements and anisotropy tensor. From the texture pattern of a deformed material, it is possible to infer active deformation systems by comparing experimental observations with corresponding polycrystal plasticity simulations (Kocks et al., 1998).

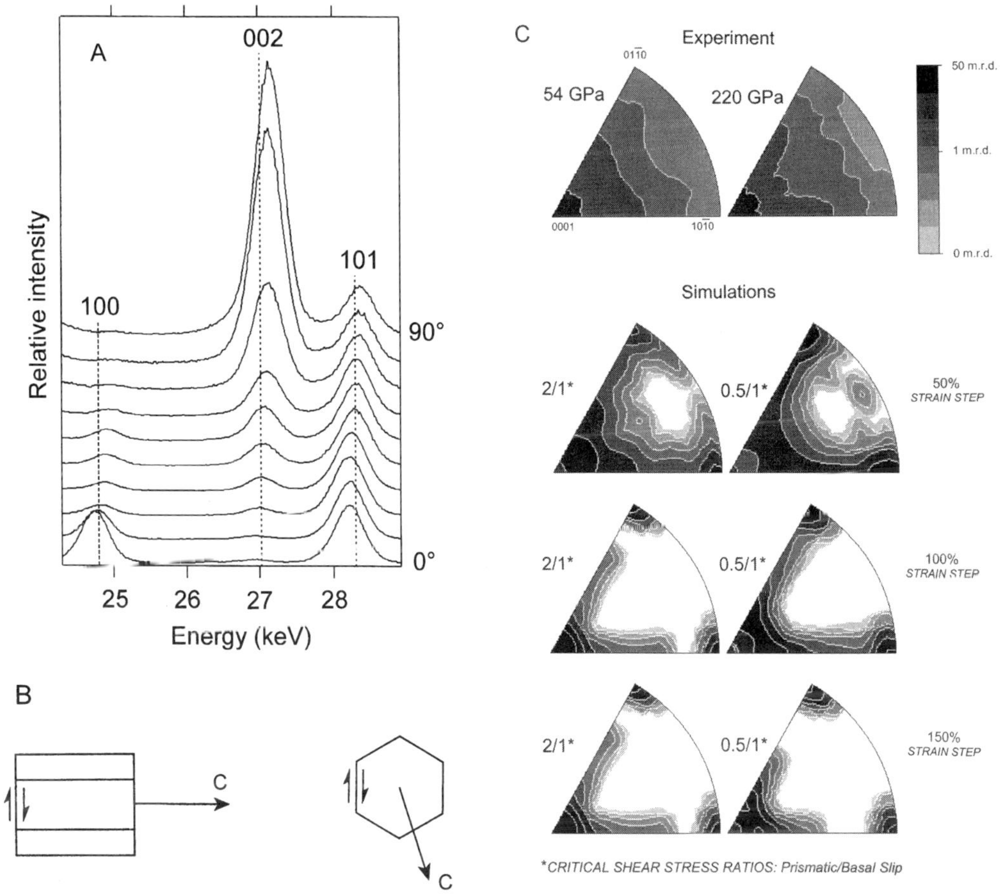

FIG. 9. A. X-ray diffraction of ε-Fe as a function of rotation, showing the strong texture that develops under non-hydrostatic compression. B. Schematic representation of basal and prismatic slip. C. Measured inverse pole figures for ε-Fe at 54 GPa and 220 GPa (300 K) and simulations of the texture assuming different values of the critical shear stress ratios for prismatic and basal slip as well as different strain steps (from Wenk et al., 2000b).

Numerous static high-pressure experiments have demonstrated that ε-Fe exhibits very strong preferred orientation when compressed under non-hydrostatic conditions (see Hemley et al., 1997). In a recent systematic study, intensities were extracted for eight diffraction peaks (100, 002, 101, 102, 110, 200, 112, and 201) for 10 sample orientations, corresponding to lattice planes oriented from parallel to perpendicular to the diamond anvil axis in 10 degree intervals (Fig. 9A) (Wenk et al., 2000b). The orientation distribution was calculated from the variations in relative intensity (Wenk et al., 1998). These can be represented by inverse pole figures for the direction of the diamond cell axis. The results show the existence of strong preferred orientations with a dominating (0001) fiber component; i.e., the *c*-axis of the crystallites tends to align parallel to the compression direction.

In the analysis of the texture pattern, calculated inverse pole figures illustrate that a fiber texture with a concentration of compression axes near *c* = [0001] develops with increasing strain. This occurs if basal slip is the primary or secondary slip system (Fig. 9B). The plastic spin due to basal slip reorients the basal poles toward the most compressed direction. The spin due to prismatic slip causes a rotation around the (0001) direction, leaving the basal pole orientation invariant. Basal slip facilitates the alignment of the basal poles with the compressive direction but does not qualitatively change the texture pattern. The analysis indicates that basal slip is an active deformation mechanism, in agreement with recent predictions (Price et al., 1999), probably in combination with other secondary slip systems (Fig. 9C). Strong textures are expected to develop if deformation in the core occurs during convection and crystals deform by basal and prismatic slip (Wenk et al., 2000b). Similar patterns are expected if deformation is caused by magnetic forces (Karato, 1999; Wenk and Buffett, 2000). Testing these hypotheses requires further theoretical refinements (Matthies et al., in prep.) and additional experiments that include effects of temperature, composition, and kinetics.

Liquid iron

Ninety-six percent of the core by volume is in a liquid state. The behavior of liquid iron at high pressures and temperatures has been investigated with shock-wave, static-compression, and theoretical techniques. A thermodynamically consistent high P-T equation of state has been determined based on the available experimental data (Anderson and Ahrens, 1994). Information on the local structure of the liquid provides a basis for understanding the origin of numerous macroscopic physical properties. However, direct experimental study of the structure of liquid iron at high pressure has long been prevented because of technical difficulties arising from the extreme conditions required (above 1810 K, melting temperature of iron at ambient pressure) (Waseda and Suzuki, 1970; Waseda and Ohtani, 1974). Information on the structure of the liquid under pressure has almost exclusively been the domain of theory (Hausleitner and Hafner, 1989; Stixrude and Brown, 1998; Alfè et al., 2000).

The structure of liquid iron at high pressures and temperatures has recently been studied directly using new synchrotron x-ray diffraction methods (Sanloup et al., 2000a) (Figs. 10A and 10B). Such measurements provide the radial distribution function $g(r)$, which gives the probability of finding atoms at a distance r from a given atom and therefore is a measure of the local structure in a disordered material (Fig. 10C). The results obtained for liquid Fe show evidence for structural changes in the liquid near the δ-γ-l triple point. At 3.0 GPa and 2300 K, and 4.8 GPa and 2100 K, the measured $g(r)$ distributions are well described by a bcc-like structure. At 3.9 GPa and 2100 K, the $g(r)$ indicates a structure intermediate between bcc-like and the less structured ambient pressure result (Waseda and Ohtani, 1974). On increasing pressure at 2300 K (4.0 and 5.0 GPa), $g(r)$ cannot be reproduced by a bcc-like structure alone, but instead can be described as a mixture of bcc-like and close-packed structures. The same effect appears at constant pressure (5.0 GPa) on increasing the temperature from 2100 K to 2300 K. These effects have not yet been examined theoretically. First-principles theoretical studies to date of the liquid appear to have been mostly restricted to the megabar pressure range (Stixrude and Brown, 1998; Alfè et al., 2000); the new experimental results should provide crucial input for furthering our theoretical understanding of dense liquid iron.

These results imply that the local order in the liquid increases in response to pressure and temperature, and this is reminiscent of the underlying crystalline phases. It is important to emphasize that the structure of the liquid should not be identified in a strict sense as that of the crystalline phases, nor is there evidence for a well-defined structural (e.g., "phase") transition in the liquid state (e.g., Brazhkin et al., 1997). Changes in liquid structure should influence macroscopic physical properties such as the viscosity, but this remains to be investigated. The viscosity of the outer core is one of the key geophysical parameters needed for modeling the dynamo, but the dependence of the viscosity of liquid iron on pressure and temperature is controversial. New data suggest systematics for the P-T dependence of viscosity for a variety of liquid metals (Brazhkin and Lyapin, 2000). These systematics have led to the prediction of a large enhancement in the viscosity of liquid iron by 5–12 orders of magnitude from ambient to core pressures, and a proposed vitreous state for the inner core; this idea is based on a proposed liquid-glass transition within the P-T stability field of the melt (Brazhkin and Lyapin, 2000), for which there is no apparent experimental evidence in any system. The conjecture of a liquid-liquid transition is more plausible on empirical grounds, having been observed in other systems under pressure, but there is no direct evidence from theory or experiment for such a transition in liquid Fe. It is important to extend the measurements to extreme conditions approaching those of the outer core to examine these and other effects. For example, simulations predict that the structure factor $S(q)$ as a function of scaled momentum transfer is constant along the melting curve (Hausleitner and Hafner, 1989); this can be checked by diffraction measurements over an extended P-T range.

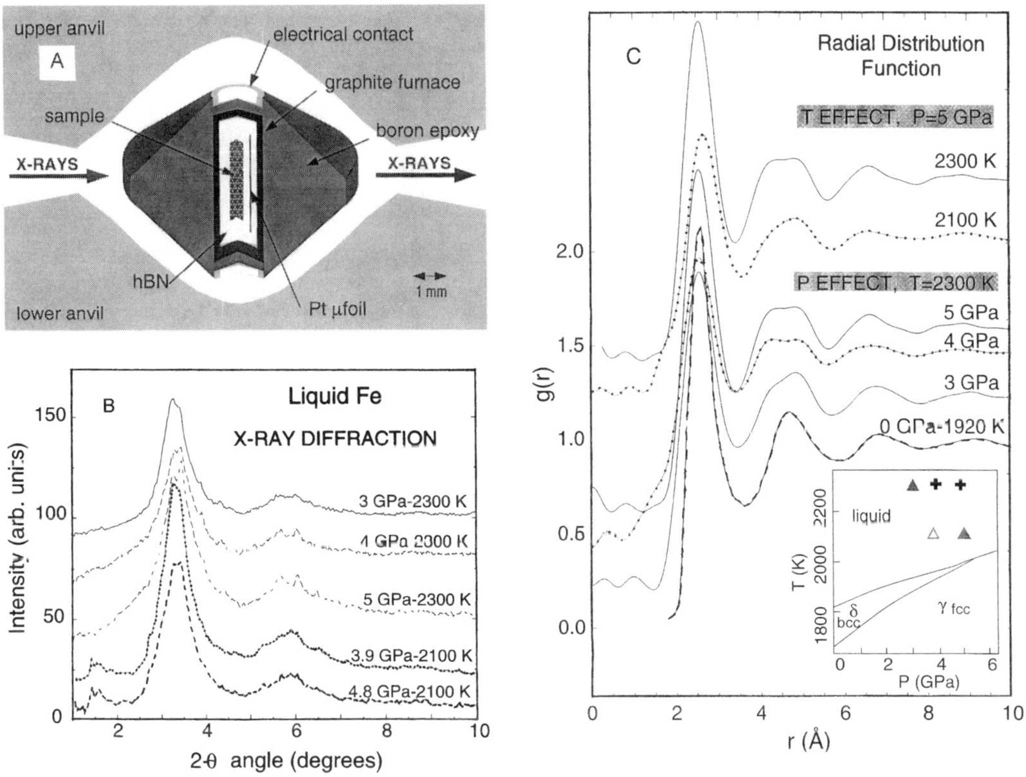

FIG. 10. High P-T x-ray diffraction measurements on liquid iron (Sanloup et al., 2000a). A. Schematic diagram of the experiment, which employs an internally heated Paris-Edinburgh cell (see Mao and Hemley, 1998) and monochromatic radiation with an area detector. The experiments were carried out at beamline ID30 of the European Synchrotron Radiation Facility (ESRF). B. X-ray diffraction patterns at selected pressures and temperatures. The weak, sharper features arise from sampling a small amount of crystalline material in addition to the melt. C. Radial distribution function $g(r)$ determined from x-ray diffraction. The explored region of the P-T phase diagram is shown in the inset.

Ultimately, all of these effects need to be addressed for core-forming liquids (i.e., alloys) at core conditions. Recent experimental data (again at low pressure) do show marked differences in the $g(r)$ and other properties as compared to pure iron (Sanloup et al., 2000b, 2001). First-principles molecular dynamics simulations carried out by Alfè et al. (2000) indicate that liquid iron is a close-packed simple liquid at outer-core conditions. Moreover, these calculations predict that the diffusion coefficient and viscosity under these conditions are similar to those of simple liquids under ambient pressure and temperature. Simulations of the dense liquid state predict that sulfur impurities (up to the 12% level) have a negligible effect on the viscosity relative to pure iron (Alfè and Gillan, 1998). Other simulations indicate that Ni should have no appreciable effect on the viscosity of iron at core conditions (Zhang and Guo, 2000; Zhang et al., 2000). All of these predictions remain to be tested experimentally.

High P-T Chemistry

The large effects of pressure and temperature on the physical properties of iron within the Earth are coupled with equally significant, and ultimately intimately related, changes in the chemistry of the element. Of paramount interest is the dissolution of light elements in the core, a major unsolved problem in geophysics. The identity of the light-element, and freezing-point depression relative to pure iron, is intimately tied to the question of the extent of light-element release at the ICB and the consequent compositional convection proposed to power the dynamo. Recent studies of end-member minerals at very high pressures and temperatures provide important new insight into this problem. As discussed above, there is compelling evidence that the inner core, as well as the liquid outer core, is less dense than pure iron, indicating the presence of a light-element component in this region as well (Jephcoat and Olson, 1987; Anderson and Ahrens, 1994; Stixrude et al., 1997). This is but a part of the larger problem of the changes in the chemistry of metallic iron induced by the combination of high pressures and temperatures encountered even at shallow depths within the planet (i.e., during core formation). Characterizing the chemistry of iron over a broad range of P-T-X conditions is crucial for solving major problems such as siderophile-element partitioning, volatile uptake in the early Earth, and chemical exchange across (as well as stability of) the core-mantle boundary region. We briefly highlight recent experimental observations and theoretical predictions that bear on these issues. Specifically, we consider reactions of iron with oxygen, sulfur, hydrogen, and additional (including heavier-element) components, as well as their mixtures.

As the most abundant element in the planet, oxygen may be considered a leading candidate for one of the light elements of the core. Originally, its role was discounted because of its low solubility in Fe at ambient pressure. However, high-pressure experiments carried out in recent years have changed this view. Early studies to 20 GPa indicated that oxygen is soluble in Fe at high P-T conditions (Ohtani et al., 1984; Kato and Ringwood, 1989; Ringwood and Hibberson, 1990). Early evidence indicated that oxygen does not significantly lower the melting point of Fe at core pressures (FeO melts at higher pressures) (see Poirier, 1994). Recent *in situ* x-ray diffraction studies of the Fe-FeO system to 100 GPa show that the eutectic shifts with pressure (Somayazulu et al., 2000). Recent analyses of thermochemical data indicate surprisingly low partial molar volumes of oxygen at high pressures that would cause a significant increase the solubility of FeO in Fe at higher pressure (Walker, 2000). Confirmation of this effect, and determination of its structural and electronic origin, requires direct *in situ* measurements (e.g., on pure oxygen at high P-T conditions).

High P-T studies of solid oxide phases provide information on the chemical affinity of Fe and O at deep-Earth conditions and a basis for evaluating candidate crystalline phases for the inner core. The polymorphism of FeO at upper- and lower-mantle conditions has been the subject of continuing study (Mao et al., 1996). Of particular interest here is the behavior at higher pressures and temperatures. Jeanloz and Ahrens (1980) found a transition near 60 GPa and high temperature; this new phase was later found to be conducting (Knittle and Jeanloz, 1986). Subsequent x-ray measurements showed that it has the hexagonal NiAs (B8) structure (Fei and Mao, 1994). Additional analysis of the diffraction data indicated the formation of a polytype or superlattice between B8 and anti-B8 (with Fe in the As-site and O in the Ni-site) (Mazin et al., 1998). On a qualitative level, the hexagonal structure suggests the possibility of significant solid-solution behavior between FeO and ε-Fe, but this remains to be observed experimentally in Fe-FeO mixtures at high P-T conditions.

The conjecture for enrichment of oxygen in the core is also supported by the evidence for high P-T reactions between molten iron and oxides (Knittle and Jeanloz, 1991; Goarant et al., 1992), i.e.,

$$(Mg_xFe_{1-x})SiO_3 + 3(1-x)Fe \leftrightarrow x\,MgSiO_3 + (1-x)FeSi + 3(1-x)FeO. \qquad (11)$$

Hillgren and Boehler (1998, 2000) report no evidence for this reaction in experiments performed under nominally similar conditions; however, differences in moisture (H_2O as catalyst or reactant) could play a role and need to be examined in detail. Nevertheless, the above reaction has figured into two recent models for core-mantle interactions. Walker (2000) proposed that the siderophile-element signals result from injection from the core to the mantle. This was based on measured Pt-Re-Os fractionation patterns (at 10 GPa), a proposal that also is able to reproduce the Os isotopic anomalies observed in plume material that were attributed to fractionation in the core. This suggests that oxygen in the core is saturated, causing the above reaction to run backwards, which could come about by core cooling (Stevenson, 1988) as well as by increasing the concentration of FeO in the outer core by crystallization of the light-element depleted inner core (Ito et al., 1995). This would cause oxide buildup in *D*" and incorporation of siderophiles into the mantle. In a related model, Buffett et al. (2000) have proposed that silicate sediments accumulate at the top of the core. These sediments would form as light elements of the core react with the mantle, entraining residual core material in the mantle, a proposal that in turn can explain the reported isotopic signatures in plumes, as mentioned above.

Recent studies have uncovered new high P-T phenomena associated with iron and sulfur, originally proposed as a light element in the core by Mason (1966) and Rama Murthy and Hall (1972). The high P-T polymorphism of FeS has been the subject of numerous recent studies (Fei et al., 1995, 1997; Kusaba et al., 1997; Nelmes et al., 1998; Kavner and Duffy, 1999), and information on both the crystal structures and phase boundaries has been obtained to above 50 GPa and 2200 K. The highest known P-T phase [FeS(V)] has the hexagonal NiAs structure, suggesting (like FeO) its possible participation in a mechanism for solid-solution behavior in the Fe-FeS system at high pressures and temperatures. The compound Fe_3S_2 has been synthesized above 14 GPa; the presence of this phase lowers the eutectic (e.g., by 500 K at 18 GPa) (Fei et al., 1997). Stixrude et al. (1997) considered the Fe-FeS (and Fe-FeO) binaries, basing their analysis on pyrite and $Fe_{0.9}S$ alloys, as well as earlier data on FeO. New results for the Fe-S system, including the implications based on the behavior of these new compounds and conclusions drawn from solubility data, have not yet been explored. The existence of other compounds, which would invalidate extrapolations of current lower pressure data, cannot be ruled out (e.g., Sherman, 1995). The recent results point to the importance of examining sulfur in combination with other elements such as oxygen (e.g., Hillgren and Boehler, 1998; McDonough, 2001), but few *in situ* high P-T studies of ternary or more complex systems have been performed.

Recent work has provided new information on the high-pressure chemistry in the Fe-H system and the role of hydrogen in the core (see Abe et al., 2000; Williams and Hemley, 2001), a proposal first examined in detail by Stevenson (1977, 1981). Iron and water react over a wide range of pressures and temperatures following

$$3Fe + H_2O \leftrightarrow 2FeH + FeO. \tag{12}$$

In situ x-ray diffraction studies showed that FeH is stable to at least 60 GPa at room temperature; that is, it does not break down but in fact becomes more stable relative to Fe and H_2O (Badding et al., 1991, 1992). Subsequent work has shown that the uptake of H is higher in liquid Fe under pressure relative to solid FeH, that this increases with temperature, and that the melting temperature of FeH_x has a minimum at 3–4 GPa, some 600 K below that of pure iron (Okuchi, 1997, 1998). Thus, small amounts of hydrogen, e.g., 3.3 mol % indicated by elemental volatility constraints (McDonough, 2001), will have an appreciable effect on Fe melting and hence the thermal regime of the core. If the core is saturated with respect to hydrogen (see Sugimoto and Fukai, 1992), hydrous phases collecting at the base of the mantle could also be plume sources (Schilling et al., 1983). There are also important implications for accretion and Earth evolution. According to the iron-water reaction, water-rich material will tend to oxidize metallic iron-bearing materials during accretion, producing FeO as well as FeH (Suzuki et al., 1984; Yagi and Hishinuma, 1995). Hydrogen strongly partitions into the metal such that ~95% of the H_2O component in the primordial Earth should have reacted with iron to form FeH. The sequestered H in this scenario requires no additional light elements for matching the density deficit for the core, including that of the inner core (Badding et al., 1991, 1992). If core formation occurs at pressures at least as high as 10 GPa (Stevenson, 1990; Karato and Rama Murthy, 1997), then FeH could be trapped and descend within the mantle. The resultant FeO formed in the iron-water reaction may in turn react with silicates, or descend with the FeH into the core (Suzuki et al., 1984). Moreover, the analysis of Righter and Drake (1999) suggests that dissolved H_2O has no appreciable effect on the siderophile-element iron-silicate partitioning. Direct measurements over a wide range of P-T-X need to be performed using new techniques to examine these proposals.

The chemical and physical properties of iron combined with other atmophiles, including both light and heavy elements, also should be mentioned. Wood (1993) found a strong enhancement of carbon solubility with pressure and further proposed that the inner core is Fe_3C. Subsequent partitioning data, however, suggest that core formation did not occur at carbon saturation (Jana and Walker, 1997). Recent high-pressure x-ray diffraction studies of the EOS of Fe_3C, however, indicate that this is not likely to be a major component of the core (Li et al., 2000; Scott et al., 2000), but its abundance as a secondary component cannot be ruled out (e.g. Fukai, 1992). Under ambient pressure, Fe and K are incompatible elements, but they are expected to form alloys (or compounds) under pressure as a result of the theoretically predicted *s-d* transition in K (Bukowinski, 1976). Although this effect has been observed in the analog system K-Ag (Parker and Badding, 1996), it has not yet been documented experimentally in K-Fe, presumably because the system has not yet been studied at sufficiently high pressures (>50 GPa estimated to be needed). Proposed sequestering of other elements such as Cs and halogens in the core (McDonough and Sun, 1995) also may be traced to pressure-induced *s-d* transitions. Another volatile is Xe, which due to its relative depletion (see Harper and Jacobsen, 1996) has been proposed to be sequestered in the core. The observation of metallic Xe near 150 GPa, first by optical studies (Goettel et al., 1989; Reichlin et al., 1989), and more recently by direct measurements of conductivity (Eremets et al., 2000), suggests the possibility of a Xe-Fe alloy at high pressure. Caldwell et al. (1997) found no evidence for compound formation in experiments to 70 GPa and in theoretical calculations for model ordered compounds, although further studies need to be performed, e.g., at higher pressure and temperatures where Xe is metallic.

Returning to the question of siderophile abundances, pressure-induced change in electronic structure (in combination with changes in melt structure) may play a role in determining the large pressure effects documented in the partitioning of Ni and Co between iron alloys and silicate melts to 80 GPa (Li and Agee, 1996; Tschanuer et al., 1999). Notably, Ni is the most abundant heavy element, with a mass fraction estimated from meteorites to be 6%. However, its equation of state is similar to that of iron (Mao et al., 1990), rendering it invisible in the core from the standpoint of density constraints. On the other hand, the elasticity (e.g., elastic anisotropy) has not been determined, nor has its effect on the solubility and compound formation with respect to light elements. Moreover, it is important to examine the possibility that the high-pressure chemistry of Fe-Ni alloys differs from that of pure Fe at core conditions (Poirier, 1994). Indeed, the role of Ni in the inner core can now be addressed with a new level of precision using the experimental techniques described above (e.g., high P-T elastic anisotropy and chemistry of Fe-Ni alloys). Once the amount of Ni in the inner core and its solid-liquid partitioning are determined from such experiments, the light-element concentration required to offset the addition of heavier Ni in the outer core can then be constrained.

Conclusions and Outlook

Recent *in situ* high-pressure measurements—from x-ray diffraction and spectroscopy, to optical and magnetic measurements, to a variety of chemical studies—have revealed a wealth of new information about the state of iron at the conditions of the deep Earth. Knowledge of the melting temperature of iron at 330 GPa is crucial for constraining the temperature distribution of the core, but despite the many debates about the melting line Fe, there appears to be an emerging degree of consensus. The new data provide an upper bound of 5800 K for the temperature at the center of the Earth; this could, however, be depressed markedly due to the presence of key light elements. As the existence of another high P-T phase (with significantly lower density) is not consistent with the most recent shock-wave data, analyses of the P-V-T EOS indicate a density deficit of 4% relative to pure iron in the inner core. Tradeoffs in the single-crystal elastic moduli appear to be such that the V_P and V_S appear to be accurate, as indicated by the excellent agreement with both ultrasonic measurements at lower pressure and the analysis of nuclear resonance inelastic scattering measurements to core pressures. The use of a thermal expansion model based on the shock-wave data gives velocities at core conditions that are somewhat higher than those determined by seismological studies of the core. The inner core is lower than the aggregate V_S of iron, suggesting the presence of low-V_S components or premelting-induced softening in the core.

Taken together with the density determinations, these results provide new constraints on the inner-core light-element abundance, and its partitioning with the outer core. Definitive determination of the anisotropy of the inner core in terms of materials properties requires an accurate treatment of the effects of texture and strength. Results to date are fully consistent with seismological observations and suggest significantly less preferred orientation than predicted on the basis of theory, a conclusion also supported by recent theoretical calculations. In addition, it is straightforward to extend the measurements to candidate core-forming alloys and simultaneous high-temperature measurements. Indeed, preliminary studies of FeO using laser heating and multi-element detectors demonstrate the feasibility of these experiments. With respect to the light-element problem and the density deficit for the core as a whole, a growing amount of data point toward the role of multiple alloying components, and the need to examine these complex alloys with the emerging array of new *in situ* high P-T techniques.

Acknowledgments

We thank the following collaborators, who contributed in various ways to the work described above: E. Alp, J. Badro. T. S. Duffy, Y. Fei, L. W. Finger, P. Gillet, A. F. Goncharov, S. Gramsch, C. C. Kao, Y. Ma, S. Merkel, C. T. Prewitt, M. L. Rivers, J. P. Rueff, G. Shen, A. K. Singh, M. Somayazulu, L. Stixrude, S. R. Shieh, V. V. Struzhkin, Y. Timofeev, L. Vocadlo, Q. Williams, H. Yang, and C. S. Zha. We thank O. L. Anderson, D. G. Isaak, and S. Gramsch for very useful reviews of the manuscript. This work was supported by the NSF, NASA, DOE, NSLS, APS, and the W. M. Keck Foundation.

REFERENCES

Abe, Y., Ohtani, E., Okuchi, T., Righter, K., and Drake, M., 2000, Water in the early Earth, *in* Canup, R. M., and Righter, K., eds., Origin of the Earth and Moon: Tucson, Univ. of Arizona Press, p. 413–433.

Alfè, D., and Gillan, M. J., 1998, First-principles simulations of liquid Fe-S under Earth's core conditions: Physical Review B, v. 58, p. 8248–8256.

Alfè, D., Gillan, M. J., and Price, G. D., 1999, The melting curve of iron at the pressures of the Earth's core from ab initio calculations: Nature, v. 401, p. 462–464.

Alfè, D., Kresse, G., and Gillan, M. J., 2000, Structure and dynamics of liquid iron under Earth's core conditions: Physical Review B, v. 61, p. 132–142.

Anderson, O. L., 1995, Mineral physics of iron and of the core: Reviews of Geophysics, v. 33, p. 429–441.

______, 1998, The Grüneisen parameter for iron at outer core conditions and the resulting conductive heat and power in the core: Physics of the Earth and Planetary Interiors, v. 109, p. 179–197.

Anderson, O. L., Dubrovinsky, L., Saxena, S. K., and Le Bihan, T., 2001, Experimental vibrational Grüneisen ratio values for e-iron up to 330 GPa at 300 K: Geophysical Research Letters, v. 28, p. 399–403.

Anderson, O. L., and Isaak, D. G., 2000, Calculated melting curves for phases of iron: American Mineralogist, v. 85, p. 376–385.

Anderson, O. L., Nelson, V., and Isaak, D. G., submitted, The solidus temperature of epsilon iron found from the experimental Grüneisen parameter in the Lindemann relationship: Physics of the Earth and Planetary Interiors.

Anderson, W. W., and Ahrens, T. J., 1994, An equation of state for liquid iron and implications for the Earth's core: Journal of Geophysical Research, v. 99, p. 4273–4284.

Andrault, D., Fiquet, G., Charpin, T., and Le Bihan, T., 2000, Structure analysis and stability field of β-iron at high P and T: American Mineralogist, v. 85, p. 364–371.

Andrault, D., Fiquet, G., Kunz, M., Visocekas, F., and Häusermann, D., 1997, The orthorhombic structure of iron: An in situ study at high temperature and high pressure: Science, v. 278, p. 831–834.

______, 1998, Structure of ß-iron at high temperature and pressure: Science, v. 281, p. 11a.

Badding, J. V., Hemley, R. J., and Mao, H. K., 1991, High-pressure chemistry of hydrogen in metals; in situ study of iron hydride: Science, v. 253, p. 421–424.

Badding, J. V., Mao, H. K., and Hemley, R. J., 1992, High pressure crystal structure and equation of state of iron hydride: Implications for the Earth's core, *in* Syono, Y., and Manghnani, M. H., eds., High pressure research: Application to Earth and planetary sciences: Tokyo/Washington, DC, Terra Scientific/American Geophysical Union, p. 363–371.

Badro, J., Struzhkin, V. V., Shu, J., Hemley, R. J., Mao, H. K., Kao, C. C., Rueff, J. P., Hu, J., and Shen, G., 1999, Magnetism in FeO at megabar pressures from x-ray emission spectroscopy: Physical Review Letters, v. 83, p. 4101–4104.

Bancroft, D., Peterson, E. L., and Minshall, S., 1956, Polymorphism of iron at high pressure: Journal of Applied Physics, v. 27, p. 291–298.

Birch, F., 1952, Elasticity and constitution of the Earth's interior: Journal of Geophysical Research, v. 57, p. 227–286.

______, 1963, Some geophysical applications of high-pressure research, *in* Paul, W., and Warschauer, D. M., eds., Solids under pressure: New York, McGraw-Hill, p. 137–162.

______, 1964, Density and composition of the mantle and core: Journal of Geophysical Research, v. 69, p. 4377–4388.

______, 1978, Finite strain isotherms and velocities for single-crystal and polycrystalline NaCl at high pressures and 300 K.: Journal of Geophysical Research, v. 83, p. 1257–1268.

Bittorf, C., Matthies, S., Priesmeyer, H. G., and Wagner, R., 1998, Diffractive determination of thermo-elastic single crystal constants: Intermetallics, v. 7, p. 1–8.

Boehler, R., 1986, The phase diagram of iron to 430 kbar: Geophysical Research Letters, v. 13, p. 1153–1156.

______, 1993, Temperatures in the Earth's core from melting-point measurements of iron at high static pressures: Nature, v. 363, p. 534536.

______, 2000, High-pressure experiments and the phase diagram of lower mantle and core materials: Reviews in Geophysics, v. 38, p. 221–245.

Bollenrath, F., Hauk, V., and Müller, E. H., 1967, Zur Berechnung der vielkristallinen Elastizitätskonstanten aus den Werten der Einkristalle: Zeitschrift fuer Metallkunde, v. 58, p. 76–82.

Boness, D. A., and Brown, J. M., 1990, Electronic band structure of iron, sulfur and oxygen at high pressure and the Earth's core: Journal of Geophysical Research, v. 95, p. 21,721–21,730.

Boness, D. A., Brown, J. M., and McMahan, A. K., 1986, The electronic thermodynamics of iron under Earth core conditions: Physics of the Earth and Planetary Interiors, v. 42, p. 227–240.

Brazhkin, V. V., and Lyapin, A. G., 2000, Universal viscosity growth in metallic melts at megabar pressures: the vitreous state of the Earth's inner core: Physics, Uspekhi, v. 43, p. 493–508.

Brazhkin, V. V., Popova, S. V., and Voloshin, R. N., 1997, High-pressure transformations in simple melts: High Pressure Research, v. 15, p. 267–305.

Bréger, L., Romanowicz, B., and Taklcic, H., 1999, PKP(BC-DF) travel time residuals and short scale heterogeneity in the deep earth: Geophysical Research Letters, v. 26, p. 3169–3172.

Brown, J. M., and McQueen, R. G., 1986, Phase transitions, Grüneisen parameter, and elasticity for shocked iron between 77 GPa and 400 GPa: Journal of Geophysical Research, v. 91, p. 7485–7494.

Brown, J. M., Fritz, J. N., and Hixson, R. S., 2000, Hugoniot data for iron: Journal of Applied Physics, v. 88, p. 5496–5498.

Buffett, B. A., 2000, Dynamics of the Earth's core, *in* Karato, S.-i., Forte, A. M., Liebermann, R. C., Masters, G., and Stixrude, L., eds., Earth's deep interior: Mineral physics and tomography from the atomic to the global scale: Washington, DC, American Geophysical Union, p. 37–62.

Buffett, B. A., Garnero, E. J., and Jeanloz, R., 2000, Sediments at the top of Earth's core: Science, v. 290, p. 1338–1342.

Bukowinski, M. S. T., 1976, The effect of pressure on the physics and chemistry of potassium: Geophysical Research Letters, v. 3, p. 491–494.

Caldwell, W. A., Nguyen, J. H., Pfrommer, B. G., Mauri, F., Louie, S. G., and Jeanloz, R., 1997, Structure, bonding and geochemistry of xenon at high pressures: Science, v. 277, p. 930–933.

Canup, R. M., and Righter, K., eds. 2000, Origin of the Earth and Moon: Tucson, Univ. of Arizona Press, 670 p.

Chen, G. Q., and Ahrens, T. J., 1996, High-pressure melting of iron: New experiments and calculations: Philosophical Transactions of the Royal Society of London A, v. 354, p. 1251–1263.

Clendenen, R. L., and Drickamer, H. G., 1964, The effect of pressure on the volume and lattice parameters of ruthenium and iron: Journal of the Physics and Chemistry of Solids, v. 25, p. 865–868.

Cohen, R. E., Gulseren, O., and Hemley, R. J., 2000, Accuracy in equation of state formulations: American Mineralogist, v. 85, p. 338–344.

Cohen, R. E., Mazin, I. I., and Isaak, D. G., 1997, Magnetic collapse in transition metal oxides at high pressure: Implications for the Earth: Science, v. 275, p. 654–657.

Dubrovinsky, L. S., Saxena, S. K., Dubrovinskaia, N. A., Rekhi, S., and Le Bihan, T., 2000a, Grüneisen parameter of β-iron up to 200 GPa from in situ x-ray study: American Mineralogist, v. 85, p. 386–389.

Dubrovinsky, L. S., Saxena, S. K., Lazor, P., and Weber, H.-P., 1998, Structure of β-iron at high temperature and pressure: Science, v. 281, p. 11a.

Dubrovinsky, L. S., Saxena, S. K., Tutti, F., Rekki, S., and Le Bihan, T., 2000b, In situ x-ray study of thermal expansion and phase transition of iron at multimegabar pressure: Physical Review Letters, v. 84, p. 1720–1723.

Duffy, T. S., and Ahrens, T. J., 1993, Thermal expansion of mantle and core materials at very high pressures: Geophysical Research Letters, v. 20, p. 1103–1106.

Duffy, T. S., Shen, G., Heinz, D. L., Ma, Y., Hemley, R. J., Mao, H. K., and Singh, A. K., 1998, Lattice strains in gold and rhenium under non-hydrostatic pressure, *in* Wentzcovitch, R., Hemley, R. J., Nellis, W. J., and Yu, P., eds., High-pressure materials research: Warrendale, PA, Materials Research Society, Materials Research Society Proceedings, v. 499, p. 145–150.

Dziewonski, A. M., 2000, Seismic tomography and physical properties of the Earth's interior: EOS (Transactions of the American Geophysical Union), v. 81, p. F1194.

Dziewonski, A. M., and Anderson, D. L., 1981, Preliminary reference Earth model: Physics of the Earth and Planetary Interiors, v. 25, p. 297–356.

Eremets, M. I., Gregoryanz, E., Mao, H. K., Hemley, R. J., Mulders, N., and Zimmerman, N., 2000, Electrical conductivity of Xe at megabar pressures: Physical Review Letters, v. 83, p. 2797–2800.

Fei, Y., Bertka, C. M., and Finger, L. W., 1997, High-pressure iron-sulfur compound, Fe_3S_2, and melting relations in the Fe-FeS system: Science, v. 275, p. 1621–1623.

Fei, Y., and Mao, H. K., 1994, In situ determination of the NiAs phase of FeO at high pressure and high temperature: Science, v. 266, p. 1678–1680.

Fei, Y., Prewitt, C. T., Mao, H. K., and Bertka, C. M., 1995, Structure and density of FeS at high pressure and high temperature and the internal structure of Mars: Science, v. 268, p. 1892–1894.

Fiquet, G., Badro, J., Guyot, F., Requardt, H., and Krisch, M., 2001, Sound velocities in iron to 110 gigapascals: Science, v. 291, p. 468–471.

Fukai, Y., 1992, Some properties of the Fe-H system at high pressures and temperature, and their implications for the Earth's core, *in* Syono, Y., and Manghnani,

M. H., eds., High pressure research: Application to Earth and planetary sciences: Tokyo/Washington, DC, Terra Scientific/American Geophysical Union, p. 373–385.

Funamori, N., Yagi, T., and Uchida, T., 1996, High-pressure and high-temperature in situ x-ray diffraction of iron to above 30 GPa using MA8-type apparatus: Geophysical Research Letters, v. 23, p. 953–956.

Glatzmaier, G. A., and Roberts, P. H., 1996, Rotation and magnetism of Earth's inner core: Science, v. 274, p. 1887–1891.

Gnäupel-Herold, T., Brand, P. C., and Prask, H. J., 1998, The calculation of single crystal elastic constants for cubic crystal symmetry from powder diffraction data: Journal of Applied Crystallography, v. 31, p. 929–935.

Goarant, F., Guyot, F., Peyronneau, J., and Poirier, J.-P., 1992, High-pressure and high-temperature reactions between silicates and liquid iron alloys, in the diamond-anvil cell, studied by analytical electron microscopy: Journal of Geophysical Research, v. 97, p. 4477–4487.

Goettel, K. A., Eggert, J. H., Silvera, I. F., and Moss, W. C., 1989, Optical evidence for the metallization of xenon at 132(5) GPa: Physical Review Letters, v. 62, p. 665–668.

Guinan, M. W., and Beshers, D. N., 1968, Pressure derivatives of the elastic constants of α-iron to 10 kbs: Journal of the Physics and Chemistry of Solids, v. 29, p. 541–549.

Harper, C. L., and Jacobsen, S. B., 1996, Noble gases and Earth's accretion: Science, v. 273, p. 1814–1818.

Hausleitner, C., and Hafner, J., 1989, A theoretical study of the melting curve of iron to very high pressure: Journal of Physics: Condensed Matter, v. 1, p. 5243–5352.

Hemley, R. J., Mao, H. K., and Cohen, R. E., 1998, High-pressure electronic and magnetic properties of minerals, *in* Hemley, R. J., ed., Ultrahigh-pressure mineralogy: Washington, DC, Mineralogical Society of America, Reviews in Mineralogy, v. 37, p. 591–638.

Hemley, R. J., Mao, H. K., Shen, G., Badro, J., Gillet, P., Hanfland, M., and Häusermann, D., 1997, X-ray imaging of stress and strain of diamond, iron, and tungsten at megabar pressures: Science, v. 276, p. 1242–1245.

Hillgren, V., and Boehler, R., 1998, High pressure reactions between metals and silicates: Implications for the light element in the core and core-mantle interactions: Mineralogical Magazine, v. 62A, p. 624–625.

Hillgren, V. J., and Boehler, R., 2000, High pressure geochemistry in the diamond cell to 100 GPa and 3300 K, *in* Manghnani, M. H., Nellis, W. J., and Nicol, M., eds., Science and technology of high pressure: Hyderabad, India, Universities Press, p. 609–611.

Howard, C. J., and Kisi, E. H., 1999, Measurement of single-crystal elastic constants by neutron diffraction from polycrystals: Journal of Applied Crystallography, v. 32, p. 624–633.

Hu, M. Y., Sturhahn, W., Toellner, T. S., Hession, P. M., Sutter, J. P., and Alp, E. E., 1999, Data analysis for inelastic nuclear resonant absorption experiments: Nuclear Instruments and Methods in Physics Research A, v. 428, p. 551–555.

Huang, E., Bassett, W. A., and Tao, P., 1987, Pressure-temperature-volume relationship for hexagonal close packed iron determined by synchrotron radiation: Journal of Geophysical Research, v. 92, p. 8129–8135.

Ito, E., Morooka, K., Ujike, O., and Katsura, T., 1995, Reactions between molten iron and silicate at high pressure: Implications for the chemical evolution of the Earth's core: Journal of Geophysical Research, v. 100, p. 5901–5910.

Jana, D., and Walker, D., 1997, The impact of carbon on element distributions during core formation: Geochimica et Cosmochimica Acta, v. 61, p. 2759–2763.

Jeanloz, R., 1979, Properties of iron at high pressures and the state of the core: Journal of Geophysical Research, v. 84, p. 6059–6069.

Jeanloz, R., and Ahrens, T. J., 1980, Equations of state of FeO and CaO: Geophysical Journal of the Royal Astronomical Society, v. 62, p. 505–528.

Jeanloz, R., and Wenk, H. R., 1988, Convection and anisotropy of the inner core: Geophysical Research Letters, v. 15, p. 72–75.

Jephcoat, A. P., Mao, H. K., and Bell, P. M., 1986, The static compression of iron to 78 GPa with rare gas solids as pressure-transmitting media: Journal of Geophysical Research, v. 91, B5, p. 4677–4684.

Jephcoat, A. P., Mao, H. K., Finger, L. W., Cox, D. E., Hemley, R. J., and Zha, C. S., 1987, Pressure-induced structural phase transitions in solid xenon: Physical Review Letters, v. 59, p. 2670–2673.

Jephcoat, A. P., and Olson, P., 1987, Is the inner core pure iron?: Nature, v. 325, p. 332–335.

Karato, S., 1999, Seismic anisotropy of the Earth's inner core caused by the Maxwell stress-induced flow: Nature, v. 402, p. 871–873.

Karato, S., and Rama Murthy, V., 1997, Core formation and chemical equilibrium in the Earth. I. Physical considerations: Physics of the Earth and Planetary Interiors, v. 100, p. 61–79.

Kato, T., and Ringwood, A. E., 1989, Melting relationships in the system Fe-FeO at high pressures: Implications for the composition and formation of the Earth's core: Physics and Chemistry of Minerals, v. 16, p. 524–538.

Kavner, A., and Duffy, T. S., 1999, Density and crystal structure of FeS to 35 GPa and 2200 K: Implications for the interior structure of Mars: EOS (Transactions of the American Geophysical Union), v. 80, p. F929.

Klotz, S., and Braden, M., 2000, Phonon dispersion of bcc iron to 10 GPa: Physical Review Letters, v. 85, p. 3209–3212.

Knittle, E., and Jeanloz, R., 1986, High-pressure metallization of FeO and implications for the Earth's core: Geophysical Research Letters, v. 13, p. 1541–1544.

______, 1991, Earth's core-mantle boundary: Results of experiments at high pressures and temperatures: Science, v. 251, p. 1438–1443.

Kocks, U. P., Tome, C. N., and Wenk, H.-R., 1998, Preferred orientation in polycrystals and their effect on materials properties: Cambridge, UK, Cambridge Univ. Press.

Krauskopf, K. B., and Bird, D. K., 1994, Introduction to geochemistry, third edition: New York, McGraw Hill, Inc., 647 p.

Kusaba, K., Syono, Y., Kikegawa, T., and Shimomura, O., 1997, Structure of FeS under high pressure: Journal of the Physics and Chemistry of Solids, v. 58, p. 241–246.

Laio, A., Bernard, B., Chiarotti, G. L., Scandolo, S., and Tosatti, E., 2000, Physics of iron at Earth's core conditions: Science, v. 287, p. 1027–1030.

Li, J., and Agee, C. B., 1996, Geochemistry of mantle-core differentiation at high pressure: Nature, v. 381, p. 686–689.

Li, J., Eremets, M. I., Gregoryanz, E., Zha, C. S., Fei, Y., and Mao, H. K., 2000, Pressure-volume relation of Fe_3C up to 30 GPa: EOS (Transactions of the American Geophysical Union), v. 81, p. S41.

Lübbers, R., Grünsteudel, H. F., Chumakov, A. I., and Wortmann, G., 2000, Density of phonon states in iron at high pressure: Science, v. 287, p. 1250–1253.

Ma, Y., Mao, H. K., Hemley, R. J., Gramsch, S. A., Shen, G., and Somayazulu, M., 2001, Two-dimensional energy dispersive x-ray diffraction at high pressures and temperatures: Review of Scientific Instruments, v. 72, p. 1302–1305.

Ma, Y., Somayazulu, M., Mao, H. K., Shu, J., Hemley, R. J., and Shen, G., submitted, High P-T x-ray diffraction of iron to conditions of the Earth's core: Journal of Geophysical Research.

Mao, H. K., Bassett, W. A., and Takahashi, T., 1967, Effect of pressure on crystal structure and lattice parameters of iron up to 300 kbar: Journal of Applied Physics, v. 38, p. 272–276.

Mao, H. K., Bell, P. M., and Hadidiacos, C., 1987, Experimental phase relations of iron to 360 kbar, 1400°C determined in an internally heated diamond-anvil apparatus, *in* Manghnani, M. H., and Syono, Y., eds., High-pressure research in mineral physics: Tokyo/Washington, DC, Terra Scientific/American Geophysical Union, p. 135–138.

Mao, H. K., and Hemley, R. J., 1998, New windows on the Earth's deep interior, *in* Hemley, R. J., ed., Ultrahigh-pressure mineralogy: Washington, DC, Mineralogical Society of America, Reviews in Mineralogy, v. 37: p. 1–32.

Mao, H. K., Shen, G., Hemley, R. J., and Duffy, T. S., 1998a, X-ray diffraction with a double hot plate laser-heated diamond cell, *in* Manghnani, M. H., and Yagi, T., eds., Properties of earth and planetary materials at high pressure and temperature: Washington, DC, American Geophysical Union, p. 27–34.

Mao, H. K., Shu, J., Fei, Y., Hu, H., and Hemley, R. J., 1996, The wüstite enigma: Physics of the Earth and Planetary Interiors, v. 96, no. 2-3, p. 135–145.

Mao, H. K., Shu, J., Shen, G., Hemley, R. J., Li, B., and Singh, A. K., 1998b, Elasticity and rheology of iron above 220 GPa and the nature of the Earth's inner core: Nature, v. 396, p. 741–743.

Mao, H. K., Wu, Y., Chen, L. C., Shu, J. F., and Jephcoat, A. P., 1990, Static compression of iron to 300 GPa and $Fe_{0.8}Ni_{0.2}$ alloy to 260 GPa: Implications for composition of the core: Journal of Geophysical Research, v. 95, p. 21,737–21,742.

Mao, H. K., Xu, J., Struzhkin, V. V., Shu, J., Hemley, R. J., Sturhahn, W., Hu, M. Y., Alp, E. E., Vocadlo, L., Alfè, D., Price, G. D., Gillan, M. J., Schwoerer-Böhning, M., Häusermann, D., Eng, P., Shen, G., Gieffer, H., Lübbers, R., and Wortmann, G., in press, Phonon density of states of iron up to 153 GPa: Science.

Mason, B., 1966, Composition of the Earth: Nature, v. 211, p. 616–618.

Mazin, I. I., Fei, Y., Downs, R., and Cohen, R., 1998, Possible polytypism in FeO at high pressures: American Mineralogist, v. 83, p. 451–457.

McDonough, W. F., 2001, Composition of the Earth, *in* Teisseyre, R., and Majewski, E., eds., Earthquake thermodynamics and phase transformations in the Earth's interior: San Diego, Academic Press, p. 3–23.

McDonough, W. F., and Sun, S.-s., 1995, The composition of the Earth: Chemical Geology, v. 120, p. 223–253.

McQueen, R. G., Marsh, S. P., Taylor, J. W., Fritz, J. N., and Carter, W. J., 1970, The equation of state of solids from shock wave studies, *in* Kinslow, R., ed., High velocity impact phenomena: San Diego, Academic Press, p. 296–417.

Merkel, S., Goncharov, A. F., Mao, H. K., Gillet, P., and Hemley, R. J., 2000, Raman spectroscopy of iron to 152 gigapascals: Implications for Earth's inner core: Science, v. 288, p. 1626–1629.

Merkel, S., Jephcoat, A. P., Shu, J., Mao, H. K., Gillet, P., and Hemley, R. J., submitted, Equation of state, elasticity, and shear strength of pyrite under high pressure: Physics and Chemistry of Minerals.

Minkiewicz, V. J., Shirane, G., and Nathans, R., 1967, Phonon dispersion relation for iron: Physical Review, v. 162, p. 528–531.

Morelli, A., Dziewonski, A. M., and Woodhouse, J. H., 1986, Anisotropy of the inner core inferred from PKIPK travel times: Geophysical Research Letters, v. 13, p. 1545–1548.

Nelmes, R. J., McMahon, M. I., Belmonte, S. A., Allan, D. R., Gibbs, M. R., and Parise, J. B., 1998, Structures of FeS polymorphs at high pressures and temperatures, *in* Nakahara, M., ed., Review of high-pressure science and technology, v. 7: Kyoto, Japan Society of High-Pressure Science and Technol, p. 202–204.

Nguyen, J., and Holmes, N. C., 1998, Iron sound velocities in shock wave experiments up to 400 GPa: EOS (Transactions of the American Geophysical Union), v. 79, p. F846.

______, 2000, Iron sound velocities in shock wave experiments, *in* Furnish, M. D., Chhabildas, L. C., and Hixson, R. S., eds., Shock compression of condensed matter—1999: Melville, NY, American Institute of Physics, p. 81–84.

Ohtani, E., Ringwood, A. E., and Hibberson, W., 1984, Composition of the core, II. Effect of high pressures on the solubility of FeO in molten iron: Earth and Planetary Sciences Letters, v. 71, p. 94–103.

Okuchi, T., 1997, Hydrogen partitioning into molten iron at high pressure: Implications for Earth's core: Science, v. 278, p. 1781–1784.

______, 1998, The melting temperature of iron hydride at high pressures and its implications for the temperature of the Earth's core: Journal of Physics: Condensed Matter, v. 10, no. 49, p. 11,595–11,598.

Olijnyk, H., and Jephcoat, A. P., 2000, Optical zone-centre phonon modes and macroscopic elasticity in hcp metals: Solid State Communications, v. 115, p. 335–339.

Parker, L., and Badding, J. V., 1996, Transition element-like chemistry for potassium under pressure: Science, v. 273, p. 95–97.

Poirier, J. P., 1994, Light elements in the Earth's outer core: A critical review: Physics of the Earth and Planetary Interiors, v. 85, p. 319–337.

Poirier, J.-P., and Price, G. D., 1999, Primary slip system of ε-iron and anisotropy of the Earth's inner core: Physics of the Earth and Planetary Interiors, v. 110, p. 147–156.

Price, G. D., Vocaldo, L., Alfe, D., Brodholt, J., and Gillan, M. J., 1999, The ab initio calculation of the structure and stability of iron [abs.], *in* International Union of Crystallography XVIII Congress and General Assembly. Collected Abstracts, Glasgow, Scotland, p. 99.

Rama Murthy, V., and Hall, H. T., 1972, The origin and chemical composition of the Earth's core: Physics of the Earth and Planetary Interiors, v. 6, p. 123–130.

Reichlin, R., Brister, K., McMahan, A. K., Ross, M., Martin, S., Vohra, Y. K., and Ruoff, A. L., 1989, Evidence for the insulator-metal transition in xenon from optical, x-ray, and band-struture studies to 170 GPa: Physical Review Letters, v. 62, p. 669–672.

Righter, K., and Drake, M. J., 1999, Effect of water on metal-silicate partitioning of siderophile elements: A high pressure and temperature terrestrial magma ocean and core formation: Earth and Planetary Sciences Letters, v. 171, p. 383–399.

Ringwood, A. E., and Hibberson, W., 1990, The system Fe-FeO revisited: Physics and Chemistry of Minerals, v. 17, p. 313–319.

Ross, M., Young, D., and Grover, R., 1990, Theory of the iron phase diagram at Earth core conditions: Journal of Geophysical Research, v. 95, p. 21,713–21,716.

Rueff, J. P., Kao, C. C., Struzhkin, V. V., J. Badro, Shu, J. F., Hemley, R. J., and Mao, H. K., 1999, Pressure induced high-spin to low-spin transition in FeS evidenced by x-ray emission spectroscopy: Physical Review Letters, v. 82, p. 3284–3287.

Sanloup, C., Guyot, F., and Gillet, P., 2001, Physical properties of liquid Fe alloys at high pressure and their bearing on the nature of metallic planetary cores: Implications for the Earth, Mars, and the galilean satellites: Journal of Geophysical Research, submitted.

Sanloup, C., Guyot, F., Gillet, P., Fiquet, G., Hemley, R. J., Mezouar, M., and Martinez, I., 2000a, Structural changes in liquid Fe at high pressures and high temperatures from synchrotron x-ray diffraction: Europhysics Letters, v. 52, p. 151–157.

Sanloup, C., Guyot, F., Gillet, P., Fiquet, G., Mezouar, M., and Martinez, I., 2000b, Density measurements of liquid Fe-S alloys at high pressure: Geophysical Research Letters, v. 27, p. 811–814.

Saxena, S. K., and Dubrovinsky, L. S., 2000, Iron phases at high pressures and temperatures: Phase transition and melting: American Mineralogist, v. 85, p. 372–375.

Saxena, S. K., Dubrovinsky, L. S., Haggkvist, P., Cerenius, Y., Shen, G., and Mao, H. K., 1995, Synchrotron x-ray study of iron at high pressure and temperature: Science, v. 269, p. 1703–1704.

Saxena, S. K., Shen, G., and Lazor, P., 1993, Experimental evidence for a new iron phase and implications for Earth's core: Science, v. 260, p. 1312–1314.

Schilling, J.-G., Zajac, M., Evans, R., Johnston, T., White, W., Devine, J. D., and Kingsley, R., 1983, Petrologic and geochemical variations along the Mid-Atlantic Ridge from 29°N to 73°N: American Journal of Science, v. 283, p. 510–586.

Scott, H. P., Williams, Q., and Knittle, E., 2000, In situ high-pressure x-ray diffraction measurements of Fe_3C: EOS (Transactions of the American Geophysical Union), v. 81, p. S41.

Shen, G., and Heinz, D. L., 1998, High-pressure melting of deep mantle and core materials, *in* Hemley, R. J., ed., Ultrahigh-pressure mineralogy: Washington, DC, Mineralogical Society of America, Reviews in Mineralogy, v. 37, p. 369–396.

Shen, G., Mao, H. K., Hemley, R. J., Duffy, T. S., and Rivers, M. L., 1998, Melting and crystal structure of iron at high pressures and temperatures: Geophysical Research Letters, v. 25, p. 373–376.

Sherman, D. M., 1995, Stability of possible Fe-FeS and Fe-FeO alloy phases at high pressure and the composition of the Earth's core: Earth and Planetary Sciences Letters, v. 132, p. 87–98.

Simmons, G., and Wang, H., 1971, Single-crystal elastic constants and calculated aggregate properties: Cambridge, MA, MIT Press.

Singh, A. K., and Balasingh, C., 1994, The lattice strains in a specimen (hexagonal system) compressed nonhydrostatically in an opposed anvil high pressure setup: Journal of Applied Physics, v. 75, p. 4956–4962.

Singh, A. K., Balasingh, A., Mao, H. K., Hemley, R. J., and Shu, J., 1998a, Analysis of lattice strains measured under non-hydrostatic pressure: Journal of Applied Physics, v. 83, p. 7567–7575.

Singh, A. K., Mao, H. K., Shu, J. F., and Hemley, R. J., 1998b, Estimation of single-crystal elastic moduli from polycrystalline x-ray diffraction at high pressure: Application to FeO and iron: Physical Review Letters, v. 80, p. 2157–2160.

Soderlind, P., Moriarty, J. A., and Wills, J. M., 1996, First-principles theory of iron up to earth-core pressures — structural, vibrational, and elastic properties: Physical Review B, v. 53, p. 14,063–14,072.

Somayazulu, M., Tschauner, O., Hemley, R. J., and Mao, H. K., 2000, Melting of Fe-FeO system at high pressures: Annual Report of the National Synchrotron Light Source, in press.

Song, X., and Helmberger, D. V., 1998, Seismic evidence for an inner core transition zone: Science, v. 282, p. 924–927.

Song, X., and Richards, P. G., 1996, Seismological evidence for differential rotation of the Earth's inner core: Nature, v. 382, p. 221–224.

Song, X. D., and Helmberger, D. V., 1993, Anisotropy of Earth's inner core: Geophysical Research Letters, v. 20, p. 2591–2594.

Steinle-Neumann, G., Stixrude, L., and Cohen, R. E., 1999, First-principles elastic constants for the hcp transition metals Fe, Co, and Re at high pressure: Physical Review B, v. 60, p. 791–799.

Stevenson, D. J., 1977, Hydrogen in the Earth's core: Nature, v. 268, p. 130–131.

______, 1981, Models of the Earth's core: Science, v. 214, p. 611–619.

______, 1988, Infiltration, dissolution, and underplating: Rules for mixing core-mantle cocktails: EOS (Transactions of the American Geophysical Union), v. 69, p. 1404.

______, 1990, Fluid dynamics of core formation, *in* Newson, H. E., and Jones, J. H., eds., Origin of the Earth: New York, Oxford University Press, p. 231–249.

Stixrude, L., and Brown, J. M., 1998, The Earth's core, *in* Hemley, R. J., ed., Ultrahigh-pressure mineralogy: Washington, DC, Mineralogical Society of America, Reviews in Mineralogy, v. 37, p. 261–282.

Stixrude, L., and Cohen, R. E., 1995, High pressure elasticity of iron and anisotropy of Earth's inner core: Science, v. 267, p. 1972–1975.

Stixrude, L., Steinle-Neumann, G., Cohen, R. E., and Gülseren, O., 2000, Elasticity of iron at high pressures and temperature—implications for the Earth's inner core: EOS (Transactions of the American Geophysical Union), v. 81, p. F1194–F1195.

Stixrude, L., Wasserman, E., and Cohen, R. E., 1997, Composition and temperature of Earth's inner core: Journal of Geophysical Research, v. 102, p. 24,729–24,739.

Sturhahn, W., Toellner, T. S., Alp, E. E., Zhang, X. W., Ando, M., Yoda, Y., Kikuta, S., Seto, M., Kimball, C. W., and Dabrowski, B., 1995, Phonon density of states measured by inelastic nuclear resonant scattering: Physical Review Letters, v. 74, p. 3832–3835.

Su, W., Dziewonski, A. M., and Jeanloz, R., 1996, Planet within a planet: Rotation of the inner core of the Earth: Science, v. 274, p. 1883–1887.

Sugimoto, H., and Fukai, Y., 1992, Solubility of hydrogen in metals under high hydrogen pressures: thermodynamical calculations: Acta Metallurgica et Materialia, v. 40, p. 2327–2336.

Suzuki, T., Akimoto, S. I., and Fukai, Y., 1984, The system iron-enstatite-water at high pressures and temperatures—formation of iron hydride and some geophysical implications: Physics of the Earth and Planetary Interiors, v. 36, p. 135–144.

Takahashi, T., and Bassett, W. A., 1964, High-pressure polymorph of iron: Science, v. 145, p. 483–486.

Timofeev, Y. A., H. K. Mao, Struzhkin, V. V., and Hemley, R. J., 1999, Inductive method for investigation of ferromagnetic properties of materials under pressure: Review of Scientific Instruments, v. 70, p. 4059–4061.

Tschanuer, O., Zerr, A., Sprecht, S., Rocholl, A., Boehler, R., and Palme, H., 1999, Partitioning of nickel and cobalt between silicate perovskite at pressures up to 80 GPa: Nature, v. 398, p. 604–607.

Uchida, T., Funamori, N., and Yagi, T., 1996, Lattice strains in crystals under uniaxial stress field: Journal of Applied Physics, v. 80, no. 2, p. 739–746.

Upadhyaya, J. C., Sharma, D. K., Prakash, D., and Upadhyaya, S. C., 1994, Three-particle forces in the lattice dynamics of some hexagonal close packed metals: Canadian Journal of Physics, v. 72, p. 61–72.

Vinet, P., Ferrante, J., Rose, J. H., and Smith, J. R., 1987, Compressibility of solids: Journal of Geophysical Research, v. 92, p. 9319–9325.

Vinnik, L., Romanowicz, B., and Breger, L., 1994, Anisotropy in the center of the inner core: Geophysical Research Letters, v. 21, p. 1671–1674.

Vocadlo, L., Brodholt, J., Alfè, D., Gillan, M. J., and Price, G. D., 2000, Ab initio free energy calculations on the polymorphs of iron at core conditions: Physics of the Earth and Planetary Interiors, v. 117, p. 123–137.

Vocaldo, L., Brodholt, J., Alfè, D., Price, G. D., and Gillan, M. J., 1999, The structure of iron under the conditions of the Earth's inner core: Geophysical Research Letters, v. 26, p. 1231–1234.

Walker, D., 2000, Core participation in mantle geochemistry: Geochemical Society Ingersoll Lecture, GSA Denver, October 1999: Geochimica et Cosmochimica Acta, v. 64, p. 2897–2911.

Waseda, Y., and Ohtani, M., 1974, Static structure of liquid noble and transition metals by x-ray diffraction: Physica Status Solidi (b), v. 62, p. 535–546.

Waseda, Y., and Suzuki, K., 1970, Atomic distribution and magnetic moment in liquid iron by neutron diffraction: Physica Status Solidi, v. 39, p. 669–678.

Wasserman, E., Stixrude, L., and Cohen, R. E., 1996, Thermal properties of iron at high pressures and temperatures: Physical Review B, v. 53, no. 13, p. 8296–8309.

Wenk, H.-R., Baumgardner, J. R., Lebensohn, R. A., and Tomé, C. N., 2000a, A convection model to explain anisotropy of the inner core: Journal of Geophysical Research, v. 105, p. 5663–5677.

Wenk, H.-R., Matthies, S., Donovan, J., and Chateigner, D., 1998, BEARTEX, a Windows-based program system for quantitative texture analysis: Journal of Applied Crystallography, v. 31, p. 262–269.

Wenk, H.-R., Matthies, S., Hemley, R. J., Mao, H. K., and Shu, J., 2000b, The plastic deformation of iron at pressures of the Earth's inner core: Nature, v. 405, p. 1044–1047.

Wenk, R., and Buffett, B., 2000, Texturing of the inner core by Maxwell stresses: EOS (Transactions of the American Geophysical Union), v. 81, p. F1195.

Williams, Q., and Hemley, R. J., 2001, Hydrogen in the deep Earth: Annual Review of Earth and Planetary Sciences, in press.

Williams, Q., Jeanloz, R., Bass, J., Svendsen, B., and Ahrens, T. J., 1987, The melting curve of iron to 250 gigapascals: A constraint on the temperature at Earth's center: Science, v. 236, p. 181–182.

Williamson, D. L., Bukshpan, S., and Ingalls, R., 1972, Search for magnetic ordering in hcp iron: Physical Review B, v. 6, p. 4194–4206.

Wilson, A., 1946, X-ray diffraction by random layers: Ideal line profiles and determination of structure amplitudes from observed line profiles: Acta Crystallographica, v. 2, p. 245–251.

Wood, B. J., 1993, Carbon in the core: Earth and Planetary Sciences Letters, v. 117, p. 593–607.

Yagi, T., and Hishinuma, T., 1995, Iron hydride formed by the reaction of iron, silicate, and water: Implications for the light element of the Earth's core: Geophysical Research Letters, v. 22, p. 1933–1936.

Yoo, C. S., Akella, J., Campbell, A. J., Mao, H. K., and Hemley, R. J., 1995, Phase diagram of iron by in situ x-ray diffracton: Implications for the Earth's core: Science, v. 270, p. 1473–1475.

______, 1997, Detecting phases of iron. Response to comment by S. K. Saxena and L. S. Dubrovinsky: Science, v. 275, p. 96.

Yoo, C. S., Holmes, N. C., and Ross, M., 1993, Shock temperatures, melting and phase diagram of iron at earth core conditions: Physical Review Letters, v. 70, p. 3931–3934.

Zhang, Y.-G., and Guo, G.-J., 2000, Molecular dynamics calculation of the bulk viscosity of liquid iron-nickel alloy and the mechanisms for the bulk attenuation of seismic waves in the Earth's outer core: Physics of the Earth and Planetary Interiors, v. 122, p. 289–298.

Zhang, Y.-G., Guo, G.-J., and Nie, G. Z., 2000, A molecular dynamics study of the bulk and shear viscosity of liquid iron using embedded-atom potential: Physics and Chemistry of Minerals, v. 27, p. 164–169.

The Statistics and Distribution of Helium in the Mantle

Don L. Anderson

California Institute of Technology, Seismological Laboratory 252-21, Pasadena, California 91125

Abstract

An unbiased estimate of the ^{3}He/^{4}He ratio (R) along the Earth's spreading ridge system is 9.14 ± 3.59 Ra (n = 503) where Ra is the atmospheric ratio (1.38×10^{-6}). By arbitrarily excluding all values with R > 11 Ra and from ocean depths less than 2500 meters, I obtain 7.91 ± 1.50 Ra (n = 212), which is close to previous "filtered" estimates based on the hypothesis that the excluded values have been influenced by plumes. These are biased estimates.

Based on unbiased statistics, many of the so-called high-^{3}He hotspot regions have isotopic ratios well within the MORB range, and all have absolute ^{3}He concentrations much less than MORB. The high variance of some oceanic-island data compared to MORB reflects, in part, the difference between a small sample and a large sample, and magma-chamber processes.

Values of 11 to 15 Ra are commonly attributed to deep mantle plumes and "indicative of lower mantle involvement," but these values are within 2σ of the mean and are not exceptional. Much higher values combined with low absolute helium concentrations are commonly associated with the onset of rifting, or volcanism, and may reflect a shallow, or lithospheric, low–^{238}U/^{3}He (LONU) source. The temporal progression to average MORB-like values suggests that the bulk of the magmas at spreading ridges and large oceanic constructs comes from below the LONU level. The termination of spreading is associated with low ratios, 6 to 7 Ra, similar to values associated with the high–^{238}U/^{204}Pb (HIMU) mantle component and some oceanic islands.

Background

Helium-3, like most of the isotopes in Earth, was brought in during accretion, some 4.5 billion years ago or, along with other volatiles, as a late veneer, mostly prior to stabilization of the crust. Some amounts of ^{3}He are still being brought in as interplanetary dust particles (IDP), which can dominate the helium budget of deep-sea sediments. Some of the highest ^{3}He concentrations to be found on Earth are in abyssal sediments. Because of its pedigree, ^{3}He is referred to as a "primordial isotope." Since the termination of accretion, and the late-veneering process, the amount of ^{3}He in the mantle has been essentially constant, except for that fraction that is vented to the surface and lost to outer space. Most ^{3}He atoms are vented from the mantle at mid-ocean ridges and island-arc volcanoes. A small fraction is lost through continents and at the so-called "hotspot" or "mid-plate" volcanoes. A small amount is made by cosmogenic and nucleogenic processes.

The number of ^{3}He atoms brought in currently by IDP is about the same as the number expelled from the interior at "hotspots" (Anderson, 1993). However, neither the influx nor the volcanic flux can be considered constant. Anderson (1993) argued that currently degassing ^{3}He at oceanic islands probably entered the mantle at least 10^9 years ago. Conditions between then and 4 Ga were more favorable for helium retention than during the bulk of the accretion process. The issue of when and how the noble gases entered the Earth, however, is different from the issue of where and how the domains of different ^{238}U/^{3}He (ν) formed. I argue in this paper and in Anderson (1998a, 1998b, 1999) that the LONU reservoir is more likely to be low [He] and shallow, rather than high [He] and deep in the convecting mantle.

Since ^{3}He is a "primordial element" and is an easily lost gas, it is often assumed that it must come from "primordial mantle," mantle that has not been subjected to melting or degassing. The assumption here is that once ^{3}He leaves its primordial parent rock, it is rapidly transferred to the atmosphere—i.e., there is no intermantle transfer of He-bearing fluids from one rock to another. The logical mistake here is confusing a *Primordial Component* with a *Primordial Reservoir*.

Most of the ^{3}He atoms that can be accounted for are in mid-ocean ridge basalts, and most of the gaseous ^{3}He escapes at mid-ocean ridges and island arcs (Sano and Williams, 1996). Nevertheless, it is a common assumption that the bulk of ^{3}He in Earth is stored in primordial undegassed lower mantle (Porcelli and Wasserburg, 1995), although the basalts presumably from this reservoir have very low ^3He concentrations compared to MORB, which is commonly thought of as coming from a degassed reservoir.

Helium-4 is a radioactive decay product of U and Th, and therefore builds up in the crust and mantle over time. The $^3He/^4He$ ratio of the crust and atmosphere is very low because the crust has high abundances of U and Th. More than 50% of these elements are in the crust (Anderson, 1989) and degassing of crustal helium is responsible for the low $^3He/^4He$ ratios of the atmosphere and seawater (1.38×10^{-6} is the atmospheric ratio, Ra). Mantle gases, and gases in mantle-derived basalts, have higher ratios, R, generally falling in the range of 5 to 20 Ra (e.g., Farley and Neroda, 1998). This reflects the low $^{238}U + ^{235}Th$ of the mantle compared to the crust.

The $^3He/^4He$ ratio [R] of a closed system (no import or export of helium atoms) depends on the time-integrated $^{238}U/^3He$ ratio, for a given U/Th ratio. This ratio is called ν, or NU. A similar ratio, $^{238}U/^{204}Pb$, called μ or MU, is important in discussions of Pb isotopes in mantle magmas. Some oceanic basalts have high MU (HIMU) ratios, attributed to recycling of ancient oceanic crust. The corresponding parts of the lithosphere, having been depleted in **U** (= U + Th) to form the crust, are LOMU, and being depleted in other large-ion incompatible (LIL) elements, such as Rb and Sm, do not contribute much to basalt chemistry, even if it is recycled along with its overlying crust.

Basalts that have low time-integrated $\mathbf{U}/^3He$ ratios (LONU), compared to the crust, will have higher R than the crust and atmosphere, the most common values lying between 7 and 9 Ra. Various reservoirs (sources of gas and magma) in the mantle can have different R if they differ in age, NU, **U** concentrations, or ^3He concentrations. The rare basalts that have R much higher than "average" MORB are widely assumed in the geochemical literature to have "excess ^3He" and therefore to come from a primordial undegassed reservoir, isolated from the convecting upper mantle. The alternatives, low **U** or different age, are generally not discussed although "high-R" basalts are low in ^3He (low [^3He]).

Helium is transferred around in the shallow mantle as a dissolved gas in magma. At shallower depth He exsolves, along with CO_2, its main carrier phase, and part is trapped in crystals or retained as CO_2-rich fluid-filled inclusions in *restite*, the residual refractory residue of basalt extraction. Restite is less dense than the original fertile (capable of producing basalt) mantle because the basaltic component at depth (eclogite) contains garnet, the densest upper mantle mineral. Restite is olivine rich and buoyant and therefore collects in the shallow mantle, below the crust and above fertile mantle. It is geochemically barren except for the compatible elements (such as Os) and trapped fluid-filled inclusions. The restite layer, when cold and strong, is called the lithosphere, the primary structural component of the plate. Old lithosphere, particularly at fracture zones, faults, and sutures can become metasomatized, as a result of transiting fluids of various sorts. Such metasomatized regions of the lithosphere are preferentially sampled by magmas since these regions are preferred conduits. The bulk of the restite in the shallow mantle, particularly the intact parts of the lithosphere, may retain ancient isotopic signatures and retain its low-**U** and low-LIL character.

Various components have been identified in ocean-island basalts, and along some ridge segments. These have been attributed to various parts of the recycled slab—EM1 and EM2 (pelagic and continental sediments) and HIMU (oceanic crust). Occasionally a component is attributed to "continental lithosphere," but the oceanic restite layer is seldom considered in mantle geochemistry. In addition to continental and oceanic lithosphere (the latter usually being assumed to be less than 200 Ma in age), there are plateaus, swells, and shallow parts of the ocean basins that may have ancient restite roots. A global restite layer, possibly about 100 km thick, is the result of accretional differentiation and billions of years of melt extraction. It is a buoyant isolated reservoir and probably is thickest under ancient cratons and some oceanic plateaus. When it is colder than some 600°C it is strong and is the "lithosphere." This part of the outer shell, plus the overlying crust, probably constitute the "slab" that recycles into the mantle. The warmer, deeper, more buoyant parts of the restite shell may remain in the shallow mantle, overlying "fertile" mantle. This "floating reservoir" is seldom considered in geochemical box models. In fact, it may be important only for the noble elements and the more compatible elements since, by definition, it has had basalts and LIL extracted.

The spatial and temporal relationships of enriched and depleted magmas, and of helium isotopic ratios, suggest that the shallow mantle is an important reservoir. Whether the inhomogeneities are arranged in layers (crust, lithosphere, restite shell, perisphere, depleted reservoir) or plums (plum pudding, marble cake) there will be similar temporal or spatial variations, and variations depending on when and how the samples are obtained. "Mid-plate" or "hotspot" volcanism is better termed "initial" or "transient" or "small-scale" magmatism, this better reflecting the temporal and tectonic contexts. The "mantle wedge" above subduction zones and metasomatized continental lithosphere (CL) are two of the more obviously enriched shallow reservoirs.

Introduction

It is usually assumed that mid-ocean ridge basalts are derived from a homogeneous, well-mixed depleted degassed reservoir, often equated with the "convecting upper mantle." Chemical heterogeneity found along the global mid-ocean ridge system is attributed to: (1) injection of enriched or "primordial" material from lower mantle plumes; (2) incorporation of delaminated continental lithosphere that somehow arrives at the ridge; (3) the presence of dispersed heterogeneities in a continually remixed convecting mantle; (4) introduction of subducted and recycled components; (5) contamination of depleted MORB by shallow layers upon eruption, or in magma chambers.

If the MORB mantle is homogeneous at all scales, then heterogeneities must be introduced from outside, as in the plume hypothesis. If the shallow mantle is heterogeneous, then the chemistry of magmas will reflect the sampling process, and history of sampling. There are many continental fragments (e.g., Seychelles, Lord Howe Rise, Jan Mayen, Broken Ridge) trapped in the ocean basins, some of which are covered with basalt. These oceanic plateaus are overlooked potential (floating) reservoirs or sources of "contamination."

The narrow range in the $^3He/^4He$ ratio in MORB is often taken as evidence for the homogeneity of the MORB source and evidence that it is a convecting well-stirred reservoir. However, the actual statistics of the $^3He/^4He$ distribution in the upper mantle are unknown since values outside of a given range are commonly deleted as being "contaminated by plumes" or recycled material. The MORB reservoir itself need not be homogeneous since the large degrees of melting, the large sampling volumes, and magma chamber processes serve to blend various components in more effective ways than available to xenoliths, melt inclusions, seamounts, and oceanic islands. The narrow range of isotopic and trace-element characteristics of MORB, compared to other magmas, may simply reflect the sampling process rather than a uniquely well mixed shallow mantle reservoir.

Anderson (1989) presented a model in which all mantle magmas, including MORB, are blends of components. Even the most depleted MORB is a hybrid, not a pure component. It simply has less of the enriched components (EM) than some smaller-scale samples, such as seamounts and islands. Niu et al. (1999) and Niu and Batiza (1997) have recently found that near-ridge seamounts exhibit a range of compositions, from more depleted to more enriched than MORB, but the average composition is N-MORB (Normal Midocean Ridge Basalts). More of the enriched component gives E-MORB. Even axial basalts have OIB-like components. This gives strong support to the idea that the upper mantle is heterogeneous, and that magma compositions depend on how the source is sampled.

The theory of statistical sampling tells us that pooled samples from a large volume of sample space will be relatively uniform and reproducible, whereas small individual samples will tend to be more diverse. Repeated small-scale sampling gives a larger variance than repeated large-scale sampling, but the means may be similar.

Islands and seamounts sample a smaller volume of the mantle than MORB and, accordingly, they should and do display a large diversity in chemistry. The question is, can MORB and island basalts be different kinds of samples from the same heterogeneous reservoir? The fact that near-ridge seamounts provide samples that are both more enriched and more depleted than axial MORB, and that there are approximately as many low–$^3He/^4He$ oceanic islands as high–$^3He/^4He$ islands suggests that sampling statistics may be responsible for the diversity of magmas and for the statistics of helium distributions.

Statistical treatment of all samples obtained on or near ridges is the best way to determine the average composition of the shallow mantle. The statistics of ridge geochemistry may approximate a Gaussian curve, while smaller subsets (e.g., seamounts, the East Pacific Rise [EPR] from 17 to 22°S, the North Atlantic Ridge from 52 to 70°N, ocean-island basalts, etc.) will approximate a t-distribution, with a similar mean but much larger variance, and more extreme values. If there is a temporal evolution, or a stratigraphy, from high R to low R, then a given ridge segment, rift, or island may yield a one-sided distribution, reflecting the stage of the volcanism. To fully define the distribution, including the high- and low-R tails, requires sampling in all tectonic environments and at various stages of rifting.

The first order of business is to compile ridge data and then determine unbiased means and standard deviations. When this is done, the various hypotheses that have been proposed for different regions can be tested. It is important not to process the data in advance, with a particular hypothesis in mind. Currently we do not know the "expected value," or what is a "significantly" different value, when we discuss helium anomalies.

$^3He/^4He$ ratios of 11 times the atmospheric ratio (Ra) or greater are now routinely referred to as "plume" components or evidence for "lower-mantle" involvement (e.g., Poreda et al., 1993; Niedermann et al., 1997). Such values are stated to be "significantly higher than the MORB average," which implies that an unbiased statistical description of MORB samples is available. Currently available compilations, for helium, have discarded values higher than 11 Ra (or 9 Ra, or 9.5 Ra, depending on the author) or "having other evidence of N-MORB [sic.; non-MORB?]...characteristics" (Hilton et al., 1993). There is thus a large degree of circular reasoning in many recent papers where a "plume" or "lower-mantle" component is invoked entirely or mainly on the basis of a few high R (and very low [^3He]) samples. Standard methods of statistical inference and hypothesis testing can be applied to the helium database, but not to a hypothesis-dependent "filtered" database.

Filtering Data

Previous estimates of helium statistics along the oceanic ridge system are "filtered" with a particular hypothesis in mind. For example:

> Samples of $^3He/^4He$ greater than 9.5 Ra were not used as they are considered to be influenced by primitive plumes from the lower mantle. (Sarda et al., 1999)

> ...MORB sample data were filtered in an attempt to include only samples uninfluenced by plumes or hotspots....Samples from less than 2500 meters depth or having "other non-MORB...characteristics"...were removed from the dataset. (Hilton et al., 1993)

Hilton et al. (1993) also discarded values higher than 11 Ra in compiling their backarc-basin statistics, obtaining (8.18 ± 0.73), thinking that Lau Basin basalts are contaminated by the "Samoan plume" (on the other side of the Tonga-Fiji slab), although samples geographically closer to Samoa, with "normal R" are retained in the compilation. Subsequently, other backarc basins have been found to include high-R samples (MacPherson et al., 1998). Backarc-basin spreading centers are isolated from the deep mantle by the underlying slab and are ideal locations for sampling the shallow mantle.

Fisher (1986) defined 8 ± 2 Ra as the range for "normal MORB,"and this provided his criteria for filtering data. Based on published compilations prior to 1976 and unpublished data (abstracts, theses, personal communications) up to 1984, the range typical for MORB was given as 7.0 to 9.0 Ra by Poreda et al. (1993). The value 8 ± 1 Ra has generally been used since that time. Subsequently, values up to 11 Ra found along ridges have been treated as "definitive evidence" for a hotspot or plume signature. Apparently, authors of recent papers are not aware of the strong a priori filtering that has been applied to the MORB dataset, or have forgotten.

Values of 11 Ra or greater are ascribed to the involvement of deep mantle plumes, implying that there is very low or zero probability of such values being available in the MORB reservoir or the result of shallow contamination. For example:

> The definitive evidence for the presence of a hotspot signature on the East Pacific Rise was the discovery of $^3He/^4He$ ratios in basalts that were greater than those of MORB (R/Ra > 8), with values ranging up to 11 Ra. (Poreda et al., 1993)

Mid-ocean ridges induce upwellings of large volumes of the mantle and generate large degrees of melting. They also process and blend magmas obtained over a large depth interval. Most ridges are mature and no longer sample (at least, without blending) the shallow parts of the mantle they may have sampled when they were younger or their ocean narrower. The North Atlantic, the Red Sea, the East African Rift, ends of propagating ridges, microplates, and backarc basin ridges are places where new ridges may be sampling the shallow mantle in a way different from that of mature ridges.

The places to search for extremes in shallow-mantle helium ratios and concentrations are at slowly spreading ridges, regions of fracture zones, seamounts, low-degree melts, initial and final melts, and xenoliths. Dense sampling along ridges may give a more representative range of helium isotopic ratios than widely spaced samples. "Spikes" of anomalous chemistry, however, often are attributed to "plumelets" or lower-mantle components rather than to intrinsic variability. The South Atlantic near the Bouvet triple junction is a region of slow spreading, numerous fracture zones, and unstable plate boundaries. One expects a wide range of $^3He/^4He$ ratios in such regions. Some of the lowest ratios in the South Atlantic are from very shallow samples, and some of the highest ratios are among the deepest samples (Kurz et al., 1998). Therefore, the criteria used by previous authors to "filter" samples from the North Atlantic does not apply. Since high $^3He/^4He$ ratios occur at various elevations and depths, since they are commonly interspersed with "normal" values on oceanic islands, and since the samples are often identical to normal MORB in other respects, there is no unbiased procedure for eliminating them from the dataset. One cannot, and should not, assume that extreme values in complex environments must be imported from the deep mantle. One also should not compare extreme values from one tectonic environment to average values in another.

Preliminaries

The $^3He/^4He$ ratio of a mantle source region depends on many factors, most of which are expected to display a broad, perhaps Gaussian, distribution: (1) U/He ratio; (2) U/Th ratio; (3) age; (4) introduced components; and (5) prior history. One expects, therefore, that basalts derived directly from any part of the mantle will have a broad distribution of $^3He/^4He$ ratios since it is unlikely that any of the above parameters, much less their integrated effects, will be single valued.

The high $^3He/^4He$ tail of the distribution is likely to be dominated by low $^{238}U/^3He$ samples and therefore will change little with time. The low $^3He/^4He$ part of the distribution may reflect a higher $^{238}U/^3He$ environment and will grow as radiogenic ^4He is added to the system. Therefore, the mean will shift to lower $^3He/^4He$ with time, and the distribution will become skewed, with a long high R tail. In fact, the R distribution of oceanic samples, both ridge and island, has the general appearance of a log-normal distribution and has more high-R members than a normal distribution (Farley and Neroda, 1998). The high-R members of the global $^3He/^4He$ ratio histogram are commonly attributed to a deep mantle, high-^3He plume. However, the evidence is that high-R samples are low in ^3He and very low in ^4He, implying a low $^{238}U/^3He$ LONU source (Anderson, 1998a, 1998b). High $^3He/^4He$, low [He] probably has more temporal, or survival, implications than implications about great depth or "primordial" mantle.

It should be noted that the refractory residual left after removal of basalt (restite) is buoyant with respect to fertile mantle. It will also be strong and of high viscosity because of the removal of water (Hirth and Kohlstedt, 1996). It serves as a carapace over fertile, more volatile-rich mantle and may collect the gases expelled by rising magmas as they pass through the saturation level of CO_2. Helium also can be partitioned into restite and olivine-rich cumulates even if the crystal-melt partition coefficient is low. The important issue regarding the $^{238}U/^3He$ ratio (NU) of the lithosphere is the relative partitioning of U, Th, CO_2, and He between crystal, melt, and gas, and the nature of the metasomatic fluids (silicate melts, carbonatites, or vapor). Buoyant restite is an isolated reservoir that can locally preserve high $^3He/^4He$ ratios if it is LONU. Subsequent magmas can be influenced by passing through the lithosphere, particularly for helium and osmium isotopes, and perhaps lead and oxygen isotopes. On the other hand, LIL contamination is more likely to occur when the magma interacts with crust.

Countering the trend toward heterogeneity in shallow reservoirs is magma mixing, magma-chamber processing, and prior removal of inhomogeneities. The sampling process, large versus small samples, can yield products (samples or subpopulations) from the same reservoir (population), with different apparent variability. Not to be overlooked is the a priori processing of the data such as discarding "anomalous" values, resulting in an artificially narrow distribution that can be "puzzling" or, statistically, "too good to be true."

If magmas are blends of various components, the $^3He/^4He$ ratio of the blend will be dominated by the high ^4He components, such as relatively undegassed MORB. If there are solubility arguments for a peak in absolute He concentrations, [He], at some particular value, then this component will dominate the ratio of any blend. If the processes of melt extraction involve degassing at relatively shallow depth, then the degassed magmas and the ultimate repository of the gases (fluid-filled inclusions, cumulates, lithosphere) will have different $^{238}U/^{3}He$, He/Ar, and He/Ne ratios and, in time, will have quite different $^{3}He/^{4}He$ ratios. Recycled crust will have different ratios than recycled depleted lithosphere. HIMU may be complementary to LONU.

It is now clear that the MORB reservoir is inhomogeneous and contains both very enriched and very depleted components, which are blended in the processes of ridge sampling and magma-chamber evolution (Anderson, 1989; Niu et al., 1999). The nature of the heterogeneity only becomes clear with small-scale sampling along the ridge and by studying xenoliths, melt inclusions, seamounts, and fracture zones. It is still commonly assumed, however, that the MORB source is intrinsically homogeneous (well-stirred "convecting" mantle) and that heterogeneities are all imported from nearby (or quite distant) hotspots, or deep mantle plumes, or delaminated continental lithosphere. This perception of "remarkable" uniformity of the upper mantle (Niedermann et al., 1997) is mainly based on the narrow distribution of helium isotopes (± 10% spread). Some incompatible elements show much larger variations—i.e., 50 to 100% variation (Ba, Rb, Cs, Th, Nb, U, Ta), and ~20% variation (K, La, Ce, Pb, Sr, Nd) (Niu and Batiza, 1997). Compatible elements (HREE, Li, Sc) tend to show variations on the order of 10%. Even major-element ratios (e.g. CaO/Al_2O_3) show scatter of 50% in ridge-derived basalts. If the $^{3}He/^{4}He$ ratio of spreading-ridge samples is as narrow as perceived in the current literature, then this would indeed be an anomalous, even remarkable, situation. I show that an unbiased estimate of the near-ridge mantle has a standard deviation of about three times the commonly assumed value, making the $^{3}He/^{4}He$ variation consistent with other measures of mantle heterogeneity.

Statistics

Unbiased statistics for the distribution of $^{3}He/^{4}He$ ratios along the global spreading ridge system are not available. The mean and standard deviation are usually quoted as 8 ± 1 times the atmospheric ratio, Ra. However, this estimate is based on a much smaller dataset than is now available and also is hypothesis dependent, the data having been "corrected" for "plume contamination." The techniques used for this "correction" are arbitrary and involve discarding all values greater than some upper cutoff, ranging from 9.5 to 11 Ra, or discarding all values from ocean depths less than some threshold, such as 2500 meters. These criteria were based on a small sample from the North Atlantic and were for the purpose of eliminating samples "too close" to Iceland. Some of the discarded values are MORB-like in all other trace-element and isotopic respects.

Subsequently, samples have been found along perfectly normal stretches of the Southern Hemisphere ridges that have values of 11 Ra or more. In the South Atlantic and near Loihi, some of the deepest samples have the highest R and, in many places, very shallow, and even emerged, samples have MORB-like values.

I have assembled a large fraction of the available data along the world's spreading systems including new ridges, near-ridge seamounts, the North and South Atlantic ridges, East Pacific Rise (EPR), Chile Rises, Indian Ocean ridges, the Azores, Galápagos, Easter, Shona, and Bouvet segments, microplates, and backarc basins. This was all analyzed together, to yield the Grand Average, and then in subsets. In contrast to previous studies, no datum was discarded because it was outside "the MORB range" or from too shallow, had a value too high, or "was influenced by plumes." The statistics are only biased by the choices of sampling areas and techniques made by individual investigators.

The results are shown in Figure 1. The black histogram excludes shallow samples, and the white histogram shows all the spreading-ridge data. There is no apparent difference between the complete dataset and the data "filtered" by depth.

Justification for Not Discarding Data

High-R basalts are found in backarc basins, at triple junctions, fracture zones, seamounts, and near microplates and propagating ridges. These usually are attributed to "hotspot influence," but they may also represent different modes of sampling normal upper mantle. Low-R basalts also are found in these places. Steady-state ridges, with shallow magma chambers, may provide a very filtered and averaged version of mantle heterogeneity.

Backarc basins are generally young, have unstable reorganizing ridges, and are cut off by slabs from the lower mantle. They provide a different sampling of the shallow mantle than do the steady-state ridges. Triple junctions often have a complex fracture system that also may provide a different kind of

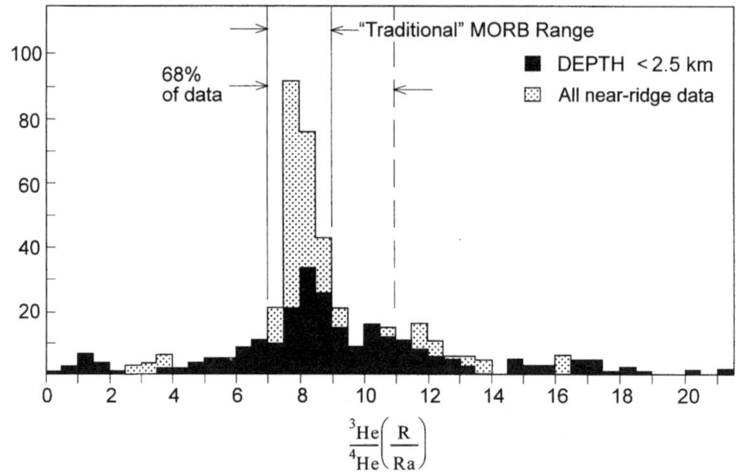

FIG. 1. Histogram showing the global spreading-ridge data and the data "filtered" by eliminating shallow data. The traditional MORB value is 8 ± 1 Ra, whereas the present results give a higher mean and a much larger standard deviation.

window into the mantle. Near-ridge seamounts may provide a more representative view of heterogeneity than ridge axes. Abandoned ridges and new ridges may provide components that are overwhelmed or diluted at normal ridges. Microplates and propagating ridges certainly utilize different sampling techniques than do steady-state ridges.

There is no a priori, hypothesis-independent reason for discarding any of this near-ridge data. Often there is no reason other than finding some samples at these locations with R > 11 Ra for invoking a plume or lower-mantle component.

Measures of Helium Chemistry

In addition to averages and standard deviations of various subsets of data, I also utilize other measures. The median is often a more robust measure of the "representative" value of a population than the average, and this is true for the helium data. The so-called "high-^{3}He" hotspot regions generally have median R values close to MORB. Unfortunately, often only the highest R values are discussed or plotted (e.g., Marty et al., 1996).

This is the inverse of the way the MORB dataset is treated. Published averages for MORB have had the high-R samples removed. Many "high-^{3}He" hotspots have means and medians within the MORB range, but only the high-R samples are compared to the "MORB average." When the highest R samples in, say, Ethiopia or Easter or the Lau Basin, are compared with the filtered MORB average, it is commonly concluded that a "lower-mantle" component is present. This is the statistical fallacy of comparing *maximum values* with *means* to conclude that two populations differ. This error is compounded by "filtering" the data prior to the comparison.

There often are orders-of-magnitude differences between the absolute helium concentrations, [He], in samples from a given area or in a single investigation. Although samples can degas, and measured [He] may not be representative of the magma at depth (or the source), there is increasing use of [He] in discussions of helium distributions (Ozima, 1994; Honda and Patterson, 1999). If the upper mantle is heterogeneous and is sampled by large-degree melts, or if magmas are blended in magma chambers, then the high [He] components will dominate. In addition to computing the averages and medians, we also calculate the [He]-weighted means, given as, e.g. {8.50}. These means, of course, will be dominated by the highest [He] samples, just as is the case in sampling the mantle or blending in magma chambers. If the shallow mantle is heterogeneous, a normal spreading ridge, with a shallow magma chamber, will yield R values that are dominated by the highest [He] samples. To get a true description of upper-mantle heterogeneity one must include samples from fracture zones, seamounts, and xenoliths and use dense profiles, without discarding values that fall outside the axial range.

TABLE 1. Summary Information on $^3He/^4He$ Ratios = R / Ra

Location	Ratio	n
All ridges	9.06 ± 3.26[1]	456
Excluding shallow depths (< 2.5 km)	8.50 ± 2.33	254
Excluding R > 9.5 Ra	7.47 ± 1.95	303
Excluding both	7.67 ± 1.35	193
Excluding d < 2.5 km, R > 11 Ra	7.91 ± 1.50	212
Excluding Easter, Galápagos, Bouvet	8.67 ± 1.88	219
North Atlantic		
All	9.37 ± 2.45	36
Excluding > 9.5 Ra	7.87 ± 0.69	24
Pacific		
All	8.84 ± 3.18[2]	269
Excluding backarc basins	8.46 ± 2.75[3]	146

[1]With slightly different datasets, I obtain R = 9.14 ± 3.59 Ra (n = 503) and R = 9.23 ± 3.35 Ra (n = 457).
[2]The alternate dataset yields 9.21 ± 3.40 (n = 210).
[3]The alternate dataset yields 9.00 ± 3.17 (n = 147).
Sources: Graham et al., 1992, 1993, 1999 and Farley and Neroda, 1998, and references therein; also Poreda et al., 1986, 1993; Moreira et al., 1995, 1996, 1999; Marty et al., 1996; Kurz et al., 1998; Nishio et al., 1999.

The various statistical measures are useful since the noble-gas literature is dominated by comparisons of OIB maximum values with MORB mean values. This is not a statistically valid procedure, even if the MORB dataset were not "filtered" by discarding all the high R values.

Results[1]

The Grand Average near-ridge mantle result is 9.14 ± 3.59 Ra (n = 503). The ±1σ range is from 5.5 to 12.7 Ra, and the ± 2σ range is 1.9 to 16.3 Ra. Slightly different datasets yield 9.06 ± 3.26 Ra (n = 456) and 9.23 ± 3.35 Ra (n = 457) (Table 1). Alternate values are given in some cases to test the robustness of the conclusions to changes in the dataset. Data selection is unbiased, but arbitrary decisions have to be made regarding how to deal with repeat measurements, samples from the same dredge haul, "oversampled" areas, and suspected air contamination. Some of the datasets differ simply because new data became available while the manuscript was being prepared.

The ± 1σ range captures all of the samples from Kerguelen, Prince Edward, Jan Mayen, Madeira, Canary Islands, St. Helena, Tristan, Gough, Haleakala, Hualalai, most of Guadalupe, Rocard, Teahitia, Mehetia, Gambiers, Pitcairn, Rurutu, Tubuai, and MacDonald, even though most of these regions are not in the spreading-center datasets used to determine the statistics. The ± 2σ level captures, in addition, Rapa, Savaii, Mas Afuera, Niihan, and Réunion (Farley and Neroda, 1998). These are all considered "hotspots," "plumes," or "samples of the lower mantle." The younger Mauna Loa and Mauna Kea samples also are in the MORB range.

If Iceland, Kolbeinsey Ridge, Reykjanes Ridge, Red Sea, and Ethiopian basalts are considered as transient and shallow products of a new or narrow ocean, or unstable ridges, then high values of R elsewhere also may sample the shallow mantle. If data for Ethiopia, Iceland, and the far north Atlantic (new ridges) are removed from the spreading-ridge dataset (arbitrarily), the result is:

$$R = 8.93 \pm 2.85 \text{ Ra } (n = 395).$$

[1]See Table 1.

For comparison, the Pacific dataset yields:

$$R = 9.20 \pm 3.40 \text{ Ra } (n = 210), \text{ or}$$
$$R = 8.84 \pm 3.18 \text{ Ra } (n = 269),$$

depending on choice of data.
If backarc basins are removed, then

$$R = 9.00 \pm 3.17 \text{ Ra } (n = 147), \text{ or}$$
$$R = 8.46 \pm 2.75 \text{ Ra } (n = 146).$$

The standard deviations (~ ± 3 Ra) are as significant as the means (which are 1 Ra unit higher than the conventional "biased" value). The values imply that many so-called "plume-influenced" or "lower-mantle" samples are unexceptional and can be drawn from the same population as MORB.

New ridges (North Atlantic, Red Sea, backarc basins) tend to have higher values than mature ridges, and abandoned ridges tend to have lower values. The statistics of R may therefore have more tectonic, local, or temporal significance than global significance. These results are tabulated in Tables 1 to 3.

The most important result is simply the statistic for the mean and standard deviation of the entire dataset, the Grand Average. The mean is significantly higher than the usually quoted 8 ± 1 Ra. The data are more normally distributed, and σ is more than three times higher than the usually quoted estimate (based on filtered and truncated data). To check the robustness of this result, I eliminated data from Ethiopia (a new spreading center), obtaining 9.22 ± 3.32 (n = 463), and then eliminated the entire region of the North Atlantic supposedly affected by the Icelandic plume, obtaining 8.93 ± 2.85 Ra. The main result survives; the mean and standard deviation are much higher than estimates based on the hypothesis-dependent "filtered," or biased, datasets. This has an immediate and dramatic implication. It means that values as high as 14 to 16 Ra are within 2σ of the "upper-mantle mean." Many recent papers use R values as low as 11 Ra as "unambiguous evidence" for a lower-mantle component. A large fraction of the so-called "hotspot" volcanics fall within the spreading-system range. The main exceptions are the more extreme values (R > 20 Ra) found in Iceland, Loihi, and Samoa. These also do not demand a lower-mantle component since the onset of rifting seems to be accompanied by high $^3He/^4He$ ratios (but extremely low [He] concentrations). A shallow U-poor region, sampled at the onset of cracking or tearing, cannot be ruled out (Anderson, 1998a, 1998b). This high-R component appears to be missing at steady-state ridges and "hotspots" that have been abandoned by ridges and HIMU islands, or the component is overwhelmed by the higher [He] later eruptives.

TABLE 2. Statistics (mean ± s.d.) of Subsets of the Data and EM, HIMU, and Low-R Islands

Location	Mean ± s.d.
Ridge abandoned islands[1]	7.10 ± 2.44
Abandoned ridges[2]	6.08 ± 1.80
"New" ridges[3]	9.93 ± 4.30
"New" ridges[4]	10.01 ± 4.67
EM islands[5]	7.89 ± 3.63
HIMU islands[6]	6.38 ± 0.94
HIMU, EM, low-R islands[7]	7.63 ± 3.33
North Chile Rise	7.78 ± 0.24
South Chile Rise	6.88 ± 1.72
Pacific Seamounts	6.58 ± 1.70
South Atlantic Seamounts	9.77 ± 1.40

[1]Gough, Azores, Tristan, Heard, Kerguelen, Réunion.
[2]Shimada, Socorro, Guadalupe.
[3]Kolbeinsey, Mohn, Easter, Galápagos, Manus, Lau, Ethiopia, Red Sea.
[4]Add Iceland.
[5]Pitcairn, Cook-Austral, Societies.
[6]St. Helena, Polynesia.
[7]Above plus Hilton et al. (1995)
Sources: Data from Farley and Neroda (1998) and references therein.

Also given in Table 1 are results for all the Pacific spreading-ridge data. This is similar to the global grand average and to subsets of the global data. There is no obvious spreading-rate dependence. The age, or maturity, of the spreading feature seems to be more important. Note that the median is a more stable estimator of the $^3He/^4He$ ratio than the mean. So-called "hotspot-influenced" sections of the global ridge system are summarized in Table 2. These are regions affected by ridge reorganization, ridge propagation, and fracture zones. They seem to be drawn from the same population.

Subsets of Data

Systematic sampling has revealed that the $^3He/^4He$ ratio can be remarkably variable over short distances (e.g., Schilling, 1986; Farley et al., 1993; Graham et al., 1993; Kurz et al., 1998). Changes apparently occur at fracture zones, in different flows in volcanoes of different ages, or near propagating rifts (Poreda et al., 1993). There are also rapid changes in R with time (Kurz et al., 1987; Farley et al., 1993).

Seawater (R = 1 Ra) and seawater-affected sediments and crust are at the top of the oceanic column. Terriginous sediments and crust are expected to have low R (high U, Th contents). Pelagic sediments, in regions of slow sedimentation, can have high R because of the influx of He-rich IDP (Anderson, 1993). The lithosphere may have low R (if metasomatized) or high R (for U, Th-poor restite). A metasomatized region, or a region that collects recycled material and residual melts (perisphere), may underlie the plate. This is expected to have low R. Some parts of the shallow mantle, above ~60 km, may collect CO_2-rich gases expelled from rising magmas, or degassing dikes. Below these residual, refractory, or enriched regions is the MORB reservoir, which itself may contain inhomogeneities of various scales and may be radially zoned in composition. One therefore expects R to evolve with time and one expects samples collected at the onset of rifting or at fracture zones, seamounts, microplates, or backarc basins to differ from samples at mature steady-state ridges, or abandoned ridges.

Once an unbiased data set for the world ridge system has been compiled, various subsets of the data (new ridges, fast-spreading ridges, backarc basins, etc.) or other datasets (oceanic-island basalts, continental flood basalts, etc.) can be tested against the Grand Average to see if they are significantly different. Tables 1 and 2 summarize various subsets of the spreading-ridge data plus oceanic-island data (data from Farley and Neroda, 1998, and references therein).

Conclusions Based on Filtered Data

Once the $^3He/^4He$ dataset is "filtered" to remove samples perceived to be contaminated by plumes, one can no longer view certain values as "significantly" higher or lower than MORB. "Significantly" is a statistical term and implies an unbiased sample. It also is common, but erroneous, to compare maximum values from some regions with the MORB average and to declare that an anomaly has been found (even if samples of similar R are found along the ridge).

For example, the average ratio for a suite of EPR glasses was determined by excluding values greater than 9 Ra, yielding 8.26 ± 0.19 Ra. Samples north of 16°S on the East Pacific Rise having values of 8.6 to 10 Ra were then viewed as "significantly higher than this 'normal' value" (Kurz et al., 1996). In the South Atlantic, near-ridge seamounts have $^3He/^4He$ up to 11 Ra, "significantly higher than adjacent ridges,"according to some authors.

Glasses from the Central Manus Basin have "typical plume $^3He/^4He$ ratios" that cluster around 12.2 Ra, "a value significantly higher than the range found in most MORB, 8 ± 1 Ra" (MacPherson et al., 1998). The Manus backarc basin spreading center yields R = 10.7 ± 3.4 Ra (n = 29) for the mean and standard deviation. This differs from the value given in the original paper because the authors eliminated MORB-like helium ratios before calculating the mean and standard deviation of the remaining data. These results are similar to new ridges elsewhere.

Continental flood basalts have very low ^3He. But some, having "$^3He/^4He$ ratios higher than MORB," confirm "the presence of plumes" in the Siberian, Deccan, and Ethiopian flood basalts (Basu et al., 1993, 1995; Marty et al., 1996, 1998).

Discussion

Previous summaries of helium data for MORB include 85 samples (Zindler and Hart, 1986) to 158 samples (Hilton et al., 1993). My result, 9.06 ± 3.26 Ra, involves 456 samples, with similar results for the alternate smaller datasets. My results for Pacific MORB (Table 1) bracket the Pacific estimate of Hilton et al. (1993), who obtained 8.68 ± 0.63 Ra (n = 30).

The mean has remained about the same, while the σ has increased because recently obtained data along the EPR are higher than previous measurements, and backarc-basin ridges and near-ridge seamounts have been included. Low values are found along the Chile Rise. The Galápagos and Easter microplate regions can be considered anomalous because of high R values, or as normal values of the shallow mantle when sampled at propagating rifts and unstable spreading boundaries rather than steady-state ridges.

A student t-test shows that the Hilton et al. (1993) estimate can be considered to be drawn from the same large population having the mean that we estimate. Interestingly, if we exclude the backarc-basin data, the Hilton mean appears, statistically, to be drawn from a different population. It errs on the high side, meaning that the inclusion of the Chile Rise and near-ridge seamount data have an important influence on estimates of shallow-mantle chemistry.

The large standard deviation associated with the present estimate merits comment since it is so much larger than the widely quoted value of ± 1 Ra based on the filtered, or "hypothesis-dependent," datasets. Even when all the shallow and high-R samples are removed from the database, I obtain ~8 ± 1.4 Ra for "filtered" MORB (Table 1). The mean is the same as the usually quoted value, but the S.D. is much larger. It turns out that the usually quoted value (±1 Ra) is based on individual ridge segments, which often tend to be uniform and have small S.D. However, the means from segment to segment or from ocean to ocean vary from ~7.8 ± 0.5 Ra to 8.7 ± 0.6 Ra for South Atlantic and Pacific Ocean MORB, respectively (Hilton et al., 1993), and some smaller subsets of ridge data vary even more. The intra-ocean variation adds considerably to the standard deviation of the MORB estimate. Even if 8 Ra is accepted as the "global mean" of MORB (based on the "filtered" dataset) a change of the σ from ±1 to ±1.5 Ra means that a value of 11 Ra is within 2σ of the mean, rather than "being significantly higher than the MORB average" and "therefore" indicative of lower mantle involvement.

The following are the [He] weighted means = {He} of some of the "anomalous" ridge systems: Azores 8.72 Ra; South Atlantic seamounts 8.68 Ra; Shona 8.67 Ra; Bouvet 8.41 Ra; Galápagos 8.79 Ra; Easter 8.80 Ra. The corresponding medians are 8.70, 10.53, 8.39, 7.88, 8.47, and 8.38 Ra. Except for the median of the South Atlantic seamounts, all of these so-called "plume-influenced" regions are very close to "normal" MORB by these measures. The mean of the seamount data is 9.77 ± 1.40 Ra, and the t-test shows that it can be of the same population as MORB. The statistics of the so-called plume-influenced regions of the global spreading-ridge system argue against the plume hypotheses. The $^3He/^4He$ ratios of many oceanic islands also are well within the MORB range.

Low $^3He/^4He$ regions include Shimada Seamount (abandoned ridge), São Miguel–St. Helena–Tristan da Cunha (previously ridge-centered), Circe, Cook-Australs, Mangaia, Tubuai, Rarotonga, Heard, Jan Mayen, Canaries, Fernando, Cameroons, Guadalupe (abandoned ridge), and Kerguelen (formerly ridge-centered). These regions are all inactive or low activity and may reflect the waning of crack- or stress-controlled volcanism. Many of these regions lie along fracture zones. Jan Mayen and Kerguelen may be continental fragments. There is little geophysical evidence that these are hot areas or are held up by plumes. The assignment of these volcanoes to "low-^{3}He" plumes is strictly ad hoc. Statistics of small samples and prior history of sampling appears to explain the variations along the global spreading-ridge system.

Plum Pudding or Perisphere

Several models have been discussed for the geometry of heterogeneity in the "upper" mantle. The plum-pudding model proposes that enriched plums are embedded in a depleted matrix, and these are preferentially removed at "hotspots" prior to sampling by the ridge system (Morgan and Morgan, 1999). In the perisphere model, parts of the mantle are preferentially enriched in incompatible and recycled material, and trapped melts, and this part of the mantle is the main contributor of geochemical "anomalies" to new rifts and at the onset of seafloor spreading and at mid-plate volcanoes (Anderson, 1998c).

In both cases there is an enriched component that is preferentially removed at the early stages of magmatism. In the statistical treatment, it does not matter whether the mantle is laminated, stirred, or permeated with plums. It is an urn, containing various balls of various properties, and is sampled by various processes (grab samples, steam shovels, tweezers—i.e. seamounts, ridges, xenoliths). The statistical question is: Are the various subsets of samples drawn from the same population?

Two of the standard tests are: (1) Is this sample more than 2 (or 3) standard deviations from the mean? What are the odds of this? (2) Do the mean and standard deviation of a population (e.g., basalts around the Easter microplate, the Azores, or the Galápagos) imply that it cannot be from the MORB, or spreading-ridge, population? I use the t-test in this regard.

"Hotspots"

This discussion uses data compiled by Farley and Neroda (1998) and references therein (see also references in the tables). The so-called "high-^{3}He" hotspots (they are actually low-^{3}He) are defined on the basis of some (usually, very few) of the samples

TABLE 3. $^3He/^4He$ ratios, Normalized to Air (1.38×10^{-6}) in Older Continental Rocks

Rock	Average + S.D.	Median	Range	*n*
Komatiites 2700 Ma	9.14 ± 13.6	0.4	.04–39	17
Carbonatites 380 Ma	11.77 ± 4.48	10.4	4.9–18.7	7
Siberia 253 Ma	6.53 ± 3.38	5.4	1.4–12.7	19
Northeast Greenland 58 Ma	6.62 ± 6.66	3.46	1.1–22	11
West Greenland 60 Ma	22.21 ± 11.46	28.05	4–31	5
Deccan 69 Ma	10.05 ± 4.65	11.45	3–14	4
Ethiopia 20 Ma	8.24 ± 5.41	8.65	.03–20	46
Heard Island	6.83 ± 5.20	7.15	3.9–18.1	16
Galápagos	12.39 ± 5.80	9.87	4.7–27	53

Sources: Basu et al., 1993, 1995; Graham et al., 1993, 1998; Hilton et al., 1995; Marty et al.,1996, 1998; Richard et al., 1996; Kurz et al., 1998; Botz et al., 1999.

(glasses, xenoliths) falling outside the perceived "normal" MORB range. Some samples (or islands) in the following regions are designated as high–$^3He/^4He$ (>11 Ra) hotspots—Hawaii, Galápagos, Iceland, Easter, Societies, Samoa, Azores, Juan Fernandez, Réunion, and Pitcairn. MORB-like He ratios also occur in most of these regions, and the ratios often exhibit dramatic temporal and spatial variations. Most of these regions are more active, and have younger volcanoes, than the low R regions. Many oceanic islands either have higher than MORB or lower-than-MORB ratios (Farley and Neroda, 1998); only a few straddle the "MORB range" (as previously defined). This may represent a temporal situation, with the high-R islands tapping a shallower, or expendable, source. The temporal evolution in both the ridge environment and the mid-plate environment is from high to low R, suggesting a shallow LONU source that is expended early or is easily overwhelmed by deeper high-[He] basalts, such as "normal" MORB.

The following islands and regions have median values of R that fall well within the MORB range—Kerguelen, Heard, Prince Edward, Jan Mayen, Canaries, Madeira, São Miguel, Gough, Haleakala, Hualalai, Galápagos, Rocard, Gambiers, Tubuai, Rurutu, Pitcairn, Mas Afuera, Ethiopia, and Yellowstone (Farley and Neroda, 1998). Thus, by this measure, many of the so-called "high-^3He" hotspots are not distinct from MORB. Even Hawaii, the classic "high-^3He" hotspot, has individual volcanoes, and flows, that are indistinguishable from MORB. It is unlikely that the Hawaiian chain, or even the island of Hawaii, is underlain by a "high-^3He" source, but quite plausible that there are LONU domains in the lithosphere or shallow mantle (LONU, low [He]). Xenoliths claimed to be Archean in age have recently been found in Hawaiian basalts (Ducea and

Sen, 1999). If this is true, an obvious source for LONU, high-R material, has been found. Many oceanic plateaus appear to be underlain by ancient continental lithosphere.

Mauna Kea, Kohala, and the past 10,000 years of Mauna Loa lavas are in the MORB range. Hualalai has higher R than these, although it is further from "the plume." This suggests that it is not motion of the plate away from "the plume" but activation of new faults and tapping of new shallow mantle that is responsible for the temporal change.

Although some basalts in the Galápagos area have $^3He/^4He$ ratios as high as 22.87 Ra, there are very low values nearby. The median value is 8.51 Ra. The samples having values near 8.5 Ra have one or two orders of magnitude more helium than the high-R samples, so any blending, as in a magma chamber, would return values near 8.5 Ra. The [He] weighted mean is {8.79} Ra. Similarly, basalts in the Easter microplate area range from 7.7 to 11.67 Ra with a median value of 8.84 Ra. The weighted mean, taking into account [He], gives {8.8} Ra. Azores samples, although having values as high as 11.3 Ra, have a median of 8 Ra and a {He} average of {8.72} Ra, the high-R basalts having an order of magnitude lower [He].

Continental Rocks; Ancient Rocks

Values of R well within the MORB range have been found in the Deccan Traps, East Greenland basalts, and Archean komatiites. The highest values have been attributed to plumes because of the perception that "they are significantly higher than the MORB average" (Basu et al., 1995; Marty et al., 1998). It is not statistically valid to compare maximum values in one population with mean values in another. There are statistical methods for determining, from the means and standard deviations, if two populations differ.

A series of recent studies on continental rocks such as flood basalts, carbonatites, and komatiites indicate extreme diversity, extraordinarily low ^3He contents, and very low values of R (Table 3). Attention has focused on the one, or few, grams that yield high $^3He/^4He$ ratios. These high values are taken as evidence of a plume component and are used to characterize "the lower mantle." The helium concentrations in these samples are measured in fractions of a picomole per gram. The standard deviations of R in these old samples are typically 50%, or more, of the mean, not much different than modern samples. Surprisingly, the maximum values are typically 20 to 30 Ra, much the same as modern values in Hawaii, Iceland, and Samoa. The standard deviation is not much greater than exhibited by fine-scale sampling of seamounts, xenoliths, or inclusions. I take the statistics to reflect the vagaries of small-scale sampling and small-sample statistics, and to reflect the variations intrinsic to the crust, lithosphere, and shallow mantle.

Ancient mantle reservoirs will have had higher $^3He/^4He$ ratios than the same reservoirs today, and if gases from these reservoirs, or their magmas, were trapped in low-U, Th lithosphere, the R will be higher than contemporary magmas. The MORB source, for example, would have had a mean R much greater than 8 to 10 Ra when the carbonatites and komatiites were emplaced. The range of values also will be different. I have shown that values of 16 Ra are not exceptional when found along the current spreading-ridge system. Much higher values would have occurred along the ancient ridge systems. One cannot compare values in ancient rocks directly with values in modern rocks, or modern reservoirs. Estimates of the rate at which R in the MORB reservoir declines with time vary widely (Anderson, 1998b) but, roughly, "hotspot"-type high R can decline to "typical" MORB-like values in 10^8 to 10^9 years. This depends, of course, on the $^{238}U/^3He$ and U/Th ratios.

After emplacement, the R of a rock also will evolve with time, if there is local U and Th, or atmospheric or seawater contamination. However, this is also the case for recycled material that enters the shallow mantle, and for magmas that ascend through the shallow mantle and crust. Surface processes apparently contribute to the diversity of components found, or inferred, in both MORB and OIB.

For comparison, in Table 3, I show results for two contemporary islands, Heard (Hilton et al., 1995) and Galápagos (Graham et al., 1993; Kurz and Geist, 1999). This shows that the statistics of ancient rocks are not much different than well-sampled modern volcanic regions. It should be pointed out that the mid-ocean ridge system is not sampled at this level of detail.

Some of the ancient rocks analyzed for helium are picrites and komatiites. Other isotopic systems place these squarely in the MORB camp (Anderson, 1994), and there is now doubt even if very high temperatures (outside the ridge range) are required to generate komatiites. There also is doubt that carbonatites are evidence for deep high-temperature mantle. Carbonatites also have been attributed to lithosphere or asthenosphere sources. In summary, the continental rocks discussed could also be termed "very low ^{3}He," or "low– ^{3}He/^{4}He" rocks. In diversity they are similar to some well-sampled modern igneous formations.

Some ancient basalts have been attributed to plumes because some specimens have high R components. For example, the 55 Ma, post-breakup basalts in Greenland have R ranging from 6.8 to 10.7 Ra, but the median is 8.3 Ra (Bernstein et al., 1998). The plume explanation is advanced since the high values are "significantly higher than present day MORB," indicating derivation from the undegassed lower mantle. This ignores the fact that values of 11 Ra are found along ridges. The Greenland values are not significantly different from the MORB mean established in this paper (8.7 ± 2 Ra, or 9.1 ± 3.3 Ra, depending on whether basalts from the Easter, Galápagos, and Bouvet regions are included). The value of 11 Ra is just about at the 2σ level, for the filtered dataset, implying that values higher than this can be expected about 16% of the time. Even with the older hypothesis-dependent value of 8 ± 1 Ra for MORB, a value of 11 Ra is at the 3σ level and is not, as implied in many papers, an impossible value for MORB. It is, in fact, a typical value in the unbiased (unfiltered) spreading-ridge dataset.

Possible Time (Depth) Sequence

If it can be shown statistically that some new ridge segment (say the Red Sea, East African Rift, Gulf of Aden, or Iceland) differs significantly from the global ridge average, then the question arises: Is this a "spatial thing" (plume), a "temporal thing" (onset of rifting), or a "depth thing" (LONU shallow mantle; recycled lithosphere)?

The life history of a mid-ocean ridge involves birth (usually by ridge propagation), mature or steady-state, and abandonment. Mature ridges, characterized by steady-state magma chambers and prior loss of easily removed or shallow components, can be expected to differ from the initial and terminal stages of spreading. The initial stage is well studied but differences from "normal MORB" are almost invariably attributed to "plumes" and "lower-mantle involvement" rather than to the normal stages of ridge development and initial or shallow-mantle sampling. The absence of precursory uplift at the site of initial rifting, or rift extension, would argue against the plume hypothesis, or any thermal explanation.

Table 2 summarizes several compilations of data that may have bearing on this. "New" ridges seem to be characterized by high R (~10 Ra) whether they grow by end-on propagation of pre-existing ridges, or in backarc basins. They are similar, in R, to the South Atlantic seamounts, which may also tap shallow mantle, undiluted by massive axial magmas.

Abandoned ridges, and islands abandoned by ridges, have low R (6 to 7 Ra) and may reflect melt-depleted (or residual) shallow-mantle or "contaminated" mantle. These values are similar to the South Chile Rise and Pacific seamounts, nominally different environments.

HIMU islands supposedly are affected by altered oceanic crust and sediments, or fluids therefrom. They have low ^{3}He/^{4}He ratios in accord with the inferred high ^{238}U/^{204}Pb ratios (high μ) and presumed high ^{238}U/^{3}He (high ν). They may tap a source similar to that tapped by abandoned ridges and ridge-abandoned islands, or the South Chile Rise (which is near the South Chile subduction zone). The so-called EM islands have R = 7.89 ± 3.63 Ra, similar to the North Chile Rise, a ridge–fracture zone system.

It appears that, initially, ridges tend toward high R, whereas abandoned ridges, and islands, tend toward low values. This suggests that the shallower, easier to remove, or initial heterogeneities have high ^{3}He/^{4}He ratios. Whether these high-R regions are shallow "lithosphere" or embedded "plums" is open for discussion. Their generally low [He] suggests that they must be shallow since they would be overwhelmed by an N-MORB component, which is probably sequestered at greater depth and certainly involves larger volumes of mantle and more extreme melting. A shallow, high-R, low-[He], and refractory source seems to be indicated, perhaps lithosphere, or recycled lithosphere (the complement to recycled HIMU crust).

The association of high ^{3}He/^{4}He ratios with early stages of magmatism (e.g., Hawaii, Juan Fernandez, Greenland, Deccan), fracture zones, young oceans (Red Sea, Gulf of Aden), rifts (Ethiopia), caldera collapse and tectonic uplift (Yellowstone), reorganizing plate boundaries (Bouvet, Shona), new and unstable backarc basins (Lau, Manus), and low ratios with abandoned ridges (Socorro, Gough, Tristan da Cunha) suggest that high ^{3}He/^{4}He ratios (but low ^{3}He absolute abundances) may be a feature of the shallow mantle, perhaps U,Th–poor lithos-

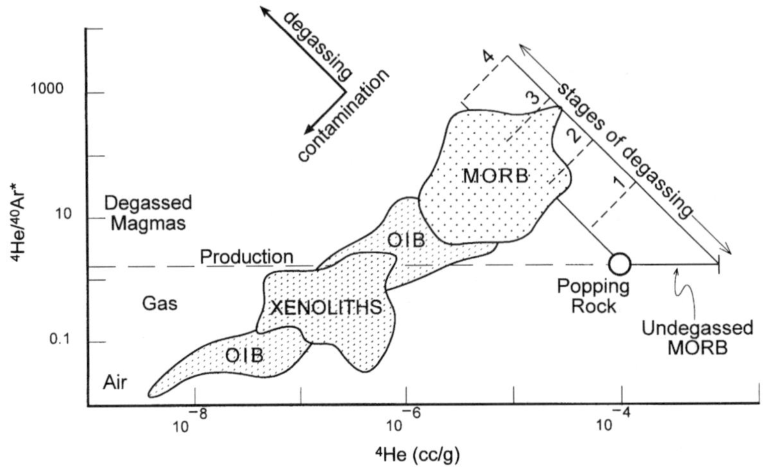

FIG. 2. ^{4}He versus ^{4}He/^{40}Ar* for mantle samples. Redrawn from Honda and Patterson (1999). "Popping rock" (Sarda et al., 1999) is the best approximation we have for a primary magma only slightly affected by degassing. A possible range of undegassed MORB is shown. Such basalts evolve as shown for one to four stages of degassing. Contamination of degassed magmas by air, seawater, xenoliths, and lithospheric vesicles will move the degassed magma toward the lower left of the diagram. Note that MORB contains much more ^{4}He (and ^{3}He) than OIB. OIB appear to be mixtures of MORB, air, and xenolith helium.

phere, or restite layer. The liberation, or incorporation, of these shallow gases by initial fracturing and magmatism seems to be indicated by the tectonic context of high ^{3}He/^{4}He localities. Once the deeper gas-rich MORB magma becomes involved, then lower ratios, ^{3}He/^{4}He of 6 to 9 Ra, will dominate over the gas-poor high ^{3}He/^{4}He ratio sources.

"High" ^{3}He/^{4}He ratios (> 9, 9.5, or 11 Ra, depending on the author) are commonly called "plume,""lower- mantle," or "primitive" helium. Because of the low [He] and low He/Ne and He/Ar ratios of these samples (Fig. 2) and because of their temporal and tectonic context, they could, with more justification, be called "lithospheric helium" (Anderson, 1998a, 1998b). There is no evidence that they are derived from a deep or an undegassed reservoir. Buoyant lithosphere, or restite, is more likely to be able to retain ancient isotopic signatures than the "convecting mantle."

Onset of Rifting

Most mid-ocean ridges are mature and steady state. One can ask: What is the R at the onset of rifting, and does this differ from mature ridges? Ethiopia is probably the newest sampled ridge and yields 8.24 ± 5.41 Ra ($n = 46$). This cannot be discarded as a separate population based on the t-test, so in some of our runs, we incorporate it into the ridge dataset.

However, we can also combine it with the Red Sea, Kolbeinsey-Mohn ridges, Easter, Galápagos, Manus, and Lau, as regions having new or propagating ridges, to obtain a "new-ridge" datum. The result is 9.93 ± 4.3 Ra ($n = 207$). If we add Iceland, because of the constant and recent ridge jumps in this area, the value only changes slightly: 10.01 ± 4.66 Ra ($n = 239$). These estimates are statistically different from the global average. Note that "new ridges" (Table 2) have a much higher variability than established ridges. This is expected since they are, in a way, "mid-plate" and involve pre-existing crust and lithosphere, and previously undrained mantle.

There is only a small amount of data on abandoned ridges, but we obtain 6.09 ± 1.80 Ra ($n = 9$) from the compilation of Graham et al. (1992). These results are consistent with a shallow LONU region and a low-R residual mantle. The LONU source seems to be exhaustible and transient.

The North Atlantic, Red Sea, Gulf of Aden, and South Pacific microplates formed by end-on, rather than pull-apart, ridge propagation. The early magmas in these regions may tap shallow mantle preferentially, and yield more diversity than is possible at a steady-state ridge. The fact that the high–^{3}He/^{4}He basalts found in these new ridge areas have low [^{3}He] is consistent with a shallow, low–U,Th source, rather than a high-[^{3}He] undegassed source.

Continental break-up (push-apart) often is attributed to deep mantle plumes, but in the above cases the processes of plate divergence involve rotation and end-on propagation and represent the present location of a propagating crack, rather than an initial break. Older ridges may well be different geochemically from early ridges or narrow oceans. Initial magmas may be contaminated by shallow sources (air, seawater, crust, lithosphere, perisphere) or by "plums" (the plum-pudding or marble-cake models). The perception that the "MORB-mantle," "Depleted Upper Mantle," or "convecting mantle" is homogeneous is based on biased statistics.

Iceland

Iceland is generally regarded as the top of a deep mantle plume, but it may also be a transient edge effect associated with the extension of the central Atlantic ridge along an intracratonic suture. This is a region of relatively new seafloor spreading, a narrow ocean, and constantly shifting plate boundaries. The magma geochemistry reflects reworking of thick crust and possibly shallow mantle. Iceland is close to the pole of rotation of the American and European plate and hence is slowly spreading. It is bordered to the north by fracture zones and islands of continental affinity. One might expect under these circumstances to be able to sample the shallow mantle in a way not available to mature, rapidly spreading ridges in the middle of wide oceans. Additionally, the highest-temperature magmas in Iceland are isotopically indistinguishable from normal MORB (Anderson, 1994). Mantle temperatures in the North Atlantic and during eruption of the Greenland basalts are not particularly high (Favela and Anderson, 1999; Korenaga, 2000). The volume of the eruptives appears to be caused by focusing, not high temperature. This is called the EDGE effect (Edge Driven Gyres and Eddies) (Anderson, 1998c).

Our view is that it is better to treat Iceland as a possible end-member of a ridge, because of its location and history, rather than eliminate its materials from the spreading-ridge database because of its diversity and preconceived notions about the allowable range of shallow-mantle geochemistry. Some low values of R are found in Iceland and along the fracture zone to the north (Condomines et al., 1983; Schilling et al., 1999). This rules out pervasive, long-distance, transport of Iceland high-R plume material, as has been invoked for the Reykjanes Ridge (Condomines et al., 1983; Schilling, 1986). An unstable system of ridges and triple junctions, and transient, local tapping of a shallow LONU source seems to be indicated. The association of geochemical anomalies with fracture zones is as defensible as their association with "hotspots," many of which are merely elevated portions of fracture zones or ridge-transform intersections. A tectonic and shallow explanation for features associated with Iceland is given in Favela and Anderson (1999) and Keller et al. (1999).

Mixing

The high-R samples are generally low [He] and would therefore be swamped if they were put into a mid-ocean ridge magma chamber or intruded by normal MORB. For example, I took a subset of the ridge data composed of 165 samples having a mean R of 8.72 Ra. To this, I added the samples from the Kolbeinsey Ridge (n = 30), Iceland (n = 32), and 16 of the highest-R basalts from Loihi and Hawaii. The composite was mixed and weighted by the [He] concentration of each sample to obtain {R}. The average changed from 8.72 Ra to {8.80} Ra for this mix, which was about two-thirds MORB and one-third "high-^{3}He hotspot" material. Except for the [He] content, all samples were assumed to be the same size. Thus it is clear that in finding the full range of R that may occur in the upper mantle, one must go to "anomalous" places, such as slowly spreading ridges, fracture zones, unstable triple junctions, seamounts, and reorganizing plate boundaries such as jumping or propagating ridge segments. Finding "anomalous" values at these locations is not evidence for a deep mantle plume. The apparent homogeneity of MORB at ridge axes may be a result of magma-chamber processing rather than an intrinsic characteristic of the upper mantle.

It is commonly assumed in geochemical box models that a homogeneous, well-stirred, depleted "reservoir" occupies the entire upper mantle, extending right up to the base of the plate. It is well known, however, that metasomatic fluids, residual melts, the mantle wedge, and recycling also affect the shallow mantle, and that N-MORB does not appear at the onset of rifting but only after a period of spreading. It is OIB-like magmas that occur in all tectonic environments, even convergence zones, and regions of localized extension. When the overlying crust and lithosphere and enriched shallow mantle are cleared away, then "normal" uncontaminated MORB makes an appearance. Because of the multiple and contradictory usages of "lithosphere" and "asthenosphere" (Anderson, 1995), I refer to the outer shell overlying the MORB reservoir as the "carapace."

Ironically, the reference state for both plume modeling and passive spreading is a static isothermal mantle. Plumes are invoked in geodynamic models to introduce a thermal anomaly, whereas they are invoked into the geochemical models to introduce chemical anomalies into "the convecting mantle." These are contradictory views. A non-static, non-isothermal, non-homogeneous evolving upper mantle should be the reference state in both dynamic and geochemical modeling. The upper mantle cannot simultaneously be static and isothermal, well-mixed and convecting.

If the carapace is under horizontal compression, it is unlikely that melts from the underlying magmasphere can escape via dikes and volcanoes. Fluids in and below the carapace can escape in regions of horizontal extension, such as near ridges, in young lithosphere, at forearcs and backarcs, and in extensional outer rises around volcanic loads. Transiting fluids can pick up gases and impurities from the various regions of the carapace. A depleted garnet-poor carapace can be considered a candidate for a long-term isolated reservoir, or source of LIL (crustal and melt-metasomatized parts) and gas (from trapped fluid-filled inclusions). Initial magmas penetrating such a carapace will differ from later magmas. Temporal and spatial variations can be expected in such an environment.

Provenance

Ocean-island and continental flood basalts are characterized by enriched or radiogenic isotopic chemistry for the heavy LIL elements and, in many cases, for helium as well. In the view of rare-gas specialists, however, the most unambiguous indicator of "undegassed plume material" is high $^3He/^4He$ ratios, inferred to require an excess of non-radiogenic helium. Since ^{206}Pb, ^{207}Pb, ^{208}Pb, and ^4He (and heat) are all daughter products of U and Th, we have the paradoxical situation that a U-rich region is required for the heavy isotopes (and implied for "hotspots"), and a U-poor region is required for helium (and neon). The "primordial undegassed lower mantle" model addresses this paradox by neutralizing the large amounts of ^4He expected in an ancient primordial reservoir with very large amounts of the primordial isotope ^3He. The amount of ^3He required is comparable to the amount in primitive carbonaceous chondrites. This brings up another paradox. The Earth is very depleted in volatiles compared to chondrites, but helium, the most volatile of all, must be an exception to this rule. This is the "undegassed lower mantle" model. A problem is that all high $^3He/^4He$ materials have very low ^3He concentrations compared to mid-ocean ridge basalts which, in this model, are from a "degassed" reservoir. The Helium-Heat Flow Paradox is a related paradox.

One could recognize, alternatively, that although He, U, Th, Rb, Sr, Pb, etc. are incompatible elements in that they enter magmas when the mantle partially melts, they differ in that gas exsolves from the magma at shallow depths, and He-CO_2 can separate from U-Th. Some fraction of the exsolved gas is stored in crystal cumulates and fluid-filled inclusions in refractory peridotite. Dike tips are low-pressure environments, and gas might exsolve from propagating dikes at greater depths than from magma chambers. In any case, magmas can retain their U and Th but can degas their complement of volatiles. Deeper exsolved gas can, in part, stay in U-Th–poor surroundings. Both the gas and the magma initially have the same $^3He/^4He$ ratio, and the same as the magma source, but as time goes on, the helium stored in a low $^{238}U/^3He$ (LONU) environment, such as fluid-filled inclusions or in olivine grains, will retain high ratios, whereas the frozen magma (crust) and original reservoir (MORB-source?) will continue to generate ^4He and will evolve to low $^3He/^4He$ ratios. The fluid-filled inclusions will have lower He/Ne (and He/Ar) ratios than the partially degassed magmas, or the source, because of the higher solubility of He in the melt. This is a property of high-R basalts (Anderson, 1998a, 1998b, 1998c). Low [He], He/Ne, He/Ar, and high but variable R can be considered as properties of lithospheric helium (Fig. 2). "Plume helium" from a hypothetical primordial undegassed reservoir should have high [He] and high He/Ne and He/Ar (as well as high He/Sr, He/Pb, He/CO_2, etc.) ratios. Such a component is not evident in high-R basalts.

Helium Incompatibility

Helium is not the same kind of incompatible element as U, Th, Pb, Sr, Nd, Rb, Sm, and Hf, the other common elements and isotopes used in discussions of mantle geochemistry. Helium is incompatible in that it prefers to enter the melt when a rock is melted. However, as a magma rises, it degasses, and at this point He can decouple from the other LIL, in particular U and Th. Magmas wet grain boundaries and can escape to the surface. Some fraction of the helium (with its CO_2 carrier) escapes the magma and ends up in magma-chamber cumulates and in the lithosphere. These are low–U,Th environments compared to magma or basaltic crust, and gas trapped in fluid-filled inclusions will evolve isotopically less rapidly than the gas in the original MORB reservoir. In the plum-pudding model, some plums may be large refractory olivine-rich domains, and the He in these domains will also retain high $^3He/^4He$ ratios compared to "enriched" (U-rich) domains. However, it is easier to envisage lithosphere, recycled lithosphere, and melt-depleted cumulates as long-term storage reservoirs for LONU material. Thus, helium is "melt incompatible" at shallow depth, and some fraction of exsolved He-CO_2 will enter the lithosphere, thereby acting as a "compatible element," along with Os. It may be that the only geochemical tracers that involve "clean" lithosphere are osmium and helium isotopes. Although there may be little helium stored in the lithosphere, very little is needed, compared to the ridge and arc flux, and a modest, exhaustible source is what is needed to explain the very low ^3He contents of mid-plate volcanoes and the temporal trends toward MORB-like $^3He/^4He$ ratios. When deeper (MORB) gas-rich magmas become involved, the lithospheric signature is quickly swamped (the same would hold true for many plume components). In order to "restart" a high-R event, a new crack (Iceland) or a new volcano (Loihi) must be initiated. A shallow high-R, low-[He] source explains the lateral and temporal variability of "hotspot" helium, plus the eventual takeover, in many cases, of gas-rich MORB geochemistry. Helium also may be more compatible, in the traditional crystal-melt sense, than U and Th (Graham, 1988). If this is the case, then residual crystals would have higher He/U ratios than the source, and these would also evolve to high-R reservoirs.

Note that the shallow LONU reservoir is not the same as the (possibly deeper) perisphere. The perisphere, like the mantle wedge, metasomatized lithosphere, and oceanic crust, is LIL rich, whereas the LONU source is LIL and volatile-poor, being characterized geochemically by small amounts of trapped (He, CO_2) gas and low $^{186}Os/^{187}Os$, and probably, unique $^{208}Pb/^{204}Pb$ ratios. The $^{238}U/^3He$ ratio may be influenced both by crystal-melt and melt-vapor partitioning.

LONU also may be thought of as complementary to HIMU (high $^{238}U/Pb$), which is widely believed to be recycled fragments of oceanic crust. LONU may be recycled fragments of oceanic (or continental) lithosphere. It is also possible that LONU may be ancient mantle trapped in the ocean basins as roots of plateaus and swells, such as has been proposed for Kerguelen, Seychelles, Lord Howe Rise, and Ontong-Fiji (Nur and Ben-Avrahem, 1982).

Arguments for a Shallow Source for LONU

There are various lines of evidence suggesting that the low $^{238}U/^3He$ (LONU) source (usually called the "high-^3He" or "undegassed" or "plume" source in the geochemical literature) is shallow: (1) the time sequence (Hawaii, Juan Fernandez, Galápagos) of high-R basalts initially, followed by MORB-like R, suggests that the high–3He/^{4}He source is shallow; (2) the rapid lateral variations, which put so-called "high-^3He" basalts (actually low-^3He, high–$^3He/^4He$ ratios) within tens of km of low–$^3He/^4He$ basalts (or geothermal areas); and (3) the existence of high-[He] and high-R lithospheric xenoliths (Farley and Neroda, 1998; Poreda and Farley, 1992) and inferred Loihi-like noble gases in continental lithosphere (Ballentine, 1997).

These observations are not predicted by the plume hypothesis. They have been "explained" by invoking a radially zoned or toroidal plume, drifting of a volcano off of a high-R plume, and/or entrainment of "normal" mantle and emplacement of plume material into the lithosphere. None of these ad hoc explanations account for the very low ^3He in high-R basalts, the rapid variations in R, and the indifference of R to magma or xenolith chemistry or location relative to a "plume axis."

Backarc basins (Lau, Manus) and geothermal systems above deep crustal batholiths (Yellowstone) commonly have variable R, but a few of the samples are high R, indicating a shallow source. Backarc basins are cut off from the deep mantle by slabs, and it is unlikely that low-[He] material could pass through U and Th-rich batholiths.

The fact that "normal" MORB values occur at spreading ridges and fracture zones adjacent to purported "plumes" (Red Sea, Wolf Darwin Rise, Azores, Shona, Bouvet, Easter, St. Helena, Amsterdam, Iceland) indicates that high-R island gas is from shallow depth and has low ^{3}He contents. Otherwise it should overwhelm the ridge signature. Proposed long-distance transport of high-R material from Iceland down the Reykjanes Ridge is countered by the presence of MORB-like helium ratios in eastern Iceland and along the fracture zones to the north.

Samoa and Juan Fernandez are located along major tears in the lithosphere having to do with dramatic changes in dip and/or strike of a nearby plate boundary (Favela and Anderson, 1999). These are high-R islands, on fracture zones.

The mantle xenolith population may not be representative of unmetamorphosed and intact lithosphere. Although many mantle xenoliths have large He contents (up to 10^{-6} cc/gm), most have R/Ra in the "MORB-range" of 7 to 9. High ^{3}He/^{4}He and high gas contents occur in xenoliths from Samoa, Hawaii, Kerguelen, and high ^{3}He/^{4}He is inferred for some continental lithosphere. Metasomatized mantle (high U, Th) and mantle in fracture zones and sutures may be most easily fragmented and brought to the surface.

The highest ^{3}He/^{4}He ratios (> 15 Ra) in Yellowstone are on the edges of a caldera overlying a silicic magma chamber, highly enriched in U and Th and which has erupted vast quantities of rhyolite. If the source of helium is deep, then the crust and magma chamber must be essentially transparent to deep helium (Craig et al., 1978). Alternatively, the rapid local uplift and the rapid lateral variations in R (Kennedy et al., 1985) suggest a fracture conduit to a shallow LONU source.

The idea that there cannot be ancient trapped gases in the oceanic "lithosphere" assumes that all material in the oceanic "crust and lithosphere" formed in the last 200 m.y. Some of the plateaus and swells in the oceans, however, may be ancient, or have continental roots (Nur and Ben-Avrahem, 1982). Evidence for isotopic isolation on the order of 1 Ga is not necessarily evidence for a deep source since buoyant refractory peridotite may form the base of the Iceland, Kerguelen, and Hawaiian swells. Archean-age xenoliths have recently been found in Hawaiian basalts (Ducea and Sen, 1999). It begs the question (and immunizes the hypothesis against falsification) to claim that all such occurrences were emplaced by deep mantle plumes.

The Common Component

The isotopic composition of most OIB can be explained in terms of mixing between extreme endmembers—EM1, EM2, HIMU, and DM. The high-R basalts are not associated with any of these components but plot between these endmembers, closer to the depleted component (DM). This indicates that high R is not carried by pelagic sediments (EM1), continental sediments or lithosphere (EM2), or altered oceanic crust (HIMU). The refractory residual left after extraction of basalt (and Rb, Sr, U, Th, Sm, Nd) is depleted and is a potential reservoir that is essentially immaterial for most isotopes and LIL but may be important for osmium and helium. On multiple-isotope plots mid-ocean ridge and island data form trends that appear to focus on a common component, similar to the component that carries a high R signature (Hanan and Graham, 1996). This suggests that the source of this component is shallow so that it can interact with magmas ascending through it, or evolving in it. A buoyant restite region, such as lithosphere, seems to be implicated.

Propagating Cracks

The tectonic context of so-called hotspots and hotspot tracks is discussed by Favela and Anderson (1999). The fan-shaped rift pattern near the Galápagos is the result of continuing fragmentation of the Farallon plate caused by divergent pulls of the Central and South American subduction zones. The Cocos, Nazca, and Juan de Fuca plates are the remnants of the fragmenting Farallon plate and the ridges separating some of them started out as incipient volcanic cracks or leaky transforms that may have resembled current island-seamount chains. The Juan Fernandez chain marks the southern border of the flat portion of the subducting Nazca plate and the volcanic gap in Chile. The plate is being torn by this change in dip angle. The Samoan chain is caused by a tear in the Pacific plate where it subducts beneath the Tonga trench to the south but continues as a surface plate to the north.

All of these regions are characterized by some high–^{3}He/^{4}He samples interspersed with samples having much lower R. The average of basalts from these regions is 11.75 ± 5.13 Ra ($n = 57$). This is similar to the value for "new ridges." The Manus Basin also could be considered a new crack or reorganizing ridge system.

The Cook-Austral island-seamount chain satisfies few, if any, of the predictions of plume theory (Dickinson and Green, 1998) and is best thought of as stress- or fracture-controlled magmatism. McNutt et al. (1997) showed that several generations of volcanoes exist in this area, and volcanism started when this section of the plate was at a ridge. In contrast to Samoa and Juan Fernandez, the Cook-Austral chain may represent a reactivated old fracture. The Cook-Austral chain is characterized by relatively low R (Table 2).

Is there any independent support for the idea that the high $^3He/^4He$ ratios found at the onset of rifting in a variety of tectonic environments, and at propagating tears, are derived from the shallow mantle? First, the samples in all these environments are extremely low in [^3He]; they are not, as commonly expressed, "high-^3He magmas." Second, high–$^3He/^4He$ xenoliths are found in the Pacific lithosphere under Samoa (Poreda and Farley, 1992). Some of these xenoliths have one or two orders of magnitude more ^3He than high-R basalts. Finally, Ballentine (1997) found indirect evidence for a Loihi-type noble-gas component in continental lithosphere. The evidence in this paper is consistent with the presence of a low [He], LONU, high-R region in the shallow mantle that is preferentially sampled at new cracks and ridges and during times of plate-boundary reorganization or activation of fracture zones.

Lassiter and Hauri (1998) argue that Kea-endmember in Hawaiian basalts represents an ultramafic source, possibly recycled lower crust or lithosphere. The key evidence was provided by Os and O isotopes. Lithospheric and ultramafic rocks are very poor in U, Th, Rb, and Sr and their isotopic tracers. The other endmember, Koolau, appears to be recycled oceanic crust and is expected to be high in $^{238}U/^3He$. The volcanoes bearing these names have helium ratios of 6 to 16 Ra. The "high R" volcanoes are between these endmembers on heavy-isotope diagrams, but they exhibit secular trends, starting high and evolving to MORB-like R with time.

In the shallow LONU–low [He] model (Anderson, 1998a, 1998b), a new crack or a new volcano can temporarily be high-R but a single crack system, or conduit, can exhaust the helium supply or be overwhelmed by the deeper gas-rich MORB magmas which always seem to show up at some point. A new crack system, or conduit, is required to maintain high-R. Regions such as Iceland or Easter, where ridges are jumping or propagating, locally have high-R magmas and gases. The amounts of ^3He coming out of so-called "hotspot" volcanoes or contained in "hotspot" rocks are extraordinarily low compared to the global ridge system and mid-ocean ridge basalts.

Marty et al. (1998) assumed that the upper mantle cannot support magmatism for as long as 30 to 60 m.y. or provide high-R basalts episodically for this period of time. But the Greenland-Iceland basalts contain orders of magnitude less [^3He] than MORB and also orders of magnitude lower [He] than some lithospheric xenoliths (e.g., Poreda and Farley, 1992). If these xenolith [He] concentrations are typical of unaltered lithosphere, the total amount of helium represented in the Greenland and Iceland basalts can be provided from a mantle lithospheric layer less than the thickness of the basalts, even if He retention in the basalts is inefficient.

Evidence That LONU Is Not a "Primordial Undegassed Reservoir"

The total ^3He flux out of "hotspot" volcanoes is much less than at ridges or island arcs (Sano and Williams, 1996). The discrepancy is about three orders of magnitude from prediction (Porcelli and Wasserburg, 1995). The ^3He content of MORB is usually more than an order of magnitude higher than OIB basalts.

The He/Ne and He/Ar ratios of MORB are higher than OIB (Fig. 2), indicating that OIB did not lose their ^3He, prior to sampling, by gas exsolving from a magma. The $^3He/^{21}Ne$ and $^4He/^{22}Ne$ ratios of MORB and OIB are complementary in the sense that OIB gases are to MORB gases as vesicles are to degassed magma. This suggests that ancient degassing events may have provided the gas (trapped in LONU refractory restite) that later contributes to mid plate volcanoes (Anderson, 1998a, 1998b).

When there are correlated He-Sr and He-Pb trends in oceanic-island basalts, indicating mixing, the He/Sr, He/Pb, etc. ratios are generally greater for the MORB component, indicating that the so-called "plume" component is from a ^3He-poor source. The assumption that $^3He/^4He$ ratios of about 8 Ra are from the Degassed Upper Mantle (DUM) while ratios of greater than 11 Ra or so are from Undegassed, or Primordial Mantle, is based on the assumption that high ratios imply excess ^3He, and that high-R, low-[He] basalts have lost substantial amounts of ^{3}He. The basic assumption and the auxiliary assumptions do not seem to be true.

Degassing

Basalts with high ^{3}He/^{4}He ratios invariably have very low absolute helium concentrations, [He], including low [^{3}He]. Although quenched glasses having R of 20% to 200% higher than the "MORB mean" are attributed to an "undegassed reservoir," the actual [^{3}He] of such glasses is typically one to three orders of magnitude lower than in the gas-rich "popping-rock" MORB samples. Gas loss is the common explanation, although some of the highest R rocks are found on the flanks of Loihi at 5 km depth, and these also have low [^{3}He]. The solubility of the noble gases in magma decreases with atomic weight. A degassed magma will accordingly develop high He/Ne and He/Ar ratios, and the vesicles, and lost gas, will have low ratios.

Figure 2 shows that MORB glasses have high [He] and high He/Ar ratios. If "popping-rock" represents average undegassed MORB, then the MORB field can be explained by about an order of magnitude range in [He] in the parent magmas and one to four cycles of degassing. As MORB-like magmas ascend at mid-plate environments they interact with olivine or trapped vesicles from prior degassing events, and with air and seawater. Figure 2 is not consistent with OIB having degassed more than MORB. The OIB field is consistent with mixtures of atmospheric gas, lithospheric gas (xenoliths), and MORB. OIB may simply be degassed and contaminated MORB.

But We Know They Are Plumes!

Some might object to the consideration of alternate explanations for many of the regions being discussed since they are well known to be "hotspots" and require, on other grounds, a lower mantle source. However, the correlation of so-called "hotspots" with fracture zones, transform faults, suture zones, new ridges, abandoned ridges, ridges in general, and continental margins is so strong that alternative mechanisms of volcanism need to be considered (Anderson, 1998c, 1998d; Favela and Anderson, 1999). Bathymetry, gravity, and heat flow considerations only require a shallow cause and often dictate against a thermal anomaly (Anderson, 1998d). The "fixed-hotspot" and "parallel-hotspot-track" predictions have not held up (Favela and Anderson, 1999) and in fact, at best, only applied to a small fraction of the extant "hotspots". In many cases a noble gas "anomaly" has been the only evidence for a plume and, as I have shown, there is circular reasoning in this identification. The only geochemical argument for "primitive" or "undegassed" or "lower-mantle" status for some basalts is the assumption that high–^{3}He/^{4}He implies high-^{3}He, and that many basalts have R values "significantly higher than the MORB mean." If "significantly" means two standard deviations, then the present study requires that R exceed about 16.3 Ra, and most of the so-called "plume" or "lower-mantle" samples can be considered to be drawn from the upper mantle, if mid-ocean ridges tap the upper mantle.

Summary

The global spreading-ridge system yields R = 9.06 ± 3.26 Ra for the mean and standard deviation. Since mid-ocean ridges blend magmas from large volumes, the upper-mantle range for ^{3}He/^{4}He may be even larger. Most so-called high-^{3}He hotspots fall within the above range. "High-R" material is probably from the shallow mantle, since it is most evident at the onset of rifting. There is no evidence that high ^{3}He/^{4}He ratios imply excess ^{3}He.

REFERENCES

Anderson, D. L., 1989, Theory of the Earth: Boston, Blackwell Sci. Publ., 366 p.

______, 1993, ^{3}He from the mantle—Primordial signal or cosmic dust: Science, v. 261, no. 5118, p. 170–176.

______, 1994, Komatiites and picrites: Evidence for a depleted "plume"source: Earth Planet. Sci. Lett., v. 128, p. 303–311.

______, 1995, Lithosphere, asthenosphere, and perisphere: Rev. Geophys., v. 33, p. 125–149.

______, 1998a, The helium paradoxes: Proc. Nat. Acad. Sci., v. 95, p. 4822–4827.

______, 1998b, A model to explain the various paradoxes associated with mantle noble gas geochemistry: Proc. Nat. Acad. Sci., v. 95, p. 9087–9092.

______, 1998c, The edges of the mantle, *in* Gurnis, M., Wysession, M. E., Knittle, E., and Buffett, B. A., eds., The core-mantle boundary region: Washington, DC: Amer. Geophys. Union, p. 255–271.

______, 1998d, The scales of mantle convection: Tectonophys., v. 284, p. 1–17.

______, 1999, The statistics of helium isotopes along the global spreading ridge system: Geophys. Res. Lett., submitted.

Ballentine, C. J., 1997, Resolving the mantle He/Ne and crustal Ne-21/Ne-22 in well gases: Earth Planet. Sci. Lett., v. 152, p. 233–249.

Basu, A. R., Poreda, R. J., Renne, P. R., Teichmann, F., Vasiliev, Y. R., Sobolev, N. V., and Turrin, B. D., 1995, High-3He plume origin and temporal-spatial evolution of the Siberian flood basalts: Science, v. 269, no. 5225, p. 822–825.

Basu, A. R., Renne, P. R., DasGupta, D. K., Teichmann, F., and Poreda, R. J., 1993, Early and late alkali igneous pulses and a high-^{3}He plume origin for the Deccan flood basalts: Science, v. 261, p. 902–906.

Bernstein, S., Kelemen, P. B., Tegner, C., Kurz, M. D., Blusztajn, J., and Brooks, C. K., 1998, Post-breakup basaltic magmatism along the East Greenland Tertiary rifted margin: Earth Planet. Sci. Lett., v. 160, p. 845–862.

Botz, R., Winckler, G., Bayer, R., Schmitt, M., Schmidt, M., Garbe-Schonberg, D., Stoffers, P., and Kristjansson, J. K., 1999, Origin of trace gases in submarine hydrothermal vents of the Kolbeinsey Ridge, north Iceland: Earth Planet. Sci. Lett., v. 171, p. 83–93.

Condomines, M., Gronvold, K., Hooker, P. J., Muehlenbachs, K., O'Nions, R. K., Oskarsson, N., Oxburgh, E. R., 1983, Helium, oxygen, strontium, and neodymium isotopic relationships in Icelandic volcanics: Earth Planet. Sci. Lett., v. 66, p. 125–136.

Craig, H., Lupton, J. E., Welhan, J., and Poreda, R., 1978, Helium isotope ratios in Yellowstone and Lassen Park volcanic gases: Geophys. Res. Lett., v. 5, 897–900.

Dickinson, W. R., and Green, R. C., 1998, Geoarchaeological context of Holocene subsidence at the Ferry Berth Lapita Site, Mulifanua, Upolu, Samoa: Geoarchaelogy: Int. Jour., v. 13, p. 239–263.

Ducea, M., and Sen, G., 1999, Archean lithosphere recycled in plume basalts: EOS (Trans. Amer. Geophys. Union), Fall Meeting, v. 80, p. F1102.

Farley, K. A., Basu, A. R., and Craig, H., 1993, He, Sr, and Nd isotopic variations in lavas from the Juan Fernandez Archipelago, SE Pacific: Contrib. Mineral. Petrol., v. 115, p. 75–87.

Farley, K. A., and Neroda, E., 1998, Noble gases in the Earth's mantle: Ann. Rev. Earth Planet. Sci., v. 26, p. 189–218.

Favela, J., and Anderson, D. L., 1999, Extensional tectonics and global volcanism: Submitted.

Fisher, D. E., 1986, Rare-gas abundances in MORB: Geochim. et Cosmochim. Acta, v. 50, p. 2531–2541.

Graham, D. W., 1988, Coupled helium-lead isotope systematics in the Earth's mantle: Chem. Geol., v. 70, p. 50.

Graham, D. W., Christie, D. M., Harpp, K. S., and Lupton, J. E., 1993, Mantle plume helium in submarine basalts from the Galápagos platform: Science, v. 262, no. 5142, p. 2023–2026.

Graham, D. W., Jenkins, W. J., Schilling, J. G., Thompson, G., Kurz, M. D., and Humphris, S. E., 1992, Helium isotope geochemistry of mid-ocean ridge basalts from the South Atlantic: Earth Planet. Sci. Lett., v. 110, p. 133–148.

Graham, D. W., Larsen, L. M., Hanan, B. B., Storey, M., Pedersen, A. K., and Lupton, J. E., 1998, Helium isotope composition of the early Iceland mantle plume inferred from the Tertiary picrites of West Greenland: Earth Planet. Sci. Lett., v. 160, p. 241–255.

Graham, D. W., Johnson, K. T. M., Priebe, L. D., and Lupton, J. E., 1999, Hotspot-ridge interaction along the Southeast Indian Ridge near Amsterdam and St. Paul islands: Helium isotope evidence: Earth Planet. Sci. Lett., v. 167, p. 297–310.

Hanan, B. B., and Graham, D. W., 1996, Lead and helium isotope evidence from oceanic basalts for a common deep source of mantle plumes: Science, v. 272, p. 991–995.

Hilton, D. R., Hammerschmidt, K., Loock, G., and Friedrichsen, H., 1993, Helium and argon isotope systematics of the central Lau Basin and Valu Fa Ridge: Evidence of crust/mantle interactions in a back-arc basin: Geochim. et Cosmochim. Acta, v. 57, p. 2819–2841.

Hilton, D. R., Barling, J., and Wheller, G. E., 1995, Effect of shallow level contamination on the helium isotope systematics of ocean-island lavas: Nature, v. 373, p. 330–333.

Hirth, G., and Kohlstedt, D. L., 1996, Water in the oceanic upper mantle: Implications for rheology, melt extraction and the evolution of the lithosphere: Earth Planet. Sci. Lett., v. 144, p. 93–108.

Honda, M., and Patterson, D. B., 1999, Systematic elemental fractionation of mantle-derived helium, neon, and argon in mid-oceanic ridge glasses: Geochim. et Cosmochim. Acta, v. 63, p. 2863–2874.

Keller, W., Anderson, D. L., and Clayton, R. W., 1999, Iceland (in prep.).

Kennedy, B. M., Lynch, M. A., Reynolds, J. H., and Smith, S. P., 1985, Intensive sampling of noble gases in fluids at Yellowstone; I. Early overview of the data; regional patterns: Geochim. et Cosmochim. Acta, v. 49, p. 1251–1261.

Korenaga, J., 2000, Magmatism and dynamics of continental breakup in the presence of a mantle plume: Unpubl. Ph.D. thesis, Massachusetts Institute of Technology.

Kurz, M. D., Garcia, M. O., Frey, F. A., and O'Brien, P. A., 1987, Temporal helium isotopic variations within Hawaiian volcanoes—Basalts from Mauna Loa and Haleakala: Geochim. et Cosmochim. Acta, v. 51, p. 2905–2914.

Kurz, M. D., and Geist, D., 1999, Dynamics and evolution of the Galápagos hotspot from helium isotope geochemistry: Geochim. et Cosmochim. Acta (Allegre vol.), submitted.

Kurz, M. D., Kenna, T. C., Lassiter, J. C., and Depaolo, D. J., 1996, Helium isotopic evolution of Mauna Kea volcano: First results from the 1 km drill core: Jour. Geophys. Res.– Solid Earth, v. 101 (B5), p. 11,781–11,791.

Kurz, M. D., Le Roex, A. P., and Dick, H. J. B., 1998, Isotope geochemistry of the oceanic mantle near the Bouvet triple junction: Geochim. et Cosmochim. Acta, v. 62, p. 841–852.

Lassiter, J. C., and Hauri, E. H., 1998, Osmium-isotope variations in Hawaiian lavas: Evidence for recycled oceanic lithosphere in the Hawaiian plume: Earth Planet. Sci. Lett., v. 164, p. 483–496.

MacPherson, C. G., Hilton, D. R., Sinton, J. M., Poreda, R. J., and Craig, H., 1998, High $^3He/^4He$ ratios in the Manus back-arc basin: Geology, v. 26, p. 1007–1010.

Marty, B., Pik, R., and Gezahegn, Y., 1996, Helium isotopic variations in Ethiopian plume lavas: Nature of magmatic sources and limit on lower mantle contribution: Earth Planet. Sci. Lett., v. 144, p. 223–237.

Marty, B., Upton, B. G. J., and Ellam, R. M., 1998, Helium isotopes in early Tertiary basalts, northeast Greenland: Evidence for 58 Ma plume activity in the North Atlantic-Iceland volcanic province: Geology, v. 26, p. 407–410.

McNutt, M. K., Caress, D. W., Reynolds, J., Jordahl, K. A., and Duncan, R. A., 1997, Failure of plume theory to explain midplate volcanism in the southern Austral islands: Nature, v. 389, no. 6650, p. 479–482.

Moreira, M., Doucelance, R., Kurz, M. D., Dupré, B., and Allègre, C. J., 1999, Helium and lead isotope geochemistry of the Azores Archipelago: Earth Planet. Sci. Lett., v. 169, p. 189–205.

Moreira, M., Staudacher, T., Sarda, P., Schilling, J.-G., and Allègre, C. J., 1995, A primitive plume neon component in MORB: The Shona Ridge anomaly, South Atlantic (51–52°S): Earth Planet. Sci. Lett., v. 133, p. 367–377.

Moreira, M., Valbracht, P. J., Staudacher, T., and Allegre, C. J., 1996, Rare gas systematics in Red Sea ridge basalts: Geophys. Res. Lett., v. 23, p. 2453–2456.

Morgan, J. P., and Morgan, W. J., 1999, Two-stage melting and the geochemical evolution of the mantle: A recipe for mantle plum-pudding: Earth Planet. Sci. Lett., v. 170, p. 215–239.

Niedermann, S., Bach, W., and Erzinger, J., 1997, Noble gas evidence for a lower mantle component in MORBs from the southern East Pacific Rise: Decoupling of helium and neon isotope systematics: Geochim. et Cosmochim. Acta, v. 61, p. 2697–2715.

Nishio, Y., Ishii, T., Gamo, T., and Sano, Y., 1999, Volatile element isotopic systematics of the Rodrigues Triple Junction Indian Ocean MORB: Implications for mantle heterogeneity: Earth Planet. Sci. Lett., v. 170, p. 241–253.

Niu, Y. L., and Batiza, R., 1997, Trace element evidence from seamounts for recycled oceanic crust in the eastern Pacific mantle: Earth Planet. Sci. Lett., v. 148, p. 478–483.

Niu, Y. L., Collerson, K. D., Batiza, R., Wendt, J. I., and Regelous, M., 1999, Origin of enriched-type mid-ocean ridge basalt at ridges far from mantle plumes: The East Pacific Rise at 11 degrees 20′N: Jour. Geophys. Res.– Solid Earth, v. 104 (B4), p. 7067–7087.

Nur, A., and Ben-Avrahem, Z., 1982, Oceanic plateaus, the fragmentation of continents and mountain building: Jour. Geophys. Res., v. 87, p. 3644–3661.

Ozima, M., 1994, Noble gas state in the mantle: Rev. Geophys., v. 32, p. 405–426.

Porcelli, D., and Wasserburg, G. J., 1995, Mass transfer of helium, neon, argon, and xenon through a steady-state upper mantle: Geochim. et Cosmochim. Acta, v. 59, p. 4921–4937.

Poreda, R. J., and Farley, K. A., 1992, Rare gases in Samoan xenoliths: Earth Planet. Sci. Lett., v. 113, p. 129–144.

Poreda, R. J., Schilling, J.-G., and Craig, H., 1986, Helium and hydrogen isotopes in ocean ridge basalts north and south of Iceland: Earth Planet. Sci. Lett., v. 78, p. 1–17.

______, 1993, Helium isotope ratios in Easter microplate basalts: Earth Planet. Sci. Lett., v. 119, p. 319–329.

Richard, D., Marty, B., Chaussidon, M., and Arndt, N., 1996, Helium isotopic evidence for a lower mantle component in depleted Archean komatiite: Science, v. 273, no. 5271, p. 93–95.

Sano, Y., and Williams, S. N., 1996, Fluxes of mantle and subducted carbon along convergent plate boundaries: Geophys. Res. Lett., v. 23, p. 2749–2752.

Sarda, P., Moreira, M., and Staudacher, T., 1999, Argon-lead isotopic correlation in Mid-Atlantic Ridge basalts: Science, v. 283, no. 5402, p. 666–668.

Schilling, J.-G., 1986, Geochemical and isotopic variation along the Mid-Atlantic Ridge axis from 79°N to 0°N., *in* Vogt, P. R., and Tucholke, B. E., eds., The Western North Atlantic Region: Washington, DC: Geol. Soc. Amer., The geology of North America, v. M, p. 137–156.

Schilling, J.-G., Kingsley, R., Fontignie, D., Poreda, R., and Xue, S., 1999, Dispersion of the Jan Mayen and Iceland mantle plumes in the Arctic: Jour. Geophys. Res., v. 104, p. 10,543–10,569.

Zindler, A., and Hart, S., 1986, Helium: Problematic primordial signals: Earth Planet. Sci. Lett., v. 79, p. 1–8.

Production of CO_2 and H_2 by Diking-Eruptive Events at Mid-Ocean Ridges: Implications for Abiotic Organic Synthesis and Global Geochemical Cycling

JOHN R. HOLLOWAY[1] AND PEGGY A. O'DAY

Department of Geology, Arizona State University, Tempe, Arizona 85287-1404

Abstract

We have calculated the amounts of CO_2 and H_2 produced by complete degassing of mid-ocean ridge basalt (MORB) magma, and by degassing during transient diking-eruptive events. Our CO_2 calculations are based on a model estimate of an initial CO_2 content of 1800 ppm in MORB magma, which is equivalent to 2.2×10^{12} mol CO_2 per year for magma production at worldwide ocean ridges. Observations indicate that many MORB magmas are emplaced in numerous small pulses of dikes and associated lava flows with very short emplacement times, which would result in release of relatively large amounts of CO_2 over short intervals. For example, a dike injected into the oceanic crust that extends from the top of its magma chamber at 2 km depth to the seafloor would degas 2.3×10^4 mol CO_2 per m^2 surface area of dike, and produce another 4.0×10^4 mol CO_2 per m^2 on complete crystallization.

Unlike CO_2, which is not strictly governed by crystallization-alteration processes, H_2 is produced from MORB by the reduction of H_2O by ferrous iron in the magma to form magnetite and H_2 as the magma cools and crystallizes. From published paired analyses of MORB glass and crystalline rock, we estimate that the amount of H_2 produced from one cubic meter of rock averages 301 mol. We suggest that the oxidizing agent is H_2O dissolved in the magma, which results in rapid generation of H_2. The amount of pre-alteration oxidation may be limited by the amount of H_2O dissolved in the magma; thus relatively water-rich magmas will undergo greater oxidation. For the case of the two-kilometer-high dike reaching the seafloor, the amount of H_2 released is 6.2×10^5 moles H_2 per m^2 surface area of the dike. This is 10 times greater than the total CO_2 released by degassing and crystallization of the dike. Assuming that the H_2 generation rate for the entire basaltic layer of the oceanic crust is the same as for MORB lavas (312 mol/m^3), then the annual global H_2 production rate is 6.3×10^{12} mol H_2 per year. This amount is about three times greater than our calculated annual CO_2 production from MORBs. Given that the annual CO_2 production rate from MORBs over 3.3 Ga can account for all CO_2 found in the Earth's crust, hydrosphere, and atmosphere, it is likely that the H_2 produced at mid-ocean ridges plays a significant role as a reducing agent in the global redox state of the Earth's surface.

In contrast to time-averaged global production rates, the rapid release of CO_2 and H_2 in diking-eruptive events may locally result in formation of a separate gas phase containing H_2-CO_2-H_2O in that order of abundance. The amounts of CO_2 and H_2 produced could provide a significant energy source for autotrophic microorganisms. It has been demonstrated that such a CO_2-H_2-H_2O gas mixture yields methanol in magnetite-surface catalyzed reactions at seafloor hydrothermal conditions. Such abiotic synthesis reactions could have been important in early Earth processes.

Introduction

CARBON DIOXIDE and hydrogen produced at mid-ocean ridges (MORs) are of primary importance in geochemical and biochemical processes on the past and present Earth. The emplacement of basaltic magma at mid-ocean ridges in diking-eruptive events provides a punctuated supply of CO_2 and H_2 over short time scales that is superimposed on a continuous background supply. The CO_2 and H_2 produced will impact the Earth at large and small scales in the following ways: (1) the annual production rate of CO_2 from MOR magmatism may be a significant contributor to the global geochemical cycle for carbon; (2) CO_2 and H_2 are primary ingredients for abiotic synthesis of organic molecules in present-day seafloor hydrothermal environments

[1] Also at Department of Chemistry and Biochemistry, Arizona State University, Tempe, AZ 85287-1404.

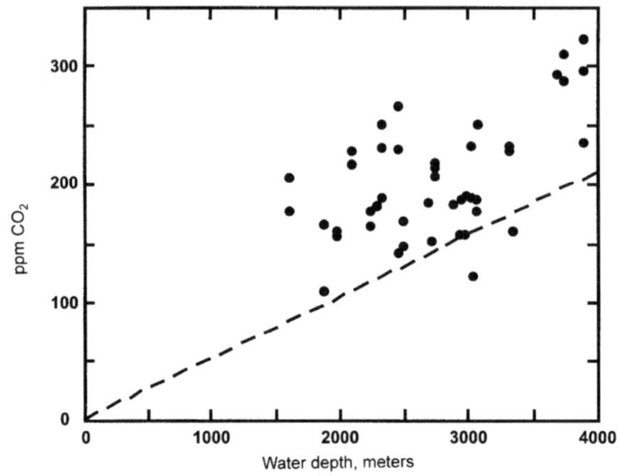

FIG. 1. Comparison of the CO_2 content in glassy MORB samples from the Juan de Fuca Ridge with the seawater depth at the sample site (from Dixon et al., 1988). The dashed line is the solubility of CO_2 with depth from Pan et al. (1991).

and possibly on the early Earth; and (3) diking-eruptive events could provide a significant nutrient supply for methanogenic microorganisms.

In this paper, we provide a summary of the field, analytical, and experimental observations that allow calculation of the average worldwide production of CO_2 and H_2 from MORB emplacement on the present-day Earth. We then discuss local temporal and spatial concentration of production resulting from diking-eruptive events at MORs.

Production of CO_2

Observations of CO_2 and H_2O in MORB magmas

Carbon dioxide in MORB magmas is present as dissolved carbonate (e.g., Stolper, 1982; Fine and Stolper, 1986; Dixon et. al, 1988) and as exsolved CO_2 within bubbles trapped in the basaltic glass (e.g., Moore et al., 1977). The presence of CO_2-rich bubbles and the observation that the carbonate concentrations determined in many MORB glasses are significantly greater than the saturation values for equilibrium degassing requires that the magmas were oversaturated in CO_2 *and were degassing* prior to eruption (Dixon et al., 1988; Pineau and Javoy, 1994). Nevertheless, the rough correlation between CO_2 and seawater depth (Fig. 1) implies that degassing commonly results in CO_2 concentrations in quenched glass that are close to equilibrium values.

In contrast, H_2O is much more soluble than CO_2 in MORB magma (e.g., Hamilton, 1964; Dixon et al., 1995) and, for the concentrations found in MORB magmas, degasses at lower pressure (Dixon and Stolper, 1995). The strong positive correlation between H_2O and K_2O shown in Figure 2 (Dixon et al., 1988; Michael, 1988; Jambon and Zimmermann, 1990) also argues strongly that little H_2O has degassed. The evidence for incomplete but substantial degassing of CO_2, and the near absence of H_2O degassing, implies that carbon (as CO_2) and hydrogen (as H_2O) are fractionated during degassing because of their differing solubilities.

CO_2 content of primary MORB magma

To estimate the amount of CO_2 released by magma degassing, the CO_2 content of the source magma must be determined. Also the intrusive stages that the magma experiences before emplacement in a diking-eruptive event or crystallization in a shallow magma chamber must be known. Gerlach (1989) considered the release of CO_2 from MORB magma in detail. The following treatment is similar to his approach; however, we use a different value for the primary CO_2 content of MORB. In Gerlach's (1989) model, he proposed initial concentrations of CO_2 in the source magma based on values measured in an alkaline basaltic magma erupting at the Earth's surface that are likely not applicable to MORB magma.

Other workers have tried to predict initial magmatic CO_2 concentrations based on experimental petrologic constraints. Mathez (1984) proposed that the mantle source region for MORB magmas contained graphite and that the magmas could be either fluid-saturated or undersaturated. The availability of more comprehensive data for the solubility of CO_2

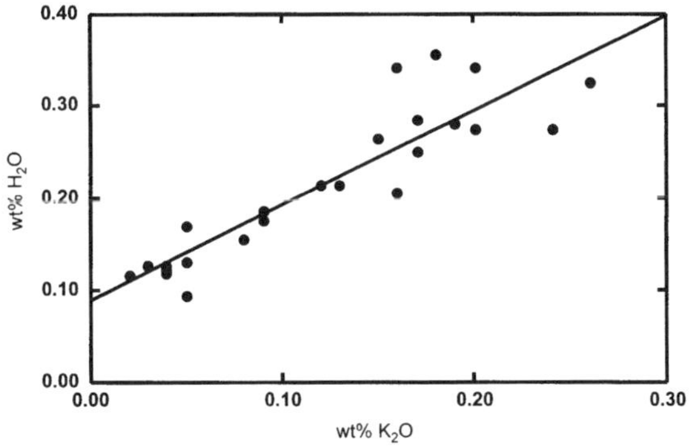

FIG. 2. Analyses of glassy MORB samples from the Juan de Fuca Ridge showing the variation of H_2O with K_2O, from Dixon et al. (1988). The line is a linear regression with a correlation coefficient of 0.90.

in basaltic magma allowed Holloway (1998) to recognize that fluid saturation at mantle pressures would result in unreasonably high initial values for dissolved CO_2. Consequently, he developed a quantitative model for fluid-absent partial melting of a graphite-saturated mantle. In this study (Holloway, 1998), the ascent trajectory of the magma was modified slightly from the adiabatic paths proposed by Langmuir et al. (1992) and Turcotte and Morgan (1992) to better match the solidus temperatures found by Baker and Stolper (1994). The MELTS experiment (MeltTeam, 1998) has since roughly confirmed that melting begins at depths in the mantle similar to those proposed in the Turcotte and Morgan model (about 75 km), and that the melt fraction remains small throughout the melting path.

A schematic view of Holloway's model is shown in Figure 3. The predicted CO_2 content of the MORB magma ranges from 900 to 1800 ppm, depending on the model parameters used (given in the caption to Fig. 3). The 1800 ppm value was considered most probable because it is based on the best estimates of source-region mineral compositions (Holloway, 1998). The 1800 ppm value is based on generation of MORB magma from 15% cumulative mantle melting. However, significantly lower degrees of melting, such as might be the case in the generation of melts associated with "popping rocks" (e.g., Javoy and Pineau, 1991; Pineau and Javoy, 1994), would result in higher CO_2 contents. For example, 5% partial melting would result in melts with ~4000 ppm CO_2. It is important to note that this model-dependent value determines the estimate of global CO_2 production. For example, if a MORB magma with an initial CO_2 content of 1800 ppm completely degassed, it would liberate 106 mol CO_2 per m^3 of magma.

CO_2 solubility in MORB

Several experimental investigations of the solubility of CO_2 in tholeiitic basaltic magmas have been made in recent years (Stolper and Holloway, 1988; Pan et al., 1991; Pawley et al., 1992; Dixon et al., 1995; Jendrzejewski et al., 1997). Such studies show that solubility is very close to a linear function of pressure up to 100 MPa. Those experimental studies yield a range in solubility coefficients from 5.0 to 5.5 ppm CO_2/MPa of CO_2 pressure. We chose a value of 5.3 ppm/MPa. Solubility in ppm can be converted to mol CO_2 per 1 m^3 of basalt by noting that 1 mol • m^{-3} = 16.5 parts per million by weight (ppmw) for a magma with a density of 2700 kg • m^{-3}. Solubility will be calculated in mol per 1 m^3 in the remainder of this paper.

At pressures below 100 MPa, the solubility of CO_2 is expressed as:

$$S_{CO_2} = 0.331 * P, \quad (1)$$

where S_{CO_2} is in mol • m^{-3} magma, and P is in MPa. At P > 100 MPa, CO_2 solubility is not linear in P, but equation (2) fits the Pan et al. (1991) thermodynamic equation to 2% relative error between 100 and 500 MPa:

$$S_{CO_2} = 0.2311 + 0.2983 * P + 2.013 \times 10^{-4} * P^2. \quad (2)$$

Release of CO_2 from MORB magma

Using a CO_2 concentration of 1800 ppm results in saturation of the rising magma at about 3000 bars,

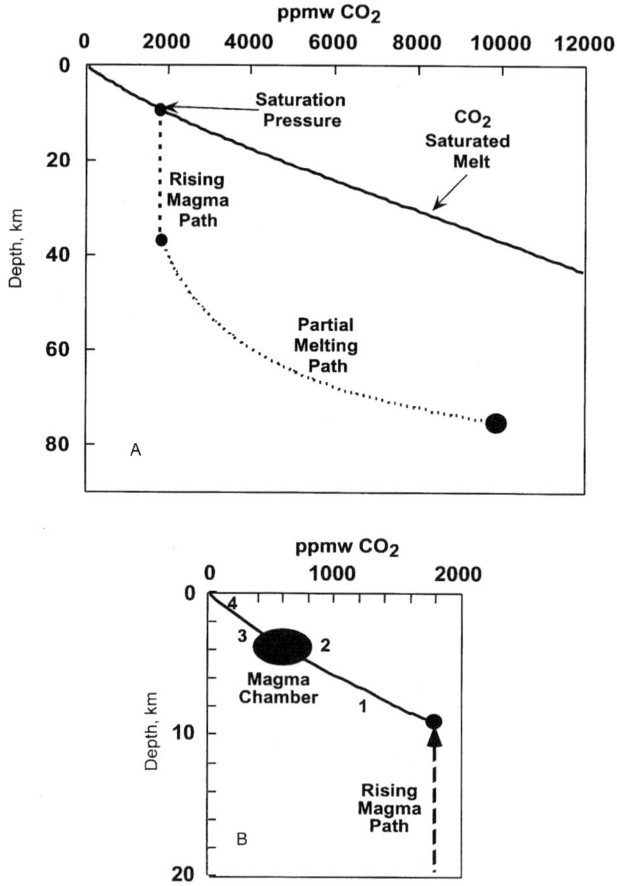

FIG. 3. The ascent trajectory used in Holloway (1998) was modified slightly from the adiabatic paths proposed by Langmuir et al. (1992) and Turcotte and Morgan (1992) to better match the solidus temperatures found by Baker and Stolper (1994). The MELTS experiment (MeltTeam, 1998) has since roughly confirmed that melting does begin at similar depths in the mantle and that the melt fraction remains small throughout the melting path. A. Path of rising magma: short-dashed line is along a partial melting path, long-dashed line is the path the magma takes after separating from its source. CO_2 saturation occurs at a pressure of 300 MPa. B. Enlargement of the low-pressure portion of Figure 3A. Numbers 1 through 4 indicate locations of steps discussed in the text.

equivalent to a depth of ~10 km (Fig. 3B). That depth is greater than is commonly assumed for axial magma chambers (≤ 6 km) (Sinton and Detrick, 1992) and there will be an interval over which the magma is degassing before reaching a typical magma chamber depth. Gerlach (1989) considered four possible modes of CO_2 degassing (Gerlach, 1989): (1) ascent of the source magma from the saturation depth to its emplacement in a sub-ridge magma chamber; (2) crystallization in the magma chamber; (3) magma injection from the chamber into the oceanic crust to form dikes, and possibly erupted as flows; and (4) crystallization of the dikes and flows. In the following discussion, we consider differences in CO_2 production from these four modes.

Degassing during ascent (Steps 1 and 3)

As a MORB magma ascends from it its source region, it remains fluid-absent until the saturation pressure is reached (Fig. 3). Further ascent will result in the exsolution of a fluid phase composed predominantly of CO_2. The rise of the magma can be separated into two idealized stages: rise in the conduit from the source region to a sub-ridge magma chamber; and rise from the magma chamber toward the seafloor in a dike, possibly after a significant residence time in the magma chamber. We recognize that the actual process is probably much more complex, but in terms of overall degassing, the amount of CO_2 degassing is the same, although the rate will depend on details of the process.

For an existing magma column of height h (in meters), the *average* pressure decrease for a packet of magma rising in the dike is given by Equation (3):

$$\Delta P = 0.030 \cdot h, \qquad (3)$$

where ΔP is in MPa and the value of 0.030 arises from the equivalence that a height of 33.3 meters of basalt generates one MPa of pressure on the Earth.

The amount of CO_2 degassed in the ascent of a packet of magma with a 1 m^2 cross section and height h is given by:

$$Mdg = \frac{(S_{P2} - S_{P1})}{2} \cdot h, \qquad (4)$$

where Mdg is moles CO_2 degassed for a 1 m^2 dike cross section, S_{P2} is the solubility at the pressure on the magma (e.g., at the initial saturation depth [step 1], or the depth to the roof of the magma chamber [step 3]), and S_{P1} is the solubility at the final pressure (e.g., the magma chamber depth [step 1], or the top of the dike [step 3], for dikes reaching the seafloor), and h is the height of the magma column in meters.

A magma initially containing 1800 ppm CO_2 rising from its source region will begin degassing at a depth of ~10 km (P_2 = 300 MPa + 25 MPa for sea water depth) and accumulate in a magma chamber with a mean depth of 4 km (Sinton and Detrick, 1992; Nicolas et al., 1998) (P_1 = 120 MPa + 25 MPa). The height h is 6000 m. In this case the values of S_{P2} and S_{P1} are 118.4 and 47.7 mol • m^{-3} magma, respectively. Substituting these values into equation (4), we find 2.1×10^5 moles CO_2 degassed for each 1 m^2 of magma body cross section. A dike extending from a magma chamber at 2 km depth would degas 2.0×10^4 mol CO_2 per m^2 area.

These processes do not result in complete degassing of the magma column because the entire column remains under some pressure. The magma column has an equilibrium CO_2 concentration profile defined by P_1 at its upper end and the magma-column-generated pressure at its lower end, P_2. The following section describes the fate of the CO_2 remaining in the magma.

Degassing during crystallization (Steps 2 and 4)

In addition to degassing of CO_2 during ascent, CO_2 will be displaced from the magma by formation of minerals crystallizing during cooling. None of the minerals contain carbon, so the CO_2 concentration in the remaining melt increases. For a particular body of magma, the amount of crystallization can vary widely, from almost none if the magma is ejected immediately from the chamber into dikes, to complete if the magma is retained in the chamber and cools below solidus temperatures. Given an equilibrium, pre-crystallization vertical CO_2 concentration profile, complete crystallization (with no CO_2 trapped in fluid inclusions) will release an amount given by:

$$Mc = \left\{\frac{(S_{P2} - S_{P1})}{2} + S_{P1}\right\} \cdot h, \qquad (5)$$

where Mc is moles CO_2 released • m^{-2} of magma cross section, S_{P2} and S_{P1} are solubilities of CO_2 for pressures at the bottom and top of the magma chamber or dike, and h is the thickness of the magma chamber or height of the dike. The calculated results for three magma column heights are given in Table 1 and shown graphically in Figure 4.

Efficiency of CO_2 production

The production of CO_2 from equilibrium degassing and crystallization and the total of both processes are shown in Figure 4 as a function of magma column height, measured from sea level downward. The amount of CO_2 produced by these two processes is roughly equal, and the total amount released is potentially large. An important question involves how much is actually released compared to the calculated equilibrium amount. The answer depends on how the magma is emplaced, either as a dike injected into the oceanic crust, or in a lava flow erupted on the seafloor.

In the case of the degassing process, we know from measurements of the CO_2 content of glassy rims of basalt lava flows that degassing in many cases is not complete (Dixon et al., 1988). Using solubility data, the extent of supersaturation is equivalent to a maximum of 700 m of excess magma pressure (Fig. 1). However, any CO_2 left from incomplete degassing could be released during the crystal-

TABLE 1. Production of CO_2 and H_2 from MORB Dikes as a Function of Dike Height

Dike height, m	P_2, MPa[1]	CO_2 solubility at P_2, mol • m^{-3}	Degassing-produced CO_2[2]	Crystallization-produced CO_2[2]	Crystallization-produced H_2[3]
2000	92	31.2	$2.3 \cdot 10^4$	$4.0 \cdot 10^4$	$0.6 \cdot 10^6$
4000	158	53.8	$9.1 \cdot 10^4$	$1.3 \cdot 10^4$	$1.2 \cdot 10^6$
6000	225	76.5	$2.0 \cdot 10^5$	$2.6 \cdot 10^5$	$1.8 \cdot 10^6$

[1]Seawater depth taken as 2500 meters; seawater pressure (P_1) is 25 MPa; CO_2 solubility at P_1 is 8.5 mol • m^{-3}. See equations 1, 3, 4, and 5 in text for calculations.
[2] In mol CO_2 • m^{-2} dike surface.
[3] In mol H_2 • m^{-2} dike surface.

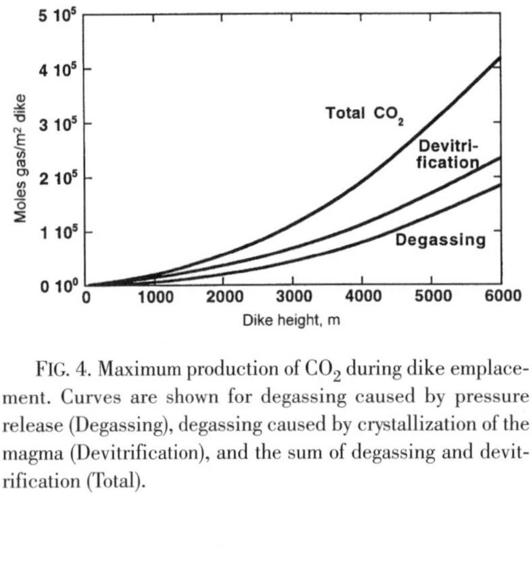

FIG. 4. Maximum production of CO_2 during dike emplacement. Curves are shown for degassing caused by pressure release (Degassing), degassing caused by crystallization of the magma (Devitrification), and the sum of degassing and devitrification (Total).

lization process. Two processes might operate to prevent the escape of the released CO_2 from dikes: entrapment of CO_2 as fluid inclusions, and formation of carbonate minerals. At seafloor pressures, carbonate minerals are only stable in basaltic compositions at temperatures below ~400°C (Slaughter et al., 1975). Carbonate minerals found in altered oceanic crust have been shown to form by precipitation from seawater, not by reaction with MORBs (Staudigel et al., 1981). The amount of CO_2 trapped in fluid inclusions in minerals is small and residual from extensive degassing (Kelley, 1996; Kelley and Fruh-Green, 1999). Taken together, these observations suggest that nearly all CO_2 transported by MORBs from the mantle to mid-ocean ridge environments is lost during degassing and crystallization of the magma.

Production of H_2

Evidence from Mid-Ocean Ridge basalt samples

Significant quantities of unaltered basalt have been collected from MORs over the last 20 years. Most samples are crystalline with glassy rims. The rims form by very rapid chilling of the magma when it contacts cold seawater, quenching the melt to a glass. The rapid cooling traps volatiles as the glass is formed, as seen in the excellent correlation between H_2O and K_2O in glassy rims shown in Figure 2. The glassy rims are thus pristine samples of the melt immediately prior to crystallization. Conversely, the more slowly cooled crystalline interiors of the samples allow determination of the effects of crystallization.

The quench crystallization of MORB magma has been shown to be an isochemical process except for loss of volatiles and a change in the Fe_2O_3:FeO ratio. Christie et al. (1986) showed by measuring Fe^{3+}/Fe^{2+} in glassy rims and their crystalline interiors of MORBs that a significant amount of oxidation of ferrous to ferric iron occurs during crystallization (Fig. 5). In the two paired samples they analyzed, the amount of ferric iron increases by 82 and 132 relative percent on crystallization.

The morphology of crystals in MORB pillows has been described in many studies (e.g., Bryan, 1972). We have carried out a petrographic examination of a sample from the Juan de Fuca Ridge. Polished thin sections were analyzed using optical microscopy, electron microprobe analysis, and scanning electron microscopy. A representative image is shown in Figure 6. A high extent of crystallization is indicated by examination of the interior of the MORB pillow, which shows the presence of magnetite grains in

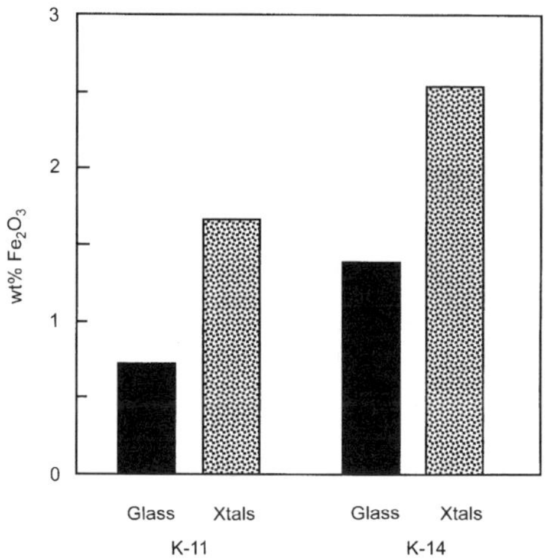

FIG. 5. Fe_2O_3 contents of two pairs of glass rims and their crystalline interiors of MORB samples (Christie et al., 1986). Crystallization results in a significant oxidation during conversion of glass to crystals.

addition to silicate and sulfide phases. The texture (indicated in Fig. 6) is characteristic of rapid ("quench") crystallization seen in high-temperature laboratory experiments (Wyllie, 1963), and is consistent with large amounts (e.g., 200°C) of undercooling (Lofgren, 1974). Lofgren's results were obtained over cooling times of three to four days, which provides an estimate of the upper time required to develop the quench texture.

The increase in the ferric/ferrous ratio and the presence of magnetite in the crystalline sample suggest a redox reaction that converts some of the ferrous iron in the MORB melt to magnetite in the crystalline rock. This process requires an oxidizing agent, and the rapid crystallization in the interior of the rock suggests that the agent was not supplied by seawater, but was present in the magma. The two most likely possibilities are CO_2 and H_2O. On a molar basis, there is not nearly enough CO_2 in the magma to account for the amount of oxidation (Fig. 1). On the other hand, there is, on average, more than enough H_2O to account for the observed increase in ferric iron (Fig. 2). Christie et al. (1986) also suggested that H_2O is the oxidizing agent. The stoichiometry of the reaction with H_2O is:

$$3\ FeO\ (magma) + H_2O\ (magma) = FeO \bullet Fe_2O_3\ (magnetite) + H_2\ (fluid). \quad (4)$$

The quench textures and fine grain size of the crystals indicate that equilibrium is not maintained during crystallization. We propose that the presence of all reactants in what was initially a single phase (melt) should allow for rapid reaction, and consequently, for rapid production of H_2.

The only nonvolatile, abundant constitutent of basalt that exhibits more than one oxidation state is iron, and so it is an important gauge of the redox state of a magma. The ferric/ferrous ratio in glassy samples of igneous rocks is a direct indicator of the redox state of magmas at high temperature (Carmichael and Ghiorso, 1986). This ratio has been used to show that the redox state of MORB magmas is significantly more reducing than that of many other igneous rocks (Christie et al., 1986; Carmichael, 1991).

Magnitude of H_2 release

The amount of H_2 produced by the reduction of H_2O by ferrous iron is easily calculated for the two samples of glassy rim and crystalline rock given in Christie et al. (1986). For the two samples, the amount of H_2 produced for 1 m^3 of rock is 246 and 355 moles. In the following discussions, an average value, 301 mol • m^{-3}, will be used. In our model, the amount of H_2 production depends only on the concentration of dissolved H_2O in the magma, and the quantity of magma crystallizing. These quantities are independent of the pressure, or the form of the magma body; thus production by crystallization of flows and dikes is the same. The amounts produced for different dike heights are shown in Table 1. The rate of H_2 production depends only on the crystallization rate, which will be different for dikes and flows. Additional H_2 production may occur if seawater reacts with ferrous iron in the rock, a potentially much slower process. Thus, an important distinction between CO_2 and H_2 is that the former is released by both degassing during ascent and by crystallization, whereas the latter is released only by crystallization and not during descent.

CO_2 and H_2 Production in Diking-Eruptive Events

A diking-eruptive event has been characterized by Delaney et al. (1998) as the quantum unit of oceanic crust formation. These events have magma volumes of a few million m^3 and are erupted in periods of hours to days (data summarized in Delaney et al., 1998). The production of H_2 and CO_2 by degassing

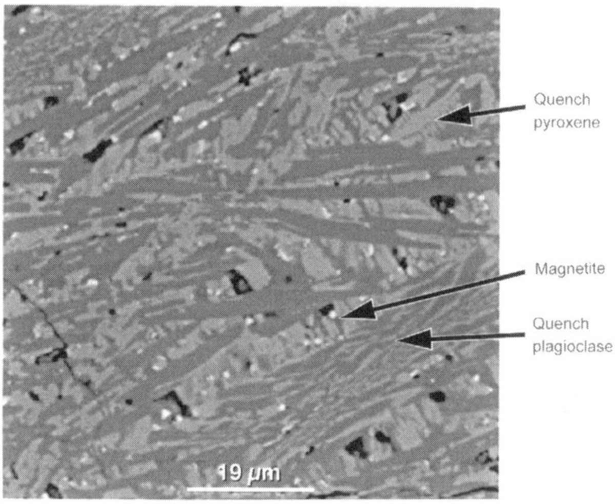

FIG. 6. Backscattered scanning electron microscope image of a portion of a polished thin-section of the crystalline interior of a MORB sample from the Juan de Fuca Ridge (TT-152 locality). Note that most of the bright grains in the image are magnetite. Also of note is the presence of quench textures in the pyroxene and plagioclase grains.

MOR basalts could provide reactants for the abiotic synthesis of organic compounds (Shock, 1996). We have demonstrated experimentally the abiotic synthesis of methanol from H_2-CO_2 gas–rich mixtures (Voglesonger et. al., in review), with yields of ~0.05 % methanol per hour. At this conversion rate, ~15 mol CH_3OH would be generated *every minute* for every square meter of cross-sectional area of a 2 km high dike.

The extent to which synthesis of organic compounds actually occurs depends in part on the local concentration of the reactants. Their concentrations depend on the rate of H_2 and CO_2 production, which depends on the rate of cooling and crystallization of the magma erupted in flows and injected in dikes. As noted above, the crystallization rate in MORB lava flows is rapid (on the order of days), and so the release of H_2 and CO_2 should be equally rapid. A basalt flow on the East Pacific Rise at 9° N emplaced $4 - 6 \times 10^6$ m^3 of lava in one to two hours (Gregg et al., 1996). The ascent and crystallization of that lava flow would produce ~10^9 mol H_2 (based on 312 mol • m^{-3}, from above) and at least 6×10^7 mol CO_2 (based only on the CO_2 contained in the lava after extrusion and before crystallization; Dixon et al., 1988).

A well-studied eruption is the 1993 emplacement event on the CoAxial segment of the Juan de Fuca Ridge. A flow with a volume of 5×10^6 m^3 was associated with a 40 km long dike that propagated over a two-day period (Baker et al., 1995). Cherkaoui et al. (1997) used data from studies of the water column above the dike and concluded that the dike lost most of its heat within three months of intrusion into permeable rock that was saturated with seawater. This implies that three months is the upper limit for the time period over which degassing would be complete in diking events. The dike had a maximum thickness of 3 to 5 m. The height of the dike is not well constrained, but Cherkaoui et al. (1997) considered heights of between 500 and 2000 meters. The quantities of H_2 and CO_2 potentially produced over this range of heights can be read from Figure 7.

Kelley et al. (1998) collected water-column samples tens of days after eruption of a lava flow on the Gorda Ridge. They found highly elevated H_2 and moderately enriched CH_4 concentrations in the event plume. The concentrations of these species rapidly declined over a one-month period. The very high H_2 levels associated with the eruption event are consistent with rapid generation of H_2 during emplacement and cooling of MORB lava flows.

Global CO_2 and H_2 Production from MORB Magmatism

The production of H_2 and CO_2 by degassing MORB magmas during emplacement into axial magma chambers and during emplacement and

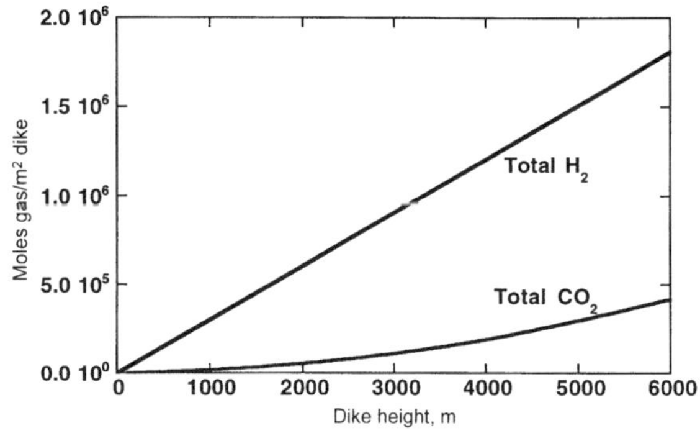

FIG. 7. Comparison of H_2 production with total CO_2 production as a function of dike height. Note that H_2 production per m^3 of magma is not dependent on pressure (which is proportional to dike height), but that CO_2 production does depend on pressure.

TABLE 2. Comparison of the Annual Production of H_2 and CO_2 at MORs

Location in MOR	Moles CO_2[1]	Moles H_2
Dikes and flows	$3.3 \bullet 10^{11}$	$9.0 \bullet 10^{12}$
All basaltic crust	$2.3 \bullet 10^{12}$	$6.3 \bullet 10^{13}$

[1]CO_2 is calculated based on complete degassing of magma with an initial content of 1800 ppmw (Holloway, 1998) and magma volumes from Crisp (1984).

crystallization in diking-eruptive events has a number of implications for local and global processes. The annual amount of magma produced by diking-eruptive events (as distinct from total MORB magma production) is estimated by Crisp (1984) to be 3 km^3. Given our estimates based on complete magma degassing, this volume could potentially produce 9.0×10^{12} mol H_2 and 3.3×10^{11} mol CO_2 at MORs annually.

Estimates of primary organic carbon production in low-nutrient open ocean regions are in the range of 0.5 to 1 g C m^{-2} d^{-1} for the subtropical North Pacific Ocean (Karl et al., 1998). If we consider a ridge area of 1 km width by 40,000 km of ridge length, gross primary organic carbon production at the surface is 0.6 to 1.2×10^{12} mol C a^{-1} using Karl et al.'s rate. Thus, our estimate of organic carbon production from magma emplacement based on 1% conversion of CO_2 is on the order of 2 to 3% of the surface organic production rate. If more CO_2 is converted to organic carbon in hydrothermal systems, then the portion of organic carbon produced by magmatic processes could be substantially higher.

On time scales of magma-chamber crystallization, most CO_2 may be released from coarse-grained gabbros that form as the magma chamber crystallizes at MORs. Kelley and Fruh-Green (1999) used carbon isotopes in fluid inclusions in MOR gabbros to show that most of the CO_2 present in the magma had escaped during crystallization. If that occurs, then the total production from MOR magma (21 $km^3 \bullet a^{-1}$; Crisp, 1984) could result in a time-averaged release rate of up to 2.3×10^{12} mol $CO_2 \bullet a^{-1}$, based on our estimate of 1800 ppm CO_2 dissolved in primary MORB magma (Table 2). Our estimates of total CO_2 production rate are similar to the recent estimate of $2.2 \pm 0.9 \times 10^{12}$ mol $CO_2 \bullet a^{-1}$ based on $^3He/CO_2$ ratios (Marty and Tolstikhin, 1998; Marty and Zimmermann, 1999). Although this agreement is exceptional, it does not imply agreement in estimated mantle carbon concentrations between those studies and Holloway's 1998 model, in which the mantle was assumed to be graphite/diamond saturated. The assumed graphite saturation requires a minimum mantle carbon concentration of ~80 ppm (Holloway, 1998).

The global annual production of CO_2 can be compared with the annual rate for hydrogen production if we assume that the H_2 generation for the entire basaltic layer of the oceanic crust is the same as it is for MORB lavas (312 mol/m^3). If that is the case, then the maximum annual global H_2 production rate at MORs would be 6.3×10^{12} mol $H_2 \cdot a^{-1}$, which is about three times greater than that of CO_2. These estimates suggest that CO_2 and H_2 production at MORs is a significant part of the long-term global cycling of carbon and hydrogen, and may have been important during the early history of the Earth. Even if only a small fraction of carbon is converted to organic forms, MOR hydrothermal systems could have served as a potential source of abiotic organic carbon throughout geologic time.

Acknowledgments

This work was supported by NASA Exobiology grant NAG5-4892, NSF grant EAR-9800147, and NASA Astrobiology grant NCC2-1051. Jim Clark helped with the electron microprobe analyses. Reviews by Jackie E. Dixon, Debbie S. Kelley, Darren Locke, and Ken Voglesonger helped greatly to improve this manuscript. John Holloway's life in science was influenced greatly by Konnie Krauskopf through the first edition of *Introduction to Geochemistry*. It was with excitement and trepidation that he taught Konnie's Geochemistry course a dozen years later while visiting Stanford. Konnie's kindness then and later is a great example for all who strive to teach and do research.

REFERENCES

Baker, E. T., Massoth, G. J., Feely, R. A., Embley, R. W., Thomson, R. E., and Burd, B. J., 1995, Hydrothermal event plumes from the CoAxial seafloor eruption site, Juan de Fuca Ridge: Geophys. Res. Lett., v. 22, no. 2, p. 147–150.

Baker, M. B., and Stolper, E. S., 1994, Determining the composition of high-pressure mantle melts using diamond aggregates: Geochim. et Cosmochim. Acta, v. 58, p. 2811–2827.

Bryan, W. B., 1972, Morphology of quench crystals in submarine basalts: Jour. Geophys. Res., v. 77, p. 5812–5819.

Carmichael, I. S. E., 1991, The redox state of basic and silicic magmas: Contrib. Mineral. Petrol., v. 106, p. 129–141.

Carmichael, I. S. E., and Ghiorso, M. S., 1986, Oxidation-reduction relations in basic magma: A case for homogeneous equilibria: Earth Planet. Sci. Lett., v. 78, p. 200–210.

Cherkaoui, A. S. M., Wilcock, W. S. D., and Baker, E. T., 1997, Thermal fluxes associated with the 1993 diking event on the CoAxial segment, Juan de Fuca Ridge: A model for the convective cooling of a dike: Jour. Geophys. Res., v. 102, p. 24,887–24,902.

Christie, D. M., Carmichael, I. S. E., and Langmuir, C. H., 1986, Oxidation states of mid-ocean ridge basalt glasses: Earth Planet. Sci. Lett., v. 79, p. 397–411.

Crisp, J., 1984, Rates of magma emplacement and volcanic output: Jour Volcanol. Geotherm. Res., v. 20, p. 177–211.

Delaney, J., Kelley, D., Lilley, M., Butterfield, D., Baross, J., Wilcock, W., Embley, R., and Summit, M., 1998, The quantum event of oceanic crustal accretion: Impacts of diking at mid-ocean ridges: Science, v. 281, p. 222–230.

Dixon, J. E., and Stolper, E. M., 1995, An experimental study of water and carbon dioxide solubilities in mid-ocean ridge basaltic liquids. Part II: Applications to degassing: Jour. Petrol., v. 36, p. 1633–1646.

Dixon, J. E., Stolper, E., and Delaney, J. R., 1988, Infrared spectroscopic measurements of CO_2 and H_2O in Juan de Fuca Ridge basaltic glasses: Earth Planet. Sci. Lett., v. 90, p. 87–104.

Dixon, J. E., Stolper, E. M., and Holloway, J. R., 1995, An experimental study of water and carbon dioxide solubilities in mid-ocean ridge basaltic liquids. Part I: Calibration and solubility models: Jour. Petrol., v. 36, p. 1607–1631.

Fine, G. J., and Stolper, E. M., 1986, Dissolved carbon dioxide in basaltic glasses: Concentrations and speciation: Earth Planet. Sci. Lett., v. 76, p. 263–278.

Gerlach, T. M., 1989, Degassing of carbon dioxide from basaltic magma at spreading centers II. Mid-ocean ridge basalts: Jour. Volcanol. Geotherm. Res., v. 39, p. 221–232.

Gregg, T., Fornari, D., Perfit, M., Haymon, R., and Fink, J., 1996, Rapid emplacement of a mid-ocean ridge lava flow on the East Pacific Rise at 9 degrees 46'–51'N: Earth Planet. Sci. Lett., v. 144, p. 1–7.

Hamilton, D. L., Burnham, C. W., and Osborn, E. F., 1964, The solubility of water and effects of oxygen fugacity and wter content on crystallization in mafic magmas: Jour. Petrol., v. 5, p. 21–39.

Holloway, J. R., 1998, Graphite-melt equilibria during mantle melting: Constraints on CO_2 in MORB magmas and the carbon content of the mantle: Chem. Geol., v. 147, p. 89–97.

Jambon, A., and Zimmermann, J. L., 1990, Water in oceanic basalts: Evidence for dehydration of recycled crust: Earth Planet. Sci. Lett., v. 101, p. 323–331.

Javoy, M., and Pineau, F., 1991, The volatiles record of a "popping" rock from the Mid-Atlantic Ridge at 14°N: Chemical and isotopic composition of gas trapped in vesicles: Earth Planet. Sci. Lett., v. 107, p. 598–611.

Jendrzejewski, N., Trull, T. W., Pineau, F., and Javoy, M., 1997, Carbon solubility in Mid-Ocean Ridge basaltic melt at low pressures (250–1950 bar): Chem. Geol., v. 138, nos. 1–2, p. 81–92.

Karl, D. M., Hebel, D. V., and Bjorkman, K., 1998, The role of dissolved organic matter release in the productivity of the oligotrophic North Pacific Ocean: Limnol. Oceanog., v. 43, p. 1270–1286.

Kelley, D. S., 1996, Methane-rich fluids in the oceanic crust: Jour. Geophys. Res., v. 101, no. B2, p. 2943–2962.

Kelley, D. S., and Fruh-Green, G. L., 1999, Abiogenic methane in deep-seated mid-ocean ridge environments: Insights from stable isotope analyses: Jour. Geophys. Res., v. 104, p. 10,439–10,460.

Kelley, D. S., Lilley, M. D., Lupton, J. E., and Olson, E. J., 1998, Enriched H_2, CH_4, and ^3He concentrations in hydrothermal plumes associated with the 1996 Gorda Ridge eruptive event: Deep-Sea Res., Part II, v. 45, p. 2665–2682.

Langmuir, C. H., Klein, E. M., and Plank, T., 1992, Petrological systematics of mid-ocean ridge basalts: Constraints on melt generation beneath mid-ocean ridges., *in* Morgan, J. P., Blackman, D. K., and Sinton, J. M., eds., Mantle flow and melt generation at mid-ocean ridges: Washington, DC, Amer. Geophys. Union, Geophys. Monog., p. 183–280.

Lofgren, G., 1974, An experimental study of plagioclase crystal morphology: Isothermal crystallization: Amer. Jour. Sci., v. 274, p. 243–273.

Marty, B., and Tolstikhin, I. N., 1998, CO_2 fluxes from mid-ocean ridges, arcs and plumes: Chem. Geol., v. 145, p. 233–248.

Marty, B., and Zimmermann, L., 1999, Volatiles (He, C, N, Ar) in mid-ocean ridge basalts: Assesment of shallow-level fractionation and characterization of source composition: Geochim. et Cosmochim. Acta, v. 63, p. 3619–3633.

Mathez, E. A., 1984, Influence of degassing on oxidation states of basaltic magmas: Nature, v. 310, p. 371–375.

MeltTeam, 1998, Imaging the deep seismic structure beneath a mid-ocean ridge: The MELT Experiment: Science, v. 280, p. 1215–1218.

Michael, P. J., 1988, The concentration, behavior and storage of H_2O in the suboceanic upper mantle: Implications for mantle metasomatism: Geochim. et Cosmochim. Acta, v. 52, p. 555–566.

Moore, J., Batchelder, J., and Cunningham, C., 1977, CO_2 filled vesicles in mid-ocean ridge basalt: Jour. Volcanol. Geotherm. Res., v. 2, p. 309–327.

Nicolas, A., Freydier, C. L., Godard, M., and Vauches, A., 1998, Magma chambers at ocean ridges: How large?: Geology, v. 21, p. 53–56.

Pan, V., Holloway, J. R., and Hervig, R. L., 1991, The pressure and temperature dependence of carbon dioxide solubility in tholeiitic basalt melts: Geochim. et Cosmochim. Acta, v. 55, p. 1587–1595.

Pawley, A. R., Holloway, J. R., and McMillan, P., 1992, The effect of oxygen fugacity on the solubility of carbon-oxygen fluids in basaltic melt at low pressure: Earth Planet. Sci. Lett., v. 110, p. 213–225.

Pineau, F., and Javoy, M., 1994, Strong degassing at ridge crests: The behavior of dissolved carbon and water in basalt glasses at 14°N, Mid-Atlantic Ridge: Earth Planet. Sci. Lett., v. 123, p. 179–198.

Shock, E. L., 1996, Hydrothermal systems as environments for the emergence of life, *in* Bock, G. R., and Goode, J. A., eds., Evolution of hydrothermal ecosystems on Earth (and Mars?): Chichester, UK, John Wiley and Sons, p. 40–60.

Sinton, J. M., and Detrick, R. S., 1992, Mid-ocean ridge magma chambers: Jour. Geophys. Res., v. 97, p. 197–216.

Slaughter, J., Kerrick, D. M., and Wall, V. J., 1975, Experimental and thermodynamic study of equilibria in the system CaO-MgO-SiO_2-H_2O-CO_2: Amer. Jour. Sci., v. 275, p. 143–162.

Staudigel, H., Muehlenbachs, K., Richardson, S. H., and Hart, R. S., 1981, Agents of low temperature ocean crust alteration: Contrib. Mineral. Petrol., v. 77, p. 150–157.

Stolper, E., 1982, The speciation of water in silicate melts: Geochim. et Cosmochim. Acta, v. 46, p. 2609–2620.

Stolper, E. M., and Holloway, J. R., 1988, Experimental determination of the solubility of carbon dioxide in molten basalt at low pressures: Earth and Planet. Sci. Lett., v. 87, p. 397–408.

Turcotte, D. L., and Morgan, J. P., 1992, The physics of magma migration and mantle flow beneath a mid-ocean ridge, *in* Morgan, J. P., Blackman, D. K., and Sinton, J. M., eds., Mantle flow and melt generation at mid-ocean ridges: Washington, DC, Amer. Geophys. Union, Geophys. Monog., p. 155–182.

Voglesonger, K. M., Holloway, J. R., Dunn, E. E., Dalla-Betta, P. J., and O'Day, P. A., in review, Experimental abiotic synthesis of methanol in seafloor hydrothermal systems: Chem. Geol.

Wyllie, P. J., 1963, The quenching technique in non-quenchable systems: A discussion concerning the alleged thermal decomposition of portlandite at high pressures: Amer. Jour. Sci., v. 261, p. 983–988.

Minerals in the Air: An Environmental Perspective[1]

PETER R. BUSECK,

Departments of Geology and Chemistry/Biochemistry, Arizona State University, Tempe, Arizona 85287

DANIEL J. JACOB,

Division of Engineering and Applied Science and Department of Earth and Planetary Sciences, Harvard University, Cambridge, Massachusetts 02138

MIHÁLY PÓSFAI[2], JIA LI,

Departments of Geology and Chemistry/Biochemistry, Arizona State University, Tempe, Arizona 85287

AND J. R. ANDERSON

Environmental Fluid Dynamics Program, Arizona State University, Tempe, Arizona 85287

Abstract

The troposphere comprises roughly the lowest 10 km of the atmosphere, the part we inhabit, that envelops us, and that we inhale and look through every day of our lives. Clouds, rain, all manner of storms, and other aspects of weather originate within it, and it contains 85% of the mass of the atmosphere, including aerosol particles. These particles, like greenhouse gases, can influence climate. Mineral particles are major constituents of this aerosol. Their sizes, shapes, compositions, and degrees of agglomeration can all be determined using transmission electron microscopy. These variables have, to a considerable extent, been overlooked in studies of atmospheric chemistry and provide an important potential area of research for geoscientists.

Introduction

THERE IS BROAD AGREEMENT that airborne particles have important consequences for health, visibility, climate, and human welfare in general. Globally, the most abundant group of aerosol particles by mass (~50%) consists of minerals (because of their relatively large sizes, their number concentration is far lower). Yet their study and environmental impacts have been relatively overlooked. Atmospheric scientists, if they even consider them, tend to group all airborne minerals into categories such as crustal, soil, or mineral dust, whereas most mineralogists and geochemists ignore them entirely. A major question we address is whether geoscientists, with their unique skills, are neglecting an important area of science where their talents and expertise are needed. For completeness, in our discussion we also include the range of common inorganic aerosol particles.

Anthropogenic emission of the "greenhouse gases" (e.g., CO_2, CH_4, N_2O, and halocarbons) is widely recognized as producing a warming effect through their absorption of radiation emitted from the Earth. Airborne particles, many of which are minerals, also have important effects on the Earth's energy balance, and thereby on global warming. For example, there is a strong correlation between the quantity of aerosol dust and recent periods of glaciation, with deposition rates between 2 and 20 times current values (Thompson and Mosley-Thompson, 1981; Hammer et al., 1985; Petit et al., 1990; Rea, 1994; Steffensen, 1997; Reader et al., 1999). It seems likely that the airborne minerals affected global climate by their interactions with solar and terrestrial radiation (Andreae, 1995, 1996; Duce, 1995; Li et al., 1996; Sokolik and Toon, 1996; Tegen and Lacis, 1996; Mahowald et al., 1999) and also by the amounts of nutrients supplied to the oceans, thereby affecting biological productivity and amounts of CO_2 released into the atmosphere (Martin and Gordon, 1988; Watson, 1997). Additionally, the large surface areas provided by airborne minerals result in abundant crystallographic sites for heterogeneous condensation and subsequent reaction

[1]Written version of presentation at a symposium honoring the contributions of Professor Konrad Krauskopf, Stanford University, December 1999.

[2]Also at Department of Earth and Environmental Sciences, University of Veszprém, Veszprém, POB 158, H8201, Hungary.

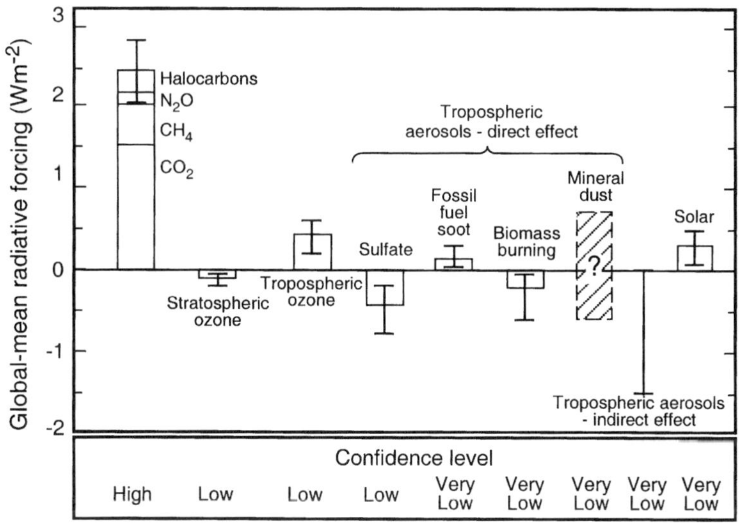

FIG. 1. Globally averaged radiative forcing at the tropopause as a result of increases in greenhouse gases, aerosols, and solar activity from pre-industrial times to the present. After IPCC (1996); the bar indicating uncertainty for mineral dust was added by the authors.

of gaseous species (Dentener et al., 1996; Zhang and Carmichael, 1999). Although not the focus of the current paper, at local and regional scales mineral aerosols also have major impacts on visibility and human health.

Although most geoscientists would have no problem agreeing that NaCl (halite) and other salts derived from the sea are minerals, atmospheric scientists tend to reserve the term for silicates. In such cases, "mineral" is commonly modified by "crustal" or used as an adjective to describe "dust" (e.g., Tegen et al., 1997; Liao and Seinfeld, 1998; Chiapello et al., 1999; Reader et al., 1999), which seems to mean what geoscientists refer to as rock-forming minerals, i.e., mainly aluminosilicates.

Despite their mass abundance among aerosol particles, little attention has been given to specific, individual rock-forming minerals. For example, the definitive reference on the effect of aerosols on climate is the report of the Intergovernmental Panel on Climate Change (IPCC, 1996). That influential publication emphasized the importance of aerosols and stimulated much of the current interest in the study of aerosols and their effects on climate. Nevertheless, the IPCC summary of radiative forcing, reproduced here as Figure 1, contains no reference to minerals; the bar for mineral dust was added by us. Moreover, none of the general circulation models (GCMs) used by the IPCC for global studies of climate change have a mineral dust feedback, implying that minerals are either inconsequential or optically inert. An interesting and, as yet, unresolved question is whether we are simply at an early stage in the recognition and evaluation of the impact of airborne minerals, or whether they are relatively unimportant. We believe the former alternative is correct. However, minerals do present special problems connected with their variety and complexity, the associated difficulties with identification, and the fact that the radiative properties of most minerals are not well known.

Although geoscientists and atmospheric scientists may use the term mineral differently, the more important point is that the tools and knowledge of mineralogy provide important information and novel perspectives for the study of airborne particles and their effects on the environment. The goals of this paper are to introduce the topic to geoscientists, to review some of the relevant literature, and to detail current issues and ways they are being addressed.

Radiative Forcing and Climate

The global energy budget of the Earth reflects a balance between absorption of solar radiation and blackbody emission of terrestrial radiation by the Earth and its atmosphere. Solar radiation (often called "short-wave") strongly peaks in the visible

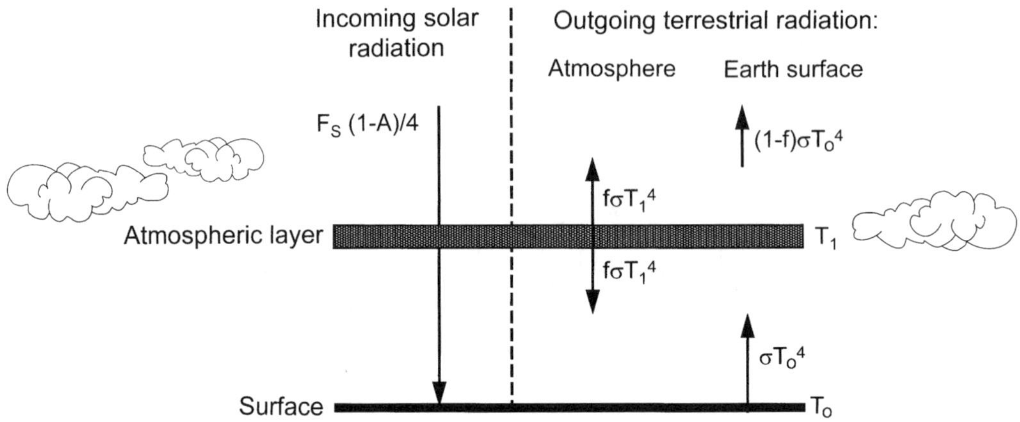

FIG. 2. Radiation balance for the atmosphere. See text for details (after Jacob, 1999).

wavelength range, whereas terrestrial radiation (often called "long-wave") is in the near-infrared 5 to 50 μm wavelength range. Gases in the atmosphere are largely transparent to visible solar radiation, but many trace gases (called greenhouse gases) efficiently absorb the long-wave terrestrial radiation emitted by the surface and re-emit it toward the Earth. The resulting warming of the Earth's surface is called the greenhouse effect.

A simple model of the greenhouse effect is shown in Figure 2 (after Jacob, 1999). In this model, we view the atmosphere as a single isothermal layer detached from the Earth's surface. The incoming solar radiation flux, $F_s = 1370$ W m^{-2}, at the top of the Earth's atmosphere is either reflected back to space (by clouds, aerosol, snow, and ice) or absorbed by the Earth's surface. We neglect the small fraction of solar radiation absorbed by the atmosphere. The fraction of incoming solar radiation reflected back to space is called the planetary albedo A; it has a global mean value of 0.28. The global mean solar radiation flux absorbed by the Earth's surface is thus $F_s(1 - A)/4 = 250$ W m^{-2}, where the geometric factor 1/4 is the ratio of the cross-sectional to the total area of the Earth.

The long-wave radiation spectrum emitted by the Earth's surface can be closely approximated as that of a blackbody. The resulting flux integrated over all wavelengths is σT_o^4, where σ is the Stefan-Boltzmann constant (= 5.67×10^{-8} W m^{-2} K^{-4}), and T_o is the global mean surface temperature of the Earth. Gases, aerosols, and clouds in the atmosphere absorb a fraction (f) of this emitted terrestrial radiation, whereas the rest is transmitted to space. The Kirchhoff law for blackbody radiation states that absorptivity equals emissivity; if the atmosphere in our simple model of Figure 2 absorbs a fraction, f, of the long-wave radiation, then its long-wave emission flux is $f\,\sigma\,T_1^4$, where T_1 is the temperature of the atmospheric layer. This emission flux from the atmospheric layer is directed both downward and upward.

As shown in Figure 2, the energy balance for the Earth system (Earth + atmosphere) is

$$\frac{F_s(1-A)}{4} = (1-f)\sigma T_o^4 + f\sigma T_1^4, \qquad (1)$$

where the left-hand side is the incoming solar radiation (F_{in}) and the right-hand side is the outgoing radiation exported to space (F_{out}). The energy balance for the atmospheric layer is

$$f\sigma T_o^4 = 2f\sigma T_1^4, \qquad (2)$$

since the atmospheric layer emits both upward and downward. From these two equations, we obtain as a solution for the surface temperature T_o of the Earth:

$$T_o = \left[\frac{F_s(1-A)}{4\sigma\left(1-\frac{f}{2}\right)}\right]^{\frac{1}{4}}. \qquad (3)$$

The observed surface temperature $T_o = 288$ K corresponds to $f = 0.72$; we reproduce the Earth's surface temperature in our model if 72% of the long-wave radiation emitted by the Earth's surface is absorbed by the atmosphere.

We see from equation (3) that T_o increases if f increases (as from increases in greenhouse gases) or if A decreases. The potential climatic effect of a perturbation to the radiative budget of the Earth is generally measured in terms of its radiative forcing, defined as the instantaneous disequilibrium, ΔF, between incoming and outgoing radiation in equation (1) as a result of a change in either A or f:

$$\Delta F = F_{in} - F_{out}. \quad (4)$$

In our model, the disequilibrium is calculated at the top of the atmosphere; in research models, it is calculated at the tropopause since the stratosphere above is only weakly coupled to the Earth's surface.

If the radiative forcing is applied as a constant perturbation, then the Earth system will eventually evolve to a new radiative equilibrium. The resulting change ΔT_o in the Earth's surface temperature can be readily calculated in our model (Jacob, 1999):

$$\Delta T_o = \lambda \Delta F, \quad (5)$$

where λ is the climate sensitivity parameter:

$$\lambda = \left\{ \frac{1}{4\sigma\left(1 - \frac{f}{2}\right)T_o^3} \right\}. \quad (6)$$

Substituting the previously given values for f, σ, and T_o, we obtain $\lambda = 0.3$ K m^2 W^{-1}.

This model has obvious weaknesses, notably in the lack of vertical resolution of the atmosphere, the lack of wavelength dependence, and the lack of climate feedbacks. Nevertheless, calculations with global three-dimensional models of the atmosphere and ocean system (general circulation models or GCMs) confirm that ΔT_o and ΔF are linearly related, with λ ranging from 0.3 to 1.4 K m^2 W^{-1}, depending on the GCM (IPCC, 1996). The value calculated in our model is at the low end of the GCM values, mostly because we did not account for the feedback from water vapor, a major greenhouse gas (as the Earth warms, increasing evaporation of water from the oceans reinforces the greenhouse effect and provides a positive feedback). But we see from the above discussion that the concept of radiative forcing can serve as a general indicator of the potential for climate change associated with a given radiative perturbation.

Aerosols and Radiation

Aerosol particles occur in a wide range of sizes, shapes, compositions, and degrees of agglomeration. Examples are given in Figures 3 to 5, which show particles collected from a varicty of tropospheric environments. Collectively, these figures illustrate features that would be taken for granted by geochemists, but that are at odds with widely accepted assumptions used among atmospheric scientists. We discuss and explain these in sequence, following a review of aerosols and radiation.

Aerosols have complicated effects on the radiation balance of the Earth. Depending on their types and sizes, they may absorb and scatter (reflect) short-wave solar radiation, and also absorb long-wave terrestrial radiation (there is no significant scatter in the IR wavelength range). Perturbations to both f and A may result in our model, as well as a vertical displacement of solar heating, since some solar radiation is absorbed by the aerosol layer instead of by the surface. The sum of these perturbations represents the direct radiative effect of aerosols.

Aerosols also affect the radiative budget of the Earth through cloud formation. They serve as condensation nuclei for clouds, so that an increase in aerosol number causes a rise in cloud albedo both by increasing the number of droplets and by increasing the lifetimes of clouds. This is called the indirect radiative effect of aerosols, and it depends on the aerosol sizes and number concentrations.

IPCC (1996) gives estimates of the cumulative radiative forcing over the past century from increasing concentrations of greenhouse gases and aerosols. The total forcing from long-lived greenhouse gases (CO_2, CH_4, N_2O, halocarbons) is 2.5 W m^{-2}, corresponding to $\Delta T_o = 1$ to 3 K (equation 5) (Fig. 1). The observed warming over the past century has been less, about 0.6 K. It is now thought that anthropogenic aerosols may be a critical compensating factor. The radiative forcing from anthropogenic sulfate and carbonaceous aerosols is estimated by IPCC (1996) to be in the range from –0.3 to –1.3 W m^{-2} (direct effect) and 0 to –1.5 W m^{-2} (indirect effect). The negative forcing arises from the increase in planetary albedo that results from scattering of solar radiation. Sulfate and carbonaceous aerosols are in the submicrometer size range, which is too small to absorb long-wave radiation and contribute to the greenhouse effect.

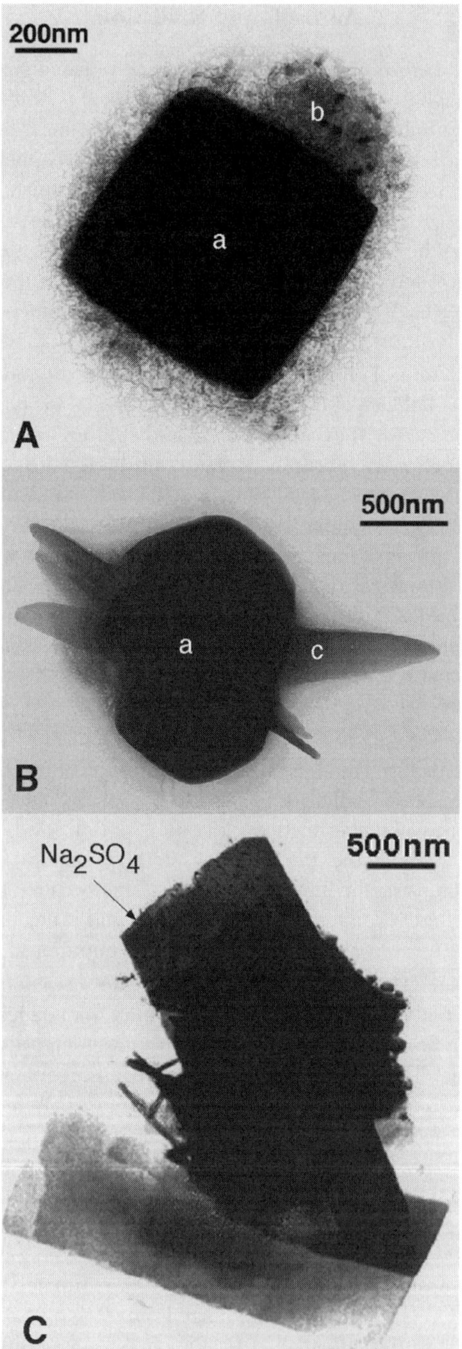

FIG. 3. Transmission electron microscope (TEM) images of reacted sea salt. A, B. Euhedral NaCl (a) with anhedral sulfate rim (b) and projecting sulfate crystals (c), Canary Islands, N. Atlantic Ocean, ACE-2. C. Tabular Na_2SO_4 particles converted from NaCl, Azores, North Atlantic Ocean, ASTEX/MAGE (from Pósfai et al., 1995).

As pointed out in the introduction, the IPCC report does not include radiative forcing from anthropogenic mineral dust. Yet dust accounts globally for over half of the total aerosol mass load in the atmosphere (Tegen et al., 1997), and its pervasive contribution to aerosol optical effects is evident from satellite observations (Herman et al., 1997; Husar et al., 1997). The largest sources of dust are in the northern tropics, although the amount of airborne dust varies considerably in space and time, reflecting the texture and dryness of soils as well as the wind speed. According to present estimates, 10 to 50% of the dust in the atmosphere originates from human influences such as desertification, cultivation, and erosion (Sokolik and Toon, 1996; Tegen et al., 1996). Clearly, the anthropogenic radiative forcing from dust needs to be included in future climate change assessments.

Dust particles in the atmosphere both scatter and absorb solar radiation (Tegen and Lacis, 1996) and also are sufficiently large to absorb long-wave terrestrial radiation and thus have a greenhouse effect. The indirect radiative forcing associated with cloud formation is generally unimportant for dust because the number concentration of dust particles is small. (Because of their relatively large sizes, they make a substantial contribution to the aerosol mass concentration but not to the aerosol number concentration.) The vertically integrated short-wave optical depth of dust, including contributions from both scattering and absorption, averages about 0.03 globally; this value can be compared to a global mean estimate of 0.10 for the total optical depth of atmospheric aerosols (National Research Council, 1996). The comparison implies that dust accounts globally for 30% of the light extinction by aerosols. Over large regions of the northern tropics, the dust optical depth can exceed 0.5 (Tegen and Fung, 1994).

The optical properties of atmospheric dust are largely determined by particles in the 0.2 to 2 μm size range; the resulting effect on the radiative budget is complicated and is highly dependent on the poorly known size distribution and refractive index of dust particles (Tegen and Lacis, 1996; IPCC, 2000). Dust increases planetary albedo over dark surfaces such as the oceans. In contrast, over bright surfaces such as deserts, absorption of solar radiation by dust results in a net decrease in albedo. On a global average, the short-wave radiative forcing from dust is negative—i.e., the cooling resulting from solar backscatter dominates over the heating arising from solar absorption. The greenhouse

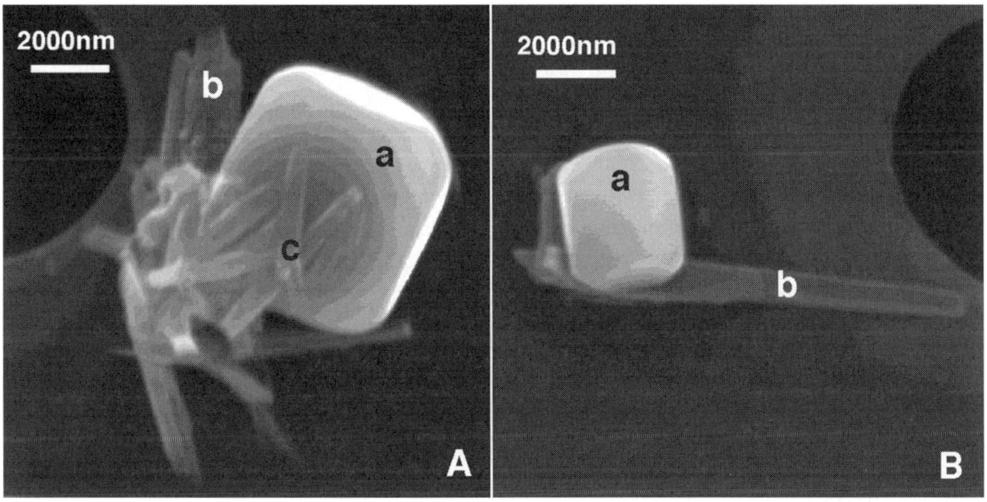

FIG. 4. Scanning electron microscope (SEM) images of partially reacted sea salt. A. Aggregation of cubic NaCl (a), tabular and rod-shaped mixed-cation sulfates (b), and Na_2SO_4 (c). Cf. Figure 2. The sulfate is a reaction product of NaCl. B. NaCl cube (a) with rod-shaped mixed-cation sulfate (b), Canary Islands, North Atlantic Ocean, ACE-2). The large black partial holes are from the SEM substrate.

warming from absorption of long-wave radiation by dust introduces further compensation in radiative forcing calculations.

Tegen et al. (1996) presented a global model study of radiative forcing at the tropopause from anthropogenic dust. They found a slight positive global forcing of 0.09 W m^{-2} at the tropopause, representing a balance between a positive long-wave forcing of 0.34 W m^{-2} and a negative short-wave forcing of –0.25 W m^{-2}. Such a calculation has large uncertainties because of our poor knowledge of the optical properties of dust particles. The latest IPCC (2000) report gives a tentative range of –0.6 W m^{-2} to +0.4 W m^{-2} for the global radiative forcing at the tropopause from anthropogenic dust. This value, while potentially large, still does not do justice to the strong regional variability of the dust radiative forcing that may have profound implications for climate; nor does it account for the climatic implications of deposition of heat in the atmospheric dust layer. Assessing the potential climatic effect of dust requires a different approach.

In a pioneering study, Miller and Tegen (1998) used a GCM to investigate the climatic consequences of the heating that results from the present-day atmospheric dust layer (as compared to an atmosphere with no dust). They found that absorption of solar radiation by the dust layer results in a net cooling of the surface of about 1 K in dusty regions of the northern tropics. The high-altitude warming resulting from the dust leads to buoyant stabilization of the lower atmospheric column, restricting convective motions and reducing precipitation. This drying effect could have a positive feedback by favoring dust generation; on the other hand, stabilization of the lower troposphere would decrease surface wind speeds and suppress dust generation (negative feedback). Considerably more work will be needed over the next decade to improve our understanding of the climatic effects of dust.

Analysis of Aerosol Particles

Aerosol particles have, with few exceptions, been neglected by geoscientists. The slack has been picked up by atmospheric scientists, primarily chemists. Most emphasis has been on bulk samples. Particles are collected through the use of what amounts to sophisticated vacuum cleaners. Aerosol concentrations in the troposphere are typically in the range of 1 to 100 cm^{-3}. Collection on a filter at a typical flow rate of 10 liters per minute yields 10^4 to 10^6 particles in one minute of sampling. These bulk samples are commonly analyzed by any of several increasingly sensitive methods. The total "mineral" components are then determined by proportioning to the amounts of Al, chosen as representative of "crustal" minerals or desert dust (e.g., Chiapello et al., 1999). The data may be processed through statistical methods, such as principal components analysis. However, no matter how the data are handled, this is clearly a crude method; it is totally unequal to the task of accurate phase identification for the range of minerals in the atmosphere.

Bulk analyses are limited in their utility. Although useful for integrating information for large numbers of particles, the effects produced by aerosol particles—whether on climate, haze, or health—depend on the characteristics of the individual particles, rather than on their averaged bulk properties. A minor or trace element might be widely dispersed, so that it is spread among all particles in tiny amounts, or it might be concentrated in a few particles in major amounts. Similarly, both speciation and whether the element(s) of interest are concentrated on particle surfaces rather than in the interiors, could have a major effect on the radiative properties.

The analysis of individual particles is much more time consuming than bulk analysis, and far fewer particles can be analyzed in given periods. However, the quality of the phase data is vastly superior. Scanning electron microscopes can be automated to produce unattended x-ray emission analyses around the clock (Germani and Buseck, 1991; De Bock et al., 1994; Katrinak et al., 1995; Anderson et al., 1996; Buseck and Anderson, 1998). Particle identifications are then based on the determined compositions. However, elements lighter than Na are not normally analyzed, and structural data are therefore not obtained. Also, as is true for all methods of chemical analysis of individual particles, the amount of water—whether structural or adsorbed—is not measured directly.

Transmission electron microscopy (TEM) has the great advantage that it provides all the necessary information (size, shape, composition, crystallographic structure, and thus speciation) for phase identification and determination of extent of aggregation, and this information can be obtained on an individual-particle basis. TEM has been used for aerosol studies to a limited, albeit increasing, extent. When performed by atmospheric chemists, it has been common practice to group samples into broad categories such as sulfates, soot, crustal minerals, and "other." In our TEM studies, we have far greater precision for mineral identification (Pósfai et al., 1994, 1995, 1998, 1999; Buseck and Pósfai, 1999), and we are able to detect and image particles as small as 30 nm. However, TEM measurements are labor intensive and must be made in the laboratory. Despite significant problems and the aforementioned limitations, they can provide "ground truth" for important selected particles that can then be used as calibration for the less specific but more rapid types of measurements.

A new method of particle analysis based on aerosol time-of-flight mass spectrometry has recently attracted much interest (Prather et al., 1994; Johnston and Wexler, 1995; Noble and Prather, 1996; Gard et al., 1998; Murphy et al., 1998). Although there are various configurations and acronyms (ATOFMS—aerosol time-of-flight mass spectrometer; PALMS—particle analysis by laser mass spectrometry), they all combine a method for introducing particles into an optical path where they are sized and then ionized with a laser. The fragments are immediately inserted into and then analyzed in a time-of-flight mass spectrometer. The resulting data include size and approximate particle compositions. Major advantages are that the measurements are sensitive and made on-line using a moderately portable instrument. Limitations are that structural (crystallographic) data are lacking and speciation must be calculated or inferred from reconstructions of the complex spectra of ionized fragments. This reconstruction becomes significantly challenging if more than one phase occurs within a given particle. Calibration of these instruments for quantitative chemical analysis is notoriously difficult.

It is clear that no one technique can provide all the information needed to completely characterize the aerosol particles in any given parcel of air. The TEM provides the most complete data per particle, but cannot be operated on-line or in automated mode. At present, no instrument or technique can provide adequate speciation of organic particles, which represent a major component of the atmospheric aerosol. Also, none of the currently available single-particle measurement methods can provide quantitative data for aerosol water.

Real versus Theoretical Aerosol Particles—Some Common Assumptions

Let us return to the figures and how they relate to several assumptions used in atmospheric studies, especially with reference to the modeling that is an essential part of predicting climate changes.

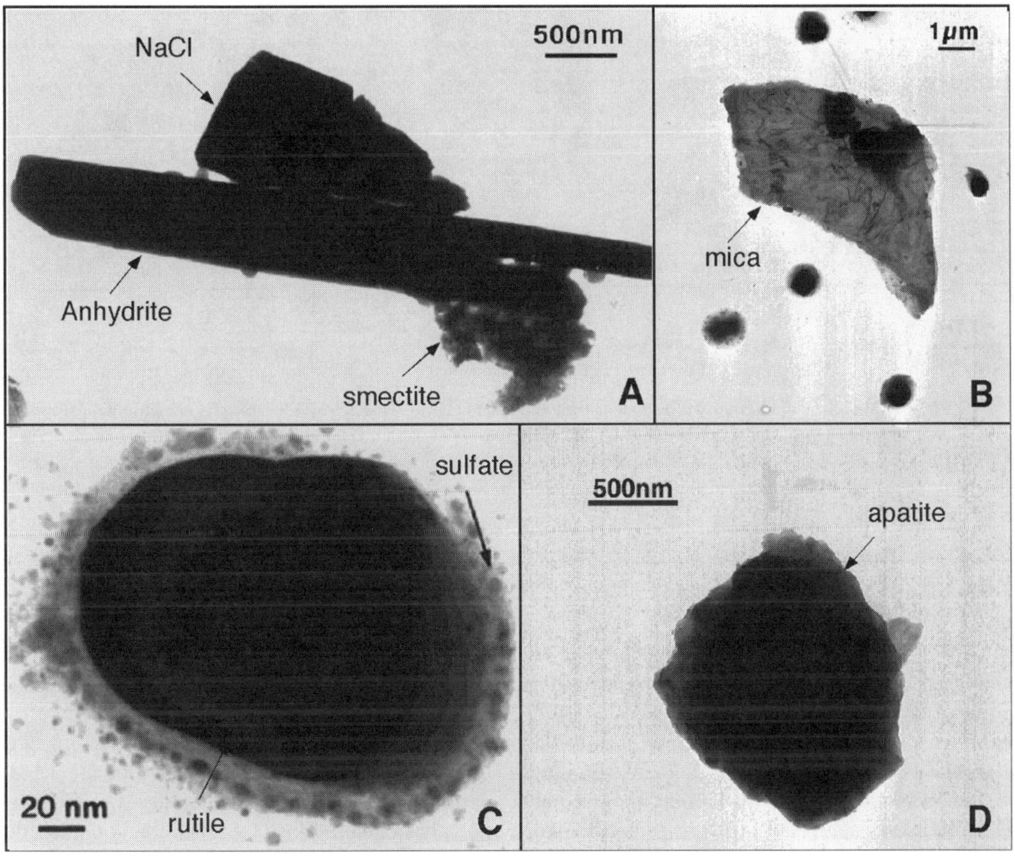

FIG. 5. TEM images of oxide and silicate mineral dust. A. Terrestrial smectite and anhydrite are internally mixed with NaCl (Azores, North Atlantic Ocean, ASTEX/MAGE; from Pósfai et al., 1995). B and D. Mica and apatite collected in the marine boundary layer (Canary Islands, North Atlantic Ocean, ACE-2). C. Twinned submicron rutile particle with sulfate coating (Equatorial Pacific, FeLINE-I cruise; from Pósfai et al., 1994).

Particle shapes and phases

Modelers commonly assume that aerosol particles are spherical. Several justifications are invoked for this assumption. Both absorption and scattering can be easily estimated for spherical shapes, and Liao and Seinfeld (1998) indicate that this assumption does not lead to significant errors. Lacis and Mischenko (1995) also report that an atmospheric radiation model that integrates over the range of particle orientations shows little difference from one based on a spherical assumption. However, other studies indicate that deviations from sphericity can have significant effects on aerosol optical properties (Bohren and Singham, 1991; Pilinis and Li, 1998; Schulz et al., 1999; Korolev et al., 1999). The supposition of sphericity is especially problematical for satellite measurements of aerosol optical properties, since the scattering geometry of irregularly shaped particles can differ greatly from that of spherical ones.

The sphericity assumption for aerosol particles is unambiguously valid only if they are liquid. Whether a particle is liquid or solid depends on its composition, on the local relative humidity (RH), and on its RH history (Seinfeld and Pandis, 1998). Most salt and organic particles deliquesce at high RH and may remain in a metastable liquid state because of lack of crystallization nuclei as the RH drops (e.g., Cziczo et al., 1997; Cziczo and Abbatt, 1999). Aluminosilicate minerals in aerosols do not normally deliquesce. However, our examinations of mineral aerosols by automated scanning electron microscopy show extensive development of hydrophilic sulfate coatings on mineral particles (Ander-

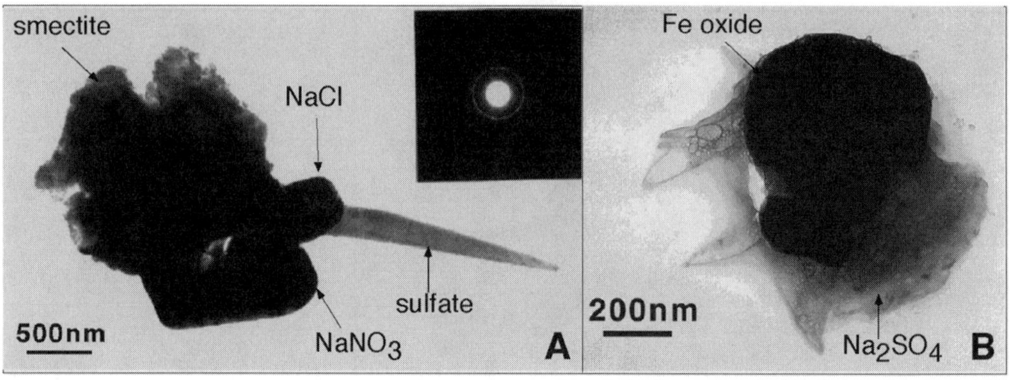

FIG. 6. TEM images of internally mixed terrestrial and marine aerosol particles. A. Terrestrial smectite with sea salt particles. Such an aggregation of mineral dust and sea salt may occur in clouds (cf. the rod-shaped sulfate crystal in Fig. 3B). The selected-area electron diffraction pattern (upper right) confirms the identification of smectite (Azores, North Atlantic Ocean, ASTEX/MAGE; after Pósfai et al., 1995). B. Fe oxide emitted from coal burning, with Na_2SO_4 (Sagres, Portugal, ACE-2).

son et al., 1992, 1996). A possible mechanism for the formation of such coatings is heterogeneous oxidation of SO_2 on the particle surfaces. Another mechanism is the scavenging of mineral particles by sulfate-containing cloud droplets. When such droplets evaporate, they leave behind mineral particles physically aggregated with or coated by sulfate. In this manner, otherwise non-deliquescent mineral particles can become cloud condensation nuclei (CCN) if a hydrophilic coating is present. Once hydrophilic, an aerosol particle can undergo many cycles of cloud processing before being removed by dry deposition or precipitation. However, unless deliquesced, their shapes will not be spherical.

In the TEM, as shown by several of the particles in Figures 3 to 5, the shapes of many particles are irregular, in some cases highly complex and even jagged. Some of these crystals may have formed by precipitation in response to evaporation of deliquesced water after sampling or in the vacuum of the TEM. Clearly, there is a need for experimental work to test that assumption, and that is an area under active study. Much work remains to adequately evaluate the complex shapes of real particles.

Mixing state

Parcels of air contain mixtures of many different particle types. Atmospheric scientists draw a sharp distinction between what they call internal and external mixtures. The distinction is important because mixing state affects radiative properties. In an external mixture, each individual particle consists of a single material, and so can be considered homogeneous. Internal mixtures, on the other hand, contain more than one material—i.e., they are polyphasic, more like a tiny rock than an individual mineral. The transition from an external to an internal mixture takes place by coagulation of particles, coalescence of particles in clouds, and gas-aerosol transfer reactions such as the uptake and oxidation of SO_2 on dust particles.

Most atmospheric radiation models assume that aerosols are externally mixed because the specification of optical properties is much easier than for internal mixtures. The assumption of external mixing has significant impacts on the calculated radiative properties. For example, the single-scattering albedo for internal mixtures of aerosols is generally lower than for external mixtures. In the next section, more details are discussed for mixtures of soot and sulfate aerosols.

Many aerosol particles are, in fact, internally mixed. We see such mixtures of different sea salts (Figs. 3 and 4), silicates and marine minerals (Figs. 5 and 6), and different types of anthropogenic species (Figs. 7 and 8). In fact, we observe a wide combination of mixture types. We will return to this subject below, in the section on soot and sulfate aerosols.

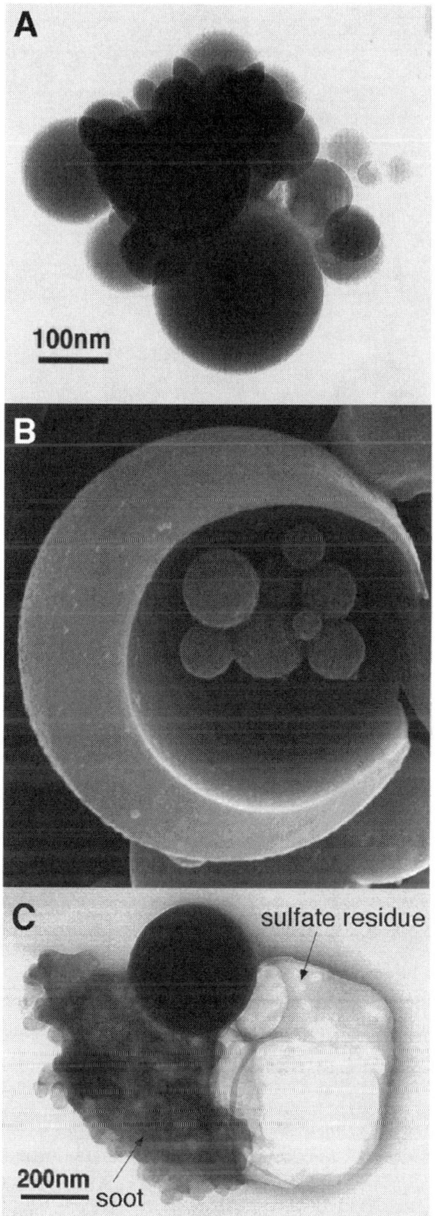

FIG. 7. SEM and TEM images of fly-ash particles. A. TEM image of an aggregate of submicron silica spheres collected in the marine boundary layer (Canary Islands, North Atlantic Ocean, ACE-2). B. SEM image of fly-ash. The hollow fly-ash particle (cenosphere) contains smaller fly-ash spheres. The largest hollow sphere is roughly 5 μm across (by Gary Aden, Arizona power plant source). C. TEM image of internal mixture of fly-ash, soot, and sulfate. The sulfate was evaporated by the TEM electron-beam, leaving behind only S-bearing residues that outline its former shape (Canary Islands, North Atlantic Ocean, ACE-2).

Phase identification/speciation

No species identification is complete without knowing its structural state. Identification of the structures of crystalline aerosol particles (such as minerals) is important for obtaining information about their refractive indices and other physical properties. Also, particles that were partly or totally liquid may crystallize during or after sampling, and their structures can provide indirect information about their compositions. The most commonly used method for mineral identification in aerosols is x-ray diffraction (e.g., Leinen et al., 1994; Caquineau et al., 1997), which is a bulk method and so provides only limited information regarding structural state and speciation for all but the most abundant particle types. Except for TEM, none of the analytical methods used by atmospheric scientists is capable of determining structural states for individual particles, and so structural states are generally ignored, as in speciation.

Soot and Sulfate Aerosols

Soot and sulfate aerosol particles are abundant, are mainly of anthropogenic origin, and have pronounced effects on radiative forcing. We discuss them here because they bear a close relation to minerals (many species in fact have mineral analogues), are important and have been studied intensively, and illustrate many of the principles discussed in this paper.

Combustion of fossil fuels releases SO_2, which is then oxidized to sulfate in the atmosphere. Aerosol sulfates are also atmospheric reaction products of natural processes such as volcanic and planktonic emissions of sulfur gases (Spiro et al., 1992). Sulfates scatter short-wave radiation and thus produce a negative forcing, i.e., cooling.

Soot is another major component of the tropospheric aerosol. It results from combustion of fossil fuels as well as from biomass burning during both controlled and wild fires (e.g., Crutzen and Andreae, 1990; Kuhlbusch et al.,1996; Kuhlbusch and Crutzen, 1996; Andreae, 1997). Soot particles absorb radiation and thereby produce a positive forcing, i.e., a net warming.

Soot and sulfate provide good examples of the assumptions discussed in the previous section. Most models assume that they are externally mixed. We have observed just the contrary. By using high-resolution TEM, we have found soot intergrown with sulfates in aerosol particles collected from marine

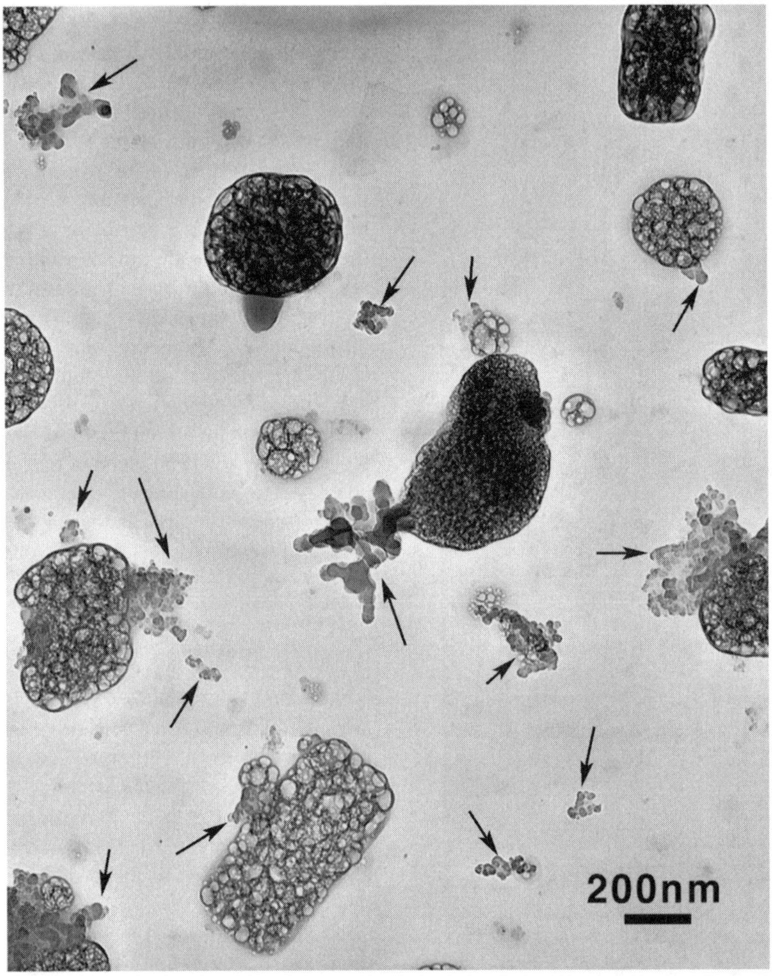

FIG. 8. TEM image of an assemblage of internally and externally mixed soot (arrowed) and ammonium sulfate. The bubbles formed as a result of the decomposition of the ammonium sulfate particles from electron-beam damage (Sagres, Portugal, ACE-2).

particles over the northern Atlantic, equatorial Pacific, Southern, and Indian oceans, and from polluted continental regions in Europe and the United States (Figs. 9 and 10). In fact, wherever we look in detail at aerosol sulfates we observe at least small amounts of soot in a significant number of particles (Pósfai et al., 1998, 1999; Buseck and Pósfai, 1999). These internal mixtures are so fine grained that they are only observable by high-resolution measurements. It remains to be determined if the reverse is true. In marine areas such as were sampled during the ACE-1 experiment, the soot particles were too small and had low contrast, so we did not attempt to count those that were not in sulfates.

The effects of such internal mixtures have been considered in several recent studies (Haywood and Shine, 1995; Pilinis et al., 1995; Andreae et al., 1986; Haywood et al., 1997; Tang, 1997). The most detailed investigation to date is that by Fuller et al. (1999), who calculated the absorption cross sections for soot/black carbon in a variety of configurations and aggregation states. Their results are complex and show that the radiative properties depend on a variety of parameters that include particle size and shape, degree of crystallinity, type of aggregation, density of aggregates, and type (internal or external) and extent of mixing with non-absorbing materials such as sulfate. The position of the soot within an

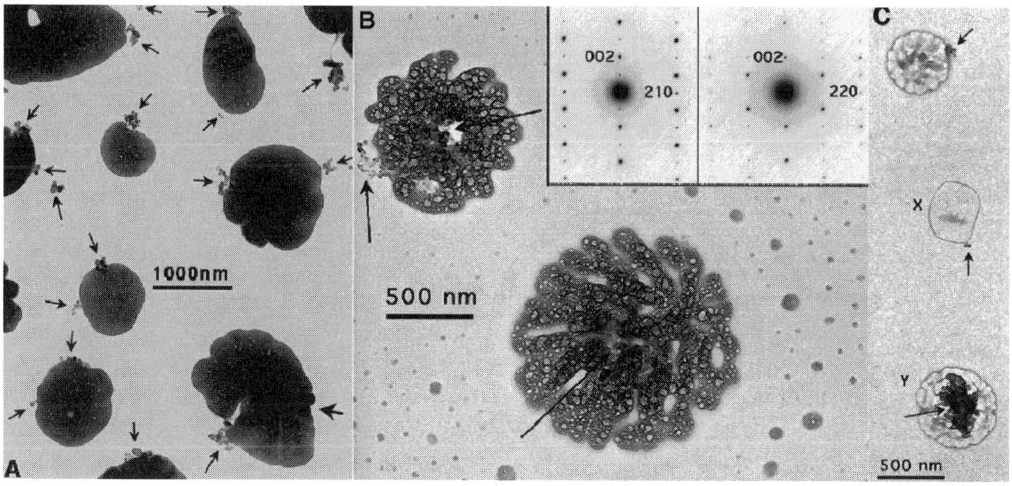

FIG. 9. TEM images of ammonium sulfate containing soot. A. Ammonium sulfate particles internally mixed with soot (small arrows) and fly-ash spheres (larger arrow in the lower-right corner), Azores, North Atlantic Ocean, ASTEX/MAGE (after Pósfai et al., 1999). B. Ammonium sulfate with soot inclusions. The rings of small ammonium sulfate particles formed during the dehydration of the large central particle. The inserted selected-area electron diffraction patterns were obtained from a similar particle and are consistent with the structure of ammonium sulfate. C. Soot inclusions (arrowed) and residues after evaporation of ammonium sulfate particles by the electron-beam. B and C are from Tasmania, Southern Ocean, ACE-1 (from Pósfai et al., 1995).

encapsulating sulfate host also affects the absorption and scattering properties. We think it is reasonable to expect effects of comparable magnitude on the seemingly abundant internal mixtures involving minerals with one another and with other species. Their results "suggest the importance of the determination of the physical state of the soot particles and their immediate environment when ascribing characteristic values for their absorption and scattering efficencies." We interpret that statement as indicating that TEM measurements are needed.

Rock-forming Minerals (RFM) in the Troposphere

Although abundant in the troposphere, no studies even closely approaching that of Fuller et al. (1999) for soot have been devoted to the RFM. Liao and Seinfeld (1998) calculated the forcing effects of minerals, although their calculations are based on integrated atmospheric optical measurements and are independent of the composition and speciation of the particular dust types. Sokolik and Toon (1996, 1999) initiated more extensive studies through their measurements of 11 important minerals; they have thereby amassed the largest data set, by far, on optical properties and radiative effects of minerals in the troposphere. They conclude that it will be highly important to determine the state of aggregation for minerals such as hematite, and also that more data are required on the regional and global scales to quantify the size-resolved composition of (mineral) dust and thus the radiative effects of minerals.

We believe that their abundance and consequent large surface areas for gas deposition and catalytic activity will bring increasing attention to the role of minerals in the troposphere. Their effects on climate will, because of their more localized distribution than gases, be more regional than global. However, some minerals will, similarly to soot, produce warming and thus could be considered "greenhouse particles" whereas others, similarly to sulfates, will produce negative forcing and thus tend toward cooling, acting like "icehouse particles." Their details and net effects on the radiative processes in the troposphere remain to be determined.

Regulatory and Health Issues

It is well known that certain mineral aerosols produce health-threatening diseases (e.g., asbestosis, silicosis, etc.), and the insult-producing materials are regulated by the U.S. Environmental

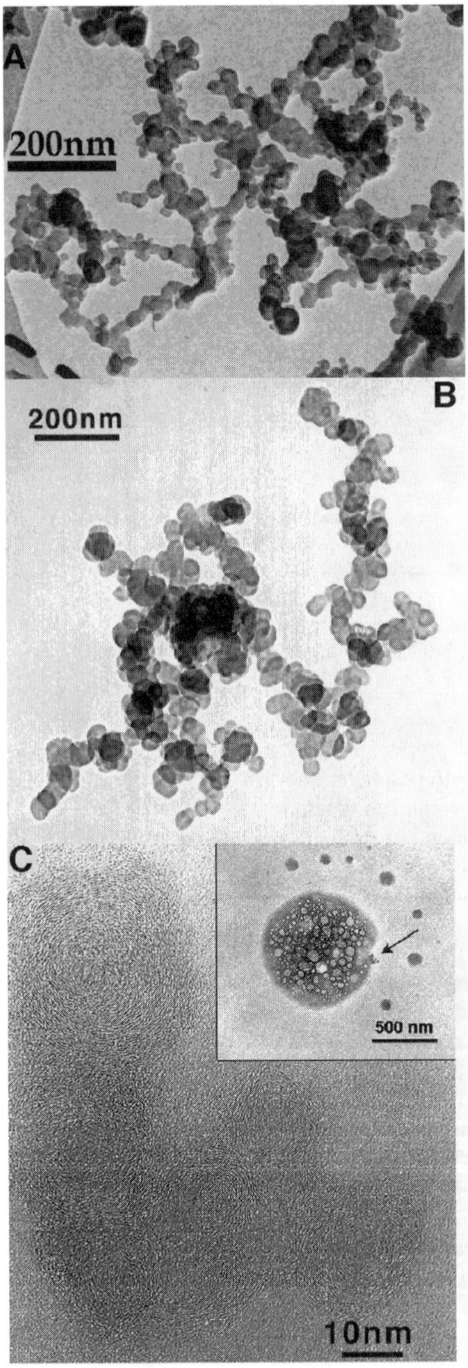

FIG. 10. TEM images of soot. A, B. Chain-like soot aggregates (A. Phoenix, Arizona after Katrinak et al., 1993; B. Sagres, Portugal, ACE-2). C. High-resolution TEM image of the arrowed soot aggregate showing the onion-like structure of soot spheres (Southern Ocean, ACE-1; after Pósfai et al, 1999).

Protection Agency (EPA). These problems tend to be localized geographically and are relatively straightforward to manage. A broader set of issues are related to ambient aerosols, and the EPA has dealt with these through the so-called PM (particulate matter) 2.5 and PM 10 standards from §109 of the Clean Air Act. These standards regulate the total mass of particles with diameters of 2.5 µm or less (with maximum allowable values of 15 and 65 $\mu g\ m^{-3}$ as annual arithmetic mean and 24-hour average concentrations, respectively). The PM 10 standard is concerned with particles 10 µm or less in diameter (with maximum allowable values of 50 and 150 $\mu g\ m^{-3}$ as annual arithmetic mean and 24-hour average concentrations, respectively).

The basis for the 2.5- and 10-µm diameter cutoffs is related to the sizes of particles that penetrate the lungs and the naso-pharyngeal systems, respectively. Although one of the goals of the Clean Air Act is to protect and improve the quality of the air we breathe, the PM standards totally ignore the compositions of the inhaled particles. With the exception of a few specified substances, there is no regulatory need or incentive to know the mineralogy of the inhaled particles. It remains for scientists and policymakers to decide whether such species- and coating-independent regulations serve the needs of society.

There is another problem related to the PM standards. The amount of mineral dust in the air exhibits great variations, both geographically and temporally. Volcanic eruptions can inject vast quantities of SO_2 (which rapidly turns into aerosol) plus particulate matter into the atmosphere (Andronova et al., 1999). Forest fires and dust storms in arid regions such as the southwestern United States also can produce sporadic events that vastly exceed EPA standards. Federal regulations accommodate such occasions by the possibility of excluding "exceptional events" produced by natural processes. Such somewhat localized events are relatively easy to identify and thus exclude.

Prospero (1999) provides an example of another type of dust-producing event that can be far more difficult to recognize because of its subtlety. At times, large quantities of minerals from northern Africa are transported across the Atlantic and deposited throughout large areas of the southeastern United States (Perry et al., 1997). These depositional events tend to be most severe during the summers. Prospero (1999) points out that, somewhat surprisingly, the highest individual PM 2.5 aerosol soil concentrations were found in the southeastern United States rather than in the arid Southwest. African dust also commonly impacts the Caribbean Islands (Arimoto et al.,1992, 1995), and there is abundant evidence that dust from Asia has important impacts over large areas in and bordering the Pacific Ocean (Tsunogai et al., 1985; Arimoto et al., 1990; Fan et al., 1996; Jaffe et al., 1999).

The evidence for long-range transport of dust complicates environmental regulations solely based on mass loadings. Identification of such distant source regions requires chemical and mineralogical fingerprinting.

Conclusion

By mass, minerals comprise the greatest part of the aerosol particles in the troposphere, the part of the atmosphere that we inhabit. They have, with prominent exceptions, been largely overlooked by atmospheric scientists, although increasing attention has been devoted to them recently. They are important as absorbers and scatterers of radiation and thus can have important effects on climate, most especially on local and regional scales because of their relatively localized spatial and temporal distributions.

Because aerosol particles are small in dimension and large in number, most analyses have been on large numbers of particles, with averaged bulk compositions and properties as the results. Such bulk measurements have important limitations. The radiative, health, and visibility effects of aerosol particles are caused by the cumulation of individual rather than averaged particles, and therefore there is much reason for studying particles individually.

Transmission electron microscopy is well suited to the study of individual aerosol particles. Many widely used assumptions in global climate models are seen to be of limited validity when individual particles are examined. Thus, spherical shapes, extent of mixing of phases, and particle speciation are all problematical, but can be resolved through individual-particle TEM studies. A difficulty is that deliquesced species lose their water in the vacuum of the electron microscope, and therefore these important aerosol types dehydrate and change their character. With this important limitation, aerosol particles can be studied on an individual basis, their most private details can be determined through TEM, and they represent a far richer and more complex world than commonly is assumed.

Acknowledgments

We thank Drs. Ina Tegen and Dante Lauretta for helpful reviews and Dr. G. Aden for use of Figure 7B. Electron microscopy was performed at the ASU Center for High-Resolution Electron Microscopy, established with support from the NSF and ASU. We acknowledge research support from the NSF and NASA.

REFERENCES

Anderson, J. R., Buseck, P. R., Patterson, T. L., and Arimoto, R. 1996, Characterization of the Bermuda tropospheric aerosol by combined individual-particle and bulk-aerosol analysis: Atmos. Environ., v. 30, p. 319–338.

Anderson, J. R., Buseck, P. R., and Saucy, D. A., 1992, Characterization of individual fine-fraction particles from the Arctic aerosol at Spitsbergen, May–June, 1987: Atmos. Environ., v. 26A, p. 1747–1762.

Andreae, M. O., 1995, Climatic effects of changing atmospheric aerosol levels, *in* Henderson-Sellers, A., ed., Future climates of the World: A modeling perspective: Amsterdam, Elsevier, World Survey of Climatology, v. 16, p. 341–392.

______, 1996, Raising dust in the greenhouse: Nature, v. 380, p. 389–390.

______, 1997, Emissions of trace gases and aerosols from southern African savanna fires, *in* van Wilgen, B. W., Andreae, M. O., Goldammer, J. G., and Lindesay, J. A., eds., Fire in southern African savannas: Ecological and atmospheric perspectives: Johannesburg, Witwatersrand Univ. Press, p. 161–183.

Andreae, M. O., Charlson, R. J., Bruynseels, F., Storms, H., Van Grieken, R., and Maenhaut, W., 1986, Internal mixture of sea salt, silicates, and excess sulfate in marine aerosols: Science, v. 232, p. 1620–1623.

Andronova, N. G., Rozanov, E. V., Yang, F., Schlesinger, M. E., and Stenchikov, G. L., 1999, Radiative forcing by volcanic aerosols from 1850 to 1994: Jour. Geophys. Res., v. 104, p. 16,807–16,826.

Arimoto, R., Duce, R. A., Ray, B. J., Ellis, W. G., Jr., Cullen, J. D., and Merrill, J. T., 1995, Trace elements in the atmosphere over the North Atlantic: Jour. Geophys. Res., v. 100, p. 1199–1214.

Arimoto, R., Duce, R. A., Savoie, D. L., and Prospero, J. M., 1992, Trace elements in aerosol particles from Bermuda and Barbados: Concentrations, sources and relationships to aerosol sulfate: Jour. Atmos. Chem., v. 14, p. 439–457.

Arimoto, R., Ray, B. J., Duce, R. A., Hewitt, A. D., Boldi, R., and Hudson, A., 1990, Concentrations, sources, and fluxes of trace elements in the remote marine atmosphere of New Zealand: Jour. Geophys. Res., v. 95, p. 22,389–22,405.

Bohren, C. F., and Singham, S. B., 1991, Backscattering by nonspherical particles: A review of methods and suggested new approaches: Jour. Geophys. Res., v. 96, p. 5269–5277.

Buseck, P. R., and Anderson, J. R., 1998, Analysis of individual airborne mineral particles, *in* Marfunin, A. S., ed., Advanced mineralogy, third ed.: Berlin, Springer-Verlag, p. 292–300.

Buseck, P. R., and Pósfai, M., 1999, Airborne minerals and related aerosol particles: Effects on climate and the environment: Proc. Nat. Acad. Sci., v. 96, p. 3372–3379.

Caquineau, S., Magonthier, M.-C., Gaudichet, A., and Gomes, L., 1997, An improved procedure for the X-ray diffraction analysis of low-mass atmospheric dust samples: Eur. Jour. Mineral., v. 9, p. 157–166.

Chiapello, I., Bergametti, G., Chatenet, B., Dulac, F., Jankowiak, L., Liousse, C., and Soares, E. S., 1999, Contribution of the different aerosol species to the aerosol mass load and optical depth over the northeastern tropical Atlantic: Jour. Geophys. Res., v. 104, p. 4025–4035.

Crutzen, P. J., and Andreae, M. O., 1990, Biomass burning in the tropics: Impact on atmospheric chemistry and biogeochemical cycles: Science, v. 250, p. 1669–1678.

Cziczo, D. J., and Abbatt, J. P. D., 1999, Deliquescence, efflorescence, and supercooling of ammonium sulfate aerosols at low temperature: Implications for cirrus cloud formation and aerosol phase in the atmosphere: Jour. Geophys. Res., v. 104, p. 13,781–13,790.

Cziczo, D. J., Nowak, J. B., Hu, J. H., and Abbatt, J. P. D., 1997, Infrared spectroscopy of model tropospheric aerosols as a function of relative humidity: Observation of deliquescence and crystallization: Jour. Geophys. Res., v. 102, p. 18,843–18,850.

De Bock, L. A., Van Malderen, H., and Van Grieken, R. E., 1994, Individual aerosol particle composition variations in air masses crossing the North Sea: Env. Sci. Technol., v. 28, p. 1513–1520.

Dentener, F. J., Carmichael, G. R., Zhang, Y., Lelieveld, J., and Crutzen, P. J., 1996, Role of mineral aerosol as a reactive surface in the global troposphere: Jour. Geophys. Res., v. 101, p. 22,869–22,889.

Duce, R. A., 1995, Sources, distributions, and fluxes of mineral aerosols and their relationship to climate, *in* Charlson, R. J., and Heintzenberg, J., eds., Aerosol forcing of climate: New York, John Wiley, p. 43–72.

Fan, X., Okada, K., Nimura, N., Kai, K., Arao, K., Shi, G. Y., Qin, Y., and Mitsuta, Y., 1996, Mineral particles collected in China and Japan during the same Asian dust-storm event: Atmos. Environ., v. 30, p. 347–351.

Fuller, K. A., Malm, W. C., and Kreidenweis, S. M., 1999, Effects of mixing on extinction by carbonaceous particles: Jour. Geophys. Res., v. 104, p. 15,941–15,954.

Gard, E. E., Kleeman, M. J., Gross, D. S., Hughes, L. S., Allen, J. O., Morrical, B. D., Fergenson, D. P., Dienes, T., Gälli, M. E., Johnson, R. J., Cass, G. R., and Prather, K. A., 1998, Direct observation of heterogeneous chemistry in the atmosphere: Science, v. 279, p. 1184–1187.

Germani, M. S., and Buseck, P. R., 1991, Automated scanning electron microscopy for atmospheric particle analysis: Anal. Chem., v. 63, p. 2232–2237.

Hammer, C. U., Clausen, H. B., Dansgaard, A., Neftel, A., Kristinsdottir, P., and Johnson, E., 1985, Continuous impurity analysis along the Dye 3 deep core, *in* Langway, C. C., Jr., Oeschger, H., and Dansgaard, W., eds., Greenland ice core: Geophysics, geochemistry, and the environment: Washington, DC, Amer. Geophys. Union, Geophys. Monog. ser. 33, p. 90–94.

Haywood, J. M., Roberts, D. L., Slingo, A., Edwards, J. M., and Shine, K. P., 1997, General circulation model calculations of the direct radiative forcing by anthropogenic sulfate and fossil-fuel soot aerosol: Jour. Climate, v. 10, p. 1562–1577.

Haywood, J. M., and Shine, K. P., 1995, The effect of anthropogenic sulfate and soot aerosol on the clear sky planetary radiation budget: Geophys. Res. Lett., v. 22, p. 603–606.

Herman, J. R., Bhartia, P. K., Torres, O., Hsu, C., Seftor, C., and Celarier, E., 1997, Global distribution of UV-absorbing aerosols from Nimbus 7/TOMS data: Jour. Geophys. Res., v. 102, p. 16,911–16,922.

Husar, R. B., Prospero, J. M., and Stowe, L. L., 1997, Characterization of tropospheric aerosols over the oceans with the NOAA advanced very high resolution radiometer optical thickness operational product: Jour. Geophys. Res., v. 102, p. 16,889–16,909.

IPCC (Intergovernmental Panel on Climate Change), 1996, Climate change 1995: The science of climate change (Houghton, J. T., Meiro Filho, L. G., Callander, B. A., Harris, N., Kattenberg, A., and Maskell, K., eds.): Cambridge, UK, Cambridge Univ. Press.

IPCC, 2000, Climate change: The scientific basis (Houghton, J. T., and Yihui, D., eds.): Cambridge, UK, Cambridge Univ. Press, in press.

Jacob, D. J., 1999, Introduction to atmospheric chemistry: Princeton, NJ, Princeton Univ. Press.

Jaffe, D., Anderson, T., Covert, D., Kotchenruther, R., Trost, B., Danielson, J., Simpson, W., Berntsen, T., Karlsdottir, S., Blake, D., Harris, J., Carmichael, G., and Uno, I., 1999, Transport of Asian air pollution to North America: Geophys. Res. Lett., v. 26, p. 711–714.

Johnston, M. V., and Wexler, A. S., 1995, MS of individual aerosol particles: Anal. Chem., v. 67, p. 721A–726A.

Katrinak, K. A., Anderson, J. R., and Buseck, P. R., 1995, Individual particle types in the aerosol of Phoenix, Arizona: Env. Sci. Technol., v. 29, p. 321–329.

Korolev, A. V., Isaac, G. A., and Hallett, J., 1999, Ice particle habits in Arctic clouds: Geophys. Res. Lett., v. 26, p. 1299–1302.

Kuhlbusch, T. A. J., Andreae, M. O., Cachier, H., Goldammer, J. G., Lacaux, J.-P., Shea, R., and Crutzen, P. J., 1996, Black carbon formation by savanna fires: Measurements and implications for the global carbon cycle: Jour. Geophys. Res., v. 101, p. 23,651–23,665.

Kuhlbusch, T. A. J., and Crutzen, P. J., 1996, Black carbon, the global carbon cycle, and atmospheric carbon dioxide, *in* Levine, J. S., ed., Biomass burning and global change: Cambridge, MA, MIT Press, v. 1, p. 160–169.

Lacis, A. A., and Mishchenko, M. I., 1995, Climate forcing, climate sensitivity, and climate response: A radiative modeling perspective on atmospheric aerosols, *in* Charlson, R. J., and Heintzenberg, J., eds., Aerosol forcing of climate: New York, John Wiley, p. 11–42.

Leinen, M., Prospero, J. M., Arnold, E., and Blank, M., 1994, Mineralogy of aeolian dust reaching the North Pacific Ocean 1. Sampling and analysis: Jour. Geophys. Res., v. 99, p. 21,017–21,023.

Li, X., Maring, H., Savoie, D., Voss, K., and Prospero, J. M., 1996, Dominance of mineral dust in aerosol light-scattering in the North Atlantic trade winds: Nature, v. 380, p. 416–419.

Liao, H., and Seinfeld, J. H., 1998, Radiative forcing by mineral dust aerosols: Sensitivity to key variables: Jour. Geophys. Res., v. 103, p. 31,637–31,645.

Mahowald, N., Kohfeld, K., Hansson, M., Balkanski, Y., Harrison, S. P., Prentice, I. C., Schulz, M., and Rodhe, H., 1999, Dust sources and deposition during the last glacial maximum and current climate: A comparison of model results with paleodata from ice cores and marine sediments: Jour. Geophys. Res., v. 104, p. 15,895–15,916.

Martin, J. H., and Gordon, R. M., 1988, Northeast Pacific iron distributions in relation to phytoplankton productivity: Deep-Sea Research, v. 35, p. 177–196.

Miller, R. L., and Tegen I., 1998, Climate response to soil dust aerosols: Jour. Climate, v. 11, p. 3247–3267.

Murphy, D. M., Anderson, J. R., Quinn, P. K., McInnes, L. M., Brechtel, F. J., Kreidenweis, S. M., Middlebrook, A. M., Pósfai, M., Thomson, D. S., and Buseck, P. R., 1998, Influence of sea-salt on aerosol radiative properties in the Southern Ocean marine boundary layer: Nature, v. 392, p. 62–65.

National Research Council, 1996, A plan for a research program on aerosol radiative forcing and climate change: Washington, DC, Nat. Acad. Press.

Noble, C. A., and Prather, K. A., 1996, Real-time measurement of correlated size and composition profiles of individual atmospheric aerosol particles: Env. Sci. Technol., v. 30, p. 2667–2680.

Perry, K. D., Cahill, T. A., Eldred, R. A., Dutcher, D. D., and Gill, T. E., 1997, Long-range transport of North African dust to the eastern United States: Jour. Geophys. Res., v. 102, p. 11,225–11,238.

Petit, J. R., Mounier, L., Jouzel, J., Korotkevich, Y. S., Kotlyakov, V. I., and Lorius, C., 1990, Paleoclimatological and chronological implications of the Vostok core dust record: Nature, v. 343, p. 56–58.

Pilinis, C., and Li, X., 1998, Particle shape and internal inhomogeneity effects on the optical properties of tropospheric aerosols of relevance to climate forcing: Jour. Geophys. Res., v. 103, p. 3789–3800.

Pilinis, C., Pandis, S. N., and Seinfeld, J. H., 1995, Sensitivity of direct climate forcing by atmospheric aerosols to aerosol size and composition: Jour. Geophys. Res., v. 100, p. 18,739–18,754.

Pósfai, M., Anderson, J. R., Buseck, P. R., Shattuck, T. W., and Tindale, N. W., 1994, Constituents of a remote Pacific marine aerosol: A TEM study: Atmos. Environ., v. 28, p. 1747–1756.

Pósfai, M., Anderson, J. R., Buseck, P. R., and Sievering, H., 1995, Compositional variations of sea-salt-mode aerosol particles from the North Atlantic: Jour. Geophys. Res., v. 100, p. 23,063–23,074.

Pósfai, M., Anderson, J. R., Buseck, P. R., and Sievering, H., 1999, Soot and sulfate aerosol particles in the remote marine troposphere: Jour. Geophys. Res., v. 104, p. 21,685–21,693.

Pósfai, M., Xu, H., Anderson, J. R., and Buseck, P. R., 1998, Wet and dry sizes of atmospheric aerosol particles: A combined AFM-TEM study: Geophys. Res. Lett., v. 25, p. 1907–1910.

Prather, K. A., Nordmeyer, T., and Salt, K., 1994, Real-time characterization of individual aerosol particles using time-of-flight mass spectrometry: Anal. Chem., v. 66, p.1403–1407.

Prospero, J. M., 1999, Long-term measurements of the transport of African mineral dust to the southeastern United States: Implications for regional air quality: Jour. Geophys. Res., v. 104, p. 15,917–15,927.

Rea, D. K., 1994, The paleoclimatic record provided by eolian deposition in the deep sea: The geologic history of wind: Rev. Geophys., v. 32, p. 159–195.

Reader, M. C., Fung, I., and McFarlane, N., 1999, The mineral dust aerosol cycle during the Last Glacial Maximum: Jour. Geophys. Res., v. 104, p. 9381–9398.

Schulz, F. M., Stamnes, K., and Stamnes, J. J., 1999, Shape dependence of the optical properties in size-shape distributions of randomly oriented prolate spheroids, including highly elongated shapes: Jour. Geophys. Res., v. 104, p. 9413–9421.

Seinfeld, J. H., and Pandis, S. N., 1998, Atmospheric chemistry and physics: New York, John Wiley.

Sokolik, I. N., and Toon, O. B., 1996, Direct radiative forcing by anthropogenic airborne mineral aerosols: Nature, v. 381, p. 681–683.

______, 1999, Incorporation of mineralogical composition into models of the radiative properties of mineral aerosol from UV to IR wavelengths: Jour. Geophys. Res., v. 104, p. 9423–9444.

Spiro, P. A., Jacob, D. J., and Logan, J. A., 1992, Global inventory of sulfur emissions with 1° × 1° resolution: Jour. Geophys. Res., v. 97, p. 6023–6036.

Steffensen, J. P., 1997, The size distribution of microparticles from selected segments of the Greenland Ice Core Project ice core representing different climatic periods: Jour. Geophys. Res., v. 102, p. 26,755–26,763.

Tang, I., 1997, Thermodynamic and optical properties of mixed-salt aerosols of atmospheric importance: Jour. Geophys. Res., v. 102, p. 1883–1893.

Tegen, I., and Fung, I., 1994, Modeling of mineral dust in the atmosphere: Sources, transport, and optical thickness: Jour. Geophys. Res., v. 99, p. 22,897–22,914.

Tegen, I., Hollrig, P., Chin, M., Fung, I., Jacob, D., and Penner, J., 1997, Contribution of different aerosol species to the global aerosol extinction optical thickness: Estimates from model results: Jour. Geophys. Res., v. 102, p. 23,895–23,915.

Tegen, I., and Lacis, A. A., 1996, Modeling of particle size distribution and its influence on the radiative properties of mineral dust aerosol: Jour. Geophys. Res., v. 101, p. 19,237–19,244.

Tegen, I., Lacis, A. A., and Fung, I., 1996, The influence on climate forcing of mineral aerosols from disturbed soils: Nature, v. 380, p. 419–422.

Thompson, L. G., and Mosley-Thompson, E., 1981, Microparticle concentration variations linked with climatic change: Evidence from polar ice cores: Science, v. 212, p. 812–815.

Tsunogai, S., Suzuki, T., Kurata, T., and Uematsu, M., 1985, Seasonal and areal variation of continental aerosol in the surface air over the western North Pacific region: Jour. Oceanog. Soc. Japan, v. 41, p. 427–434.

Watson, A. J., 1997, Volcanic iron, CO_2, ocean productivity and climate: Nature, v. 385, p. 587–588.

Zhang, Y., and Carmichael, G. R., 1999, The role of mineral aerosol in tropospheric chemistry in East Asia—A model study: Jour. Appl. Meteorol., v. 38, p. 353–366.

PHYSICS AND CHEMISTRY OF CRUST MANTLE

Water as a Petrologic Catalyst Driving $^{18}O/^{16}O$ Homogenization and Anatexis of the Middle Crust in the Metamorphic Core Complexes of British Columbia

GREGORY J. HOLK

Department of Geological Sciences and Geological Engineering, Queen's University, Kingston, Ontario K7L 3N6, Canada

AND HUGH P. TAYLOR, JR.

Division of Geological and Planetary Sciences, California Institute of Technology, Pasadena, California 91125

Abstract

In the Thor-Odin and Valhalla metamorphic core complexes, we have documented a remarkable uniformity of mineral $\delta^{18}O$ values in the middle continental crust beneath the detachment faults. For example, in the Thor-Odin Complex, throughout an 8 km thick section of metasedimentary rocks and early Tertiary leucogranites in the hanging wall of the Monashee décollement (MD), quartz $\delta^{18}O$ = 12.3 ± 0.5‰ (1σ S.D.) for metapelite (N = 11), 12.0 ± 0.1‰ for quartzite (N = 2), 12.6 ± 0.6‰ (N = 4) for < 1 m thick amphibolite layers, and 12.1 ± 0.4‰ (N = 24) for the concordant leucogranites. No exceptions have been found to this remarkable $^{18}O/^{16}O$ homogeneity except locally in a couple of thick amphibolites and within a ductile, relatively impermeable, marble-rich section. Similar zones of $^{18}O/^{16}O$ homogeneity associated with leucogranite genesis are observed throughout the mid-crustal section of the Valhalla Complex and just beneath the MD in the Monashee Complex, the only difference being that those rocks are overall 0.5 to 1.5‰ lower in $\delta^{18}O$ than in the middle crust at Thor-Odin. These zones of pervasive homogenization in $^{18}O/^{16}O$ must be a result of exchange with magmatic or metamorphic H_2O, and these same volatiles appear to have been responsible for the leucogranite anatexis. A wide range in quartz $\delta^{18}O$ from +8 to +16 within and below the MD suggests that this major thrust fault was impermeable to aqueous fluid flow during the partial melting stage; at that time, the basement appears to have been isolated from the mid-crustal metamorphic-hydrothermal system.

LITHOPROBE crustal seismic profiles establish the MD as a W-dipping, crustal-scale ramp with 20 km of vertical relief, and Carr (1992) proposed an anatectic origin for the leucogranites during decompression melting associated with tectonic shortening as the mid-crustal section moved up this thrust ramp. Partial melting of metapelites and feldspathic grits from the Late Precambrian Windermere Supergroup began in response to influx of metamorphic H_2O, aided by internal muscovite dehydration at ≈8 kbar and ≈750°C at the base of the Monashee ramp. Metapelites are volatile rich, but feldspar poor, whereas the opposite is true for the grit lithologies. Thus, at the base of the Monashee ramp large-scale (≈30%) partial melting of the metapelites produced magmas near H_2O saturation (10 to14 wt%), whereas the intercalated arkosic grit-derived magmas were undersaturated (5 to 6 wt%). As these H_2O-rich, pelite-derived leucogranite melts moved upward to shallower depths, they cooled adiabatically and underwent decompressive exsolution of H_2O. The released H_2O was then able to exchange oxygen with lithologies infertile to melting as it concurrently migrated through the section toward the feldspathic grit layers, where it could act as a catalyst and be re-used, promoting further hydrothermal melting of the arkosic grits. Continued decompression melting and exsolution occurred simultaneously in different parts of the section during uplift, tectonic shortening, and buoyant uprise of the magma bodies, until final crystallization of all of the leucogranites took place much higher in the crust, where almost all of the H_2O was released and again re-used for a final episode of $^{18}O/^{16}O$ exchange with the unmelted metamorphic lithologies. In addition to the direct $^{18}O/^{16}O$ exchange that takes place between the metamorphic rocks and the migrating leucogranite magmas, this use and re-use of the same H_2O during repeated episodes of partial melting and exsolution in different parts of the section seems adequate to explain the pervasive oxygen isotopic homogenization of these metasedimentary rocks. It is estimated that 25 to 30% partial melting of a typical section of the Windermere Supergroup occurred as a result of these cumulative processes, and this probably played a pivotal role in determining the susceptibility of this orogen to subsequent extensional collapse along the detachment faults.

Introduction

METAMORPHIC CORE COMPLEXES represent ductilely deformed portions of the middle continental crust uplifted and exposed in an extensional setting (Armstrong, 1982). In western North America (Fig. 1), these features are associated with extensive magmatism (Armstrong and Ward, 1991) in crust thickened during Mesozoic orogenesis (Coney and Harms, 1984; Anderson et al., 1988; Sevigny et al., 1989, 1990; Liu and Furlong, 1993). Extension may be compensated by flow of rock material from unextended areas (Block and Royden, 1990; Wernicke, 1990), possibly driven by the weakening of the middle and/or lower crust by anatexis (Patiño Douce et al., 1990; Hollister, 1993) and/or injection of basaltic magma into the lower crust (Gans, 1987; Grunder and Wickham, 1991).

It is now clear that processes involving H_2O play a prominent role in the development of metamorphic core complexes at all structural levels, and also that stable isotope studies are essential in understanding these kinds of water/rock interactions. The influence of H_2O is clearly paramount in the production of crustal melts (Tuttle and Bowen, 1958; Wyllie, 1977; Taylor, 1988). These melts are a likely agent of H_2O transport and $^{18}O/^{16}O$ exchange during the early events that affected the ductile, deeper portions of metamorphic core complexes (Wickham and Peters, 1990; Wickham et al., 1991; Morrison, 1994; Peters and Wickham, 1995; Holk and Taylor, 1997a; 1997b; 1999). At later stages, it is well known that large-scale meteoric-hydrothermal systems have penetrated the upper plates of these complexes along major detachment faults (Kerrich, 1988; Beaudoin et al., 1991, 1992; Fricke et al., 1992; Morrison, 1994; Nesbitt and Muehlenbachs, 1995; Holk and Taylor, 1997a; 1997b; 1999).

At least two modes of hydrothermal activity are active during detachment faulting (e.g., see Fig. 2; Reynolds and Lister, 1987): (1) a convective system involving large quantities of heated meteoric waters or formation waters at hydrostatic pressure at shallow crustal levels within the brittle regime (Kerrich, 1988; Nesbitt and Meuhlenbachs, 1989, 1991, 1995); and (2) a low water/rock ratio system involving metamorphic and/or magmatic water at lithostatic pressures active at deep crustal levels in the ductile regime (Wickham and Peters, 1990, 1992; Holk and Taylor, 1997a). Hydrothermal fluids from the deep system may rise upward and mix with the meteoric-hydrothermal fluids (Beaudoin et al., 1992; Holk and Taylor 1997b) and/or the deeper system may later be overprinted by meteoric-hydrothermal alteration as these formerly ductile rocks enter the brittle regime during denudation (Reynolds and Lister, 1987; Morrison, 1994; Holk and Taylor, 1999). This overprinting may extend throughout the brittle portion of the crust during extension, perhaps to depths of 15 km (Nesbitt and Muehlenbachs, 1989, 1991, 1995). Mixing between the two types of aqueous fluids may also play a role in the formation of ore deposits in the upper plate (Spencer and Welty, 1986; Beaudoin et al., 1991, 1992; Beaudoin and Taylor, 1997; Holk and Taylor, 1997b; 1999).

In this paper we quantitatively discuss the hydrothermal activity associated with mechanism (1) above, namely the metamorphic and anatectic processes that produced the homogenization of $\delta^{18}O$ values in deep portions of the Thor-Odin and Valhalla metamorphic core complexes in southern British Columbia (Holk and Taylor, 1997a; 1999). We specifically evaluate the couplings between protolith stratigraphy, thermotectonic evolution, and aqueous fluid-rock interaction in the formation of the leucogranites. These terranes in southern British Columbia (Fig. 2) arguably represent the optimum location for addressing the role of H_2O in the tectonic and petrologic evolution of Cordilleran metamorphic core complexes, because: (1) the core complexes in this area are parts of very large systems in which deep-seated pre-extensional thrust faults are exposed along a strike length of as much as 200 km at the surface (Fig. 2; Brown and Journeay, 1987); (2) major uplifts have provided excellent exposures of a complete mid-crustal section below the detachment faults, which are well mapped and dated (Reesor and Moore, 1971; Carr et al., 1987; Parrish et al., 1988; Carr, 1991, 1992, 1995); and (3) stratigraphic units are well defined (Reesor and Moore, 1971) and reliably correlated (Hoy, 1977; Carr, 1991). Nonetheless, the most compelling reason for singling out these areas for isotopic study is (4) their high-latitude location relative to other metamorphic core complexes in North America (Fig. 1); meteoric water is known to have been

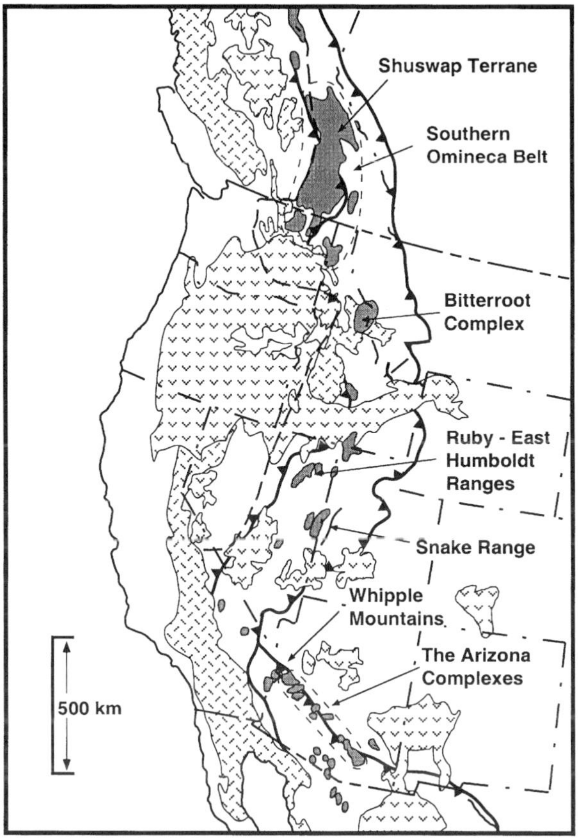

FIG. 1. Generalized map of Cenozoic metamorphic core complexes in western North America (after Armstrong, 1982). The various core complexes, denoted by the black shading, formed as the result of the exhumation of the middle crust during detachment faulting. Tertiary volcanic rocks are shown with a "V" pattern. The Cordilleran batholith belt is shown with a cross-hatch stipple pattern.

very low in ^{18}O and deuterium during the Late Cretaceous and Early Tertiary in British Columbia (Magaritz and Taylor, 1986; Nesbitt and Muehlenbachs, 1989, 1991, 1995; Beaudoin et al., 1991, 1992; Criss et al., 1991; Holk and Taylor, 1997a; 1999), providing a very clear-cut isotopic contrast between the local meteoric waters and either metamorphic or magmatic water (e.g., Sheppard, 1986) This allows us to quantitatively sort out the extent of the meteoric-hydrothermal overprint that is ubiquitous in all of these tectonic systems, particularly in the vicinity of the detachment faults. Detailed discussions of the effects of the meteoric-hydrothermal systems in the upper plates and along the detachment faults associated with the Thor-Odin and Valhalla complexes are given in a series of other papers (Holk and Taylor, 1997a, 1997b, 1999; Holk et al., 1997).

Basement Rocks of the Monashee Complex

Geologic setting

The Monashee Complex, which occurs in the footwall of the 58 to 59 Ma Monashee décollement (Carr, 1992), represents the largest and best-documented exposure of basement rocks in the metamorphic core complexes of southern British Columbia (Fig. 3). This basement terrane is comprised of ~2200 Ma, Lower Proterozoic paragneiss together with multiple intrusive bodies of highly deformed, locally migmatized, 1934 ± 6 Ma Proterozoic orthogneiss (Reesor and Moore, 1971; Wanless and Reesor, 1975; Armstrong et al., 1991; Parkinson, 1991); these units are unconformably overlain by an up to 2300 m thick mantling gneiss sequence correlated with the Upper Proterozoic Windermere Supergroup (Hoy, 1988; Scammel and Brown, 1990). Migmatiza-

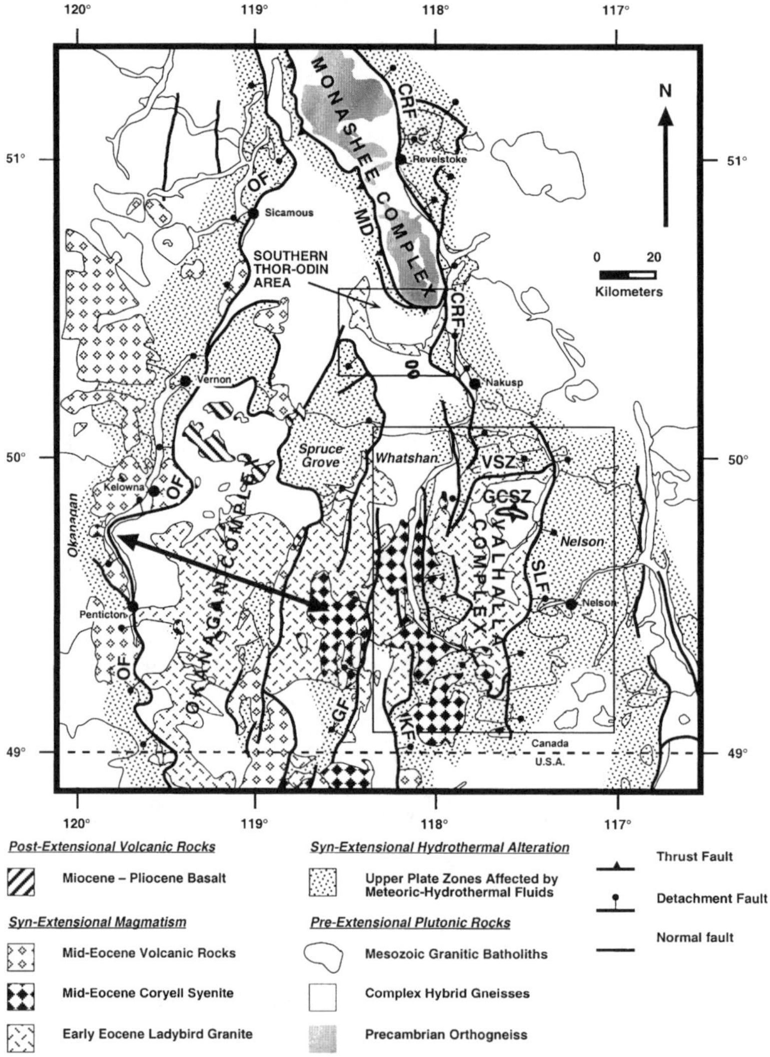

FIG. 2. Simplified geologic map of the southern Omineca belt (after Parrish et al., 1988), identifying our study areas in the vicinity of the southern Thor-Odin (small rectangle) and Valhalla (large rectangle) metamorphic core complexes. Deep, ductile Jurassic to Paleocene compressive structures in the lower plate form the boundaries between the underlying basement and the mid-crustal Selkirk allochthon; two of these are shown—the large Monashee décollement (MD) in the north and a small remnant of the Gwillim Creek shear zone (GCSZ) in the south, both indicated with teeth in the hanging wall. Extensional detachment faults form the boundaries between the upper and lower plates, with barbs pointing into the upper plate. The Early Eocene (58 to 52 Ma) E-dipping detachment faults—such as the Columbia River fault (CRF), Slocan Lake fault (SLF), and Kettle fault (KF)—are coeval with the synorogenic lower-plate Ladybird leucogranites. The Mid-Eocene (52 to 45 Ma) W-dipping detachment faults, such as the Okanagan fault (OF) and Granby fault (GF), are coeval with the syn- to post-tectonic Coryell intrusions. Extensional displacements along the Okanagan fault may have approached 100 km, as indicated by the large, double-barbed arrow linking two Coryell intrusions thought to have originally been part of the same 51 Ma pluton (Tempelman-Kluit and Parkinson, 1986; Carr and Parkinson, 1989). The Early Eocene Valkyr shear zone (VSZ) is locally cut by both the Ladybird leucogranites and the Coryell syenites. The dotted pattern denotes the broad areas of the upper plates above the detachment faults that have experienced $^{18}O/^{16}O$ and/or D/H changes as a result of meteoric-hydrothermal activity (Magaritz and Taylor, 1986; Beaudoin et al., 1992; Kyser et al., 1994; Nesbitt and Muehlenbachs, 1995; Holk and Taylor, 1997a, 1999; Holk et al., 1997). Such meteoric-hydrothermal effects also occur locally just above the MD.

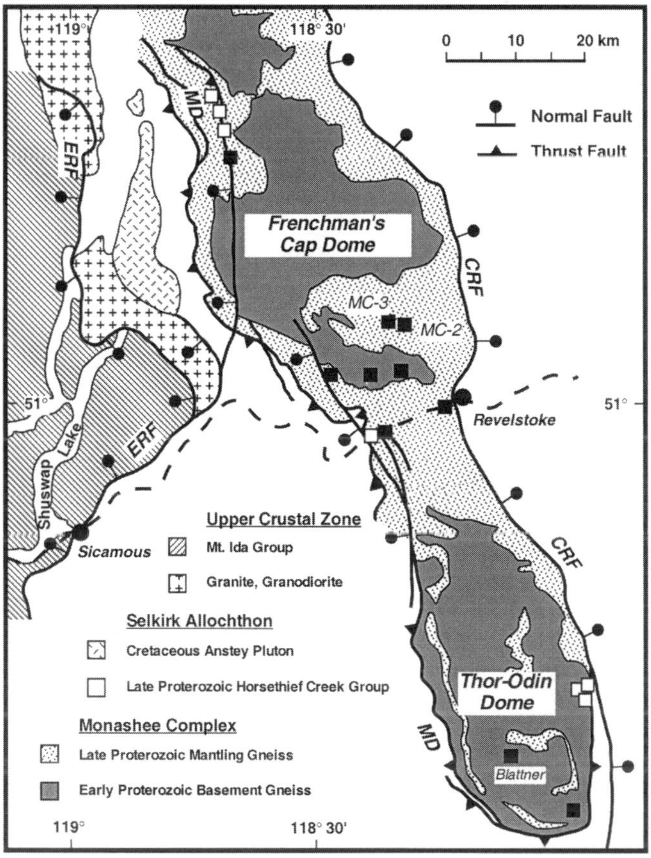

FIG. 3. Simplified geology (modified from Brown and Journeay, 1987; Parkinson, 1991) of the Monashee Complex showing sample locations. The thick dashed curve is the Trans-Canada Highway. The Monashee décollement (MD) separates the basement from the Selkirk allochthon to the west and south. The Eagle River fault (ERF) and the Columbia River fault (CRF) are extensional detachment faults that separate the upper-crustal zones from the Monashee Complex and the mid-crustal Selkirk allochthon. The seven sample localities <200 m below the Monashee décollement (MD) are denoted by open squares. All other samples are indicated by solid squares, including the outcrop locality in the Thor-Odin dome studied by Blattner (1971). MC-2 and MC-3 are localities where $^{18}O/^{16}O$ heterogeneity was observed at the outcrop scale.

tion and metamorphism of these older gneiss units took place before deposition of the mantling gneiss sequence.

The mantling gneiss is composed dominantly of quartzite with interlayered metapelite, quartzofeldspathic gneiss, amphibolite, hornblende gneiss, pyroclastic carbonatite, and calcareous schist (Hoy, 1988; Scammell and Brown, 1990). These paragneisses record a two-stage, Mesozoic to early Cenozoic history of prograde metamorphism related to the eastward thrusting of the Selkirk allochthon (Brown and Journeay, 1987; Journeay, 1988; Lane et al., 1989). In the footwall of the Monashee décollement, the metamorphic grade increases from greenschist facies (450°C, 3.4 kbar) to upper amphibolite facies (650 to 680°C, 2.5 kbar) in an *upward* direction, indicating an inverted thermal gradient during the latest stages of movement along this thrust zone during the Paleocene (Journeay, 1988). Immediately below the décollement, Nyman et al. (1995) reported granulite-facies conditions (720–820°C, 7.5 to 9.0 kbar) in which the migmatitic mantling gneiss experienced as much as 50% partial melting.

Isotopic results

Excluding the rocks that lie immediately beneath the Monashee décollement, our data demonstrate that the basement orthogneiss and para-

TABLE 1. $^{18}O/^{16}O$ Data on Rocks and Minerals from the Monashee Complex[1]

Sample	Rock type	Qz[2]	Fs[2]	Gar[3]	Am[2]	Bi[2]	WR[2]
	Early Proterozoic basement gneiss						
3099[4]	Granodiorite gneiss				3.8		
3100[4]	Pegmatoid vein	8.0	6.5			2.5	
3117[4]	Granite	8.4	6.7		4.3	2.7	
3118[4]	Amphibolite				4.1		
? ? ?[4]	Metaconglomerate		8.8				
GH-49	Metapelite gneiss	10.2	7.7				
GH-669	Granodiorite gneiss	9.7	8.1	7.1			
MC-4	Granite gneiss	9.3	2.7				
MC-5	Quartzite	12.0					
	Late Proterozoic mantling gneiss > 200 m below Monashee décollement						
GH-76	Metapelite	14.0					
GH-556	Calc-silicate gneiss	16.4					
MC-2	Pelitic calc-silicate	14.7					
MC-2a	Metapelite	13.6	12.1				
MC-2b	Calc-silicate	13.4					
MC-3a	Psammite	12.2					
MC-3b	Quartzite	12.2					
MC-3c	Metapelite	14.7					
	< 200 m below Monashee décollement						
GH-75	Migmatitic metapelite	11.0	9.0				10.7
GH-552	Metapelite	10.6					
GH-553	Metapelite	11.6	9.6				
GH-555	Metapelite	11.3					
GH-841	Granodiorite gneiss	10.6	9.0				
GH-842	Leucogranite gneiss	11.0	8.4				
GH-843	Migmatitic leucogranite	11.1	9.1				

[1]Mineral separates were hand-picked under the microscope. Some quartz separates were treated with HF to remove impurities. Oxygen was liberated from quartz and feldspar by reaction with F_2 in Ni reaction vessels at 550°C, converted to CO_2, and analyzed on a Finnegan MAT 252 mass spectrometer (Taylor and Epstein, 1962).

[2]The oxygen isotope data are from sample localitites shown on Figure 3 for whole rocks (WR), quartz (Qz), feldspar (Fs), almandine garnet (Gar), amphibole [Am], biotite [Bi], and calcite (Cc), all reported in the δ notation, where δ is the relative difference in parts per thousand in $^{18}O/^{16}O$ between the sample and a standard. The standard is V-SMOW (Vienna standard mean ocean water). NBS-28 has a $\delta^{18}O$ = +9.60 on this scale. Replicate analyses of samples and the Caltech rose quartz working standard ($\delta^{18}O$ = +8.45) have an analytical error of ± 0.2‰.

[3]The garnet oxygen isotope data were also acquired by a fluorination procedure, but one using the Caltech oxygen isotope laser extraction system (e.g., Sharp, 1990), instead of externally heated Ni reaction vessels.

[4]These data are from Blattner (1971).

gneiss of the Monashee Complex are very heterogeneous in ^{18}O (Table 1). Samples of Early Paleozoic basement gneiss have quartz $\delta^{18}O$ = +8.0 to +12.0 (N = 6) and metasediments from the mantling gneiss > 200 m below the Monashee décollement have quartz $\delta^{18}O$ = +12.2 to +16.4 (N = 8). Isotopic heterogeneity of up to 3‰ in $\delta^{18}O$ quartz has been observed in a single outcrop of the mantling gneiss. This very large 8.4‰ range in quartz $\delta^{18}O$ contrasts sharply with the rocks in the overlying middle-crustal units from the Thor-Odin and Valhalla complexes, as described below.

Also, in sharp contrast to the isotopic heterogeneities in most of the basement rocks of the Monashee Complex, the rocks <200 m below the Monashee décollement were found to be remarkably uniform in $^{18}O/^{16}O$ (quartz $\delta^{18}O = +11.0 \pm 0.3‰$, $N = 7$; feldspar $\delta^{18}O = +9.0 \pm 0.4‰$, $N = 5$). Isotopic equilibrium at magmatic temperatures is indicated by $\Delta_{Q\text{-}F} = 1.6$ to $2.6‰$. Water in equilibrium with this quartz at ~700°C would have $\delta^{18}O \sim +10.5$ (Clayton et al., 1972). The isotopic homogenization of these rocks is most likely associated with the anatectic event described by Nyman et al. (1995), who showed that the rocks directly underlying the décollement underwent as much as 50 vol% melting during thrusting.

Local $^{18}O/^{16}O$ homogenization beneath the Monashee décollement

The heterogeneous $^{18}O/^{16}O$ results from basement rocks of the Monashee Complex indicate relatively low water/rock ratios for this deep section of the southern Omineca belt at the time of prograde metamorphism. However, there must have been enough H_2O locally in the system to drive anatexis and $^{18}O/^{16}O$ homogenization in the immediate footwall of the Monashee décollement. The most likely source of aqueous fluids in this system is from the dehydration of hydrous minerals as the mantling gneisses rocks are heated during overthrusting of the Selkirk allochthon. During this short-lived thermal event, metamorphic dehydration fronts (muscovite-out, chlorite-out) migrate downward (e.g., England and Thompson, 1984), whereas the buoyant aqueous fluids released during dehydration move upward (e.g., Walther and Orville, 1982) and up temperature toward the Monashee décollement. The impermeable ductile rocks of the décollement are inferred to have focused fluid flow in an up-dip direction along the sole of the shear zone, resulting in relatively high water/rock ratios and $^{18}O/^{16}O$ homogenization just beneath the décollement.

The amount of water introduced into the system from dehydration of chlorite and muscovite of metapelite during the thermal inversion event can be estimated as follows. Journeay (1988) reported a minimum temperature of ~450°C, a temperature well within the stability field of chlorite, for the deepest parts of this inverted sequence. A metapelite containing 35 vol% muscovite and 20 vol% chlorite contains about 4.5 wt% H_2O. Fluids released during dehydration of these minerals are enough to drive ~35 vol% melting of the pelites (assuming the melt contains 12 wt% H_2O). The remaining 15% partial melting, needed to match the estimate of Nyman et al. (1995) for the sole of the Monashee décollement, is assumed to be fluxed by the addition of H_2O from deeper, lower-grade sequences of mantling gneiss. These quantities of aqueous fluids seem adequate to explain the degree of $^{18}O/^{16}O$ exchange and homogenization in the gneisses just beneath the Monashee décollement (see Fig. 6 below).

These pockets of melt along the sole of the Monashee décollement may function as an impermeable barrier inhibiting the transport of aqueous fluids from the basement zone across and into the overlying Selkirk allochthon. Also, the lubricating efects of these shear zone–related melts could facilitate rapid movement along the Monashee décollement (e.g., Hollister, 1993), providing a mechanism for the adiabatic decompression of the Selkirk allochthon as it is transported up the thrust ramp.

The Southern Thor-Odin Complex

Geologic setting

The Thor-Odin Complex, which lies structurally above the Monashee Complex (Figs. 2 and 4), is the product of late Mesozoic to Paleocene compression followed by Eocene extension (Parrish et al., 1988; Carr, 1991, 1992, 1995; Vanderheaghe and Teyssier, 1997). It consists of three fault-bounded zones (Figs. 4 and 5) that are typical of core complexes in the southern Omineca belt (e.g., Parrish et al., 1988). Going from deepest to shallowest, these include: Precambrian crystalline basement; mid-crustal, ductilely deformed, upper amphibolite-facies metamorphic rocks; and a brittle upper plate that lies above the detachment faults.

The mid-crustal zone and adjoining bounding structures preserve evidence of Mesozoic to Early Tertiary deformation, metamorphism, and anatexis. This zone consists of a 6000 m thick sequence of metapelite, quartzite, psammite, amphibolite, calc-silicate, and marble (Reesor and Moore, 1971; Carr, 1991, 1992), portions of which are correlated with (1) the Late Precambrian to Mississippian clastic, post-rift, pericratonic sequence of the Kootenay terrane, and (2) the Permian to Triassic, deep-ocean basin and island-arc sediments of the Slide Mountain and Quesnellia terranes (Hoy, 1977; Carr, 1991). Interlayered within this sequence are cm to km thick sheets of Eocene (Ladybird) leucogranite.

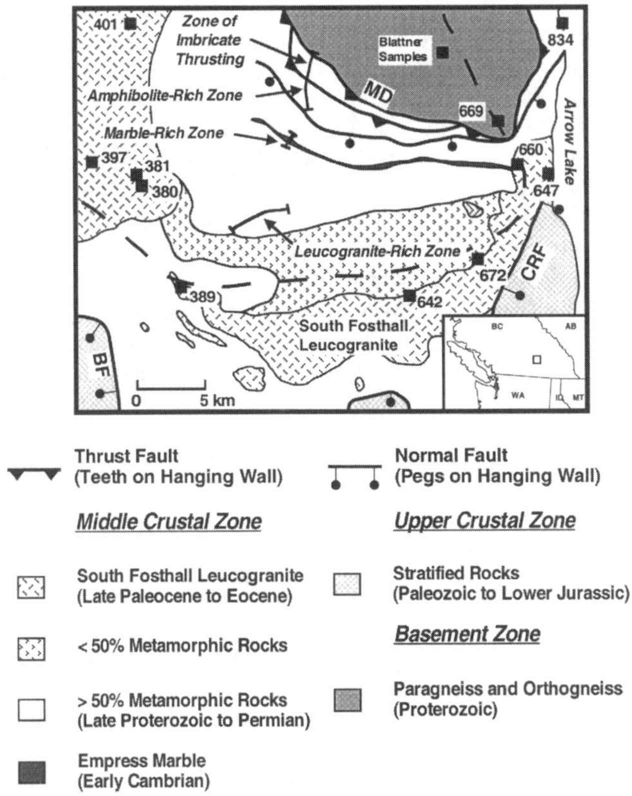

FIG. 4. Map of the southern Thor-Odin metamorphic core complex (modified from Carr, 1992) showing the locations of Traverses 1, 2, and 3, and other sample locations (squares). The Monashee décollement (MD) separates the basement from the middle crustal zone of the Selkirk Allochthon. The Columbia River fault (CRF) and Beavan fault (BF) are Eocene detachment faults that separate the middle crustal zone from the overlying upper plate.

Cross-cutting dikes, stocks, and plutons of this leucogranite also are common.

The Monashee décollement (Figs. 3, 4, and 5) is a ductile shear zone, active up to 58–59 Ma (Brown et al., 1992; Carr, 1992), responsible for the northeastward transport of mid-crustal rocks of the Selkirk allochthon over the basement zone of the Monashee Complex. In the southern Thor-Odin area, this shear zone is a duplex containing imbricate slices of basement and supracrustal gneisses (McNicoll and Brown, 1995). Lithoprobe crustal seismic profiles show that this structure is a W-dipping crustal-scale ramp with 20 km of vertical relief (Cook et al., 1992). This served as the root thrust of the Rocky Mountain Foreland Thrust Belt, and is proposed to have had at least 200 km displacement (Brown et al., 1992). Just to the north of the Thor-Odin Complex, apparent temperature and pressure conditions during shearing were >650° C and >6 kbar (Lane et al., 1989; Nyman et al., 1995). According to Carr (1995), the thermal maximum in this terrane occurred at ~60 to 70 Ma, and this was associated with anatexis and formation of the Ladybird leucogranite suite during the waning stages of compressional deformation. Rapid cooling from >700°C to ~300°C accompanied the extensional denudation of the complex during the internal 58 to 45 Ma.

Mylonitic and brittle deformation along the low-angle Columbia River and Beavan detachment faults (Fig. 2) accommodated the respective eastward and westward movement of the upper crustal zone during the Eocene exhumation of the mid-crustal zone (Parrish et al., 1988; Carr, 1992). The Columbia River fault was definitely active at 55 Ma, but may have been active as early as 59.7 Ma (Parrish et al., 1988). Displacement along the Columbia River fault at the latitude of the

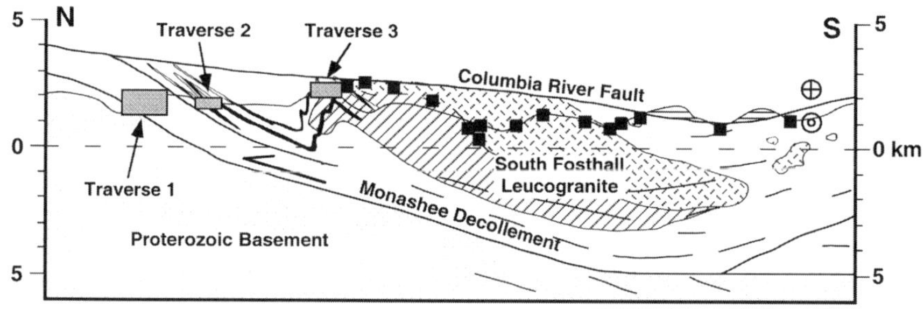

FIG. 5. N-S cross section through the southern Thor-Odin metamorphic core complex, modified after Carr (1992). Sample traverses and sample locations from Figure 4 are projected onto this cross section at their correct structural locations. Traverse 1 includes the zone of imbricate thrusting and an overlying amphibolite-rich zone. Traverse 2 corresponds to a portion of the marble-rich zone. Traverse 3 is the pegmatite-rich and leucogranite-rich zone beneath the Ladybird granite pluton. Standard symbols indicating the direction of movement along the Columbia River fault are shown on the right-hand side of the cross section (plus sign for movement away from viewer and open circle for movement toward the viewer).

Thor-Odin Complex is estimated to be 20–30 km (Read, 1979; Parrish et al., 1988).

In the southern Thor-Odin Complex, the Ladybird leucogranite sheets are synkinematic with deformation associated with the latest stages of upper amphibolite-facies metamorphism (Reesor and Moore, 1971; Duncan, 1984; Carr, 1991, 1992). The outcrop abundance of leucogranite increases from 30% immediately above the Monashee décollement to nearly 100% in the vicinity of the 3 km thick, sill-like South Fosthall pluton (Carr, 1991). U-Pb zircon ages (55 to 62 Ma) constrain the time of crystallization of these granites to be concurrent with the latest deformation along the Monashee décollement and the earliest stages of movement along the Columbia River fault (Carr, 1992, 1995). Carr (1992) cited Wyllie (1977) and Whitney (1988) in proposing an anatectic origin for these leucogranites during decompression melting associated with tectonic shortening as the Selkirk allochthon was carried up the Monashee thrust ramp. Anatexis resulted in the mechanical weakening of the middle crustal zone, facilitating the initiation of extension during the collapse of the overthickened crust (e.g., Vanderheaghe and Teyssier, 1997).

Isotopic results

Sample localities and sample traverses are shown on the simplified geologic map and cross section of the southern Thor-Odin Complex (Figs. 4 and 5), and the stable isotope data for rocks and coexisting minerals are given in Tables 2 and 3 and Fig. 6. Traverse 1 is a suite of samples collected across the imbricate thrust zone and into an overlying amphibolite-rich zone in which about 20% of the section is composed of amphibolite layers of variable thickness; this zone is correlative with a Proterozoic rift sequence, namely the semipelitic amphibolite division of the Horsethief Creek Group, a subdivision of the Windermere Supergroup (Carr, 1991). This amphibolite-rich zone is lithologically heterogeneous, consisting of 0.1 m to 40 m thick layers of metapelite, calc-silicate, quartzite, marble, and amphibolite intruded by 0.2 m to 30 m thick leucogranite sheets. Traverse 2 crosses an equally heterogeneous, but more marble rich zone comprised of thin layers (<1 m) of all of the above lithologies except amphibolite; these are intercalated with marble units that are up to 50 m thick (Cambrian Empress Marble). This marble-rich zone is correlated with intercalated marble and quartzite of the Cambrian Mohican and Badshot formations (Carr, 1991). Traverse 3 passes through a leucogranite-rich zone just beneath a large Ladybird Series leucogranite body (the South Fosthall pluton; Fig. 4); here, sequences of metapelite, amphibolite, and calc-silicate correlated with the Cambrian–Ordovician Lardeau Group occur as discrete layers within the dominant leucogranite.

$^{18}O/^{16}O$ homogenization of the middle crust in the Thor-Odin Complex

Throughout the Selkirk allochthon, but particularly within the uppermost 4 km thick leucogranite-rich zone, minerals in the metasediments and leucogranites are characteristically uniform in ^{18}O (Fig. 6 and Table 2), and in approximate isotopic equilibrium at high temperatures. For example, in

TABLE 2. $^{18}O/^{16}O$ Data on Rocks and Minerals from the Thor-Odin Middle Crustal Zone

Sample[1]	Rock type	Qz	Fs	Gar	Other[2]	Cc[3], $\delta^{18}O$	Cc[3], $\delta^{13}C$
		Amphibolite-rich zone					
MD-33a	Leucogranite sheet	12.1	10.8				
MD-35	Pelitic gneiss	12.5		10.1			
MD-35a	Leucogranite dike	13.2	–4.0				
MD-36a	Calc-silicate					15.2	–3.5
MD-36b	Marble					18.8	8.5
MD-37	Pelitic gneiss	11.6		8.9			
MD-38	Leucogranite sheet	11.9	5.2				
MD-38a	Quartzofeldspathic gneiss	13.1	11.1		7.3a 7.2b		
MD-38b	Amphibolite	12.8			10.1a 7.8b		
MD-39a	Leucogranite sheet	12.0	10.6				
MD-40	Amphibolite		–1.0		4.1a		
MD-40a	Leucogranite sheet	11.5	10.9				
MD-40b	Pelitic gneiss	11.6	9.4		4.0b		
MD-41	Quartzite	12.1					
MD-41a	Leucocratic sheet	12.7	11.3	9.5			
MD-41b	Amphibolite	13.4		9.8	10.5a 9.5b		
MD-50	Pelitic gneiss	12.2	10.7				
MD-50a	Leucocratic sheet	12.5	10.8	8.4			
MD-49	Pelitic gneiss	12.6	6.8	10.3	3.9b		
MD-49a	Quartz vein	11.9					
MD-48	Feldspathic gneiss		10.7	8.5			
MD-48a	Amphibolite	11.7		7.9	7.9a 6.2b		
MD-48b	Leucocratic sheet	12.1	9.8	8.1	7.1b		
		Thick amphibolite outcrops					
MD-47	Pelitic gneiss	10.5	8.2	6.5			
MD-47a	Amphibolite	9.5		6.4			
MD-46	Amphibolite	9.5	7.9	5.9	7.6a 2.9b		
MD-46a	Pelitic gneiss	10.9	9.6	7.8			
MD-46b	Leucogranite sheet	11.5		7.6			
MD-45	Amphibolite	12.8		10.2	9.4a		
MD-45a	Leucocratic sheet	12.5	11.0	10.0	8.8b		
MD-44	Calc-silicate	12.7		10.2			
MD-44a	Leucocratic sheet	11.9	10.2				
MD-43	Leucocratic sheet	12.1	11.3				
MD-43a	Calc-silicate	13.6					
MD-42	Amphibolite		8.9	6.6	6.1a		
NF-3	Amphibolite	12.1		10.3			
		Marble-rich zone					
MD-57	Quartzite	13.4					
MD-57a	Semi-pelitic gneiss	13.7	13.1				
MD-57b	Psammitic gneiss	13.8					
MD-58	Pelitic gneiss	14.3	13.5		9.4b		
MD-58a	Leucogranite sheet	14.1	13.7				
MD-58b	Psammitic gneiss	14.9	13.5				
MD-58c	Marble					21.0	–1.9
MD-58d	Calc-silicate	12.2			12.3p		

(continued)

TABLE 2. (*continued*)

Sample[1]	Rock type	Qz	Fs	Gar	Other[2]	Cc[3], $\delta^{18}O$	Cc[3], $\delta^{13}C$
Marble rich zone (*continued*)							
MD-58e	Marble					18.2	0.9
MD-58f	Leucogranite sheet	14.8	13.1				
MD-58g	Semi-pelitic gneiss	15.1		13.0	11.6b		
MD-58h	Quartzite	13.4					
MD-56a	Leucogranite sheet	18.6	16.8				
MD-56b	Marble					20.6	0.3
MD-56c	Leucogranite dike	17.2	16.6				
MD-55	Pelitic gneiss	13.7	13.0	12.3			
MD-54	Marble					21.7	–0.4
MD-53	Quartzite	15.9					
MD-53a	Pelitic gneiss	14.9	13.7	12.2			
MD-53b	Marble					19.3	1.9
MD-51	Calc-silicate	15.5				15.2	–4.0
MD-51W	Marble: whole rock					21.5	2.3
MD-51C	Marble: coarse fraction					22.0	2.4
MD-51M	Marble: matrix fraction					21.5	2.2
MD-51a	Calc-silicate	15.7				15.5	–5.2
MD-51b	Leucocratic sheet	16.1	14.0				
MD-51c	Marble					15.9	2.0
MD-51d	Pelitic gneiss	16.0	15.1	13.3			
Leucogranite-rich zone							
MD-59	Leucogranite dike	10.9	10.1				
MD-60a	Leucogranite sheet	12.6	10.5				
MD-63	Pelitic gneiss	12.9	11.5				
MD-63a	Amphibolite		11.5	10.1	9.9a		
MD-66	Leucogranite sheet	11.6	10.9				
MD-68	Leucogranite sheet	12.7	11.0				
MD-70	Leucogranite sheet	12.2	11.2				
MD-70a	Amphibolite	13.1	11.4	10.3	8.8b		
MD-71a	Pelitic gneiss	12.1		9.6			
MD-73	Leucogranite	11.7	10.4				
MD-74a	Marble					12.4	0.2
MD-74b	Calc-silicate					9.6	–6.4
MD-77a	Pelitic gneiss	11.7		9.4			
MD-78	Pelitic gneiss	12.8		9.6			
MD-80a	Pelitic gneiss	12.1		9.8			
MD-81	Leucogranite	11.7	10.4				
South Fosthall pluton							
GH-380	Leucogranite	12.0	10.3				
GH-381	Leucogranite	12.3		9.0			
GH-389	Leucogranite	12.4	10.6				
GH-397	Leucogranite	12.3	10.3				
GH-401	Leucogranite	11.7	9.9				
GH-612	Leucogranite	11.6	9.9				

[1]Numbers refer to samples archived at the Division of Geological and Planetary Sciences, California Institute of Technology, Pasadena, California. Data are from sample localities and traverses shown in Figure 4. See footnotes to Table 1 for an explanation of symbols.

[2]The symbols a and b in the column labeled "Other" represent, respectively, analyses of amphibole (hornblende) and biotite.

[3]The calcite stable isotope data were obtained using the standard analytical procedure involving reaction with H_3PO_4 (McCrea, 1950), and corrected using the fractionation factor of 1.01008 (Sharma and Clayton, 1965).

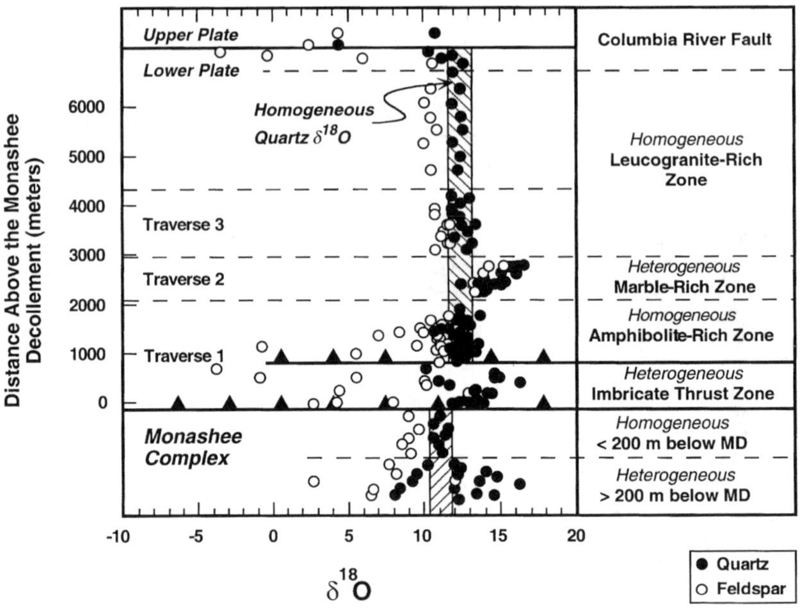

FIG. 6. Quartz and feldspar $\delta^{18}O$ vs. structural height relative to the Monashee décollement for 115 rock specimens analyzed from the southern Thor-Odin complex, excluding three samples from thick, low-^{18}O amphibolite units and five late-stage leucocratic dikes (sample numbers MD-35a, MD-17a, MD-19b, MD-56c, and MD-59). Note that quartz and feldspar are each very uniform in ^{18}O and in apparent isotopic equilibrium at near-magmatic temperatures throughout much of the mid-crustal zone, indicating pervasive homogenization with an aqueous fluid having $\delta^{18}O \approx +11.0$. The high $\delta^{18}O$ values in the marble-rich zone represent a lack of complete isotopic homogenization with the initially extremely ^{18}O rich ($\delta^{18}O > 20$), relatively impermeable marble layers (see text). Similarly, the low $\delta^{18}O$ values of quartz from two pelites and a leucogranite collected in the amphibolite-rich zone are attributable to buffering by the directly adjacent, thick low-^{18}O amphibolite layers. Feldspar ^{18}O depletions (down to $\delta^{18}O = -4.0$) within the imbricate thrust zone and in the vicinity of the Columbia River fault are the result of late-stage meteoric-hydrothermal exchange associated with detachment faulting (see Holk and Taylor, 1997a; 1999). Marked variability of quartz $\delta^{18}O$ within the basement and the imbricate thrust zone indicates that these deeper parts of the complex did not undergo oxygen isotope homogenization, except for a zone of anatexis within 200 m of Monashee décollement. There, the $\delta^{18}O$ of quartz is uniform at a level 1.0 to 1.5‰ lower than homogeneous quartz in the overlying allochthon, as shown by the offset of the two patterned vertical bands. The plotted samples of deeper basement are from a wider area than shown on Figure 4, and two of the plotted quartz-feldspar pairs are from Blattner (1971).

the leucogranite-rich zone and in most of the amphibolite-rich zone, quartz $\delta^{18}O = 12.3 \pm 0.5$‰ (1σ S.D.) for metapelite ($N = 11$), 12.0 ± 0.1‰ for quartzite ($N = 2$), and 12.1 ± 0.4‰ ($N = 24$) for the concordant leucogranites. Within the leucogranite-rich zone, no exceptions have been found to this remarkable $^{18}O/^{16}O$ homogeneity that is delineated by the patterned vertical band in the upper part of Figure 6; this includes a 1 m thick marble unit, which has calcite $\delta^{18}O = 12.2$, as well as even the very refractory garnets (see Holk and Taylor, 1997a). Ignoring a few low-^{18}O feldspars that are obviously the result of local, later-stage meteoric-hydrothermal alteration of this easily exchanged mineral during detachment faulting (Holk and Taylor, 1997a, 1999), the quartz-feldspar $^{18}O/^{16}O$ fractionations (i.e., Q-F values) from all igneous and metamorphic lithologies within this zone are typical of granitic plutons throughout the world ($\Delta_{Q\text{-}F} = 1.5 \pm 0.6$‰, $N = 28$). Using the calibrations of Clayton et al. (1972) and Clayton and Kieffer (1991), the above data imply apparent equilibrium temperatures of 550 to 650°C and a coexisting magmatic-hydrothermal fluid with $\delta^{18}O \approx +11.0$, which is at least 20‰ higher than the coeval meteoric surface waters associated with detachment faulting (Magaritz and Taylor, 1986).

Quartz associated with the amphibolites is isotopically more variable then the quartz in the litholo-

TABLE 3. $^{18}O/^{16}O$ Data on Minerals from the Zone of Imbricate Thrusting Associated with the Monashee Décollement at Thor-Odin

Sample[1]	Rock type	Qz	Fs	Other[2]	Cc, $\delta^{18}O$	Cc, $\delta^{13}C$
		Lower-plate basement				
MD-1	Basement gneiss	13.2	2.5			
		Zone of imbricate thrusting				
MD-2	Quartzite	11.7				
MD-3	Pelitic schist	13.7				
MD-4	Quartzite	11.9				
MD-5	Pelitic schist	12.9	4.0			
MD-6	Pelitic schist	12.1	7.8			
MD-6a	Aluminous schist	12.7				
MD-7	Basement gneiss	12.7				
MD-14	Quartzite	13.3				
		Thrust				
MD-17	Calc-silicate				15.7	–7.6
MD-17a	Leucogranite dike	14.2	12.7			
MD-17b	Quartz vein	14.0				
		Thrust				
MD-19b	Leucogranite dike	13.0	4.2			
MD-20a	Quartz vein	13.3				
		Thrust				
MD-24	Basement gneiss	11.5	10.0			
		Thrust				
MD-26a	Calc-silicate	16.2			15.4	–4.9
MD-27	Pelitic schist	10.8	9.8			
MD-29	Quartzite	14.7				
MD-29a	Pelitic schist	14.4	–1.1	4.2b		
MD-29b	Amphibolite		5.3	8.4g 8.6a		
		Thrust				
MD-30	Quartzite with late calcite veins	14.5			12.8	–1.7
MD-31	Pelitic schist with late calcite	10.0	–4.0		–2.1	–7.9
MD-32	Pelitic schist	11.9		9.1g		

[1]Numbers refer to samples archived at the Division of Geological and Planetary Sciences, California Institute of Technology, Pasadena, California. Data are from sample localities and traverses shown on Figure 4. See footnotes to Table 1 for an explanation of symbols.

[2]Note that g in the column labelled "other" represents garnet (almandine).

gies described above, but quite systematic in the sense that relatively low $\delta^{18}O$ values (+9.5) are found only in two very thick layers (10 to 40 m) of undeformed amphibolite within a narrow interval in the upper part of the amphibolite-rich zone. Elsewhere in this zone, homogenized quartz $\delta^{18}O$ values of +12.6 ± 0.6‰ ($N = 4$) are observed in all of the numerous thin (<1 m) amphibolite layers; in contrast to the thick amphibolites, these thin layers typically contain abundant quartzofeldspathic veins or melt segregations. Quartz in metapelitic paragneiss directly adjacent to the thick amphibolite units is 0.5 to 1.5‰ lower in ^{18}O than the characteristic homogenized quartz found in all lithologies throughout the rest of the 900 m thick amphibolite-rich zone (Fig. 6); this indicates local $^{18}O/^{16}O$ buffering as a result of exchange with the nearby reservoir of lower-^{18}O amphibolite.

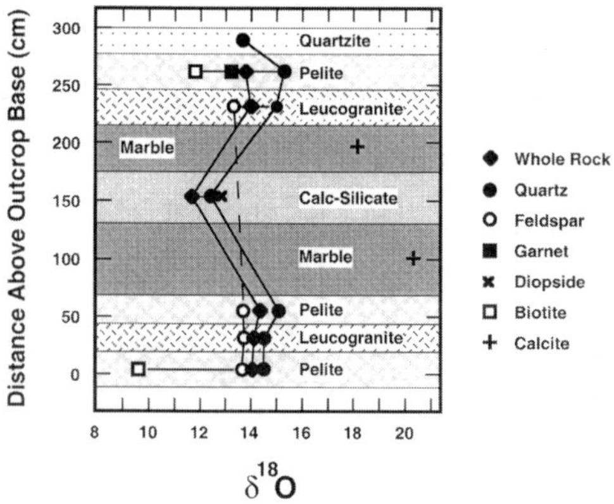

FIG. 7. Plot showing outcrop-scale variations in mineral and whole rock $\delta^{18}O$ at site MD-58 in the marble-rich zone. For the metapelites and leucogranites, $\delta^{18}O$ values of quartz (+14.8; $N = 5$), feldspar (+13.7; $N = 4$), and whole rock (+14.2; $N = 5$) are homogeneous, even across the up to 0.5 m thick, high-$\delta^{18}O$ marbles ($\delta^{18}O$ = +18.2 and +21.0).

Quartz $\delta^{18}O$ values in the marble-rich zone (Fig. 6) are higher and more variable than in any other part of the allochthon—13.7 to 16.0 ($N = 8$) for metapelite, 13.4 to 16.0 ($N = 3$) for quartzite, 12.2 to 15.7 ($N = 3$) for calc-silicate, and 14.1 to 18.6 ($N = 5$) for leucogranite. Quartz-feldspar $^{18}O/^{16}O$ fractionations are typically lower than 1‰ and indicative of isotopic disequilibrium. These relict high $\delta^{18}O$ values and anomalous Δ_{Q-F} values are readily attributable to local exchange with the adjacent intercalated ^{18}O-rich marble layers (calcite $\delta^{18}O$ = +15.9 to +22.0, $N = 9$), whose protoliths probably started out with $\delta^{18}O$ = +20 to +30 (Veizer and Hoefs, 1976). This lack of ^{18}O homogenization was not unexpected, because such ductile, pure marble layers are known to be relatively impermeable to metamorphic pore waters (Rye et al., 1976; Wickham and Peters, 1992; Bickle et al., 1995). A detailed transect through a portion of Traverse 2 is shown on Figure 7. The preservation of relatively high $\delta^{18}O$ values in this part of the section indicates that these ductilely deformed, marble-rich sections underwent incomplete $^{18}O/^{16}O$ homogenization with the aqueous fluids derived from the surrounding metasedimentary units during anatexis and high-grade metamorphism. However, quartz from a calc-silicate layer between the two thick marbles has $\delta^{18}O$ = +12.4, identical to the range of homogeneous quartz $\delta^{18}O$ found throughout the rest of the Selkirk Allochthon; this contrasts sharply with the effects observed around the pure marbles, but is consistent with the relatively high local permeabilities that are commonly produced during decarbonation of such calc-silicate lithologies (e.g., Rumble and Spear, 1983). In any case, although these sedimentary carbonate units did not become completely isotopically homogenized with the pelitic rocks and leucogranite, they certainly underwent very significant ^{18}O depletion relative to their initial protoliths, which originally would have had $\delta^{18}O$ > +20.

In the imbricate thrust zone, quartz $\delta^{18}O$ values of metapelite, quartzite, gneiss, and calc-silicate display a wide range, 10.0 to 14.4 ($N = 7$), 11.7 to 14.8 ($N = 7$), 10.8 to 13.2 ($N = 4$), and 16.1, respectively. Some of these lower $\delta^{18}O$ values are clearly attributable to the late-stage meteoric-hydrothermal activity associated with detachment faulting, but all quartz $\delta^{18}O$ values significantly higher than +12.1 *must* represent relict metasedimentary values. Concordant leucogranite sheets are totally absent in the imbricate thrust zone; the only leucogranitic materials in that zone are rare, late-stage dikes that have undergone meteoric-hydrothermal ^{18}O depletion (with feldspar $\delta^{18}O$ as low as –4.0). Note that these post-kinematic leucogranite magmas originally must have been more ^{18}O rich than the typical concordant leucogranites in the allochthon, because they retain quartz $\delta^{18}O$ values as high as +13.0 to +14.2 ($N = 3$).

In contrast to quartz, coexisting almandine garnet is isotopically variable in all parts of the allochthon except the leucogranite-rich zone, where garnet $\delta^{18}O = +9.6 \pm 0.5$ ($N = 8$) in all lithologies (Holk and Taylor, 1997a). In samples from homogeneous parts of the amphibolite-rich zone, garnet $\delta^{18}O = +9.3 \pm 1.0$‰ ($N = 13$), whereas coexisting quartz is more uniform at $\delta^{18}O = +12.4 \pm 0.5$‰. Garnets from metapelites in the marble-rich zone have $\delta^{18}O$ = +12.2 to +13.3 (N = 4). Quartz-garnet fractionations (Δ_{Q-G} values) are relatively constant at 2.7 ± 0.4‰ ($N = 7$) in all lithologies of the leucogranite-rich zone. Using the calibration of Javoy (1977), these values indicate equilibrium at 700 ± 100°C, compatible with the estimates of Carr (1995). Inasmuch as all these rocks were metamorphosed at similar grade, the variable Δ_{Q-G} values in the amphibolite-rich (1.8 to 4.1‰) and marble-rich zones (1.4 to 2.7‰) clearly indicate isotopic disequilibrium. We interpret the variable $\delta^{18}O$ values of these relatively refractory garnets as being a relict of the original $^{18}O/^{16}O$ heterogeneity of this Proterozoic sedimentary section (see below and Holk and Taylor, 1997a).

Stable-isotope analyses of carbonates from marble and calc-silicate units in the Selkirk allochthon exhibit systematic variations indicative of H_2O-driven decarbonation. In the zoisite-bearing calc-silicates from the imbricate thrust zone, the amphibolite-rich zone, and the marble-rich zone, these carbonates are very uniform in ^{18}O ($\delta^{18}O$ = +15.4 ± 0.2; N = 5) but variable in ^{13}C ($\delta^{13}C$ = –3.5 to –7.6); this probably indicates decarbonation in the presence of an aqueous fluid having a fairly uniform oxygen isotope composition. Calcite from marble of the marble-rich zone has $\delta^{18}O$ = +15.9 to +22.0 (N = 8) and $\delta^{13}C$ = –1.9 to +2.4; these values indicate that the marbles that did not undergo significant decarbonation underwent variable degrees of $^{18}O/^{16}O$ exchange with lower-^{18}O metamorphic-hydrothermal fluids. One marble from the amphibolite-rich zone has $\delta^{18}O$ = +18.8 and $\delta^{13}C$ = +8.5; this high-^{13}C marble is typical of carbonates of Late Precambrian age (Knoll et al., 1986; Wickham and Peters, 1993), and rocks such as these are widespread in the Windermere sequence in southeastern British Columbia (Ghent and O'Neil, 1985). The only marble analyzed from the leucogranite-rich zone was found to have a $\delta^{18}O$ = +12.4 and $\delta^{13}C$ = +0.2; this indicates complete $^{18}O/^{16}O$ homogenization of this <1 m thick marble layer with the surrounding metasedimentary values and leucogranites. A carbonate from the calc-silicate zone adjacent to this homogenized marble has $\delta^{18}O$ = +9.6 and $\delta^{13}C$ = –6.4; the low ^{18}O and low ^{13}C characteristics of this sample are readily explained by Rayleigh fractionation during localized decarbonation of this calc-silicate (e.g., Valley, 1986).

The Valhalla Complex

Geologic setting

The Valhalla Complex is another of several large metamorphic core complexes exposed in the southern Omineca belt of British Columbia (Fig. 2). The geology of this complex (Fig. 8) is in many respects similar to that of the Thor-Odin Complex. However, it is more difficult to correlate the metasedimentary units in the lower plate with known stratigraphic sections, because this complex experienced a somewhat greater degree of anatexis than is observed at Thor-Odin. Another difference is that the lowermost portion of the crustal section analogous to the Selkirk allochthon is composed of Cretaceous Mulvey orthogneiss, which is in turn overlain by a thick hybrid anatectic gneiss unit and a 1 to 3 km thick sill-like body of Paleocene to Eocene Ladybird-series leucogranite quite analogous to the leucogranite body at Thor-Odin (Reesor, 1965; Carr et al., 1987; Parrish et al., 1988). Another similarity is that deep in the lower plate is the Cretaceous to Paleocene Gwillim Creek shear zone (Figs. 8 and 9), an E-directed compressional mylonite zone analogous to the Monashee décollement (Carr et al., 1987; Spear and Parrish, 1996).

The lower-plate hybrid gneiss in the Valhalla Complex is a highly deformed, mylonitized unit of unknown protolith age. The intensity of mylonitic deformation in this paragneiss unit increases upward in the section as one approaches the detachment faults (Carr et al., 1987). Dominant lithologies include quartzofeldspathic gneiss and metapelitic schist and gneiss, with minor amphibolite, calc-silicate gneiss, and minor quartzite (Reesor, 1965). These rocks are highly migmatitic, containing up to 75% lit-par-lit pegmatitic leucogranite (Reesor, 1965; Carr et al., 1987). Zircon and monazite U-Pb ages of the leucogranites indicate that the peak of metamorphism and anatexis was reached at 67 to 72 Ma (Spear and Parrish, 1996). Garnet-biotite Fe-Mg exchange thermometry and dehydration melting phase equilibria indicate a peak metamorphic temperature of 820 ± 30°C and pressure of 8 ± 1 kbar for the deep Valhalla lower plate (Spear and Parrish, 1996). In other words, peak metamorphic conditions were somewhat more extreme than at Thor-Odin, and they may have occurred very slightly earlier.

Lower-plate plutonic gneiss units are sheetlike (Fig. 9) and become younger in an upward direction (Carr et al., 1987). The Mulvey gneiss is a 110 Ma (Spear and Parrish, 1996) melanocratic, megacrystic, K-feldspar–hornblende–biotite granodiorite gneiss cut by leucocratic veins. The oldest and deepest phase of the Ladybird suite, the Airy quartz monzonite, is a 62 Ma (Carr et al., 1987) megacrystic, K-feldspar–biotite–hornblende granitic gneiss. Variably deformed and mylonitized leucocratic biotite quartz monzonites belonging to the uppermost and youngest phase of the Ladybird suite give U-Pb zircon ages of 56 to 59 Ma (Carr et al., 1987; Parrish et al, 1988; Spear and Parrish, 1996).

The Valkyr shear zone forms the roof of this domal complex (Fig. 9). This gently dipping, amphibolite-grade ductile shear zone makes up the western, southern, and northern boundaries of the lower plate (Fig. 8). Movement of the hanging wall was to the east (Carr et al., 1987; Simony and Carr, 1997). The footwall is composed of variably

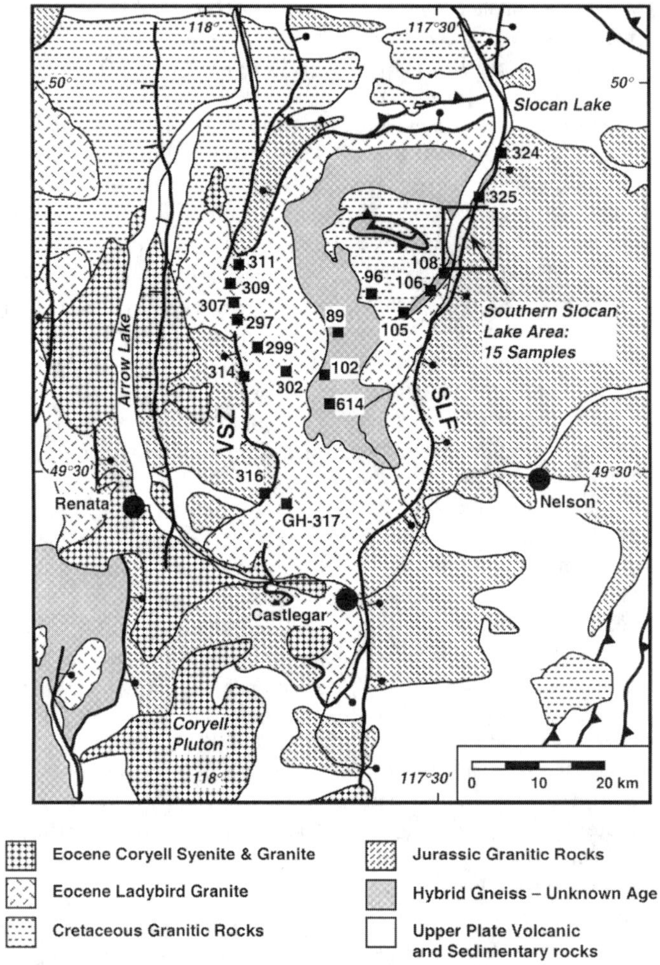

FIG. 8. Simplified geologic map of the Valhalla Complex and surrounding areas (modified after Carr et al., 1987) showing sample locations studied in this work (this map area is indicated by a large rectangle on Fig. 2). Sample locations are numbered and the $\delta^{18}O$ and $\delta^{13}C$ values are keyed to these samples in Table 4. Legend: SLF = Slocan Lake fault; VSZ = Valhalla shear zone.

deformed, synkinematic bodies of Ladybird leucogranite, and the hanging wall consists of Jurassic granitoids similar to those in the Nelson batholith, as well as Cretaceous granitoids and low-grade metasedimentary and metavolcanic rocks of the Quesnellia terrane (Carr et al., 1987). Undeformed pegmatitic Ladybird leucogranites that crosscut this shear zone date its latest movement at just before 56.5 Ma (Carr et al., 1987). To the east, the Valkyr shear zone merges into and is deformed by the younger, more brittle Slocan Lake fault, an indication that these two faults were synkinematic during the earliest stages of extension. Carr et al. (1987) suggested that the Valkyr shear zone was deactivated while still in the ductile regime, in response to arching of the Valhalla Complex.

Isotopic results

As shown in Figure 10 and Table 4, a diverse group of rocks from the hybrid gneiss unit of the lower plate, including metapelite, calc-silicate, amphibolite, and leucogranite, all display uniform quartz $\delta^{18}O$ values of +11.2 ± 0.9‰ (N = 6) and feldspar $\delta^{18}O$ = +10.0 ± 0.2‰ (N = 3). An amphib-

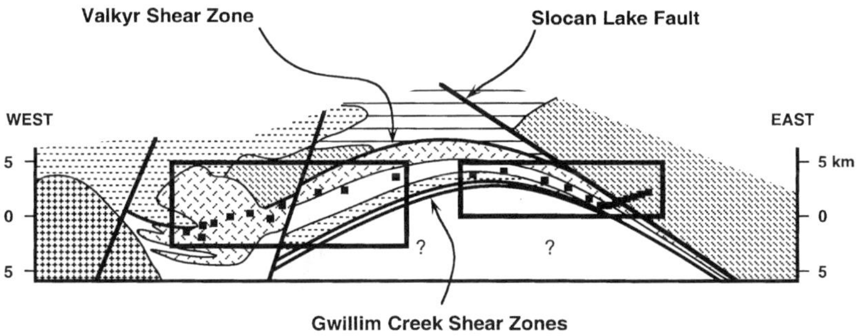

FIG. 9. ENE-trending cross section (no vertical exaggeration, modified after Spear and Parrish, 1996) across the Valhalla metamorphic core complex (Fig. 8), showing the layered and domed structure and the relative structural positions of samples studied in this work (solid squares). As indicated by the two heavy rectangles, these samples are divided into an eastern group that straddles the Slocan Lake detachment fault and a western group in the vicinity of the Valkyr shear zone. The indicated patterns are the same as on Figure 8 except that, for clarity, the hybrid gneiss is shown without any pattern and the the upper plate Slocan Group is shown by the horizontal lined pattern.

olite and a leucogranite sheet collected within 1 m of each other have slightly lower whole-rock $\delta^{18}O$ = +8.8 and +9.1, respectively, probably indicative of local $^{18}O/^{16}O$ exchange with the lower-^{18}O protolith of the amphibolite (and similar to effects described above for the Thor-Odin Complex). One very highly strained graphitic, siliceous calc-silicate collected just beneath the mylonitic shear zone associated with the Slocan Lake fault has a much higher quartz $\delta^{18}O$ = +16.3, indicative of partial preservation of the original high-^{18}O signature of its carbonate protolith, again analogous to features locally observed at Thor-Odin.

Quartz-feldspar oxygen isotope fractionations (0.8 to 2.0‰) in all but one of these samples are consistent with the preservation of $^{18}O/^{16}O$ equilibrium at magmatic temperatures. At a temperature of 750 to 800°C, H_2O in equilibrium with the Valhalla mid-crustal section would have $\delta^{18}O \sim +10.0$. Clearly, we can rule out the possibility that any significant quantity of low-^{18}O meteoric-hydrothermal fluid exchanged with these rocks from deep in the Valhalla lower plate, despite the massive evidence for such effects in the upper plate and along the Slocan Lake fault (Holk and Taylor, 1997b, 1999). Our whole-rock $\delta^{18}O$ values for these Valhalla mid-crustal, lower-plate rocks are similar to the range ($\delta^{18}O$ = +8 to +11) observed by Nesbitt and Muehlenbachs (1995). We analyzed only one sample of dolomitic marble ($\delta^{18}O$ = +20.6 and $\delta^{13}C$ = +2.2), but its isotopic composition is typical of marbles throughout the region (Ghent and O'Neil, 1985).

$^{18}O/^{16}O$ homogenization of the middle crust in the Valhalla lower plate

Quartz and feldspar $\delta^{18}O$ values from the mid-crustal, lower-plate rocks of the Valhalla Complex display a remarkable degree of oxygen isotope homogeneity, very similar to what is observed in analogous rocks of the Selkirk allochthon in the southern Thor-Odin Complex (see Fig. 6 and also Holk and Taylor, 1997a). However, as shown by the position of the patterned vertical band on Figure 10, the $\delta^{18}O$ values of quartz in the homogenized rocks at Valhalla are about 1‰ lower than in similar rocks of the southern Thor-Odin Complex; this probably reflects an originally lower bulk average $\delta^{18}O$ value for the Valhalla protolith, compatible with the fact that marbles and calc-silicate rocks are much less abundant in the Valhalla section (Reesor, 1965) as compared to the one at Thor-Odin (Reesor and Moore, 1971). Quartz $\delta^{18}O$ values of the deep-seated Mulvey orthogneiss and the Airy quartz monzonite are, respectively, 1.3‰ and 1.6‰ lower in ^{18}O than the average hybrid gneiss quartz $\delta^{18}O$ value (+11.2), another indication of an overall lower $\delta^{18}O$ value for the Valhalla mid-crustal section.

Except for the fact that the mid-crustal section is overall about 1‰ lower in ^{18}O at Valhalla than at Thor-Odin (compare Figs. 6 and 10), all of the above discussion of the southern Thor-Odin data concerning $^{18}O/^{16}O$ homogenization of the middle crust and the relationship of these effects to the isotopic uniformity of the abundant associated anatectic leucogranites (Holk and Taylor, 1997a) also basically

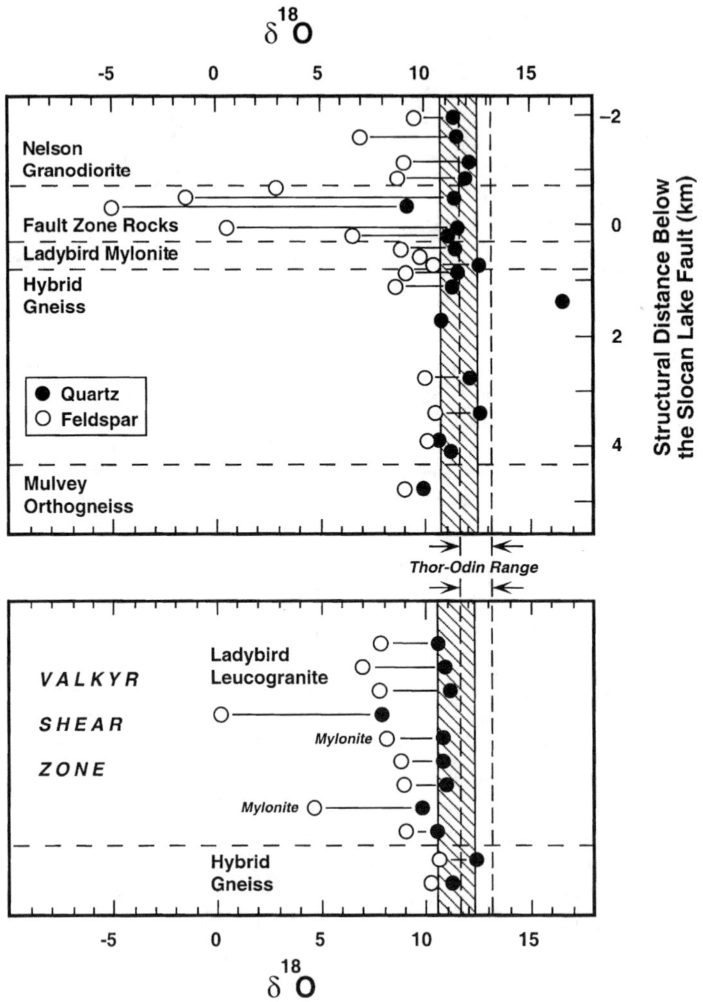

FIG. 10. Quartz and feldspar $\delta^{18}O$ versus measured structural height relative to the Slocan Lake fault (upper part of diagram) and versus the unmeasured structural height relative to the Valkyr shear zone (lower part of diagram). The striking homogeneity of quartz $\delta^{18}O$ values from the lower plate is shown by the diagonal pattern. Note that this range is also nearly identical to the $\delta^{18}O$ values of quartz from the upper-plate Nelson batholith (Holk and Taylor, 1999). The somewhat higher range of homogeneous $\delta^{18}O$ values for quartz from the analogous mid-crustal zone of the southern Thor-Odin Complex (see Fig. 6) is indicated by the two dashed lines. Very large ^{18}O depletions in feldspar (down to a minimum $\delta^{18}O = -5.0$) are observed in most samples collected within the 1 km thick zone that encompasses the brittle part of the Slocan Lake detachment fault zone; this includes the upper 300 m of quartz monzonite mylonite beneath the fault, and the lower 700 m of Nelson granodiorite above the fault, clearly showing that the meteoric-hydrothermal exchange effects are related to the episode of detachment faulting.

applies to the Valhalla lower plate. Also, as was the case in the marble-rich zone at Thor-Odin, $^{18}O/^{16}O$ homogenization at Valhalla was locally incomplete; for example, this event does not seem to have markedly affected the extremely rare samples of dolomitic marble ($\delta^{18}O = +20.6$), and it has played only a minor role in such samples as the highly strained calc-silicate ($\delta^{18}O = +16.3$). Again, as discussed above and by Holk and Taylor (1997a), these carbonate-rich lithologies were probably ductile and relatively impermeable to aqueous fluids during metamorphism.

TABLE 4. $^{18}O/^{16}O$ Data from the Valhalla Metamorphic Core Complex

Sample[1]	Rock type	WR	Qz	Fs	Cc[2], $\delta^{18}O$	Cc[2], $\delta^{13}C$
	Upper plate					
	Jurassic Nelson batholith					
GH-118	Slightly altered granodiorite	5.3	11.3	6.8	2.0	–4.8
GH-120	Fresh granodiorite	8.2	11.9	8.8	1.2	–5.8
MT-325	Fresh granodiorite		11.4	9.6		
GH-122	Fresh granodiorite	7.7	11.7	8.5	2.1	–5.5
GH-123	Altered granodiorite			2.7		
GH-124a	Highly altered granodiorite	–0.4	11.1	–1.6	–3.0	–4.9
MT-324	Highly altered granodiorite		8.9	–5.0		
	Zone of brecciation					
GH-125	Carbonate-rich breccia	0.0			–2.1	–5.6
GH-126	Very altered granitoid	0.7			–3.2	–2.9
GH-150	Chloritic granitoid	3.1	11.2	0.3	2.5	–1.1
	Lower-plate mylonite					
GH-151	Ladybird mylonite with muscovite	7.4	10.8	6.4		
GH-154	Brittle shear zone in granite	5.1				
GH-155	Leucogranite mylonite	9.8	11.1	8.7		
GH-159	Mylonitic paragneiss		12.4	10.1		
GH-140	Mylonitic leucocratic sheet	9.4	11.4	8.9		
GH-134	Mylonitic augen gniess	9.9	11.0	8.4		
MT-326	Migmatite in hybrid gneiss			9.8		
	Deep lower plate–Valhalla core					
	Hybrid gneiss					
GH-89	Amphibolite	8.8	11.0			
GH-89a	Leucogranite sheet	9.1	10.2	10.0		
GH-102a	Calc-silicate		12.4	10.3		
GH-105	Quartzofeldspathic gneiss		11.8	9.8		
GH-106	Metapelite		10.5			
GH-108	Graphitic calc-silicate		16.3			
GH-614	Marble				20.6	2.2
	Paleocene Airy quartz monzonite					
GH-302	Quartz monzonite	9.0	9.9	8.8		
	Cretaceous Mulvey granodiorite gneiss					
GH-96	Melanocratic gneiss	8.2	9.6	8.8		
	Valkyr shear zone					
	Eocene Ladybird quartz monzonite					
GH-297	Protomylonite granite		10.6	7.9		
GH-307	Undeformed granite	0.8	7.6	0.0		
GH-309	Quartz monzonite		11.0	7.7		
GH-309a	Quartz + feldspar vein		10.7	6.8		
GH-311	Quartz monzonite		10.3	7.5		
GH-314	Quartz monzonite		10.7	8.6		
GH-316	Undeformed granite		10.6	8.6		
GH-317	Mylonitic quartz monzonite		9.5	4.4		

[1]Numbers refer to samples archived at the Division of Geological and Planetary Sciences, California Institute of Technology, Pasadena, California. Data are from sample localities on Figure 8. For explanation of symbols, see footnotes in Table 1.
[2]Regarding calcite (Cc), see notes in Table 2.

Based on the absence of late-stage muscovite throughout much the Valhalla Complex, during peak metamorphism, this terrane appears to have been thoroughly dewatered at temperatures above the wet granite solidus (Spear and Parrish, 1996). Thus, it apparently experienced somewhat higher anatectic temperatures (>800°C) than did the Thor-Odin Complex (~750°C), consistent with observed mineralogical differences in these Paleocene–Eocene leucogranites (i.e., hornblende is absent from the Ladybird suite at Thor-Odin, but is present at Valhalla).

$^{18}O/^{16}O$ Homogenization of the Selkirk Allochthon during Anatexis

The profound $^{18}O/^{16}O$ homogeneity of large volumes of the middle continental crust in both the Thor-Odin and the Valhalla core complexes could be the result of: (1) initial protolith homogeneity, (2) influx and exchange with large amounts of external oxygen-bearing fluids, and/or (3) an internal closed-system homogenization process of some type, perhaps involving multiple stages of $^{18}O/^{16}O$ exchange involving leucogranite melt, internally derived aqueous fluid, and unmelted rock. Similar statements also apply to the widespread, but limited, section (<200 m thick) of isotopically homogenized rocks in the anatectic zone lying just beneath the Monashee décollement.

Mechanism (1) almost certainly cannot represent a general explanation of these phenomena, because of the known $^{18}O/^{16}O$ heterogeneity observed in similar metasedimentary sections elsewhere on Earth (e.g., Burt and Taylor, 1989; 1993), particularly noting the extreme $\delta^{18}O$ values exhibited by typical protoliths of amphibolite ($\delta^{18}O$ ~ +6; Schwarcz and Clayton, 1965) and marble ($\delta^{18}O$ ~ +25 to +30), both of which are common rock types in the mid-crustal sections of southern British Columbia. In the leucogranite-rich and amphibolite-rich zones, quartz $\delta^{18}O$ values in calc-silicates and thin amphibolites are almost as homogeneous as those in the intercalated leucogranites and metapelites. Note that mechanism (1) would also require that both the Proterozoic and the Paleozoic sections by coincidence must have started out with almost exactly the same, uniform, bulk $\delta^{18}O$ values; however, that possibility can be refuted because of the variable quartz $\delta^{18}O$ values of +12.7 to +15.6 (N = 14) in analogous (but lower grade) Upper Proterozoic metapelites 180 km to the north (O'Neil and Ghent, 1975; Bowman and Ghent, 1986). Those rocks appear to be stratigraphically correlative with metapelites from the Thor-Odin amphibolite-rich zone (where quartz $\delta^{18}O$ = +12.5 ± 0.5, N = 6). Furthermore, quartz ($\delta^{18}O$ = +10.2 and +10.5) and feldspar ($\delta^{18}O$ = +15.8 and +13.0) from greenschist-grade arkosic and pebble conglomerates from the Purcell Range (Holk, 1997) are additional evidence of initial isotopic heterogeneity.

Our own data set itself provides evidence of original $^{18}O/^{16}O$ heterogeneity. Zones that were a priori expected to be relatively impermeable or refractory (the marble-rich zone, the thick amphibolites, and the garnets) are in fact the only parts of the allochthon that now display any significant $^{18}O/^{16}O$ heterogeneity. In fact, these rocks themselves appear to have been very significantly homogenized, because they are markedly more uniform in $\delta^{18}O$ than their lower-grade protoliths or counterparts elsewhere (i.e., these thick marbles are much lower in ^{18}O than limestones, and these thick amphibolites are much higher in ^{18}O than basalts). Also, the $\delta^{18}O$ heterogeneity in quartz from the deepest and most deformed allochthonous assemblages (i.e., the imbricate thrust zone) is similar to that displayed by garnets coexisting with homogeneous quartz higher in the section, implying that prior to anatexis these other sections must have originally exhibited a whole-rock $\delta^{18}O$ variation of 4.5‰ or more.

Despite the above arguments, there is a remote, but unlikely, possibility that the protolith rocks of the mid-crustal sections at both Thor-Odin and Valhalla (and the basement rocks directly beneath the Monashee décollement?) may indeed have originally been relatively uniform in $^{18}O/^{16}O$. For example, Bowman and Ghent (1986) interpreted their own $^{18}O/^{16}O$ data, as well as those of O'Neil and Ghent (1975) from lower-grade rocks belonging to the Windermere Supergroup, as being the product of a protolith that was fairly uniform in ^{18}O. However, note that the oxygen isotope homogeneity that does exist in these staurolite-to sillimanite-grade rocks in the footwall of the Columbia River fault at Mica Creek might equally well be interpreted to be the product of widespread exchange with a metamorphic pore fluid having a fairly uniform $\delta^{18}O$ ~ +11.0 at T = 500°C. Quartz $\delta^{18}O$ values from rocks of similar metamorphic grade in the hanging wall of the Columbia River fault (upper crustal layer) are indeed quite variable in $^{18}O/^{16}O$ (Bowman and Ghent, 1986), consistent with the concept that the metamorphic/magmatic hydrothermal event associ-

TABLE 5. Calculation of Bulk $\delta^{18}O$ for the Selkirk Allochthon

Lithology	$\delta^{18}O$	Horsethief Creek Group[1]	Hamill Group	Mohican-Badshot formations[2]	Lardeau Group[3]
Arkosic grit	+12[4]	30			35
Pelite	+16[5]	30	35	59	30
Quartzite	+11	19	65	14	16
Amphibolite	+9	19			17
Marble	+22	2		27	2
Bulk $\delta^{18}O$		+12.6	+12.7	+16.9	+12.6

[1]Amphibolite-rich zone.
[2]Marble-rich zone.
[3]Leucogranite-rich zone.
[4]The $\delta^{18}O$ value used for arkosic grit in the material balance calculation is based on $^{18}O/^{16}O$ results on quartz ($\delta^{18}O$ = +10.5) and feldspar ($\delta^{18}O$ = +13.0) from low-grade arkosic grits of the Windermere Supergroup in the Purcell Mountains (Holk, 1997).
[5]The model pelite $\delta^{18}O$ value is taken from the approximate average whole-rock $\delta^{18}O$ (+15.8) of argillites from the Belt (Purcell) Supergroup, Montana (Eslinger and Savin, 1973).
Sources: Lithologic proportions are from the following sources: Horsethief Creek Group from Pell and Simony, 1986; Hamill Group and Mohican-Badshot formations from Fyles and Eastwood, 1962; and Lardeau Group from Fyles (1964).

ated with ^{18}O homogenization and anatexis was restricted to the deeper, mid-crustal layer. Furthermore, the data cited above are all from similar rock types having similar bulk compositions and stratigraphic positions, making the conclusion concerning original homogeneity perhaps likely for those particular units, but not necessarily for the entire section. Based on the available data, we believe it is very improbable that the entire Precambrian to Cambrian post-rift, clastic sequence could have initially been so uniform in $^{18}O/^{16}O$ at the onset of metamorphism. Nevertheless, this is a sufficiently important problem to justify checking it in the future by doing more $^{18}O/^{16}O$ studies of lower-grade rocks from these correlative units elsewhere in British Columbia.

TABLE 6. Bulk $\delta^{18}O$ Calculation for the Mantling Gneiss[1]

Lithology	$\delta^{18}O$	Percent of section
Arkosic grit	+12	8
Pelite	+16	17
Quartzite	+11	40
Amphibolite	+9	35
Bulk $\delta^{18}O$		+11.2

[1]Lithologic proportions are from Scammell and Brown (1990). See Table 5 for sources of $\delta^{18}O$ values.

In addition to almost certainly ruling out mechanism (1), we can also rule out mechanism (2) as a sole explanation, because there is no obvious source for the very large amounts of external hydrothermal fluids that would be required. Also, it seems unlikely that the hypothesized external H_2O would by coincidence already have had just the appropriate $\delta^{18}O$ to be in equilibrium with the bulk average material of the allochthon. Excluding the marble-rich zone, rough material-balance calculations utilizing $^{18}O/^{16}O$ data on analogous and/or correlative lower-grade protoliths (Tables 5 and 6) show that, overall, there cannot have been any more than about a 0.5 to 1.0‰ lowering in the average $\delta^{18}O$ of either the mid-crustal metasedimentary section or the basement mantling gneiss. This contrasts with other localities on Earth where, on a scale of several km^3, the $\delta^{18}O$ of the bulk continental or oceanic crust has indeed been changed during interactions with externally derived aqueous fluids during metamorphism (e.g., Shieh and Schwarcz, 1974; Gregory and Taylor, 1981; Wickham and Taylor, 1985; Peters and Wickham, 1995). However, in those cases the large quantities of H_2O are typically derived from the Earth's surface waters, and the effects generally occur in the brittle portion of the crust, not in the ductile regime.

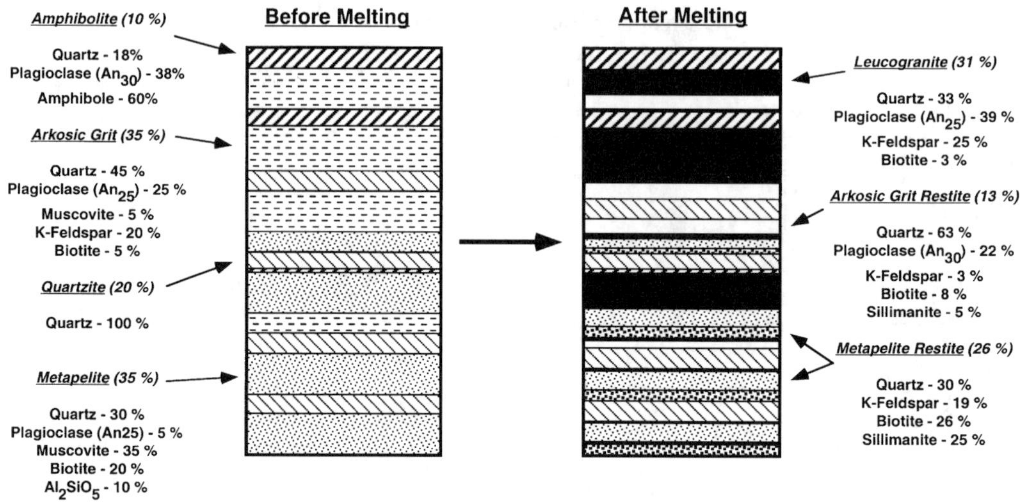

FIG. 11. Comparison of the characteristics of the schematic stratigraphic section before and after melting. The premelting stratigraphic section is a simplified representation of the semipelite-amphibolite division of the Horsethief Creek Group of the Windermere Supergroup as documented by Pell and Simony (1987). The model operates by partial melting of metapelite and arkosic grit layers in response to the influx of small amounts of external H_2O, muscovite dehydration, and decompression, together with cycling of H_2O out of H_2O-saturated melts and into newly formed melts. After melting, leucogranite comprises 32% of the section, while metapelite restite and paragneiss together comprise 38% of the section. Overall, the arkosic grit is completely removed from the column, while the proportions of amphibolite and quartzite are unchanged.

The portions of the core complexes in British Columbia that are most homogenized in $^{18}O/^{16}O$ are either those that are most leucogranite rich or the ones that have undergone the most extensive anatexis (and vice versa); this strongly suggests that the anatectic melts themselves facilitated $^{18}O/^{16}O$ exchange with the unmelted rocks, and/or that they served as a source of the hydrothermal fluids that promoted such exchange. Thus, together with a partial contribution from mechanism (2), the multistage exchange model (3) is preferred, because in all three of the isotopically homogenized zones under discussion, most non-leucogranite lithologies have mineral $\delta^{18}O$ values identical to those of the leucogranites, and local $^{18}O/^{16}O$ variability is only observed in the parts of the section that a priori would be expected to be relatively impermeable (e.g., marbles and thick amphibolite layers). Apparently, only a limited amount of aqueous fluid was present in these systems at any one time (i.e., they were essentially rock-buffered systems).

In these metamorphic core complexes, particularly along the detachment faults and within the upper crustal zone, there is overwhelming evidence for large-scale, meteoric-hydrothermal water-rock interaction during the Eocene (see Fig. 2, and Magaritz and Taylor, 1986; Nesbitt and Muehlenbachs, 1989, 1995; Holk and Taylor, 1999); these effects are also observed along the Monashee décollement. These meteoric-hydrothermal events had local impact on the basement and mid-crustal zones along late-stage brittle normal faults associated with the extensional event related to detachment faulting (see Holk and Taylor, 1997a, 1999; Holk at al., 1997). However, it has been clearly demonstrated that no significant amount of meteoric water can have infiltrated more than a few hundred meters into the lower plates beneath the detachment faults, probably because of the very low permeabilities of the plastically deforming mylonites (Fricke et al., 1992; Holk et al., 1997; Holk and Taylor, 1999). It is thus clear that meteoric-hydrothermal fluids cannot be responsible for the observed degree of $^{18}O/^{16}O$ homogenization of the three zones documented in Figures 6 and 10. Also, the type of massive lowering of $\delta^{18}O$ associated with large-scale infiltration of more deep-seated low-^{18}O waters (Wickham and Taylor, 1987; Wickham and Peters, 1990; Peters and Wickham, 1995) is not observed in the mid-crustal zones of any of these core complexes in

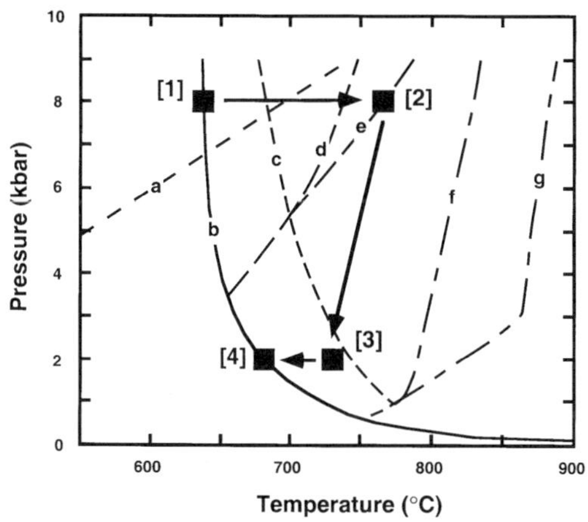

FIG. 12. Pressure-temperature plot (modified from Vielzeuf and Holloway, 1988) showing phase relations for the (a) kyanite-sillimanite phase boundary (Holdaway et al., 1971), (b) the solidus curve for the minimum melt eutectic (Wyllie, 1977), (c) the dehydration melting curve for wet pelite (Vielzeuf and Holloway, 1988), (d) the H_2O-absent breakdown of muscovite (Althaus et al., 1970), (e) the H_2O-present breakdown of muscovite (Althaus et al., 1970), (f) the H_2O-absent breakdown of biotite, and (g) the H_2O-present breakdown of biotite (Vielzeuf and Holloway, 1988). The inferred P-T path for the amphibolite-rich zone of the Selkirk allochthon during thrust ramping and decompression-related anatexis is shown by the arrows from [1] initial melting to [2] end of muscovite dehydration, [3] end of decompression, and [4] final crystallization (see text).

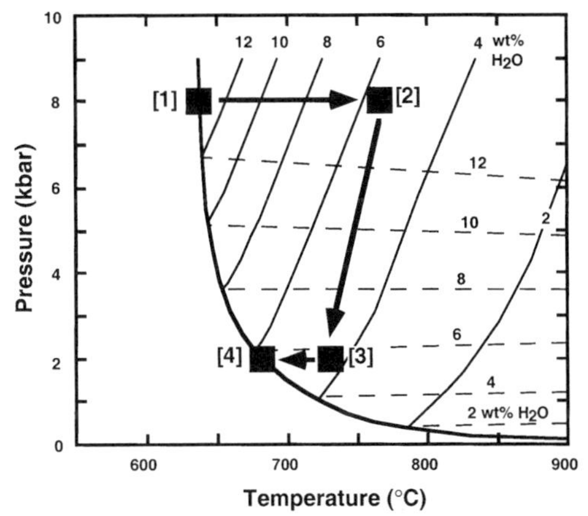

FIG. 13. Plot showing liquidus phase relations and H_2O solubilities in the system Qz-Ab-Or-H_2O for minimum and eutectic compositions from Holtz and Johannes (1994), together with the approximate P-T path from Figure 12 for the amphibolite-rich zone of the Selkirk allochthon. The heavy solid curve is the solidus of the minimum melt eutectic (Wyllie, 1977). The lighter solid lines denote liquidus curves at various values of wt% H_2O from 12% to 2%. The subhorizontal dashed curves represent the solubilities of H_2O (wt%) in the various granitic melts; these H_2O solubilities decrease with decreasing pressure. The H_2O contents on the liquidus decrease with increasing temperature and decreasing pressure. The presence of H_2O-undersaturated melt in the arkosic grit units keeps H_2O activity low because all locally available H_2O is assumed to be incorporated into the melt. At a given P and T, the H_2O contents on the liquidus curves provide minimum possible H_2O concentrations for the melt, whereas the intersecting solubility curves provide maximum constraints on this parameter. The intersections of these two sets of curves are used to constrain the H_2O contents of the hypothetical melts utilized in the forward progress material-balance melting model (see text) that is constructed in Figures 11 to 14 to account for the $^{18}O/^{16}O$ homogeneity observed in the Selkirk allochthon.

southern British Columbia. Therefore, the source of any such external aqueous fluid must be: (1) volatiles produced during dehydration of the mid-crustal layer at depth; and/or (2) H_2O-rich, basement-derived leucogranite melt that intruded through the Monashee décollement during the thrusting stage.

$^{18}O/^{16}O$ Homogenization Using Recycled Magmatic H_2O as a Catalyst

Pressure-temperature paths and H_2O contents of melts

Carr (1991) has correlated the stratigraphic section of the Thor-Odin amphibolite-rich zone (Reesor and Moore, 1971) with the semipelite-amphibolite subdivision of the Horsethief Creek Group (Pell and Simony, 1987). Based on this correlation, and to address the $^{18}O/^{16}O$ homogenization problem described above, we have formulated a material-balance model that follows the evolution and recycling of metamorphic and magmatic H_2O during progressive partial melting of a representative 1 km thick section of this metasedimentary column. Before anatexis, the model stratigraphic section (Fig. 11) is assumed to be composed of 35 vol% arkosic grit, 35% pelite, 20% quartzite, and 10% amphibolite, dominantly of pelite and quartzite in its lower part, and mainly arkosic grit and amphibolite in its upper part. While the metapelite and meta-arkose are undergoing anatexis the quartzite and amphibolite units are assumed to be inert to partial melting.

The pressure-temperature paths for the mid-crustal portion of the Selkirk allochthon (Figs. 12 and 13) were constructed from field and petro-

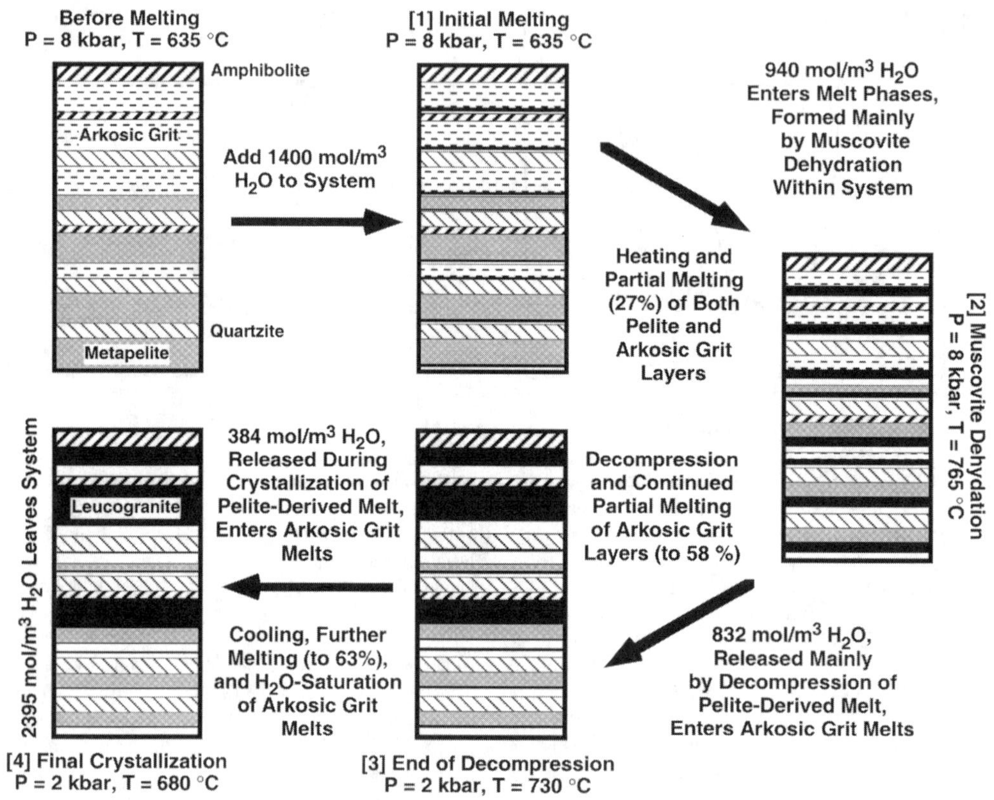

FIG. 14. The influx-exsolution-transfer-release model for the partial melting of arkosic grit and pelite layers and H_2O-melt-rock interaction of the middle crust of the southern Thor-Odin metamorphic core complex. This material-balance model follows the forward progress of the partial melting (solid black zones) that occurs in response to the initial influx of 1400 mol/m^3 of H_2O into the system, followed by muscovite breakdown. This dissolved H_2O is then released from the pelite-derived melts as they reach H_2O saturation during decompression, as the system moves up the Monashee ramp. These advecting H_2O-rich pore fluids exchange $^{18}O/^{16}O$ with adjacent restites, quartzites, and amphibolites, and they also drive further melting of nearby arkosic grit layers. During post-decompression cooling there is continued release of H_2O from all of these anatectic melts as they finally crystallize at about 680°C, allowing this H_2O to be used once again to imprint a terminal episode of $^{18}O/^{16}O$ exchange and homogenization upon the section.

graphic observations (Reesor and Moore, 1971; Duncan, 1984; Carr, 1991; Holk, 1997), and from geochronological studies (Parrish et al., 1988; Carr, 1992, 1995). The presence of kyanite (as inclusions in garnet) in the imbricate thrusting zone, together with the lack of primary muscovite in the amphibolite-rich zone constrain $P_{max} \geq$ 8kbar and $T_{max} \geq$ 765°C (Althaus et al., 1970; Holdaway, 1971), whereas the presence of stable biotite indicates T_{max} < 800°C (see Vielzeuf and Holloway, 1988, and references within). Overall, 6 kbar of decompression is inferred from the 20 km of vertical relief on the Monashee ramp (Cook et al., 1992), and from the obvious fact that these core complexes have all experienced major uplifts. The presence of retrograde cordierite (Reesor and Moore, 1971) in the vicinity of the Monashee décollement suggests rapid, nearly isothermal decompression (Hollister, 1977; Selverstone and Chamberlain, 1990) at least to pressures as low as 4 kbar. Melt H_2O contents are constrained using the model of Holtz and Johannes (1994). When H_2O-undersaturated melt is present, it is assumed that all locally available metamorphic pore H_2O will migrate toward, and be incorporated into, these melt layers.

Assumptions and controls on the progressive influx-exsolution-release model are mainly a function of the H_2O contents of the melts, the amounts of

TABLE 7. Tabulation of H_2O Contents of Unmelted Rocks and Leucogranite Melts[1]

	Before melting	Initial melting	Muscovite dehydration	Final decom- pression	Final crystalization	After crystallization
Stage:		Stage 1	Stage 2	Stage 3	Stage 4	
Temperature:	<635°C	635°C	765°C	730°C	680°C	<680°C
Pressure:	8 kbar	8 kbar	8 kbar	2 kbar	2 kbar	<2 kbar
		External H_2O				
	400 in	0	0	0	0	2395 out
		Arkosic grit				
Grit restite-muscovite	123	123	0	0	0	0
Grit restite-biotite	115	110	96	75	72	72
"Total" H_2O-grit melts[2]	0	700 8%	819 27%	1630 58%	2014 63%	0
Biotite H_2O-grit melts	0	5	19	40	43	0
Grit-xtallized biotite in leucogranite	0	0	0	0	0	43
		Pelite				
Pelite restite-muscovite	872	790	0	0	0	0
Pelite restite-biotite	459	453	440	440	440	440
"Total" H_2O-pelite melts[3]	0	782 9%	1576 27%	765 24%	381 12%	0
Biotite H_2O-pelite melts	0	6	19	17	8	0
Pelite-xtallized biotite in leucogranites	0	0	0	2	11	19
Total H_2O	2969	2969	2969	2969	2969	2969

[1]In mol/m^3 relative to the volume of the entire 1 km thick section.
[2]These percentage values represent the fraction of the arkosic grit layers that are melted at each stage. The "total" H_2O (mol/m^3) is all of the dissolved H_2O in the grit-derived melts, except that contributed by the original biotite in the grit.
[3]These percentage values represent the fraction of the pelite layers that are melted at each stage. The "total" H_2O (mol/m^3) is all of the dissolved H_2O in the pelite-derived melts, except that contributed by the original biotite in the pelite.

partial melting of each unit, the melt compositions, and the thermal history during decompression. The controlling agents for the production of granitic melt in such rocks are the supply of volatiles and the amount and distribution of feldspars in the source rock (e.g., Clemens and Vielzeuf, 1987). Metapelites are initially volatile rich (~3.5 wt% H_2O, mainly as OH in muscovite), but feldspar poor (<5 vol%); the converse applies to the grit lithologies (<1 wt% H_2O, mainly as OH in biotite, but containing as much as 50 vol% feldspar). At any given pressure and temperature, the H_2O content of the grit-derived melt is the value on the liquidus for a haplogranitic melt having a composition near the eutectic. The H_2O contents of melts on the liquidus decrease with increasing temperature; at 8 kbar the H_2O content at the liquidus is 14 wt% at 630°C, decreasing to 5.5 wt% at 765°C (see Fig. 13, and Holtz and Johannes, 1994, and the references therein). The maximum solubility of H_2O in such melts decreases with decreasing pressure from 14 wt% at 8 kbar to 5.5 wt% at 2 kbar (see Fig. 13). Water-saturated melting occurs (e.g., see Wyllie, 1977) when the water-saturated solidus temperature is attained for a given lithology (~650°C for most lithologies). Melt compositions produced in this system are constrained by experimentally derived phase equilibria (Tuttle and Bowen, 1958; Luth et al., 1964; Vielzeuf and Holloway, 1988) and petrographic observations (Reesor and Moore, 1971; Holk, 1997). Pre-melt mineralogical compositions of pelite and arkosic grit are based on average compositions given by Shaw (1956) and Raeside and Ghent (1983). Restite compositions are calculated using material balance.

Material-balance model

Our model assumes that all available H_2O goes into the melt phase when melt is present, unless the melt is H_2O saturated. The presence of H_2O-undersaturated melt (e.g., grit-derived melt) keeps the H_2O activity low in portions of the system in communication with those melts. Once some melt portions become saturated, however, there can be transport of H_2O from saturated to unsaturated melts through the intervening lithologies that are infertile to melting, particularly as the H_2O-saturated melts undergo decompresssion, or as they crystallize. Since this is the case envisioned in this general scenario, some H_2O exsolved from the pelite-derived melts will indeed eventually end up in nearby grit-derived melts. At these temperatures, any exsolved waters will certainly rapidly exchange $^{18}O/^{16}O$ with all lithologies infertile to melting as the H_2O is transported through them (either by diffusion or by advection along fractures and grain boundaries). On Figure 14 and in Table 7, we describe the basic elements of the material-balance model, at successive stages and at various positions along the P-T trajectories of Figures 12 and 13. Excluding the 30% of the section composed of amphibolite and quartzite, the initial H_2O content of the stratigraphic section is assumed to be 1569 mol/m^3, all as OH in biotite and muscovite in the pelites and arkosic grits (note that although the H_2O contents of the amphibolites and quartzites are ignored in these calculations, the volumes of those units are included in all the tabulations). This initial 1569 mol/m^3 mainly resides in the pelites (1331 mol/m^3); only 238 mol/m^3 is in the arkosic grits (Table 7).

A plausible way of supplying the Selkirk allochthon with the necessary amount of aqueous fluid to flux the observed amount of partial melting in that metasedimentary section (Fig. 11) is from crystallization and/or decompression of leucogranite melts formed within the zone of isotopic homogenization in the basement just beneath the Monashee décollement (see Figs. 3, 5, and 6). With a melt H_2O content of 12 wt% (14,600 mol/m^3), a leucogranite melt layer only 100 m thick could provide 1460 mol/m^3 to the overlying 1 km thick section of pelites and arkosic grits that is discussed and modeled below.

The initial partial melting of metapelites and arkosic grits is assumed to occur at 8 kbar and 630°C (point 1 in Fig. 12) in response to an addition to this 1 km thick section (from below?) of a modest amount of external H_2O (1400 mol/m^3 relative to the entire section). This raises the total H_2O content of the section up to a value of 2969 mol/m^3, which represents the minimum amount of H_2O required to produce the observed amount of leucogranite in the model. In all subsequent stages of the model, we assume that the 1 km thick section is a closed system, and all we do is redistribute the 2969 mol/m^3 H_2O between different parts of the section. Initial melt composition is taken to be the minimum melt eutectic composition at 8 kbar (i.e., quartz 24%, plagioclase (An_{25}) 52%, K-feldspar 24%; Luth et al., 1964). All of the available H_2O is assumed to enter the initially water-saturated melt (~14 wt% H_2O; Holtz and Johannes, 1994), and the extent of melting is limited solely by the available H_2O. Under these constraints the pelites lose almost all their plagioclase (from 5% down to less than 0.1%) while they undergo 9 vol% partial melting and the grits 8 vol% partial melting. The Windermere section at this stage would overall contain about 6% leucogranite melt.

Muscovite dehydration is probably complete by the time these rocks reach the base of the Monashee ramp (Point 2 in Figs. 12 and 13), with this reaction having produced 950 mol/m^3 H_2O. Melt H_2O contents are still mainly dependent upon the amount of available water, but in combination with the internally derived water we have added just enough external water to cause complete loss of plagioclase from the pelite restite. The melt is now much more potassic than at stage 1, and has the following normative chemical composition: quartz 30%, plagioclase (An_{25}) 17%, K-feldspar 50%, and "biotite" 3%. This composition is taken from data of Reesor and Moore (1971) and our own petrographic observations, but it is also very close to compositions determined experimentally by Thompson and Tracy (1979). These stage 2 percent-melting values are now basically plagioclase limited, and for the assumed amounts of H_2O in the system the pelite-derived melts are no longer H_2O saturated. In

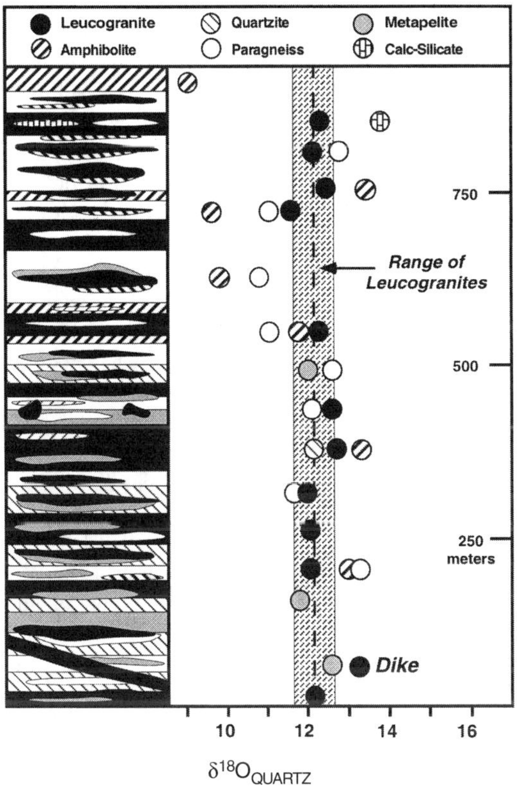

FIG. 15. Quartz $\delta^{18}O$ versus stratigraphic position in the amphibolite-rich zone. The generalized stratigraphy is from measured stratigraphic sections described in Reesor and Moore (1971). This section is comprised of ~30% leucogranite as sheets, pods, and lit-par-lit segregations; ~10% amphibolite; ~20% quartzite; ~30% paragneiss; ~20% metapelite; and <1% calc-silicate. This section was correlated by Carr (1991) with the Late Precambrian semipelite-amphibolite division of the Horsethief Creek Group (Windermere Supergroup). Except for the thick amphibolite outcrops, quartz is homogeneous in $\delta^{18}O$ (+12.4) for essentially all lithologies. The patterns of layers on the stratigraphic column are the same as those indicated for the data points. The vertical band indicates the range in $\delta^{18}O$ of quartz in the isotopically homogenized rocks.

our model these pelite-derived, peraluminous granitic magmas contain about 10 wt% H_2O.

Thus, at the base of the Monashee ramp, the muscovite dehydration reaction triggers large-scale (27%) partial melting of the muscovite-rich metapelite units. However, because they start out chemically much more similar to granites, the muscovite-poor feldspathic grits coincidentally undergo a similar degree of partial melting even with much less available H_2O. At this temperature (~750°C) and pressure (8 kbar), the grit-derived melts are less potassic and have much lower H_2O contents (i.e., the minimum value at point 2, namely 5.5 wt%). The calculated extent of partial melting of the grit (27%) is based upon the chemical composition of this type of melt at the stage of muscovite breakdown, as inferred from the actually observed end-member leucogranite compositions of the Ladybird suite at Thor-Odin (Reesor and Moore, 1971). In our model, this arkosic grit-derived melt at stage 2 has the following normative chemical composition: quartz 30%, plagioclase(An_{25}) 28%, K-feldspar 39%, and "biotite" 3%. These melts are within the compositional range of possible melts in the KNASH system (Thompson and Algor, 1977). At this stage, about 19 vol% of the Windermere section is leucogranite melt, with equal amounts derived from the pelites and grits.

Because the solubility of H_2O in granitic melt decreases during its ascent (Whitney, 1988), H_2O is released as these peraluminous melts reach saturation during their migration to shallower crustal depths, either passively because of displacements along the underlying décollement and/or during extension-related isostatic rebound (e.g., Wernicke and Axen, 1988), or actively in response to buoyant forces. Saturation should occur at pressures lower than 5 kbar (Fig. 13), and as the pressure is reduced from 5 to 2 kbar there should be continuous release of H_2O from these peraluminous magmas. Such H_2O-rich melts are probably less viscous than the drier, grit-derived melts, and these magma bodies may become quite mobile, cooling adiabatically and producing decompressive exsolution of H_2O. This 811 mol/m^3 of released H_2O is assumed to migrate through the section toward adjacent pockets of still undersaturated arkosic grit melt, where it can be re-dissolved, more than doubling the amount of arkosic grit melt (to 58% of the grit layers). As this H_2O passes through adjacent lithologies infertile to melting (paragneiss, amphibolite, quartzite, calc-silicate) it promotes $^{18}O/^{16}O$ homogenization of these lithologies. However, this H_2O does not leave the system; it continues to drive hydrothermal melting of any feldspathic grit layers that it encounters as it moves through the system. Aqueous fluid transport is assumed to be generally upward and/or subparallel to the leucocratic melt layers because the H_2O-saturated magmas would tend to act as impermeable barriers to the passage of aqueous fluids. Eventually, partial melting of the grit units exceeds the critical melt fraction (~30%; Arzi, 1978) and these mobilized packets of melt also begin migrating up the Monashee ramp. The movement of these leucogranite magmas provides another mechanism for promoting $^{18}O/^{16}O$ exchange and homogenization with any rocks that these magmas come into contact with. Thus, in our model, passage of both the H_2O and the leucogranite melts through the rocks contributes to the overall $^{18}O/^{16}O$ homogenization of the stratigraphic section, with release of latent heat of crystallization also contributing energy for continued melting and hydrothermal circulation. The post-decompression H_2O contents at 2 kbar and 730°C are calculated to be 5.5 wt% for the pelite-derived melt and 4.5 wt% for the grit-derived melt. In other words, the pelite-derived melts are still slightly more H_2O rich than the grit-derived melts, but the two types of leucogranites are now much more similar than they were at greater depths. At the end of decompression, 29 vol% of the Windermere section is leucogranite melt.

During cooling at 2 kbar, the post-decompression compositional evolution of the two types of melt follow the phase relations for the Qz-Ab-Or-H_2O system given in Tuttle and Bowen (1958). Crystallization of the H_2O-saturated (5.5 wt%) pelite-derived melt reduces its volume from 24% of the pelite section to 12%, releasing another 384 mol/m^3 of H_2O, which can be used to drive further melting of nearby grit layers (to 63% of the grit section), as well as to promote further $^{18}O/^{16}O$ exchange. The grit-derived melts become H_2O saturated (5.5 wt%) at T = 730°C and P = 2 kbar near the time of final crystallization, at which stage the Windermere section is composed of 26% leucogranite. This process goes on until final crystallization of the leucogranites (i.e., 680°C at 2 kbar; Wyllie, 1977), which releases all of the remaining melt H_2O (2395 mol/m^3), except the tiny amount that is ultimately incorporated into the leucogranite biotite (62 mol/m^3). This large outflow of H_2O at 680°C produces the final, and probably the most significant, episode of $^{18}O/^{16}O$ exchange among the various igneous and metamorphic lithologies, "locking in" the final $\delta^{18}O$ values in the 1 km thick section as we see it today (Fig. 15). Although this H_2O is then assumed to leave our model section, it should be noted that such a pore fluid would obviously be available for continued $^{18}O/^{16}O$ exchange with overlying lithologies, and just such a scenario is shown in Figure 16.

Summing up, the model shown on Figure 14 and in Table 7 indicates that the following amounts of H_2O were transported through, and exchanged with, the 1 km thick section at various stages: Prior to stage 1, 1400 mol/m^3 of H_2O moved into and through the stratigraphic column, where it was dissolved in the early leucogranite melts of stage 1. Between stages 1 and 2, another 913 mol/m^3 of H_2O was generated locally and moved into the leucogranite melts. Between stages 2 and 3, still another 832 mol/m^3 of H_2O moved through the section and entered the H_2O-undersaturated, grit-derived melts; most of this H_2O was derived from decompression of the H_2O-saturated, pelite-derived melts. Between stages 3 and 4, a much smaller amount of H_2O (384 mol/m^3) is released to enter the grit-derived melts, but following stage 4 a very significant quantity of H_2O is released during final crystallization of both kinds of melts, and this 2395 mol/m^3 of H_2O also

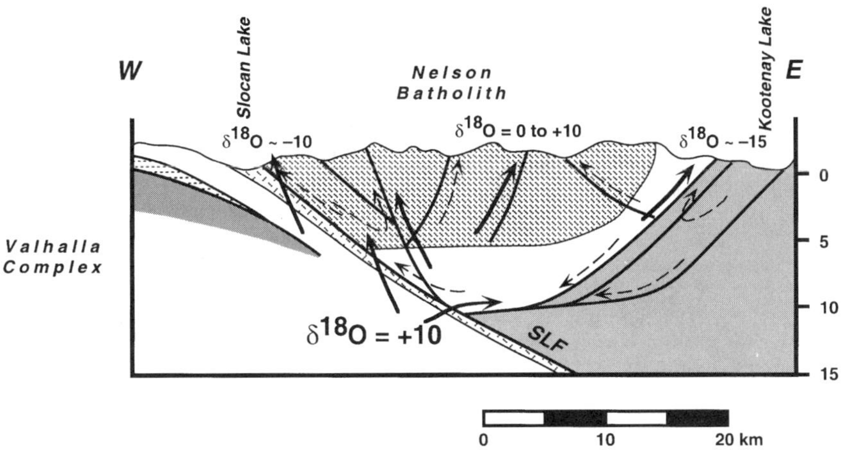

FIG. 16. Schematic cross section through the terrane on the eastern side of the Valhalla Complex (see Fig. 2), illustrating the hypothetical upward flow of H_2O (solid arrows, $\delta^{18}O = +10$) derived from final crystallization of the isotopically homogenized leucogranites in the Selkirk allochthon. This high-^{18}O H_2O is considered to be the deep seated component required by Beaudoin et al. (1992) and Holk and Taylor (1997b, 1999) to mix with the overlying low-^{18}O meteoric-hydrothermal fluids (dashed arrows) in the zone of brittle fracturing that overlies the Slocan Lake detachment fault (SLF); such mixed hydrothermal fluids are necessary to explain the isotopic characteristics of the mineral deposits and associated hydrothermal alteration phenomena in the upper plates of the metamorphic core complexes in southern British Columbia.

has to move through the section, triggering an additional major episode of $^{18}O/^{16}O$ homogenization.

These repeated episodes of H_2O influx, dehydration, exchange, incorporation, and exsolution in leucogranite melts and rocks in the hanging wall of the Monashee décollement, together with the effects of direct interaction and $^{18}O/^{16}O$ exchange between the metasedimentary rocks and the migrating partial melts, appear to be adequate to explain the pervasive $^{18}O/^{16}O$ homogenization observed in these core complexes. In this model, we have ignored the CO_2 released from calc-silicate rocks (e.g., see Fig. 7), but the passage of that oxygen-bearing fluid through the system could further contribute to the $^{18}O/^{16}O$ homogenization of the section. The amounts of H_2O utilized in the model are also sufficient to produce the amount of leucogranite melt observed in the actual section, including the melts that migrated upward and coalesced to form separate leucogranite plutons (such as the South Fosthall pluton). The above model readily accounts for the observed 30% partial melting of this section of the Windermere Supergroup, and we believe that the cyclic use and re-use of H_2O as an exchange catalyst, together with processes of direct oxygen communication between the rocks and the migrating leucogranite melts, are sufficient to impose the observed degree of $^{18}O/^{16}O$ homogenization, including the effects observed at higher levels in this mid-crustal section. These processes of large-scale hydrothermal anatexis of thick sequences of clastic metasediments during crustal-scale thrust ramping are likely to play a pivotal role in determining the susceptibility of such an orogen to subsequent wholesale extensional collapse along detachment faults.

Acknowledgments

It is a great honor to present this paper on the occasion of the 89th birthday of Konrad Krauskopf, who has spent much of his career studying the volatile contents of granitic plutons and the fluids that formed their associated mineral deposits. Also, his early mapping studies of the Colville batholith in northern Washington (Krauskopf, 1941; Waters and Krauskopf, 1941) provided the first comprehensive documentation of the southward extensions of the metamorphic core complexes in British Columbia studied in this work, and of the protoclastic deformation associated with what came to be known as the Okanagan detachment fault. Our research was supported by National Science Foundation grants EAR-90-19190, EAR-93-17036, and EAR-97-25811, and Geological Society of America grant

4951-92. We thank S. Carr, D. Clemens-Knott, C. Gazis, B. Nesbitt, M. Palin, R. Parrish, M. Peters, L. T. Silver, B. Taylor, B. Wernicke, D. J. Wood, and S. Wickham for discussions, and in particular Prof. Peter J. Wyllie for a very helpful and thorough review of this paper. Contribution No. 8559, Division of Geological and Planetary Sciences, California Institute of Technology.

REFERENCES

Althaus, E., Karotke, E., Nitsch, K. H., and Winkler, H. G. F., 1970, An experimental re-examination of the upper stability limit of muscovite plus quartz: N. Jahrbuch fur Mineralogie, v. 112, p. 324–336.

Anderson, J. L., Barth, A. P., and Young, E. D., 1988, Mid-crustal Cretaceous roots of Cordilleran metamorphic core complexes: Geology, v. 16, p. 366–369.

Armstrong, R. L., 1982, Cordilleran metamorphic core complexes from Arizona to southern Canada: Ann. Rev. Earth Planet. Sci., v. 10, p. 129–154.

Armstrong R. L., Parrish, R. R., Van der Heyden, P., Scott, K., Runkle, D., and Brown, R. L., 1991, Early Proterozoic basement exposures in the southern Canadian Cordillera: Core gneiss of Frenchman Cap, Unit I of the Grand Forks gneiss and the Vaseaux Formation: Can. Jour. Earth Sci., v. 28, p. 1169–1201.

Armstrong, R. L., and Ward, P., 1991, Evolving geographic patterns of Cenozoic magmatism in the North American Cordillera: The temporal and spatial association of magmatism and metamorphic core complexes, Jour. Geophys. Res., v. 96, p. 13,201–13,224.

Arzi, A. A., 1978, Critical phenomena in the rheology of partially melted rocks: Tectonophys., v. 44, p. 173–184.

Beaudoin, G., and Taylor, B. E., 1997, $^{18}O/^{16}O$ homogenization of the middle crust during anatexis: The Thor-Odin metamorphic core complex, British Columbia: Comment: Geology, v. 25, p. 1053–1054.

Beaudoin, G., Taylor, B. E., and Sangster, D. F., 1991, Silver-lead-zinc veins, metamorphic core complexes, and hydrologic regimes during crustal extension: Geology, v. 19, p. 1217–1220.

______, 1992, Silver-lead-zinc veins and crustal hydrology during Eocene extension, southeastern British Columbia, Canada: Geochim. et Cosmochim. Acta, v. 56, p. 3513–3529.

Bickle, M. J., Chapman, H. J., Wickham, S. M., and Peters, M. T., 1995. Strontium and oxygen isotope profiles across marble-silicate contacts, Lizzies Basin, East Humboldt Range, Nevada: Constraints on metamorphic permeability contrasts and fluid flow: Contrib. Mineral. Petrol., v. 121, p. 400–413.

Blattner, P., 1971, Migmatite by partial fusion and short range hydrothermal transfer, British Columbia: Schweiz. Min. Petr. Mitt., v. 51, p. 155–177.

Block, L. H., and Royden, L. H., 1990, Core complex geometries and regional-scale flow in the lower crust: Tectonics, v. 9, p. 557–567.

Bowman, J. R., and Ghent, E. D., 1986, Oxygen and hydrogen isotope study of minerals from metapelitic rocks, staurolite to sillimanite zones, Mica Creek, British Columbia: Jour. Metamor. Geol., v. 4, p. 131–141.

Brown, R. L., and Journeay, J. M., 1987, Tectonic denudation of the Shuswap metamorphic terrane of Southeastern British Columbia: Geology, v. 15, p. 142–146.

Brown, R. L., et al., 1992, The Monashee décollement of the southern Canadian Cordillera: A crustal-scale shear zone linking the Rocky Mountain foreland belt to lower crust beneath accreted terranes, *in* McClay, K. R., ed., Thrust tectonics: London, Chapman and Hall, p. 357–364.

Burt, E. A., and Taylor, H. P., Jr., 1989, A comparison of whole-rock oxygen isotopic variations in regionally metamorphosed and unmetamorphosed sedimentary rocks [abs.]: Geol. Soc. Amer. Abs. Prog., v. 21, p. 274.

______, 1993, Remarkably uniform bulk silicate $\delta^{18}O$ values of terrigenous sedimentary rocks from the Central Appalachain and Ouachita geosynclines [abs.]: Geol. Soc. Amer. Abs. Prog., v. 25, p. 203–204.

Carr, S. D., 1991, Three crustal zones in the Thor-Odin-Pinnacles area, Southern Omineca Belt, British Columbia: Can. Jour. Earth Sci., v. 28, p. 2003–2023.

______, 1992, Tectonic setting and U-Pb geochronology of the early Tertiary Ladybird leucogranite suite, Thor-Odin-Pinnacles area, Southern Omineca belt, British Columbia: Tectonics, v. 11, p. 258–278.

______, 1995, The Southern Omineca belt, British Columbia: New perspectives from the Lithoprobe geoscience program: Can. Jour. Earth Sci., v. 32, p. 1720–1739.

Carr, S. D., and Parkinson, D., 1989, Eocene stratigraphy, age of the Coryell batholith, and extensional faults in the Granby Valley, southern British Columbia, *in* Current Research, Part E, Geol. Surv. Can. Pap., 89-1E, p. 79–87.

Carr, S. D., Parrish, R. R., and Brown, R. L., 1987, Eocene structural development of the Valhalla complex, Southeastern British Columbia: Tectonics, v. 6, p. 175–196.

Clayton, R. N., and Kieffer, S. W., 1991, Oxygen isotopic thermometer calculations, *in* Stable isotope geochemistry: A tribute to Samuel Epstein: Geochem. Soc. Spec. Publ., no. 3, p. 3–10.

Clayton, R. N., O'Neil, J. R., and Mayeda, T. K., 1972, Oxygen isotope exchange between quartz and water: Jour. Geophys. Res., v. 77, p. 3057–3067.

Clemens, J. D., and Vielzeuf, D., 1987, Constraints on melting and magma production in the crust: Earth Planet. Sci. Lett., v. 86, p. 287–306.

Coney, P. J., and Harms, T., 1984. Cordilleran metamorphic core complexes: Cenozoic extensional relics of Mesozoic compression: Geology, v. 12, p. 550–554.

Cook, F. A., et al., 1992, Lithoprobe crustal reflection cross section of the Southern Canadian Cordillera, 1, Foreland thrust and fold belt to Fraser River fault: Tectonics, v. 11, p. 12–35.

Criss, R. E., Fleck, R. J., and Taylor, H. P., Jr., 1991, Tertiary meteoric hydrothermal systems and their relation to ore deposition, Northwestern United States and Southern British Columbia: Jour. Geophys. Res., v. 96, p. 13,335–13,356.

Duncan, I. J., 1984, Structural evolution of the Thor-Odin gneiss dome: Tectonophys., v. 101, p. 87–130.

England, P. C., and Thompson, A. B., 1984, Pressure-temperature-time paths of regional metamorphism: I. Heat transfer during the evolution of regions of thickened continental crust: Jour. Petrol., v. 25, p. 929–955.

Eslinger, E. V., and Savin, S. M., 1973, Oxygen isotope geothermometry of the burial metamorphic rocks of the Precambrian Belt Supergroup, Glacier National Park, Montana: Geol. Soc. Amer. Bull., v. 84, p. 2549–2560.

Fricke, H. C., Wickham, S. M., and O'Neil, J. R., 1992, Oxygen and hydrogen isotope evidence for meteoric water infiltration during mylonitization and uplift in the Ruby Mountains–East Humboldt Range core complex, Nevada: Contrib. Mineral. Petrol., v. 111, p. 203–221.

Fyles, J. T., 1964, Geology of the Duncan Lake area, Lardeau District, British Columbia: Brit. Col. Dept. Mines Petrol. Resources Bull., v. 49.

Fyles, J. T., and Eastwood, G. E. P., 1962, Geology of the Ferguson area: Brit. Col. Dept. Mines Petrol. Resources Bull., v. 45.

Gans, P. B., 1987, An open-system, two-layer crustal stretching model for the Eastern Great Basin: Tectonics, v. 6, p. 1–12.

Ghent, E. D., and O'Neil, J. R., 1985, Late Precambrian marbles of unusual carbon-isotope composition, southeastern British Columbia: Can. Jour. Earth Sci., v. 22, p. 324–329.

Gregory, R. L., and Taylor, H. P., Jr., 1981, An oxygen isotope profile in a section of Cretaceous oceanic crust, Samail Ophiolite, Oman: Evidence for $\delta^{18}O$ buffering of the oceans by deep (>5 km) seawater-hydrothermal circulation at mid-ocean ridges: Jour Geophys. Res., v. 86, p. 2737–2755.

Grunder, A. L., and Wickham, S. M., 1991, Homogenization and lowering of $^{18}O/^{16}O$ in mid-crustal rocks during extension-related magmatism in Eastern Nevada: Earth Planet. Sci. Lett., v. 107, p. 416–431.

Holdaway, M. J., 1971, Stability of andalusite and the aluminum silicate phase diagram: Amer. Jour. Sci., v. 271, p. 97–131.

Holk, G. H., 1997, The role of water in the magmatic and tectonic evolution of metamorphic core complexes: A stable isotope study of the Southern Omineca crystalline belt, British Columbia, Canada: Unpubl. Ph.D. thesis, Calif. Inst. Technol., Pasadena, California, 444 p.

Holk, G. H., Taylor, B. E., and Taylor, H. P., Jr., 1997, Influx of low D/H meteoric water into the lower plates of metamorphic core complexes, southeastern British Columbia [abs.]: Geol. Soc. Amer. Abs. Prog., v. 29, p. 449.

Holk, G. H., and Taylor, H. P., Jr., 1997a, $^{18}O/^{16}O$ homogenization of the middle crust during anatexis: Thor-Odin metamorphic core complex, British Columbia: Geology, v. 25, p. 31–34.

______, 1997b, $^{18}O/^{16}O$ homogenization of the middle crust during anatexis: Thor-Odin metamorphic core complex, British Columbia: Reply: Geology, v. 25, p. 1054–1055.

______, 1999, $^{18}O/^{16}O$ studies of regional metamorphism, anatexis, extensional magmatism, meteoric-hydrothermal activity, and ore deposition in the Valhalla metamorphic core complex, British Columbia: Jour. Volcanol. Geotherm. Res. (submitted).

Hollister, L. S., 1977, The reaction forming cordierite from garnet, the Khtada Lake Metamorphic complex, British Columbia: Can. Mineral., v. 15, 217–219.

______, 1993, The role of melt in the uplift and exhumation of orogenic belts: Chem. Geol., v. 108, p. 31–48.

Holtz, F., and Johannes, W., 1994, Maximum and minimum water contents of granitic melts: Implications for chemical and physical properties of ascending magmas: Lithos, v. 32, p. 149–159.

Hoy, T., 1977, Big Ledge, *in* Geology in British Columbia 1975: Brit. Col. Dept. Mines Petrol. Resources, G12–G18.

______, 1988, Geology of the Cottonbelt lead-zinc-magnetite layer, carbonatites, and alkalic rocks in the Mount Grace area, Frenchman Cap dome, southeastern British Columbia: Brit. Col. Dept. Mines Petrol. Resources Bull., v. 80.

Javoy, M., 1977, Stable isotopes and geothermometry: Jour. Geol. Soc. Lond., v. 133, p. 609–636.

Journeay, J. M., 1988, Relic P-T regimes of the Monashee Complex: A record of repeated imbrication and heat transfer, S.E. Omineca belt [abs.]: Geol. Soc. Amer. Abs. Prog., v. 20, p. 19.

Kerrich, R., 1988, Detachment zones of Cordilleran metamorphic core complexes: Thermal, fluid and metasomatic regimes: Geol. Rundschau, v. 77, 157–182.

Knoll, A. H., Hayes, J. M., Kaufnam, A. J., Swett, K., and Lambert, I. B., 1986, Secular variation in carbon isotope ratios from Upper Proterozoic successions of Svalbard and East Greenland: Nature, v. 321, p. 832–838.

Krauskopf, K., 1941, Intrusive rocks of the Okanogan Valley and the problem of their correlation: Jour. Geol., v. 49, p. 1–53.

Kyser, T. K., Ghent, E. D., and Sevigny, J. H., 1994, Petrogenesis of two-mica granites, Northern Monashee Mountains, British Columbia: Evidence from oxygen, hydrogen, and strontium isotopic compositions: Geol. Soc. Amer. Bull., v. 106, p. 1531–1537.

Lane, L. S., Ghent, E. D., Stout, M. Z., and Brown, R. L., 1989, P-T history and kinematics of the Monashee décollement near Revelstoke, British Columbia: Can. Jour. Earth Sci., v. 26, p. 231–243.

Liu, M., and Furlong, K. P., 1993, Crustal shortening and Eocene extension in the southeastern Canadian Cordillera: Some thermal and rheological considerations: Tectonics, v. 12, p. 776–786.

Luth, W. C., Jahns, R. H., and Tuttle, O. F., 1964, The granite system at pressures of four to ten kilobars: Jour. Geophys. Res., v. 69, p. 759–773.

Magaritz, M., and Taylor, H. P., Jr., 1986, Oxygen 18/Oxygen 16 and D/H studies of plutonic granitic and metamorphic rocks across the Cordilleran batholiths of Southern British Columbia: Jour. Geophys. Res., v. 91, p. 2193–2217.

McCrea, J. M., 1950, On the isotopic chemistry of carbonates and a paleotemperature scale: Jour. Chem. Phys., v. 18, p. 849–857.

McNicoll, V. J., and Brown, R. L., 1995, The Monashee décollement at Cariboo Alp, southern flank of the Monashee complex, Southern British Columbia, Canada: Jour. Struct. Geol., v. 17, p. 17–30.

Morrison, J., 1994, Meteoric water-rock interaction in the lower plate of the Whipple Mountain metamorphic core complex, California: Jour. Metamor. Geol., v. 12, p. 827–840.

Nesbitt, B. E., and Muehlenbachs, K., 1989, Origins and movement of fluids during deformation and metamorphism in the Canadian Cordillera: Science, v. 245, p. 733–736.

______, 1991, Stable isotopic constraints on the nature of the syntectonic fluid regime of the Canadian Cordillera: Geophys. Res. Lett., v. 18, 963–966.

______, 1995, Geochemical studies of the origins and effects of synorogenic crustal fluids in the Southern Omineca Belt of British Columbia: Geol. Soc. Amer. Bull., v. 107, 1033–1050.

Nyman, M. W., Pattison, D. R. M., and Ghent, E. D., 1995, Melt extraction during formation of K-feldspar + sillimanite migmatites, west of Revelstoke, British Columbia: Jour. Petrol., v. 36, p. 351–372.

O'Neil, J. R., and Ghent, E. D., 1975, Stable isotope study of coexisting metamorphic minerals from the Esplanade Range, British Columbia: Geol. Soc. Amer. Bull., v. 86, p. 1708–1712.

Parkinson, D., 1991, Age and isotopic character of early Proterozoic basement gneisses in the Southern Monashee Complex, southeastern British Columbia: Can. Jour. Earth Sci., v. 28, p. 1159–1168.

Parrish, R. R., Carr, S. D., and Parkinson, D. L., 1988, Eocene extensional tectonics and geochronology of the Southern Omineca belt, British Columbia and Washington: Tectonics, v. 7, p. 181–212.

Patiño Douce, A. E., Humphreys, E. D., and Johnston, A. D., 1990, Anatexis and metamorphism in tectonically thickened continental crust exemplified by the Sevier hinterland, Western North America: Earth Planet. Sci. Lett., v. 97, p. 290–315.

Pell, J., and Simony, P. S., 1987, New correlations of Hadrynian strata, south-central British Columbia: Can. Jour. Earth Sci., v. 24, p. 302–313.

Peters, M. T., and Wickham, S. M., 1995, On the causes of ^{18}O-depletion and $^{18}O/^{16}O$ homogenization during regional metamorphism, the East Humboldt range core complex, Nevada: Contrib. Mineral. Petrol., v. 119, p. 68–82.

Raeside, R. P., and Ghent, E. D., 1986, The migmatization of Hadrynian sedimentary rocks, Mica Creek, British Columbia: Can. Mineral., v. 24, p. 547–556.

Read, P. B., 1979, Relationship between the Shuswap metamorphic complex and Kootenay Arc, Vernon east-half, southern British Columbia: Geol. Surv. Can. Pap., 79-1A, p. 37–40.

Reesor, J. E., 1965, Valhalla gneiss complex: Geol. Surv. Can. Bull., no. 129.

Reesor, J. E., and Moore, J. M., 1971, Petrology and structure of the Thor-Odin gneiss dome, Shuswap metamorphic complex, British Columbia: Geol. Surv. Can. Bull., no. 197.

Reynolds, S. J. and Lister, G. S., 1987, Structural aspects of fluid-rock interactions in detachment zones: Geology, v. 15, p. 362–366.

Rumble, D., and Spear, F. S., 1983, Oxygen isotope equilibrium and permeability enhancement during regional metamorphism: Jour. Geol. Soc. Lond., v. 140, p. 619–628.

Rye, R. O., Schuiling, R. D., Rye, D. M., and Jansen, J., 1976, Carbon, hydrogen, and oxygen isotope studies of the regional metamorphic complex at Naxos, Greece: Geochim. et Cosmochim. Acta, v. 40, p. 1031–1049.

Scammell, R. J., and Brown, R. L., 1990, Cover gneisses of the Monashee terrane: A record of synsedimentary rifting in the North American Cordillera: Can. Jour. Earth Sci., v. 27, p. 712–726.

Schwarcz, H. P., and Clayton, R. N., 1965, Oxygen isotopic studies of amphibolites: Can. Jour. Earth Sci., v. 2, p. 72–84.

Selverstone, J., and Chamberlain, C. P., 1990, Apparent isobaric cooling paths from granulites: Two counterexamples from British Columbia and New Hampshire: Geology, v. 18, p. 307–310.

Sevigny, J. H., Parrish, R. R., Donelick, R. A., and Ghent, E. D., 1990, Northern Monashee Mountains, Omineca Crystalline belt, British Columbia: Timing of metamorphism, anatexis, and tectonic denudation: Geology, v. 18, p. 103–106.

Sevigny, J. H., Parrish, R. R., and Ghent, E. D., 1989, Petrogenesis of peraluminous granites, Monashee

Mountains, southeastern Canadian Cordillera: Jour. Petrol., v. 30, p. 557–581.

Sharma, T., and Clayton, R. N., 1965, Measurement of $^{18}O/^{16}O$ ratios of total oxygen of carbonates: Geochim. et Cosmochim. Acta, v. 29, p. 1347–1353.

Simony, P. S., and Carr, S. D., 1997, Large lateral ramps in the Eocene Valkyr shear zone: Extensional ductile faulting controlled by plutonism in southern British Columbia: Jour. Struct. Geol., v. 19, p. 769–784.

Sharp, Z. D., 1990, A laser-based microanalytical method for the in situ determination of oxygen isotope ratios of silicates and oxides: Geochim. et Cosmochim. Acta, v. 54, p. 1353–1357.

Shaw, D. M., 1956, Major elements and general geochemistry (Part 3) *in* Geochemistry of pelitic rocks: Geol. Soc. Amer. Bull., v. 67, p. 919–934.

Sheppard, S. M. F., 1986, Characterization and isotopic variations in natural waters, *in* Valley, J. W., Taylor, H. P., Jr., and O'Neil, J. R., eds., Stable isotopes in high temperature geological processes: Rev. Mineral., v. 16, p. 165–181.

Shieh, Y. N., and Schwarcz, H. P., 1974, Oxygen isotope studies of granite and migmatite, Grenville province of Ontario, Canada: Geochim. et Cosmochim. Acta, v. 38, 21–45.

Spear, F. S., and Parrish, R. R., 1996, Petrology and cooling rates of the Valhalla complex, British Columbia, Canada: Jour. Petrol., v. 37, p. 733–765.

Spencer, J. E., and Welty, J. W., 1986, Possible controls on base- and precious-metal mineralization associated with Tertiary detachment faults in the lower Colorado River trough, Arizona and California: Geology, v. 14, p. 195–198.

Taylor, H. P., Jr., 1988, Oxygen, hydrogen, and strontium isotope constraints on the origin of granites: Trans. Roy. Soc. Edinburgh, Earth Sci., v. 79, p. 317–388.

Taylor, H.P., Jr., and Epstein, S., 1962, Relationship between $^{18}O/^{16}O$ ratios in coexisting minerals of igneous and metamorphic rocks, Part I. Principles and experimental results. Geol. Soc. Amer. Bull., v. 73, p. 461–480.

Templeman-Kluit, D., and Parkinson, D., 1986, Extension across the Eocene Okanagan crustal shear in southern British Columbia: Geology, v. 14, p. 318–321.

Thompson, A. B., and Algor, J. R., 1977, Model systems for anatexis of pelitic rocks: Contrib. Mineral. Petrol., v. 63, p. 247–269.

Thompson, A. B., and Tracy, R. J., 1979, Model systems for anatexis of pelitic rocks: II. Facies series melting and reactions in the system $CaO-KAlO_2-NaAlO_2-SiO_2-H_2O$: Contrib. Mineral. Petrol., v. 70, p. 429–438.

Tuttle, O. F., and Bowen, N. L., 1958, Origin of granite in the light of experimental studies in the system $NaAlSi_3O_8-KAlSi_3O_8-SiO_2-H_2O$: Geol. Soc. Amer. Memoir, no. 74.

Valley, J. W., 1986, Stable isotope geochemistry of metamorphic rocks, *in* Valley, J. W., Taylor, H. P., Jr., and O'Neil, J. R., eds., Stable isotopes in high temperature geological processes: Rev. Mineral., v. 16, p. 445–490.

Vanderhaeghe, O., and Teyssier, C., 1997, Formation of the Shuswap metamorphic core complex during late-orogenic collapse of the Canadian Cordillera: Role of ductile thinning and partial melting of the mid-to lower crust: Geodinamica Acta, v. 10, p. 41–58.

Veizer, J., and Hoefs, J., 1976, The nature of $^{18}O/^{16}O$ and $^{13}C/^{12}C$ secular trends in sedimentary carbonate rocks: Geochim. et Cosmochim. Acta, v. 40, p. 1387–1395.

Vielzeuf, D., and Holloway, J. R., 1988, Experimental determination of the fluid-absent melting relations in the pelitic system: Contrib. Mineral. Petrol., v. 98, p. 257–276.

Walther, J. V., and Orville, P. M., 1982, Volatile production and transport in regional metamorphism: Contrib. Mineral. Petrol., v. 79, p. 252–257.

Wanless, R. K., and Reesor, J. E., 1975, Precambrian zircon ore of orthogneiss in the Shuswap metamorphic complex, British Columbia: Can. Jour. Earth Sci., v. 12, p. 326–331.

Waters, A. C., and Krauskopf, K., 1941, Protoclastic border of the Colville Batholith: Bull. Geol. Soc. Amer., v. 52, p. 1355–1418.

Wernicke, B., 1990, The fluid crustal layer and its implications for continental dynamics, *in* Salisbury, M. H., and Fountain, D. M., eds., Exposed cross-sections of the continental crust: Dordrecht, Kluwer, p. 559–544.

Wernicke, B., and Axen, G. J., 1988, On the role of isostacy in the evolution of normal fault systems: Geology, v. 16, p. 848–851.

Whitney, J. A., 1988, The origin of granite: The role and source of water in the evolution of granitic magmas, Geol. Soc. Amer. Bull., v. 100, p. 1971–1980.

Wickham, S. M., and Peters, M. T., 1990, An oxygen isotope discontinuity in high-grade rocks of the East Humboldt Range, Nevada: Nature, v. 345, p. 150–153.

______, 1992, Oxygen and carbon isotope profiles in metasediments from Lizzies Basin, East Humboldt Range, Nevada: Constraints on mid-crustal metamorphic and magmatic volatile fluxes: Contrib. Mineral. Petrol., v. 112, p. 46–65.

______, 1993, High $\delta^{13}C$ Neoproterozoic carbonate rocks in Western North America: Geology, v. 21, p. 165-168.

Wickham, S. M., and Taylor, H. P., Jr., 1985, Stable isotopic evidence for large-scale seawater infiltration in a regional metamorphic terrane; the Trois Seigneurs Massif, Pyrenees, France: Contrib. Mineral. Petrol., v. 91, p. 122–137.

______, 1987, Stable isotope constraints on the origin and depth of penetration of hydrothermal fluids associated with Hercynian regional metamorphism and crustal anatexis in the Pyrenees: Contrib. Mineral. Petrol., v. 95, p. 255–268.

Wickham, S. M., Taylor, H. P., Snoke, A. W., and O'Neil, J. R., 1991, An oxygen and hydrogen isotope study of high-grade metamorphism and anatexis in the Ruby Mountains–East Humboldt Range core complex, Nevada, *in* Stable isotope geochemistry: A tribute to Samuel Epstein: Geochem. Soc. Spec. Publ. no. 3, p. 373–389.

Wyllie, P. J., 1977, Crustal anatexis: An experimental review: Tectonophys., v. 43, 41–71.

Composition and Petrogenesis of Ti-Oxides Associated with Diamonds

N. V. SOBOLEV AND E. S. YEFIMOVA

Institute of Mineralogy and Petrography, Siberian Branch, Russian Academy of Sciences, Novosibirsk 630090, Russia

Abstract

Among Ti-bearing oxides, rutile is common in eclogitic (E-type) diamonds and diamondiferous eclogites of varied bulk chemistry, but rarely occurs in association with ultramafic (or peridotitic) (U-type) diamonds, especially in lherzolitic assemblages. Roughly 100 rutile samples associated with diamonds as inclusions (71 samples), intergrowths with polycrystalline diamond aggregates (20 samples), and extracted from diamondiferous eclogites (11 samples) were studied by electron microprobe techniques for Fe, Al, Mg, Ca, Mn, Cr, Nb, and Zr contents. Rutiles included in diamonds from Yakutia and Venezuela dominate the study. Compositions of rutiles associated with diamond and/or coesite from crustal ultrahigh-pressure (UHP) metamorphic rocks from the Kokchetav massif, northern Kazakhstan, and the Dora Maira massif in the Italian Alps are presented for comparison.

Titanates of the LIMA (lindsleyite-mathiasite) series and perovskite are rare accessories in U-type diamonds. Ilmenite of a wide range of compositions, containing up to 80% geikielite, is closely associated with rutile in bimineralic intergrowths in a number of diamonds and diamondiferous eclogites. Diamond-associated rutiles contain up to 0.8 wt% Al_2O_3, 2.5 wt% Fe_2O_3, 1.4 wt% Nb_2O_5, 0.45 wt% ZrO_2, and 2.3 wt% Cr_2O_3 (in the U type). The majority of rutiles, including diamond inclusions, are heterogeneous with widely varying Al, Fe, and Mg contents. These heterogeneities are mainly caused by the presence of sigmoidal oriented lamellae of ilmenite, picroilmenite (geikielite), and corundum; these are documented for the first time in diamond-associated rutiles and may represent exsolution from a UHP precursor. These features reflect the complex P-T and $f(O_2)$ history of the rutiles. The presence of complex titanates of the LIMA series and perovskite in diamond-bearing assemblages reflects the metasomatizing conditions of diamond formation.

Introduction

TITANIUM-BEARING OXIDES, especially rutile and ilmenite, play an important role in many ore deposits within crustal rocks, and typically occur as accessory minerals in a wide range of igneous and metamorphic rocks, including mantle-derived rocks, represented by xenoliths of varying compositions in alkali basalts and kimberlites. Ti-bearing oxides in these rocks are chiefly accessory minerals amounting to no more than 1 to 3 vol%, but petrogenetically they are very important (Sobolev et al., 1972, Haggerty, 1991a, 1991b; Rudnick et al., 2000). Although rutile + ilmenite have been identified as inclusions in diamonds and in high-pressure rocks that include diamond, the oxides have not been systematically studied in these rocks (e.g., Sobolev, 1974; Meyer and Svisero, 1975; Prinz et al., 1975; Mvuemba Ntanda et al., 1982; Bulanova, 1985; Meyer and McCallum, 1986; Meyer, 1987). Rutile was among the first identified inclusions in natural diamonds, based on x-ray diffraction methods (Futergendler, 1956; Harris, 1968; Orlov, 1973). Accessory rutile was also documented in a number of xenoliths of diamondiferous eclogites, including kyanite- and corundum-bearing eclogites from kimberlites (Sobolev et al., 1972; Sobolev, 1974; Hills and Haggerty, 1989).

The present paper reports on the composition of rutile, associated ilmenite, and some other rare Ti-bearing oxides (titanates of the LIMA series, perovskite) from the ultrahigh-pressure mineral assemblages containing natural diamond. These assemblages are known in both ultramafic (peridotitic) and eclogitic geochemical types (U and E types, respectively) of diamond forming environments (Sobolev, 1974; Meyer, 1987), represented by xenoliths of diamondiferous rocks in kimberlites and by inclusions in diamonds. However, titanate (Sobolev et al., 1997) and especially perovskite are among the rarest diamond inclusions. Rutiles occur dominantly in E-type diamondiferous environments in all types of eclogites, including ilmenite-, corundum-, kyanite-, and coesite-bearing samples that may con-

TABLE 1. Composition of Rutiles Associated with Picroilmenite (geikielite), and Diamonds and Perovskite Included in Diamond[1]

Sample:	Mr-20			Mr-553			Mr-574			Mr-562		Mr-25/15	S-28/87
	Rutile	1σ	Ilmenite	Rutile	1σ	Ilmenite	Rutile	1σ	Ilmenite	Rutile	1σ	Rutile	Perovskite
n:	2			9			6			3		I	I
TiO_2	98.4	0.21	49.7	96.0	0.25	63.1	98.9	0.85	61.8	97.0	0.46	97.1	56.9
Al_2O_3	0.27	0.17	0.67	0.54	0.02	0.2	0.11	0.05	0.17	0.00	0.00	0.11	0.19
Cr_2O_3	0.12	0.01	1.1	0.21	0.01	0.09	0.09	0.01	0.06	1.67	0.06	2.25	0.63
FeO			36.9			10.5			14.80				1.08
Fe_2O_3	0.65	0.07		1.40	0.06		0.29	0.11		0.07	0.05	0.32	
MgO	0.02	0.01	10.5	0.10	0.32	25.5	0.07	0.04	23.30	0.02	0.03	0.06	0.05
CaO													36.7
Nb_2O_5	0.37	0.01	0.27	1.02	0.20	0.43	0.12	0.14	0.08	0.04	0.03		0.48
ZrO_2	0.15	0.07		0.12	0.04		0.03	0.03		0.19	0.02		0.07
NiO			0.12			0.12			0.12				0.01
Total	99.92		99.14	99.39		99.82	99.55		100.21	98.99		99.84	96.10

[1]Samples with symbol Mr, M, BM, and TM are from the Mir pipe, Yakutia; S = Sytykanskaya pipe, Yakutia. Inclusions in diamonds = Mr-574, Mr-25/15, S-28/87; intergrowths with polycrystalline diamond aggregates = Mr-20, Mr-553, Mr-562.

tain rutile as an accessory mineral. In comparison, rutiles from ultrahigh-pressure crustal rocks that contain diamond and/or coesite also were investigated. These include metasedimentary diamond-bearing felsic gneiss of the Kokchetav massif in northern Kazakhstan, where intimate rutile-diamond intergrowths were observed, garnet-pyroxene rock (Sobolev and Shatsky, 1990), and coesite-bearing pyrope quartzite of Dora Maira in northwestern Italy (Chopin, 1984).

In this paper, the compositions of diamond-associated rutile are newly reported for diamonds and diamondiferous eclogites from Yakutia, South Africa, and Venezuela (Sobolev et al., 1972, 1997, 1998), and these results along with earlier unpublished data by the authors are compared with rutiles occurring as diamond inclusions from other regions (Meyer, 1987) and presented in other publications (Smith and Dawson, 1975; Kovalsky et al., 1979; Bulanova, 1985; Haggerty, 1991a, 1991b; Kopylova et al., 1997; Spetsius and Griffin, 1997; Rudnick et al., 2000).

Rutile is an accessory in xenolithic eclogites, including diamondiferous ones, and also appears as inclusions in a small proportion of E-type diamonds. Systematic investigations of E-type diamonds from Venezuelan kimberlites (Sobolev et al., 1998) show that only 7% contain rutile in two-mineral assemblages with garnet, magnetite, and omphacite as well as in three-mineral associations, containing clinopyroxene, ilmenite, coesite, and magnetite. A similar picture (6% of diamonds contain rutile inclusions) is observed in the E-type diamonds from the placers north of the Siberian Platform (Sobolev et al., 1999).

Analytical Techniques

Rutiles and associated ilmenites were analyzed by electron microprobe (EMP) using a four-spectrometer CAMEBAX-micro at the United Institute of Geology, Geophysics, and Mineralogy, Novosibirsk, employing an accelerating voltage of 20 kV

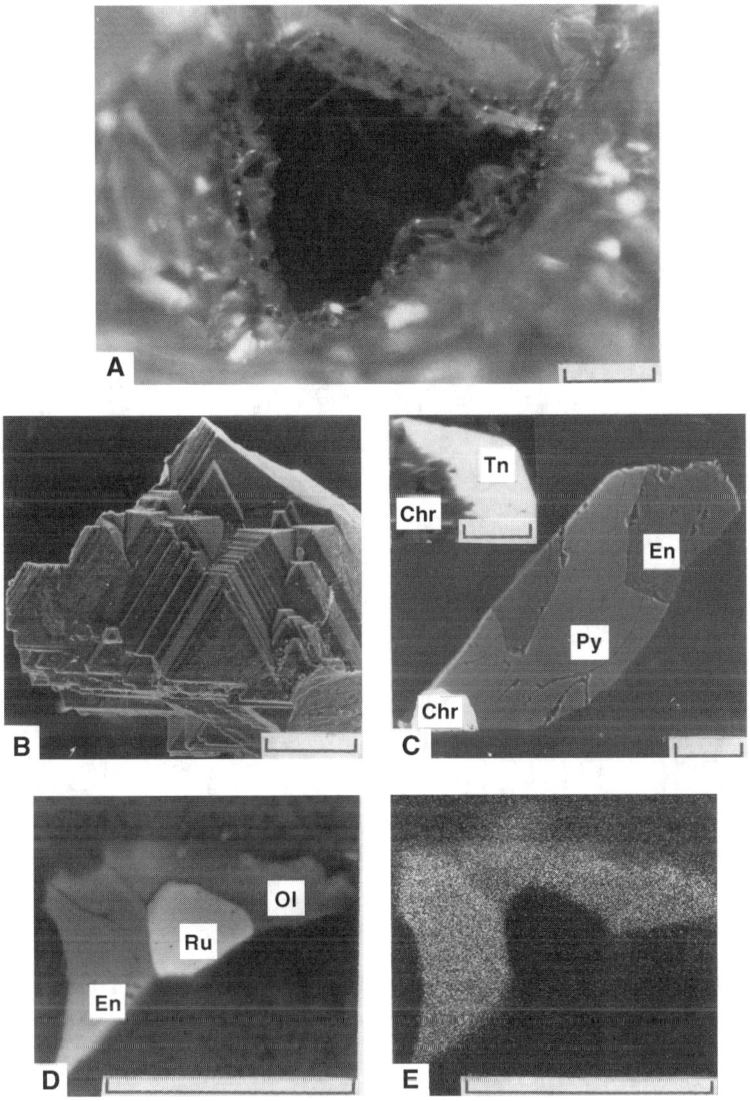

FIG. 1. Morphological features and polymineralic assemblages of Ti-bearing oxides associated with diamonds (A–L). Rutile inclusion in diamond (A) and rutile extracted from diamond (B), samples Mr-782 and Mr-567, respectively. Both samples are characterized by negative diamond morphology. Polymineralic inclusion of chromite (Chr) and titanite (Tn) of the LIMA series, pyrope (Py) and enstatite (En) from the Sputnik pipe (Yakutia) diamond, sample Sp-728, backscattered electron image (C); polymineralic inclusion of enstatite, rutile, and olivine from the Mir pipe diamond, sample Mr-25/15, backscattered electron image (D); the same in Si k_{α} radiation (E); rutile (light) included in omphacite from Venezuelan diamond, sample Gm-97-16, backscattered electron image (F); rutile (light) intergrown with geikielite (dark), from the Mir pipe diamond, sample Mr-574, backscattered electron image (G); rutiles containing oriented exsolution-type ilmenite lamellae, including sample Ud-2023,B from diamond polycrystalline aggregate of the Udachnaya pipe (H) and diamondiferous eclogite from the Mir pipe, sample TM-180 (I); rutile from diamondiferous eclogite of the Roberts Victor mine (sample RV-15) containing oriented corundum (dark) and ilmenite (light) sigmoidal rods, general view (J) and enlarged portion of a grain (K), backscattered electron image. Scale bars are 100 μm for all photos except: 10 μm for C (upper left corner) and F, and 25 μm for K. (*Figure continues on following page.*)

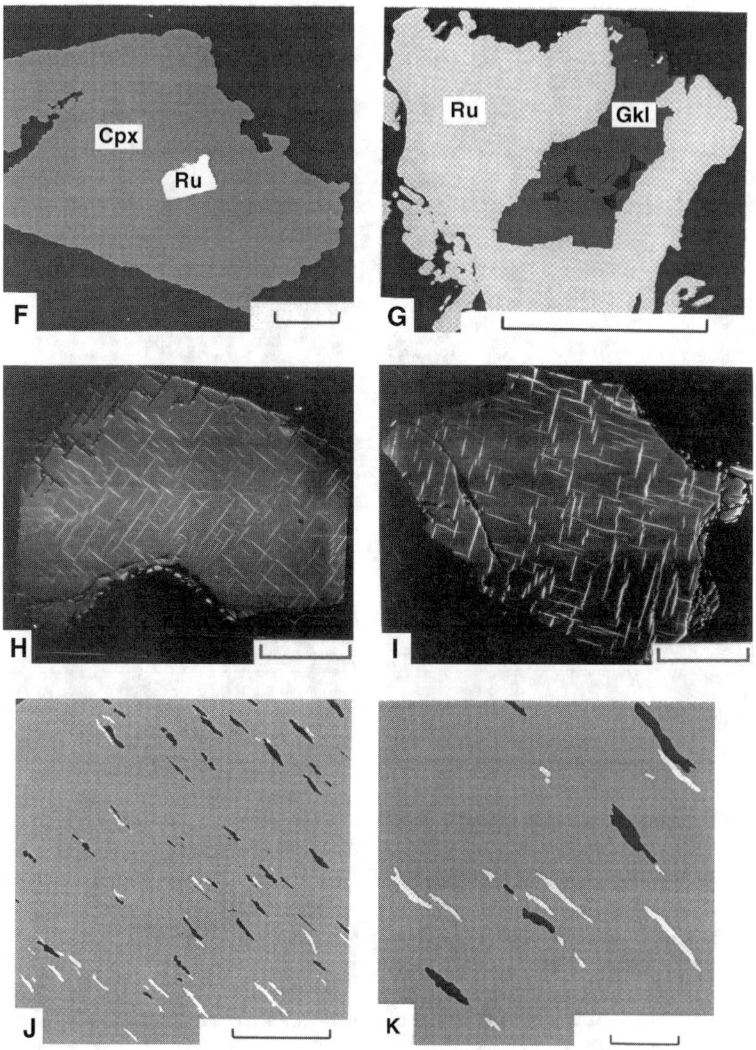

FIG. 1. Continued (*caption on preceding page*)

and a beam current of 50 nA. The beam size was 3 to 4 μm, with a counting time of 10 seconds for all elements (Ti, Cr, Al, Ca, Mg, Mn, Fe, Nb, and Zr). Several samples also were analyzed using a defocused beam with a diameter of ~50 μm. All data were corrected using ZAF procedures. Limits of detection for Nb and Zr were 0.05 wt% and 0.02 wt% for all other elements recalculated as oxides. Special attention was paid to the homogeneity of rutile grains. No fewer than five points were analyzed for each sample and even more were performed when heterogeneities were observed. Only minimal values of Mg, Al, and Fe were accepted for individual inhomogeneous samples.

Results

Ca and Mn abundances did not exceed the detection limit (0.02 wt% of CaO and MnO), and thus are not discussed. The results of major- and minor-element analyses of representative samples are presented in Tables 1, 2, and 3. Most rutile samples occur as inclusions in diamonds; some are present in polymineralic assemblages, in particular as intergrowths with clinopyroxenes. Rutile inclusions in pyroxenes from Venezuelan diamonds are among them. Some samples of rutile were extracted from polycrystalline aggregates of diamond (boart, framesite). Accessory rutiles from 11 samples of diamondiferous eclogites from Yakutian and South African kimberlites also were investigated for comparison.

Rutiles associated with diamonds are variable in color, from almost opaque to reddish brown and moderately transparent grains. Rutile inclusions in diamonds have a diamond-negative crystal morphology (Sobolev, 1974) (Figs. 1A and 1B). Both rutile and other Ti-oxide inclusions are found in some polymineralic diamond inclusions (Figs. 1C–1G). A typical feature of most rutiles is their heterogeneity, reflected by widely varying Al, Fe, and Mg contents. (Despite these heterogeneities, it is possible to find relatively homogeneous portions in all studied samples.) Only a few rutiles occurring as diamond inclusions and from ultrahigh-pressure metamorphic rocks are relatively homogeneous. All others, including rutiles exposed in the outer portions of diamonds, as well as rutiles from polycrystalline diamond aggregates and diamondiferous eclogites, are heterogeneous. Minimal Al values for homogeneous sections are accompanied by the absence of Mg, and by minimal Fe contents. The impurity concentrations vary from sample to sample and are rather monotonous within individual samples. Fe and Cr are reported as trivalent cations in accordance with evidence from electron paramagnetic resonance (e.g., Smith and Dawson, 1975). Al_2O_3 correlates positively with Fe_2O_3 (Fig. 2), with a significant correlation coefficient (r = 0.44 for 103 samples), whereas Nb_2O_5 does not (Fig. 3). It is interesting to note the significant Nb_2O_5 contents in rutiles from UHP rocks (see Fig. 3 and Table 2). Most rutiles have relatively low ZrO_2—<0.3 wt% to 0.45 wt%; Fig. 4).

Only one rutile sample (Table 1, Mi-25/15) from a polymineralic diamond inclusion (Fig. 1D) contains relatively high Cr_2O_3, reaching 2.25 wt%, which is similar to rutile included in pyrope from Colorado (McGetchin and Silver, 1970). Cr_2O_3 contents of all studied rutiles does not exceed 0.2 wt% in the majority of the analyzed samples (see Fig. 5).

Corundum lamellae in rutile

The overwhelming majority of rutiles show a heterogeneous distribution of Al, typically ranging between 0.2–0.3 to 2 wt% Al_2O_3. When considering only minimal and reproducible values of Al_2O_3 during analysis of each sample, the maximal contents of Al_2O_3 of all studied samples does not exceed 0.8 wt% (Fig. 2). However, some earlier work reports up to 1.9 wt% Al_2O_3 (Prinz et al., 1975; Mvuemba Ntanda et al., 1982; Meyer, 1987). Several rutile samples in the present study contain up to 15 wt% Al_2O_3 for separate analyzed points without visible heterogeneities. However, one sample of diamondiferous kyanite eclogite from the Roberts Victor mine, South Africa (RV-15) revealed much higher Al_2O_3 contents for several points, reaching 85 wt%. Backscattered electron imagery (see Figs. 1J and 1K) shows a number of oriented dark rods that are most likely corundum. The electron beam size exceeds the width of such rods, and hence a mixture of corundum and rutile was analyzed at several points (from 15 to 85 wt% Al_2O_3, reflecting a wide range in their relative abundances).

Detailed EMP study of RV-15 rutile was performed both with standard conditions for 6 points (see Table 3, RV-15) and with a defocused beam for 59 points (Table 3, RV-15a, RV-15b, RV-15c, and RV-15d). Several parts of this sample are free of lamellae, and this is confirmed from a comparison of RV-15 and RV-15a data—being almost identical except for a heterogeneous Fe distribution. However, sample portions containing from 1 to 2 wt% Al_2O_3 (RV-15b), from 2 to 4 wt% (RV-15c), and from 5 to 6.5 wt% (RV-15d) show a positive correlation of Al and Fe. The major variables are Al_2O_3 and Fe_2O_3 at negligibly low MgO. These relationships reflect differences in densities of oriented lamellae, and probably signify the presence of ilmenite or magnetite (white rods in Fig. 1J and 1K) closely associated with corundum (black rods).

Ilmenite and magnetite lamellae in rutile

A typical feature of some samples of rutile (see Table 1) is the presence of regular lamellae of other mineral phases, which form a kind of grid on the backscattered electron images (Figs. 1H–1K). These lamellae are very thin and in nearly all cases are considerably smaller than the electron beam. As a result, their identification is very difficult and they may be detected unambiguously only by TEM techniques (Banfield and Veblen, 1991). The recognized heterogeneities, coupled with increased Mg and Fe concentrations, suggest that these lamellae are

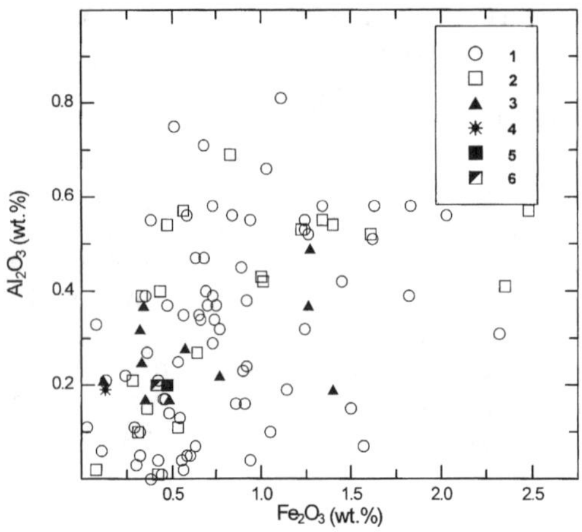

FIG. 2. Al_2O_3 versus Fe_2O_3 in rutiles included in diamonds (1), intergrown with diamond polycrystallure aggregates (2), from xenoliths of diamondiferous eclogites (3), from coesite-bearing pyrope quartzite, Dora Maira massif (4), from felsic diamondiferous gneisses (5), and from garnet-pyroxene diamondiferous rock, Kokchetav massif, northern Kazakhstan (6). All data are from Sobolev et al., 1997, 1998 and the present study (Tables 1–3).

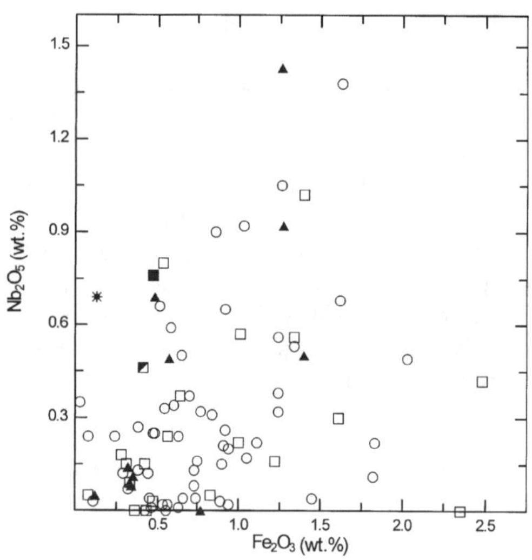

FIG. 3. Nb_2O_5 versus Fe_2O_3 in rutiles associated with diamonds and coesite. For symbols and data sources, see Figure 2.

probably picroilmenite (geikielite), ilmenite, and/or magnetite (Fig. 6).

Titanate of the LIMA Series

Titanate of a unique composition, related to the LIMA series (Haggerty, 1991a, 1991b), and containing 22.1 wt% TiO_2, 48.8 wt% Cr_2O_3, and Mg, Fe, V, Sr, K, and Ba was detected as part of a polymineralic inclusion of a harzburgitic assemblage in a diamond from the Sputnik pipe, Yakutia (Sobolev et al., 1997). The position of this Ti-phase (Tn) in relation to other coexisting phases is shown in Figure 1C. The titanate along the contact between the chromite and pyrope probably formed through a reaction relationship. Another example of a titanate included in diamond of the Internatsional'naya pipe, Yakutia, is discussed by Sobolev et al. (1997) and Haggerty (1991b), and may also be related to the LIMA series.

Perovskite

Perovskite typically occurs in the groundmass of many kimberlite bodies as a late-stage mineral (Mitchell, 1972). The U-Pb age of perovskite, along with corresponding zircon dating, allows the absolute age of the kimberlite intrusion to be determined (Kinny et al., 1997). In mantle-derived xenoliths, perovskite is very rare. Perovskite inclusions in diamonds are even rarer. We know of only two examples; it occurs as part of a polymineral inclusion of perovskite-rimmed chromite in a diamond from the River Ranch kimberlite, Zimbabwe (Kopylova et al., 1997), and as an inclusion in a diamond from the Sloan pipe, Colorado (Meyer and McCallum, 1986). Both occurrences have been suggested to be secondary in origin, based on the similarity of perovskite compositions with that in the groundmass and a lack of Cr_2O_3.

We have observed perovskite as an isolated discrete grain present together with a series of chromite inclusions in a diamond from the Sytykanskaya kimberlite pipe, Yakutia. Similar sizes of this euhedral perovskite and chromite inclusions (about 50 to70 μm each), as well an elevated content of Cr_2O_3 in perovskite and TiO_2 in the associated chromite suggests that this perovskite is primary.

Discussion

The results presented above confirm earlier studies showing significant Al, Fe, Cr, Nb, and Zr concentrations in rutiles from high-pressure mantle rocks (e.g., Rudnick et al., 2000). However, the present study draws attention to compositional features of rutiles solely from mantle and crustal rocks, compared with previously published data that are related to general features of rutiles from both mantle and crustal eclogites (Fett and Brey, 1991; Rudnick et al., 2000). Each kimberlitic pipe may

TABLE 2. Representative Analyses of Rutiles Associated with Diamonds and/or Coesite[1]

Sample:	— Mr-565 —		— Mr-567 —		— Mr-554 —		— Mr-4K —		— Mr-564 —		— Mr-819 —		— BM-56 —	
Location:	Yakutia	1σ	Yakutia	1σ	Yakutia	1σ	Yakutia	1σ	Yakutia	1σ	Yakutia	1σ	Yakutia	1σ
n:	6		4		3		8		5		4		8	
TiO_2	98.1	1.08	96.5	0.34	98.2	0.32	95.5	0.31	98.3	0.74	97.3	0.39	97.3	0.77
Al_2O_3	0.45	0.08	0.38	0.07	0.35	0.05	0.58	0.02	0.37	0.08	0.53	0.10	0.58	0.02
Cr_2O_3	0.05	0.04	0.41	0.03	0.36	0.01	0.04	0.03	0.02	0.03	0.13	0.02	0.17	0.02
Fe_2O_3	0.89	0.21	0.92	0.10	0.65	0.20	1.62	0.04	0.75	0.20	1.24	0.27	1.35	0.13
MgO	0.06	0.05	0.02	0.02	0.01	0.01	0.08	0.01	0.07	0.04	0.02	0.00	0.03	0.03
Nb_2O_5	0.03	0.02	0.65	0.06	0.51	0.03	1.38	0.09	0.16	0.03	0.38	0.06	0.56	0.07
ZrO_2	0.27	0.02	0.14	0.02	0.03	0.02	0.08	0.04	0.10	0.01	0.11	0.03	0.10	0.02
Total	99.79		99.00		100.07		99.23		99.77		99.64		100.12	

Sample:	– Ud-12/86 –		– Ud-1/90 –		– Ud-238K/1 –		– Sp-734 –		– Mr-550 –		– Ud-2023 B –		– Ur-77-76 –	
Location	Yakutia	1σ	Yakutia	1σ	Yakutia	1σ	Yakutia	1σ	Yakutia	1σ	Yakutia	1σ	Ural	1σ
n:	2		2		4		9		10		6		9	
TiO_2	97.2	0.21	98.1	0.57	98.9	0.70	96.5	0.75	95.5	0.32	97.6	1.15	97.7	0.53
Al_2O_3	0.16	0.00	0.17	0.07	0.24	0.18	0.50	0.05	0.78	0.23	0.69	0.16	0.75	0.23
Cr_2O_3	0.10	0.01	0.20	0.01	0.01	0.02	0.02	0.02	0.13	0.02	0.02	0.03	0.02	0.02
Fe_2O_3	0.83	0.03	0.45	0.01	0.17	0.06	1.26	0.42	2.01	0.18	0.83	0.11	0.51	0.14
MgO	0.01	0.01	0.00	0.00	0.01	0.02	0.04	0.05	0.05	0.05	0.02	0.02	0.01	0.01
Nb_2O_5	0.90	0.01	0.04	0.03	0.25	0.04	1.08	0.04	0.42	0.04	0.04	0.02	0.66	0.07
ZrO_2	0.11	0.01	0.29	0.02	0.22	0.02	0.07	0.03	0.13	0.04	0.17	0.02	0.07	0.02
Total	99.24		99.24		99.80		99.48		99.00		99.33		99.74	

Sample:	— TM-180 —		— M-75 —		— M-29 —		— F-1508 —		— K-83 —		— K 2-4 —		— DM-15 —	
Location:	Yakutia	1σ	Yakutia	1σ	Yakutia	1σ	S. Africa	1σ	Kazakh-stan	1σ	Kazakh-stan	1σ	Dora Maira	1σ
n:	6		4		6		3		5		6		18	
TiO_2	97.2	0.50	99.0	0.36	96.9	1.31	98.7	1.37	98.1	0.06	98.5	0.40	98.9	0.65
Al_2O_3	0.28	0.03	0.25	0.02	0.36	0.02	0.17	0.02	0.23	0.01	0.20	0.03	0.2	0.03
Cr_2O_3	0.02	0.04	0.04	0.02	0.06	0.02	0.20	0.02	0.03	0.01	0.03	0.01	0.02	0.01
Fe_2O_3	0.57	0.04	0.33	0.03	1.26	0.47	0.48	0.22	0.48	0.04	0.41	0.02	0.13	0.04
MgO	0.01	0.01	0.00	0.00	0.03	0.03	0.01	0.01	0.00	0.00	0.00	0.00	0.01	0.01
Nb_2O_5	0.49	0.03	0.09	0.03	1.41	0.10	0.69	0.03	0.77	0.08	0.46	0.04	0.7	0.06
ZrO_2	0.07	0.02	0.23	0.04	0.17	0.02	0.12	0.01	0.09	0.03	0.22	0.04	0.04	0.03
Total	98.64		99.90		100.18		100.41		99.70		99.86		100.00	

[1]Samples with symbol Ud are from Udachnaya pipe, Yakutia; with symbol Sp from the Sputnik pipe, Yakutia; F-1508 = diamondiferous eclogite, Newlands pipe, South Africa. Samples Mr-550 and Ud-2023B are intergrowths with polycrystalline diamond aggregates; TM-180, M-75, M-29, F-1508 = rutiles from diamondiferous eclogites. All other samples are inclusions from diamonds.

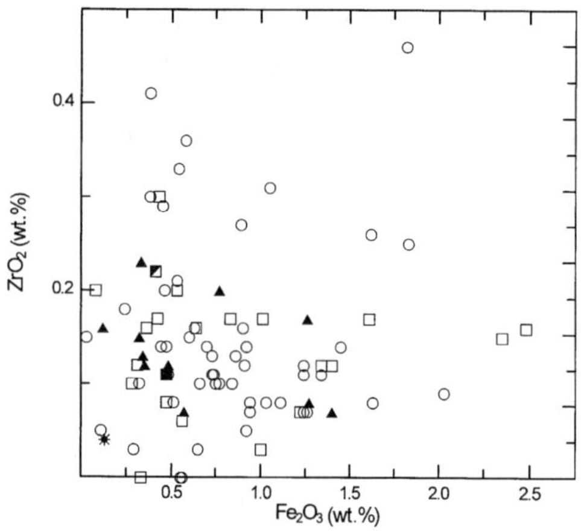

FIG. 4. ZrO_2 versus Fe_2O_3 in rutiles associated with diamonds and coesite. For symbols and data sources, see Figure 2.

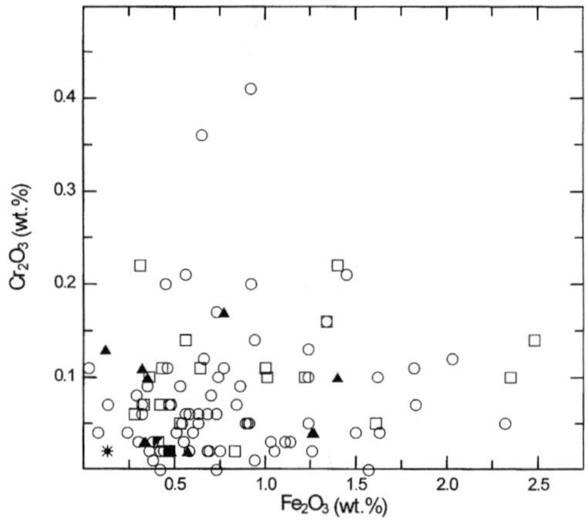

FIG. 5. Cr_2O_3 versus Fe_2O_3 in rutiles associated with diamonds and coesite. For symbols and data sources, see Figure 2.

TABLE 3. Composition of Rutile from Kyanite Eclogite Xenolith, Roberts Victor Mine[1]

Sample:	RV-15		RV-15a		RV-15b		RV-15c		RV-15d	
Location:	S. Africa	1σ	S.Africa	1σ	S. Africa	1σ	S. Africa	1σ	S.Africa	1σ
n :	6		16		16		17		10	
TiO_2	98.1	1.16	98.8	1.27	97.4	1.79	95.8	1.65	94.4	1.42
Al_2O_3	0.37	0.02	0.35	0.06	1.36	0.37	2.91	0.49	5.07	0.83
Cr_2O_3	0.03	0.04	0.05	0.01	0.06	0.01	0.05	0.01	0.05	0.01
Fe_2O_3	0.34	0.07	0.51	0.46	0.54	0.20	0.95	0.32	1.44	0.55
MgO	0.00	0.00	0.03	0.04	0.06	0.05	0.07	0.06	0.09	0.07
Nb_2O_5	0.08	0.03	0.01	0.03	0.03	0.04	0.05	0.05	0.07	0.04
ZrO_2	0.13	0.02	0.14	0.02	0.14	0.02	0.13	0.03	0.15	0.03
Total	99.00		99.85		99.55		99.98		101.28	

[1]At standard conditions (RV-15) and with defocused beam (RV-15 a-d).

contain eclogites derived from a wide range of depths (e.g., Sobolev, 1974) and only the presence of diamond (or coesite) may provide a lower limit of possible ultrahigh pressure. The same is true for crustal UHP metamorphic rocks (Chopin, 1984; Sobolev and Shatsky, 1990).

Electron microprobe studies of rutiles from crustal and mantle eclogites and diamond inclusions suggested that Al and Fe contents in rutile are a function of P-T conditions and $f(O_2)$, and that Cr reflects bulk chemistry (Sobolev et al., 1972). Significant hydroxyl content (to 0.8 wt% H_2O) was detected in a rutile from an eclogite xenolith from the Roberts Victor mine, South Africa by infrared spectroscopy (Rossman and Smyth, 1990).

A wide range in different minor impurities in rutiles from crustal and mantle eclogites was demonstrated by Smith and Dawson (1975), Bulanova (1985), Fett and Brey (1991), and Babich and Sobolev (1995). This range, especially for Al and Fe, exceeds that found in the present study (see Figs. 2–6) and probably reflects the sparse presence of corundum and ilmenite lamellae, which affect the rutile bulk compositions.

Conclusions

Rutile is the most important Ti-bearing oxide associated with E-type diamonds, diamondiferous eclogites of varied bulk compositions, and UHP metamorphic rocks associated with diamond and/or

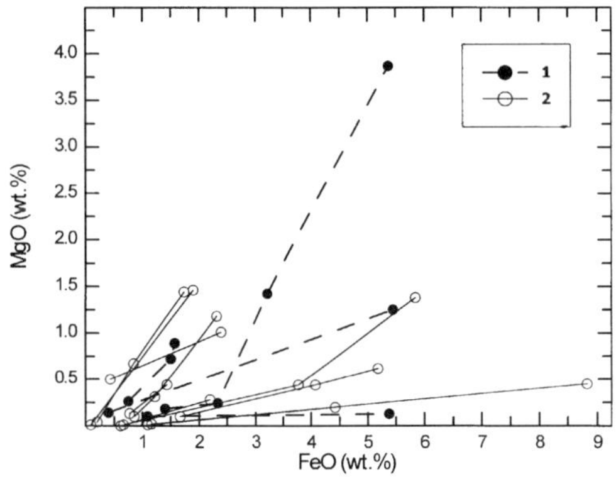

FIG. 6. MgO versus FeO in some heterogeneous rutiles intergrown with polycrystalline diamond aggregates (1) and included in diamonds (2). Tie lines connect several (up to four) different points analyzed within the same rutile grain, demonstrating the positive correlation of MgO and FeO contents.

coesite. In some cases, it occurs in lherzolitic U-type diamond-inclusion assemblages. Titanates of the LIMA series and perovskite are rare accessory phases in U-type diamonds. Ilmenite of wide compositional range, containing up to 80% geikielite, is associated with rutile in a number of diamonds. Rutile inclusions in diamond contain up to 0.8 wt% Al_2O_3, 2.5 wt% Fe_2O_3, 1.5 wt% Nb_2O_5, and 0.45 wt%, ZrO_2. Al contents generally are a function of P-T conditions; Fe content depends on $f(O_2)$ (Babich, 1999), P-T conditions, and bulk chemistry; and Cr in rutile correlates only with bulk-rock chemistry. Unusual ilmenite and, especially, corundum lamellae, previously unknown in diamond-associated rutiles (ilmenite lamellae) and in rutiles in general (corundum lamellae), expand the list of exsolution lamellae in minerals from ultrahigh-pressure rocks that require detailed characterization and experimental investigation (Zhang and Liou, 1999).

Rutiles from UHP metamorphic rocks are relatively homogeneous, and are similar to diamond-associated kimberlitic (U-type) rutiles in their high Al contents. Nb contents of these rutiles are variable and sometimes also high, but this is true of many types of rutiles from crustal and mantle metamorphic rocks (Rudnick et al., 2000). Negligibly low Fe_2O_3 in the Dora Maira rutile (0.13 wt%) reflects the very low Fe content of the bulk-rock composition. The presence of complex titanates of the LIMA series, and perovskite in the diamond-bearing assemblages reflect the metasomatizing conditions of diamond formation.

Acknowledgments

Detailed reviews by Roberta Rudnick and Vladislav Shatsky are greatly appreciated. The authors also thank Larissa Usova for assistance with the EMP analyses. This research was partially supported by RFBR grant 96-5-98547 to N. V. Sobolev.

REFERENCES

Babich, Y. V., 1999, Experimental study of rutile composition at variable oxygen fugacity in the Mg-Fe-Ti-Si-O system: Geochem. Int., v. 37, p. 683–687.

Babich, Y. V., and Sobolev, N. V., 1995, Rutiles from eclogite xenoliths of Roberts Victor kimberlite pipe (S. Africa) [ext. abs.]: Sixth Int. Kimberlite Conf., Novosibirsk, p. 23–24.

Banfield, J. F., and Veblen, D. R., 1991, The structure and origin of Fe-bearing platelets in metamorphic rutiles: Amer. Mineral., v. 76, p. 113–127.

Bulanova, G. P., 1985, Syngenetic inclusions in Yakutian diamonds: Unpubl. Ph.D. thesis, Leningrad Univ., 142 p. (in Russian).

Chopin, C., 1984, Coesite and pure pyrope in high-grade blueschists in the Western Alps: A first record and some consequences: Contrib. Mineral. Petrol., v. 86, p. 107–118.

Fett, A., and Brey, G., 1991, Significance of aluminiurn, calcium, chromium, zirconium, niobium, and iron concentrations in rutile from high pressure rocks [ext. abs.]: Fifth Int. Kimberlinte Conf., Araxa, Brazil, p. 92–93.

Futergendler, S. I., 1956, The study of inclusions in diamonds using X-ray diffraction method: Zap. Vses. Mineral. Obshch., v. 85, no. 2, p. 568–569 (in Russian).

Haggerty, S. E., 1991a, Oxide textures—a mini-atlas: Rev. Mineral., v. 25, p. 129–219.

______, 1991b, Oxide mineralogy of the upper mantle: Rev. Mineral., v. 25, p. 355–416.

Harris, J. W., 1968, The recognition of diamond inclusions. Part 1: Syngenetic inclusions: Indust. Diam. Rev., v. 28, p. 402–410.

Hills, D. V., and Haggerty, S. E., 1989, Petrochemistry of eclogites from the Koidu kimberlite Complex, Sierra Leone: Contrib. Mineral Petrol., v. 103, p. 397–422.

Kinny, P. D., Griffin, B. J., Heaman, L. M., Brakhfogel, F. F., and Spetsius, Z. V., 1997, SHRIMP U-Pb ages of perovskite from Yakutian kimberlite: Russ. Geol. Geophys., v. 38, p. 97–105.

Kopylova, M. G., Gurney. J. J., and Daniels, L. R. M., 1997, Mineral inclusions in diamonds from the River Ranch kimberlite, Zimbabwe: Contrib. Mineral. Petrol., v. 129, p. 366–384.

Kovalsky, V. V., Bulanova, G. P., Nikishov, K. N., Botkunov, A. I., Makhotko. V. F., Shestakova, O. E., and

Gotovtsev, V. V., 1979, Composition of garnets, chromites and rutiles associated with diamonds from kimberlite pipes of Yakutia: Dokl. Akad. Nauk SSSR, v. 247, p. 946–951 (in Russian).

McGetchin, T. R., and Silver, L. T., 1970, Compositional relations in minerals from kimberlite and related rocks in the Moses Rock dike, San Juan County, Utah: Amer. Mineral., v. 55, p. 1738–1771.

Meyer, H. O., 1987, Inclusions in diamond, *in* Nixon, P. H., ed., Mantle xenoliths: New York, John Wiley and Sons, p. 501–523.

Meyer, H. O. A., and McCallum, M. E., 1986, Mineral inclusions in diamonds from the Sloan kimberlites, Colorado: Jour. Geol., v. 94, p. 600–612.

Meyer, H. O., A., and Svisero, D. P., 1975, Mineral inclusions in Brazilian diamonds: Phys. Chem. Earth, v. 9, p. 785–795.

Mitchell, R. H., 1972, Composition of perovskite in kimberlite: Amer. Mineral., v. 57, p. 1748–1753.

Mvuemba Ntanda, F., Moreau, J., and Meyer, H. O. A., 1982, Particularites des inclusions cristallines primaries des diamants du Kasai, Zaire: Can. Mineral., v. 20, p. 217–230.

Orlov, Y. L., 1973, The mineralogy of the diamond: Moscow, Nauka Press, 222 p. (in Russian) (Engl. transl., 1977, New York, John Wiley, 235 p.).

Prinz, M., Manson, D. V., Hlava, P. F., and Keil, K., 1975, Inclusions in diamonds: Garnet lherzolite and eclogite assemblages: Phys. Chem. Earth, v. 9, p. 797–815.

Rossman, G. R., and Smyth, J. R., 1990, Hydroxyl contents of accessory minerals in mantle eclogites and related rocks: Amer. Mineral., v. 75, p. 775–780.

Rudnick, R. I., Barth, M., Horn, I., and McDonough, W. F., 2000, Rutile-bearing refractory eclogites: Missing link between continents and depleted mantle: Science, v. 287, p. 278–281.

Smith, J. V., and Dawson, J. B., 1975, Chemistry of Ti-poor spinels, ilmenites, and rutiles from peridotite and eclogite xenoliths: Phys. Chem. Earth., v. 9, p. 309–322.

Sobolev, N. V., 1974, Deep-seated inclusions in kimberlites and problem of upper mantle composition: Novosibirsk, Nauka Press, 263 p. (in Russian) (Engl. trans., 1977, Washington, DC, Amer. Geophys. Union, Geophys. Union, 279 p.).

Sobolev, N. V., Kaminsky, F. V., Griffin, W. L., Yefimova, E, S., Win, T. T., Ryan, C. G., and Botkunov, A. I., 1997, Mineral inclusions in diamonds from the Sputnik kimberlite pipe, Yakutia: Lithos, v. 39, p. 135–157.

Sobolev, N. V., Lavrent'yev, Yu. G., and Usova, L. V., 1972, Minor elements in rutiles from eclogites: Geol. i Geofiz., v. 13, no. 11, p. 108–112 (in Russian).

Sobolev, N. V., and Shatsky, V. S., 1990, Diamond inclusions in garnets from metamorphic rocks: A new environment for diamond formation: Nature, v. 343, p. 742–746.

Sobolev, N. V., Yefimova, E. S., Channer, D. M. DeR., Anderson, P. F. N., and Barron, K. M., 1998. Unusual upper mantle beneath Guaniamo, Guyana Shield, Venezuela: Evidence from diamond inclusions: Geology, v. 26, p. 971–974.

Sobolev, N. V., Yefimova, E. S., and Koptil, V. I., 1999, Mineral inclusions in diamonds in the northeast of the Yakutian diamondiferous province: Proc. Seventh Int. Kimberlite Conf., Capetown, South Africa, v. II, p. 816–822.

Spetsius, Z. V., and Griffin, W. L., 1997, Trace elements in minerals from eclogites from the Udachnaya kimberlite pipe, Yakutia, *in* Proc. Sixth Int. Kimberlite Conf., Geol. i Geofiz., v. 38, p. 226–232 (in Russian) (Engl. transl. in Russ. Geol. Geophys., v. 38, p. 240–246).

Zhang, R. Y. , and Liou, J. G., 1999, Exsolution lamellae in minerals from ultrahigh-pressure rocks: INT. GEOL. REV., v. 41, p. 981–993.

Geospeedometry and Time Scales of High-Pressure Metamorphism

A. L. PERCHUK[1]

Institute of Geology of Ore Deposits, Petrography, Mineralogy, and Geochemistry (IGEM), Russian Academy of Sciences, Staromonetny 35, Moscow, 109017, Russia

AND P. PHILIPPOT

CNRS-ESA 7058, Laboratoire de Pétrologie, Université Paris-Jussieu, Case 110, T26-E3, 4 place Jussieu, 75252 Paris cedex 05, France

Abstract

Kinetic theory allows the calculation of a time scale for metamorphic events using the extent of relaxation of garnet growth zoning along a particular P-T trajectory. Eclogitic garnets from the Kokchetav Massif (Kazakhstan), the Great Caucasus (Russia), and the Yukon Tanana terrane (Canada) experienced different metamorphic P-T histories and display different types of zoning patterns, which allowed testing of a variety of geospeedometric procedures. In all cases, the preservation of sharp compositional gradients and hence the limited degree of diffusive modification of garnet compositions can be explained if associated tectono-metamorphic processes were of very short duration. Results of diffusion modeling indicate rates of temperature and pressure change on the burial and/or the exhumation path on the order of several hundreds of °C/m.y. and several cm/yr, respectively. These extreme exhumation and cooling rates apply for rocks buried to a depth greater than, for example, 20 km, thus arguing for the existence of contrasted velocity fields for eclogitic block exhumation from deep versus shallow levels of the lithosphere.

Introduction

SINCE THE PIONEERING WORK of Korzhinkii (1959) and Thompson (1959), a major focus of geochemical and petrological studies has been the application of classical thermodynamics to natural mineral systems. One of the most active areas of investigation concerned the quantification of physical-chemical conditions of rock equilibration through mineral geothermobarometry (Perchuk, 1969, 1991; Ellis and Green, 1974) and/or employing petrogenentic grids (Berman, 1988; Gottschalk, 1997; Holland, and Powell, 1998). Although the application of thermodynamics has influenced our fundamental understanding of the metamorphic pressure-temperature evolution of rocks, recent efforts has been devoted to better constraining the temporal aspect of geological systems (Lasaga, 1983; Muncill and Chamberlain, 1988, Gerasimov, 1992; O'Brien and Vrana, 1995; Perchuk et al., 1996; Duchene et al., 1998; Perchuk and Philippot, 2000). One of the most complex geological systems is located along convergent plate junctions (Ernst, 1971; Sobolev and Shatsky, 1990; Ernst et al., 1994; Peacock et al., 1994; Dobretsov et al., 1995; Liou et al., 1997). Constraining temporal changes of P-T conditions during high-pressure metamorphism provides considerable insight into tectonic processes operating in subduction zones and collisional orogens.

Our understanding of the time scales and rates of subduction-zone metamorphism relies primarily on two main approaches. The first one concerns geochronological investigations based on the time-dependent decay of radiogenic isotopes in minerals. Despite the obvious progress in chemical and mass spectrometric procedures, as well as in the development of high–spatial resolution analysis, geochronology still encounters the problem of establishing the relationship between isotopic age derived from minerals and the P-T evolution of particular rocks (see Cliff, 1985; von Blakenburg et al., 1989; Mezger et al., 1992; Vance et al., 1998). The problem arises from the fact that isotopic age does not always relate to the temperature of crystal growth but to some "closure" temperature, at which time diffusional loss of the isotope ceases and the mineral begins to record age information. A basic theo-

[1]Also at CNRS-ESA 7058, Laboratoire de Pétrologie, Université Paris 6, Case 110, 4 place Jussieu, 75005 Paris, France.

retical approach for deciphering closure temperature (T_c) in radiogenic systems was developed by Dodson (1973). Its broad application to metamorphic rocks is commonly limited, however, because of a lack of constraints on the cooling rates (Burton et al , 1995) and isotope diffusivities (Chakraborty and Ganguly, 1991; Burton et al., 1995).

The second approach for evaluating the time scale of a metamorphic event is geospeedometry (Lasaga, 1983). The basic principle of geospeedometry uses the extent of diffusional modification of compositional zoning of minerals as a means of determining the duration of the metamorphic event. Originally developed to extract information on cooling rates of metamorphic rocks (Walker et al.,1977; Lasaga, 1983; Powell and White, 1995), it has been extended recently for evaluating the time scale of both temperature and pressure changes endured by high-pressure (HP) metamorphic rocks (Perchuk et al., 1996, 1998, 1999; Perchuk and Philippot, 1997, 2000; Duchene et al., 1998).

Two main approaches in geospeedometry exist—exchange geospeedometry and monocrystalline geospeedometry. Exchange geospeedometry is based on the cationic exchange between different phases (Lasaga, 1983; Ozawa, 1984; Gerasimov, 1992). For example, compositional retrograde zoning in garnet and adjoining minerals (cordierite, biotite, olivine) is a widespread phenomenon in high-grade rocks (Perchuk and Lavrent'yeva, 1983). This so-called diffusive zoning has been used for several comprehensive geospeedometric studies (Medaris et al.; 1990; Lindstrom et al.,1991; Gerasimov and Savko, 1995). With respect to high-pressure rocks, application of exchange geospeedometry is limited, however, because of a lack of evidence of cation exchange between garnet and clinopyroxene in natural (Perchuk et al., 1996) and experimental (Aranovich and Pattison, 1995) systems. Recent numerical simulations of Duchene et al. (1998) applied to solid garnet-clinopyroxene couples provide additional support for this observation. Moreover, Aranovich and Pattison (1995) showed experimentally that clinopyroxene coexisting with garnet tends to recrystallize rather than to participate in the exchange reaction. In contrast, numerical modeling of Fe-Mg exchange between garnet and clinopyroxene via the medium of an intergranular fluid (aggregate model, analogue of the Fast Grain Boundary model for stable isotopes of Eihler et al., 1992) failed to reproduce the amplitude of zoning and thickness of boundary layers in garnet (Duchene et al., 1997).

The second approach (monocrystalline geospeedometry) is based on the diffusional relaxation of growth-zoning in a crystal (Walker et al., 1977; Muncill and Chamberain, 1988; Jiang and Lasaga, 1990; Chakraborty and Ganguly, 1991; Ganguly et al., 1996) when the interior of the crystal is effectively isolated from the rest of the rock (i.e., closed-system conditions). The common observation of marked growth zoning of eclogite garnets (Ghent, 1988; Perchuk and Varlamov, 1995; Carswell et al., 1997) indicates that this approach is particularly well suited for evaluating the thermal and baric history of high-pressure rocks. In addition, because most garnets grow during prograde metamorphism, garnet growth patterns permit investigation of the early stages of metamorphic evolution (Perchuk et al., 1999).

Most geospeedometric data published so far derive from the mineral exchange approach (see Jaoul and Sautter, 1999, and references therein). In constrast, as discussed by Chakraborty and Ganguly (1991, p.144), the monocrystalline approach is critically dependent on two assumptions that are difficult to constrain and therefore are generally considered to be of limited geological application. The two assumptions are the initial garnet composition and closed-system behavior. In order to evaluate the effect of post-growth histories on garnet chemical patterns, it is necessary to refer to a starting profile unmodified by diffusional homogenization. Several strategies have been elaborated to minimize this first source of uncertainties. Jiang and Lasaga (1990) calculated the degree of garnet homogenization using T-t history extracted from single-grain petrologic and geochronologic data. A crucial limitation to this approach is that the time-scale parameter is not extracted from the calculation itself but is known from isotopic dating. Muncill and Chamberlain (1988) considered that garnet-growth compositional profiles from low-grade rocks of New Hampshire may be regarded as unmodified vestiges of higher-grade garnets, owing to the low kinetics of cation diffusivities at low temperatures. As shown by Ayres and Vance (1997), comparing different garnet populations of different origins and P-T histories may be severely in error, as minor variations in bulk-rock chemistry and P-T paths can have strong controls on garnet zoning patterns. In order to counteract this problem, Ayres and Vance (1997) proposed the use of Mn-zonation profiles; here, compositional evolution can be expressed in terms of a simple Rayleigh fractionation process. However, diffusion processes in garnets are known to be mainly controlled by other divalent cations (Ca, Mg, Fe), which have higher compositional gradients and lower diffusion coefficients.

With respect to closed-system behavior, Chakraborty and Ganguly (1991) showed that even in the most favorable circumstances, such as garnet crystals enclosed in an inert matrix (e.g., quartz), the matrix could behave as an open system as a result of the possible appearance of an intergranular fluid phase that facilitates exchange of diffusive species. Although there exists no irrevocable argument to describe boundary conditions for natural systems, several textural observations including stair-like core-mantle interfaces (Ganguly et al., 1996; Perchuk et al.,1998) or the preservation of delicate garnet inclusions in the cores of eclogitic garnet porphyroblasts (Perchuk et al., 1999) offer strong textural support in favor of closed-system behavior during post-growth processes.

The objective of the present paper is to present a review and some new results obtained using the monocrystalline geospeedometry approach and to examine in some detail the possibilities and limitations of the method in evaluating the duration of metamorphic events along particular P-T trajectories (prograde, retrograde, linear, nonlinear). Examples treated in this study are eclogitic garnets from the Kokchetav Massif (Kazakhstan), the Great Caucasus (Russia), and the Yukon-Tanana terrane (Canada). These rocks experienced different metamorphic P-T histories and display different types of zoning patterns, which allowed the testing of a variety of geospeedometric procedures.

Monocrystalline Geospeedometry of High-Pressure Rocks

Garnets from high-pressure rocks are characterized by different types of growth zoning, each of which requires a specific solution of the diffusion problem. Appropriate mathematical treatments of the relaxation processes affecting the different growth-zoning patterns are presented below.

Concentric growth zoning

Concentric growth zoning is a widespread phenomenon in metamorphic garnets. This type of zoning originates from continuous or discontinuous supply of material to the garnet growth site. The compositional changes are controlled by different factors, including: (1) a change in P-T conditions, (2) a modification of the effective bulk-rock composition, and (3) a transformation of the mineral assemblage (Tracy, 1982; Loomis, 1983; Chakraborty and Ganguly, 1991; Spear, 1993). Understanding the processes operating during garnet growth provides an opportunity for evaluating the conditions attending garnet initial growth zoning and hence for quantifying the degree of post-growth modification through intracrystalline diffusion.

Homogenization of concentric growth zoning. Homogenization of garnet displaying concentric growth zonation can be treated in a first approximation as a sphere. The mathematical treatment of diffusion in a sphere has been addressed by Anderson and Olimpio (1977), Walker et al. (1977), Yardley (1977), Muncill and Chamberlain (1988), Jiang and Lasaga (1990), Chakraborty and Ganguly (1991), Ganguly et al. (1996), Perchuk and Philippot (1997, 2000), and Perchuk et al. (1996, 1998, 1999), and will be briefly introduced here.

The differential equation used to describe the evolution of an initial profile of an isotropic spherical crystal of radius (a) as a function of time (t) is given by:

$$\frac{\partial C}{\partial t} = D\left(\frac{\partial^2 C}{\partial r^2} + \frac{2}{r}\frac{\partial C}{\partial r}\right), \tag{1}$$

where C is the concentration of the element (in mole%), and r the radial position of interest. The diffusion coefficient, D, is given in the Arrhenius form

$$D = D_0 \exp\left(\frac{-E_a}{RT}\right), \tag{2}$$

where D_o is the pre-exponential factor, Ea is the activation energy, R is the gas constant, and T is the temperature.

For garnet with an initial distribution function $g(r)$ of the element of interest that evolved under closed system conditions, one can write the following initial and boundary conditions, respectively:

$$C(r, t = 0) = g(r), \text{ and} \tag{3}$$

$$\left.\frac{\partial C}{\partial r}\right|_{r=0} = 0, \quad \left.\frac{\partial C}{\partial r}\right|_{r=a} = 0. \tag{4}$$

The general solution of differential equation (1) that satisfies both initial and boundary conditions

(3,4) described above can be written as follows (Crank, 1975)

$$C(r',t') = 3\int_0^1 r^2 g(r)dr + \frac{2}{r'}\sum_{n=1}^{\infty}$$

$$\exp(-\beta_n^2 t')\frac{\sin(\beta_n r')}{\sin^2\beta_n}\int_0^1 r\,g(r)\sin(\beta_n r)dr\,, \quad (5)$$

where βn is the n-th solution of the equation, βn ctg $(\beta n) = 1$, $g(r)$ is the initial initial growth-zoning profile, and $r' = r / a$ is the dimentionless radius. The dimensionless parameter t′ for the spherical coordinates can be expressed as

$$t' = \int_0^t \frac{D(u)}{a^2}du\,, \quad (6)$$

where u is a constant of integration. It is worth noticing that equation (6) incorporates a temperature-activated diffusion coefficient, and hence permits one to examine the thermal history of a rock in terms of the parameter t' (e. g., Jiang and Lasaga, 1990). Equations (5) and (6) form the basis of the kinetic treatment to follow.

Types. Two main types of garnet concentric growth zoning have been described in nature (Tracy, 1982). Continuous growth zoning is the most common. It is devoid of gaps or breaks in the zoning trends and is generally attributed to a continuous change in composition of the material supplying the growth. In contrast, discontinuous growth is characterized by abrupt change in mineral composition. This "stair-like" geometry is commonly attributed to a change in the reaction producing garnet, either because of a change in metamorphic conditions and/or to modification of the bulk-rock composition (Rosenfeld, 1970; Rumble and Finnerty, 1974; Tracy, 1982). The particular solution of equation (5) is a function of the type of growth zoning considered, and therefore will be treated separately.

Discontinuous growth zoning

Discontinuous growth zoning can be expressed mathematically as

$$C(0 \le r < h, t=0) = C_1,\ C(h \le r \le a, t=0) = C_2, \quad (7)$$

where h and a stand for the radial position of the compositional break, and the radii of garnet grain, respectively, and C_1 and C_2 denote the initial concentrations of the component of interest near the two sides of the compositional break. A particular solution of the differential equation (5) that satisfies the initial (7) and boundary (3) conditions can be written as follows (see Perchuk et al., 1998).

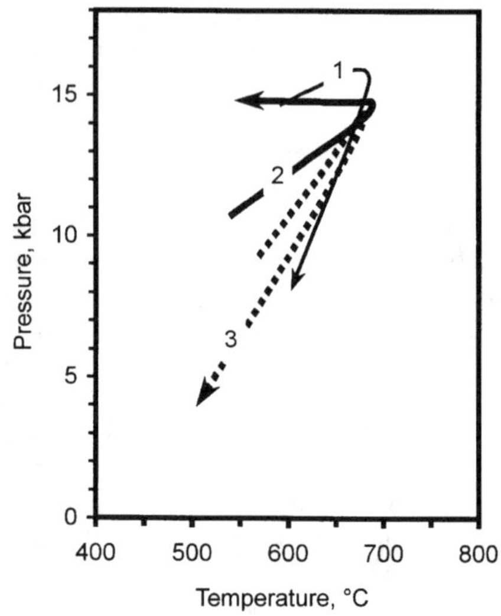

FIG. 1. Pressure-temperature diagram showing P-T paths of the eclogitic rocks studied. Legend: 1 = Red Cliff area, Great Caucasus, Russia (after Perchuk, 1993); 2 = Faro area, Yukon, Canada (after Perchuk et al., 1999); 3 = Sulu Tjube area, Kokchetav massif, Kazakhstan (after Perchuk et al., 1998).

$$C(r',t') = b^3C_1 + (1-b^3)C_2$$

$$+\frac{2}{r'}\sum_{n=1}^{\infty}\exp(-\beta_n^2 t')\frac{\sin(\beta_n r')}{\sin^2\beta_n}A_n$$

$$*\frac{1}{\beta_n^2}[(C_2-C_1)(b\beta_n\cos b\beta_n - \sin b\beta_n)$$

$$-C_2(\beta_n\cos\beta_n - \sin\beta_n)] \quad , \quad (8)$$

where $b = h/a$.

Yukon eclogite. A representative illustration of diffusion modeling of discontinuously zoned garnets comes from the Yukon-Tanana eclogites, of the western Canadian Cordillera (Erdmer and Helmstaedt, 1983; Erdmer et al.,1998). Recently, Perchuk et al. (1999) reported that the eclogites near Faro experienced a counterclockwise P-T trajectory (Fig. 1,

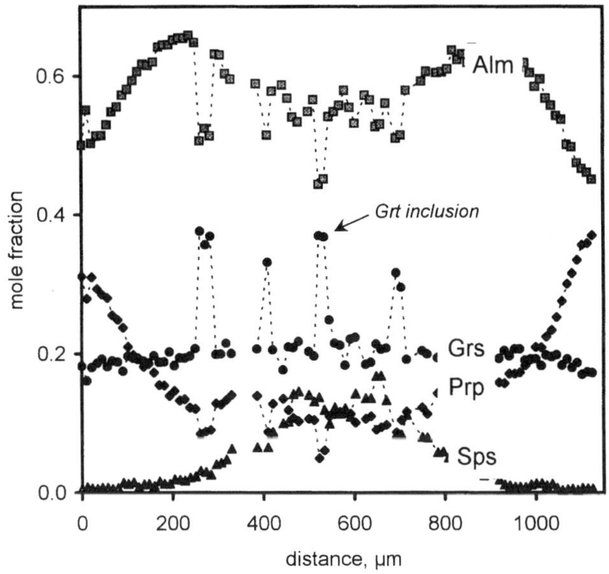

FIG. 2. Composition profile of zoned garnet from the Faro area, Yukon, plotted in mol fraction of major garnet components. Abbreviations: Alm = almandine; Prp = pyrope; Grs = grossular; Grt = garnet; Sps = spessartine. Grossular-rich peaks relate to minute garnet inclusions preserved in eclogitic garnet cores.

curve 2; see Perchuk et al., 1999 for discussion) and contain remarkable, minute garnet "inclusions" preserved in eclogitic garnet porphyroblasts. The garnet inclusions and garnet hosts show distinctly different compositions (Fig. 2). Host garnets show prograde chemical zoning characterized by a marked increase in Mg, and decrease in Fe, from core to rim. In contrast, garnet inclusions are markedly enriched in Ca and depleted in Fe compared to host-garnet core compositions. Preservation of a sharp compositional gradient along the garnet host–garnet inclusion interface and the results of diffusion modeling along the P-T trajectory and use of the diffusion dataset of Schwandt et al. (1996) indicate that the prograde evolution and near-isobaric cooling event took place over the very short time scale of less than 0.2 Ma ($\log t = -0.7 \pm 1.4$) (Fig. 3). Details of the procedure for establishing rates of metamorphism are presented elsewhere (Perchuk and Philippot, 2000).

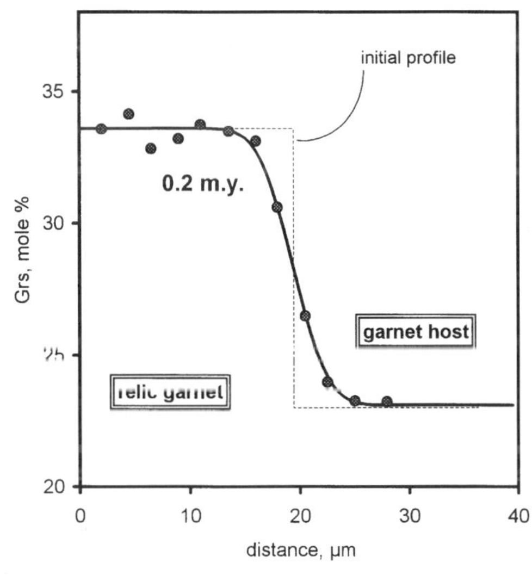

FIG. 3. Relaxation profile of garnet inclusion–garnet host interface calculated along the counterclockwise P-T trajectory shown in Figure 1 (path 2, Yukon eclogite) using a time scale of 0.2 m.y. Initial (t = 0 m.y.) growth zoning profile was assumed as a perfect step.

Kokchetav eclogite. The studied rocks are from the Sulu-Tjube area of the diamond-bearing Kokchetav massif. Two main stages of P-T equilibration under eclogite- and amphibolite-facies conditions have been documented (see Perchuk et al, 1998, for details; P-T curve 3 in Fig.1). Eclogitic garnet shows a concentric zoning pattern characterized by an abrupt increase in Ca and decrease in Fe from core to rim (Fig. 4). This compositional change delineates a "stair-like" boundary located ~600 μm

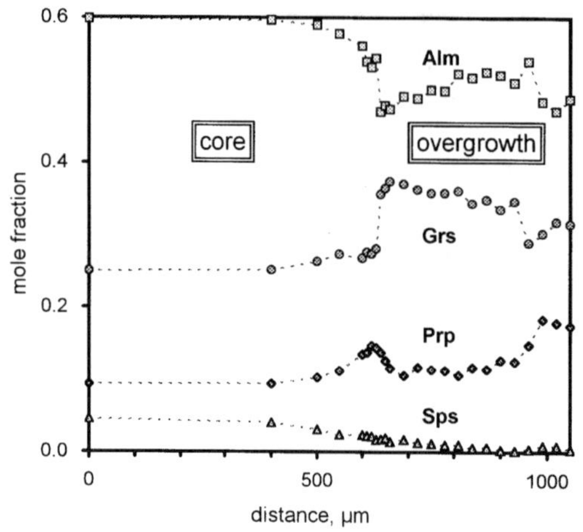

FIG. 4. Divalent cation zonation data for garnet from the Sulu-Tjube area, Kokchetav massif, Kazakhstan. The zonation is characterized by an abrupt increase in grossular content from core to rim that delineates a "stair-like" boundary between the core and its overgrowth.

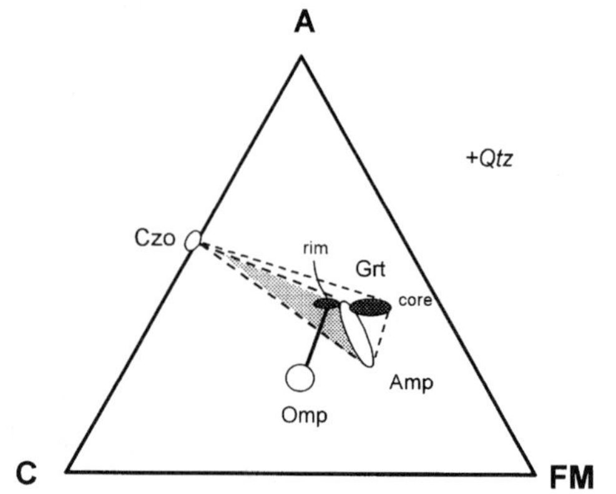

FIG. 5. ACFM phase diagram showing the relationships between omphacite (Omp), garnet (Grt) rim, and quartz (Qtz) and pre-eclogite facies mineral inclusions of amphibole (Amp) and epidote (Ep) located in the garnet rim. The intersection between Grt_{rim}-Omp and Amp-Ep tie-lines attest to the enhancement of the reaction Amp + Ep→ Grt_{new} + Omp + H_2O on the prograde path, and hence for the transition from the epidote-amphibolite to the eclogite metamorphic facies during burial.

from the garnet rim, thus satisfying closed-system boundary conditions. The outer zoning pattern is interpreted as an overgrowth zone that developed around a pre-eclogitic garnet core during prograde metamorphism. Phase equilibrium analysis indicates that the sharp compositional heterogeneity resulted from discontinuous garnet growth associated with the continuous reaction amphibole + epidote → $garnet_{new}$ + omphacite + H_2O (Fig.5). This reaction may be viewed as testimony to a prograde transition from epidote-amphibolite- to eclogite-facies conditions, thus implying that the garnet cores remained unaffected by the eclogitization process. Accordingly, we assumed that the initial garnet profile had a "stair-like" geometry that has been slightly attenuated during exhumation of the rock. Diffusion modeling of the stair-like initial compositional profile along the reconstructed P-T path and use of the diffusion coefficient of Schwandt et al. (1996) indicate that the eclogitic block experienced a depth-interval change of ~35 km (from 50 km to 15 km depth) in 0.6 m.y. ($\log t = -0.2 \pm 1.0$) (Fig. 6).

Continuous growth zoning

Theory. A main limitation associated with continuous growth zoning is the absence of marked compositional gradients, and hence difficulties in defining a starting profile unmodified by the diffusional processes. One way to counteract this problem is to compare zoning profiles of grains of different sizes. The reason for this is that homogenization rates are primarily dependent on the effective distance of diffusion (see equations 1 and 6), thus arguing for the presence of a direct correlation between compositional gradients and the size of the grains.

Growth zoning trends in high-pressure garnets are very different and hence can be approximated by different functions, each of which requires particular solution of the differential equation (5). For example, in order to describe the initial distribution of Fe and Mg in garnet, Chakraborty and Ganguly (1991) proposed a trigonometric function of the form

$$g(r') = C_1 + [C_0 - C_1]\sin^2\left(\frac{\pi r'}{2}\right),$$

where C_0 and C_1 are the concentrations in the core ($r' = 0$) and in the rim ($r' = 1$), respectively. Substituting this expression into equation (5) leads to the following particular solution:

$$C'(r', t') = 3C_0 + \frac{3}{2}C_1 + \frac{2}{r'}\sum_{n=1}^{\infty} \exp(-\beta_n^2 t')\frac{\sin(\beta_n r')}{\sin^2\beta_n}A_n,$$

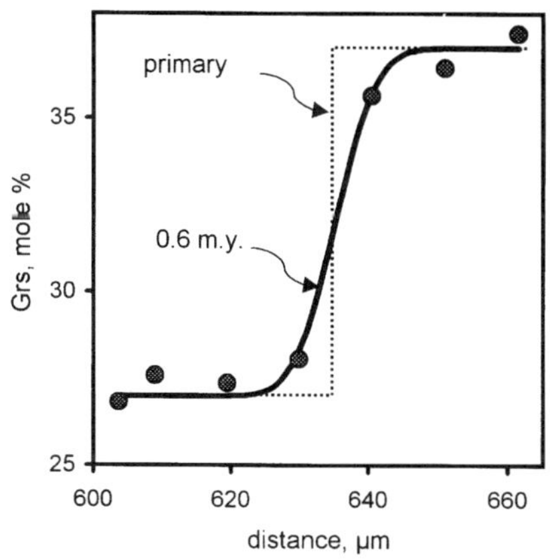

FIG. 6. Variation of grossular content in garnet inclusion–garnet host interface of the Kokchetav eclogite as a function of time. The curve was calculated using equations (6) and (8) and the retrograde P-T trajectory shown in Figure 1 (path 3). The initial (t = 0 m.y.) growth zoning profile was defined as a perfect step.

where

$$A_n = [C_0((2\beta_n^4 - 2\pi^4 - 2\pi^2\beta_n^2)\sin\beta_n - (2\beta_n^5 + 2\beta_n\pi^4 - 4\pi^2\beta_n^3)\cos\beta_n) + \pi^2 C_1((\pi^2 - 3\beta_n^2)\sin\beta_n - (\beta_n\pi^2 - \beta_n^3)\cos\beta_n)]\frac{1}{2\beta_n^2(\beta_n + \pi)^2(\beta_n - \pi)^2}.$$

Applying the linear initial distribution of diffusive components (e.g., Muncill and Chamberlain, 1988):

$$g(r,0) = a + b\,r',$$

where a and b are constants, leads to the equation of

$$C'(r',t') = a + \frac{3}{4}b + \frac{2}{r'}\sum_{n=1}^{\infty}\exp(-\beta_n^2 t')\frac{\sin(\beta_n r')}{\sin^2\beta_n}A_n,$$

where $$A_n = a\left(\frac{\sin\beta_n}{\beta_n} - \cos\beta_n\right) + b\left(\frac{2\sin\beta_n}{\beta_n^2} - \frac{\beta_n^2 - 2}{\beta_n^2}\cos\beta_n - \frac{2}{\beta_n^2}\right)$$

Strong compositional zoning can be adequately expressed as the polynomial function (e.g., Perchuk et al.,1996):

$$f(r') = a + b\,(r')^2 + c\,(r')^4, \tag{9}$$

where a, b, and c are constants. Inserting the last equation into equation (5), and carrying out the integration results in the expression below describing relaxation profiles with the polynomial form of zoning:

$$C'(r',t') = a + \frac{3}{5}b + \frac{3}{7}c + \frac{2}{r'}\sum_{n=1}^{\infty}\exp(-\beta_n^2 t')\frac{\sin(\beta_n r')}{\sin^2\beta_n}A_n, \tag{10}$$

where

$$A_n = \frac{a}{\beta_n}\left(\frac{\sin\beta_n}{\beta_n} - \cos\beta_n\right) + b\left(-\frac{\cos\beta_n}{\beta_n} + \frac{3}{\beta_n^2}\left(\sin\beta_n - \frac{2\sin\beta_n}{\beta_n^2} + \frac{2\cos\beta_n}{\beta_n}\right)\right) + \frac{c}{\beta_n}\left(\left(\frac{5}{\beta_n} - \frac{60}{\beta_n^3} + \frac{120}{\beta_n^5}\right)\sin\beta_n - \left(1 - \frac{20}{\beta_n^2} + \frac{120}{\beta_n^4}\right)\cos\beta_n\right).$$

Great Caucasus eclogite. The diffusion model for closed-system homogenization of spherical grains has been applied to eclogitic garnets from the Red Cliff area, Great Caucasus, Russia (Perchuk et al., 1996; Perchuk and Philippot, 1997). P-T conditions were estimated at 680°C and 16 kbar and 600°C and 8 kbar for the high-pressure and retrograde stage of equilibration, respectively. The effects of changing the shape of the retrograde P-T path on garnet zoning patterns and hence calculated cooling and exhumation rates have been examined in some detail by Perchuk et al. (1996) and Perchuk and

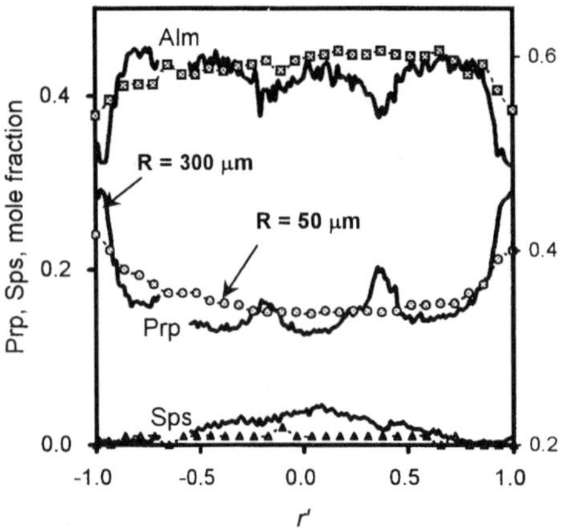

FIG. 7. Superposition of concentration profiles across garnet crystals of radius 50 μm (small) and 300 μm (large) from the Great Caucasus eclogite (see text).

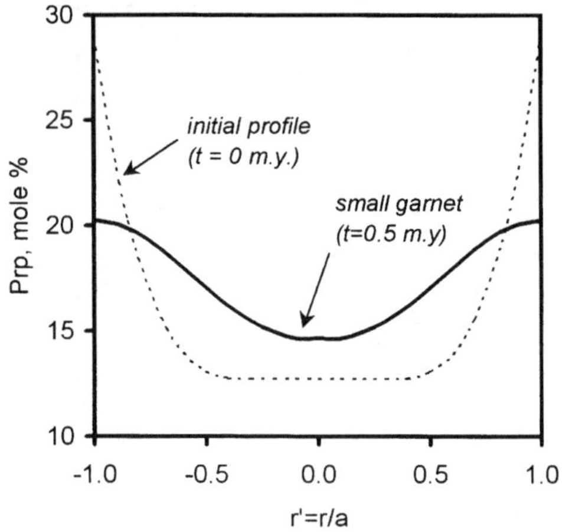

FIG. 8. Relaxation of concentric growth zoning in garnet from the Great Caucasus as a function of time. The initial distribution profile (t = 0 m.y.) corresponds to the microprobe profile of the large garnet (Fig. 6B). The line at t = 0.5 m.y. for the regular P-T trajectory (Fig. 1, path 1) was chosen as the boundary above which the zoning profile of the smallest garnet is no longer preserved.

Philippot (1997). For simplicity, in this article we will discuss only the trajectory reflecting a regular decrease of both temperature and pressure (Fig. 1, path 1). To convert the P-T path to a P-T-t path, the kinetics of diffusion in a spherical garnet must be treated explicitly.

As shown in Figure 7, small garnet porphyroblasts exhibit a smaller range of pyrope concentrations from core to rim than do larger grains. This, together with recognition that the large garnets display similar compositional gradients at their rims independently of the size of the grains (300 μm to 1 mm), indicates that the larger grains are likely to retain the original growth zonation. In contrast, because homogenization rates are primarily dependent on the size of the grains, one would expect to find a correlation between the size of the grains and the ranges of compositional gradients. Accordingly, we assumed that the initial growth zoning profile is represented by profiles of the large garnets.

Relaxation of the concentric growth zoning profiles was modeled using equations (6) and (10) and the Mg self-diffusion dataset of Chakraborty and Rubie (1996) along the regular P-T path shown in Figure 1. The calculated contour presented in Figure 8 corresponds to a time scale of 0.5 m.y. and characterizes more advanced stages of homogenization than the initial profile of small garnets. Hence the zoning patterns of the small garnet can be preserved if the exhumation history does not exceed 0.5 m.y. (see also Perchuk and Philippot, 1997 for details). Although better constraints on the initial growth zoning in the small garnet are necessary, even the upper limit of the time scale presented provides unambiguous confirmation of the high rate of HP metamorphism.

Stringers

Definition, origin, and initial profile

Although described in a variety of HP occurrences (Erambert and Austrheim, 1993; Perchuk, 1993; Perchuk and Varlamov, 1995: Cliff et al., 1998), stringers (terminology of Cliff et al., 1998) or microchannels (terminology of Perchuk and Varlamov, 1998) remain relatively poorly documented in most eclogitic rocks. Prior to developing the mathematical treatment associated with diffusion modeling of stringers, a brief review of the possible mechanisms involved in their formation is presented below.

Stringers refer to more or less radially oriented linear zoning that emanates from the rim of garnet and converges toward the core (Fig. 9). Perchuk and Varlamov (1995) developed a numerical simulation for the generation of stringers using the grain-boundary diffusion theory of Fisher (1951), and showed that stringers can be associated with a stage of fluid-controlled garnet growth along pre-existing microcracks during prograde metamor-

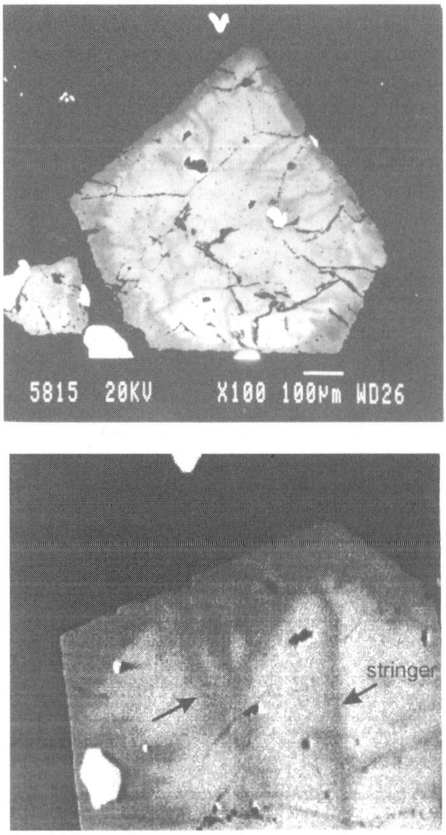

FIG. 9. Backscattered electron images showing Mg-rich stringers (dark linear patterns) located in the core (light grey) of eclogitic garnet, Red Cliff area, Great Caucasus.

phism. However, because of the absence of constraints on the grain-boundary diffusion coefficient of garnet and on the relative effect of H_2O on this parameter (Rubie, 1986; Liou et al., 1997; Yund, 1997), alternative mechanisms of transport also can be considered.

For example, in addition to the pure grain-boundary–diffusion mechanism, Walther and Wood (1984) proposed that transport of material along grain boundaries to a new site of mineral growth can be controlled by the flow of fluid containing dissolved solutes or by diffusion of solutes in a stationary fluid when the grain-boundary spacing is too thin for adequate fluid circulation.

Lack of the stringers with so-called "V-shape" compositional contours clearly indicate that rates of penetration of material within the microcracks were much higher than rates of lateral spreading via volume diffusion into the adjacent garnet (see Joesten, 1991; Perchuk and Varlamov, 1995 for details). Pure grain-boundary diffusion would favor transfer of material normal to the stringer walls, thus producing a smoothed profile across the stringer–garnet host interface. Alternatively, two other mechanisms (bulk fluid flow or diffusion in the fluid) could operate without lateral spreading of elements into the adjacent garnet, thus producing the "plateau-like" shape of the profile.

Homogenization of stringers

Stringers are characterized by a plane sheet geometry that can be expressed as a finite region of the form $-h < x < h$ penetrating a semi-infinite crystal. Owing to their location in the core of the garnet, stringers can be considered as a closed system with negligible mass flux across the stringer–garnet core interface. Considering Fe and Mg as the main interdiffusion cations, the evolution of the initial zoning profile of a stringer is governed by the equation

$$\frac{\partial C}{\partial t} = D\frac{\partial^2 C}{\partial x^2}, \quad (11)$$

with the following initial and boundary conditions:

$$C(x, t = 0) = \varphi(x), \quad (12)$$

$$\left.\frac{\partial C}{\partial t}\right|_{x = \pm\infty} = 0, \quad (13)$$

where C is the concentration of pyrope in garnet (in mole%), t the time, x the distance, and $\varphi(x)$ the initial distribution function of pyrope.

The general solution to equation (11) with conditions (12) and (13) is given by

$$C(x, t) = \frac{1}{2\sqrt{\pi D t}}\int_{-\infty}^{\infty}\exp\left(\frac{|x-\xi|^2}{4Dt}\right)\varphi(\xi)d\xi, \quad (14)$$

where ξ is the parameter of integration.

At this point we can reduce the problem of a variable coefficient of diffusion (D) by introducing the dimensionless variable

$$t' = \int_0^t \frac{D(\tau)}{h^2}d\tau, \quad (15)$$

where u is a constant of integration.

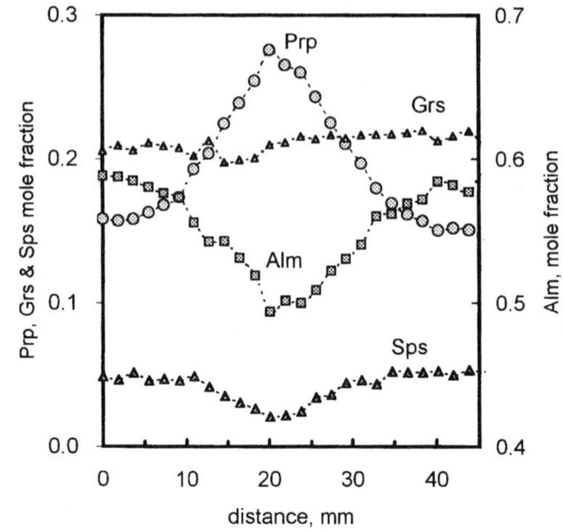

FIG. 10. Microprobe profile across the stringer expressed in mole fractions of spessartine, almandine, pyrope, and grossular.

Great Caucasus

Stringers represent an important textural feature of eclogitic garnets from the Red Cliff area, Great Caucasus (Perchuk and Varlamov, 1995; Perchuk and Philippot, 1997). Stringer compositional zoning is mainly expressed in terms of Fe-Mg exchange, with a marked Mg enrichment and Fe depletion toward the stringer (Fig. 10).

Because of very high compositional gradients across the stringer (Fig. 10), we consider a "plateau-shaped" initial zoning profile, which can be defined as $\varphi(|x| \geq d) = C_1$; $\varphi(|x| < d) = C_2$. Integrating equation (15) yields a particular solution of the form

$$C(x', t') = C_1 + \frac{C_2 - C_1}{2}$$

$$\left[\operatorname{erf}\left(\frac{1-x'}{2\sqrt{t'}}\right) + \operatorname{erf}\left(\frac{1+x'}{2\sqrt{t'}}\right)\right], \quad (16)$$

where $x' = x/h$, and C_1 and C_2 refer to the concentration of pyrope in the host garnet and in the rim, respectively. Note that equation (16) has a similar form as the equation governing cooling of a dike intruded into country rocks (see Lasaga, 1998).

Expression (16) allows the examination of relaxation of the initial "plateau-like" Mg zoning profile across the stringer as function of t' and distance. Results of the calculations performed for the stringer with an initial pyrope content C_2 = 32 mole%, width $2h$ = 12 μm, and initial concentration of the adjacent garnet core C_1 = 15 mole% are shown in Figure 11. The plot adequately describes a partly relaxed compositional profile across the stringer, when t' = 0.5. Accordingly, in the following treatment, we will use this value as a quantitative degree of stringer homogenization. Also note that homogenization of the stringer is completed at t' = 10.

In order to relate t' data for our stringer to the timing of metamorphism, we should quantify the metamorphic P-T path followed by the Great Caucasus eclogite in terms of the t' parameter. For the present calculation we have employed the expression (15) and the dataset of Chakraborty and Rubie (1996) for self diffusion of magnesium in garnet. Results of the calculations for temperature and pressure change of 80°C and 8 kbar along the P-T path 1 in Figure 1 are shown in a $t'-t$ diagram (Fig. 12). The plot shows that preservation of the observed profile across the stringer could be achieved along regular P-T paths for very short time scales of ~0.1 m.y. ($\log t < -1.0 \pm 1.3$).

Uncertainties of the Method

Uncertainties inherent in the geospeedometric approach arise from two main sources—the diffusion coefficient data and the thermal history of rock. With regard to the first source of uncertainty, the problem lies in a choice of a particular experimental dataset. Perchuk and Philippot (1997, 2000) discussed in detail appropriate experimental datasets containing diffusion coefficients for Ca in garnet (Schwandt et al., 1996) and Mg (Chakraborty and Rubie, 1996) in terms of geospeedometry. Recent experimental data of Ganguly et al. (1998) on cation diffusion in pyrope-almandine couples are in good agreement with the data of Chakraborty and Rubie (1996). Accordingly, $t'-t$ dependence of the stringer homogenization is insignificantly affected by the choice of diffusion datasets, hence yielding very close time-scale estimates (Fig. 12). In contrast, Ca-(Fe, Mg) interdiffusion coefficients obtained in the new experimental study by Freer and Edwards (1999) are almost three orders of magnitude higher than Ca tracer diffusion data (Schwandt et al., 1996) used in this study. Thus using the new dataset one obtains even higher rates of metamorphism than those we presented in the study for Yukon and Kokchetav eclogites.

The second source of uncertainties concerns the thermal history of particular rocks. The Cpx-Grt

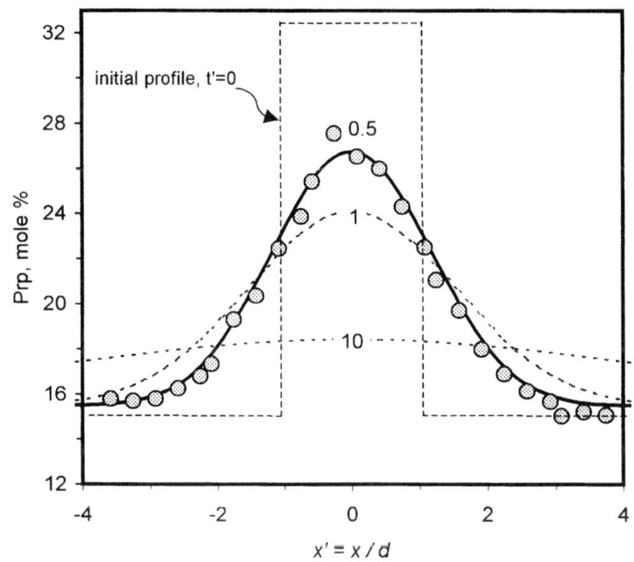

FIG. 11. Compositional profiles illustrating the progressive homogenization of a stringer in a garnet from the Great Caucasus eclogite. Initial (t' = 0) parameters of the growth zoning profile were defined as: C_0 = 15 mol% pyrope; C_1 = 32 mol% pyrope; h = 6 μm.

geothermometer of Krogh (1988) used in our studies provides quite realistic temperatures for crustal eclogites (e.g., Carswell et al., 1997). Statistically, the deviation of temperature estimates using different equations present in the literature do not exceed 50°C. Figure 12 illustrates the effect of this uncertainty on the time scale calculated using the degree of stringer homogenization. Using higher temperatures, such as those estimated using the Ellis and Green (1974) or Powell (1985) equations, does not change the overall conclusions of the derived T-t history, as all results are shifted to higher rates of temperature change. The shape of the T-t path may influence the geospeedometric estimates (Perchuk et al., 1996; Perchuk and Philippot, 1997). Special additional calculations show that very rapid cooling (1500°C/m.y.) prior to decompression leads to some increase of the time scale (from 0.1 to 0.3 Ma) of metamorphism, but this does not change our overall conclusions about the very high rates of metamorphism.

A simple assumption involving the ideal behavior of garnet solid solutions has been employed in diffusion models throughout this paper. Although it is not generally valid (e.g., Berman and Aranovich, 1996), because of the scales of compositional variations within diffusion zones, the considered non-ideality effects of garnet solid solutions may be neglected.

Rates of High-Pressure Metamorphism

The diffusion models presented above provide quantitative constraints on the rates of P and T change experienced by high-pressure rocks. As shown above, the duration of prograde and early isobaric cooling events in the Faro eclogite should not have exceeded 0.2 m.y. Although the relative duration of these two events is unknown, it is possible to determine critical values of burial/heating and cooling rates above which the relict garnet-interface chemical profile cannot be preserved. By assuming that the Faro eclogite experienced a continuous P-T evolution, and considering a time scale of 0.2 Ma, and a P-T change of 4 kbar and 170°C during the prograde metamorphic event, we obtain minimum rates of burial and heating of 7 cm/yr and 850°C/m.y., respectively. Conversely, a temperature change of 170°C and a duration of 0.2 m.y. for the isobaric cooling event results in a minimum cooling rate of 750°C/m.y.

Geospeedometric study of the eclogites from the Kokchetav massif (Sulu-Tjube area) yielded a very short time scale (–0.6 m.y.) for the decompression

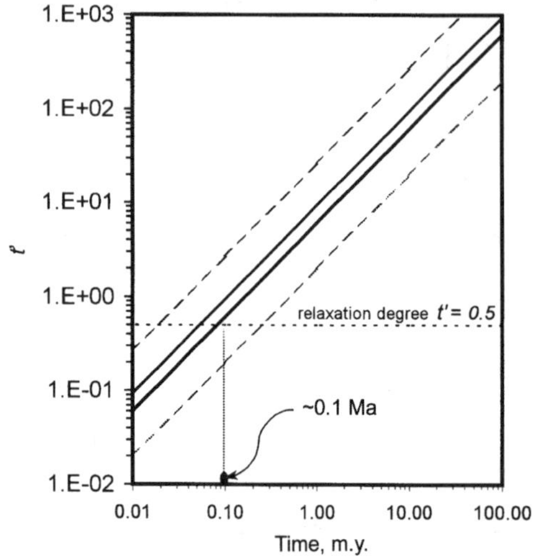

FIG. 12. Dimensionless parameter t' - time (t) diagram for a stringer of dimension h = 6 μm along the retrograde P-T path of the Great Caucasus eclogite (curve 1, Fig. 3). The heavy solid line was calculated using the diffusion data of Chakraborty and Rubie (1996) and the Grt-Cpx theromometer of Krogh (1988); the heavy dashed line illustrates the effects of changing the peak temperature by 50°C; the dotted line was calculated using the diffusion dataset of Ganguly et al. (1998) and the Grt-Cpx thermometer of Krogh (1988). The time estimate of 0.1 m.y. corresponds to the degree of stringer homogenization of t' = 0.5.

event. Considering a pressure change of 10 kbar and a temperature change of 180°C occurring in 0.6 m.y. result in average exhumation and cooling rates of ~7 cm/yr and 320°C/m.y., respectively.

With respect to the Great Caucasus eclogites, we used two different textural features—"concentric zoning" and "stringer"—to extract rates of cooling and exhumations on the same samples. Both methods provide comparable time scales of the decompression stage (0.1 m.y. versus <0.5 m.y). The difference between the estimates originates from the overestimated relaxation degree of the concentric growth zoning pattern. In turn, the stringer provides the most accurate value of the time. Considering a time scale of 0.1 Ma, a P-T change of 8 kbar and 80°C along the regular P-T path shown in Figure 1 yields minimum rates of exhumation and cooling of the Great Caucasus eclogites of ~28 cm/yr and 800°C/m.y., respectively.

Cooling and exhumation rates presented above considerably exceed most of the estimates derived from high-pressure (HP) and ultrahigh-pressure (UHP) rocks worldwide by combination of geothermobarometric results with radiometric data (see summaries in the works of Perchuk et al.,1996; Ernst and Peacock, 1996; Duchene et al., 1997). Some results (e.g., Zeck et al., 1992; Gebauer et al., 1995) seem comparable with our estimates.

The minimum rate of burial (7 cm/yr) determined above for the Yukon eclogites is consistent with rates of plate convergence of modern subduction zones. The heating rate (850°/C/m.y.) is, to our knowledge, the first estimate obtained from high-pressure rocks, which compares with values extracted from thermal models of newly initiated subduction zones (e.g., Peacock, 1987).

Closure Temperature

Geospeedometry represents a tool of interest for interpreting the results of isotopic dating. The interpretation of chronological information requires that the isotopic compositions of minerals dated were not significantly modified by diffusion after growth. The problem of losing (by diffusion) isotopes in the crystal was solved by Dodson (1973), who developed the following mathematical treatment that relates the closure temperature (T_c) of the mineral to the parameters of cooling rate (s), activation energy (E_a), and length scale for diffusion (a)

$$T_c = \frac{E_a/R}{\ln[AD_oRT^2c/a^2sE_a]}, \qquad (17)$$

where R is the gas constant, A is a geometric factor (for a plane sheet, A = 8.7, a is a half width; for a cylinder, A = 27 and a is the radius; for a sphere, A = 55 and a is the radius).

Considering that geospeedometry provides information on cooling rates, information on closure temperatures can be extracted from the equation above if appropriate sets of experimentally determined diffusion data are available. Doing this, we recognize that boundary conditions and/or isotope behavior in minerals can be complex (e.g., Harrison et al., 1991; Hodges et al., 1994; Reddy et al., 1996), implying that the applicability of Dodson's formalism should be regarded with some caution (see Ehlers and Powell, 1994). Specifically, pressure effects on retention of Ar in micas (e.g., Dahl, 1996) may considerably complicate the situation.

Extrapolating high-temperature Sm and Nd diffusivities of Harrison and Wood (1980) and Coghlan (1990) to the temperature region of interest, it is

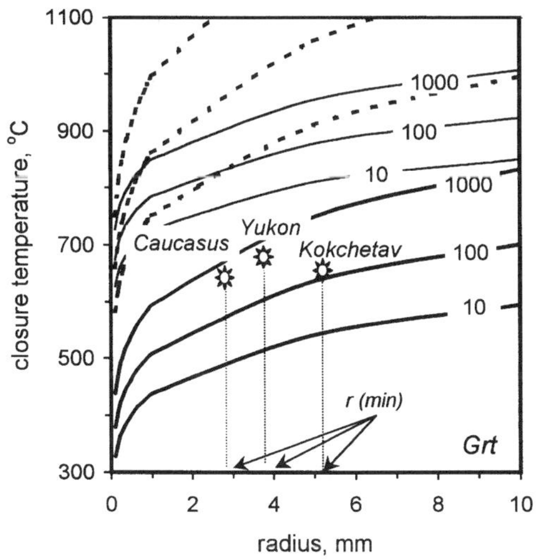

FIG. 13. Relationships between mineral size, calculated cooling rate, and closure temperature (T_c) of the Sm-Nd system calculated using the Dodson (1973) formalism. Diffusion coefficient data for Sm in grossular (heavy solid line) and pyrope (solid line) are from Harrison and Wood (1980), and the diffusivity of Nd in Fe-Mn garnet is taken from Coghlan (1990). Cooling rates and peak metamorphic temperatures established for the Great Caucasus, Yukon, and Kokchetav eclogites were used to evaluate critical garnet radii required for the relevant Sm-Nd dating.

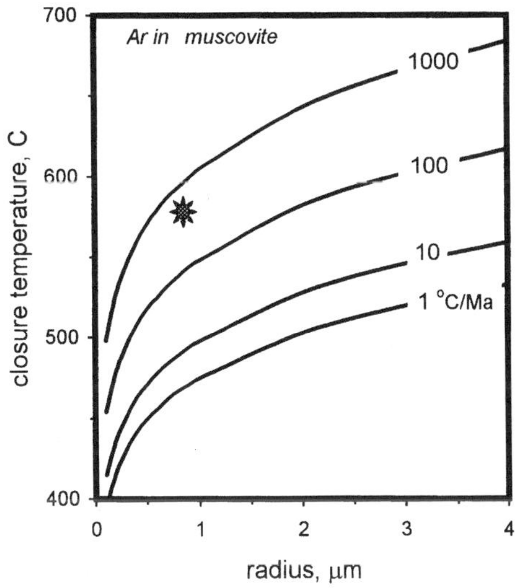

FIG. 14. The effect of the radius and the cooling rate on the closure temperature (T_c) of Ar in muscovite, according to Dodson (1973). The D_{Ar} values were retrieved by Kirschner et al. (1996). The filled symbol corresponds to the phengite used by Erdmer et al. (1998) for Ar-Ar dating (260 Ma) of the Yukon eclogite with a geospeedometrically obtained cooling rate of 750°C/Ma.

possible to quantify the dependence of the Sm blocking temperature using cooling rates deduced from geospeedometry and measured garnet size (Fig. 13). Although much more experimental data are needed to establish the diffusivities in garnet under metamorphic conditions (Cohen et al., 1988; Chakraborty and Ganguly, 1991), this diagram shows the important role of cooling rate for quantifying the closure temperature of the Sm-Nd isotopic system in garnet. As an example, considering a garnet of radius >5 mm and a range of cooling-rate values of about two orders of magnitude (see Perchuk et al., 1996, Table 2) results in a possible uncertainty regarding the closure temperature of more than 200°C.

In addition, Figure 13 allows the calculation of a critical size of garnet above which the initial distribution of isotopes is preserved, thus providing realistic age estimates. For example, considering a peak temperature of equilibration of 690°C and a subsequent cooling rate of 750°C/m.y., the minimum garnet grain size retaining Sm in the crystal structure is 3.5 μm. Similar reasoning applied to the Great Caucasus and Kokchetav eclogites yields a minimum garnet size of of ~2 and ~5 μm, respectively. In other words, Figure 13 provides an elegant means for determining critical garnet sizes above which the closed-system behavior of isotopes is guaranteed.

Ar-Ar dating studies have been successful in setting age constraints on many HP and UHP occurrences worldwide. The recent development of microanalytical $^{40}Ar/^{39}Ar$ laser-probe dating techniques has highlighted the heterogeneous distribution of argon isotopic compositions on a grain scale (Scaillet et al., 1992; Hodges et al., 1994; Reddy et al., 1996), which require extending Dodson's formalism to particular situations. In contrast, the stepped-heating technique (e.g., Camacho et al., 1997), which is based on the argon release spectra of the bulk sample, provides no information about the grain scale distribution of isotopes. This implies that a plot of closure temperature versus mica radius derived from Dodson's equation can be used for the interpretation of Ar-Ar age spectra.

Using equation (17) and the recent empirical calibration of Ar diffusivity in muscovite of Kirschner et al. (1996), Figure 14 shows the closure temperature of the Ar system in muscovite for different cooling rates and grain sizes. Calculated curves indicate that the closure temperature of the Ar system in phengite can vary by as much as 300°C over a wide range of equilibration temperatures. For example, considering a 1.5 mm phengite grain from the Faro eclogite dated at 261 ± 2 Ma using the $^{40}Ar/^{39}Ar$ method (Erdmer et al., 1998) that experienced a cooling rate of 750°C/m.y. (Perchuk et al., 1999) yields a closure temperature of 570°C. Interestingly, note that this temperature is close to the peak temperature experienced by the rock in the case studied, thus implying that caution must be used when interpreting Ar-Ar dates as late-stage cooling ages (typically below 400°C).

Conclusions

Although important work remains to be done using the geospeedometry approach, the data presented above provide crucial information on the velocity field of lithospheric processes at convergent plate junctions. Calculated exhumation/burial and cooling/heating rates are extracted directly from natural rocks, thus representing the first "objective" dataset to be compared with predictions of thermal and mechanical simulations.

An important aspect of the diffusion modeling is that different geospeedometric approaches, using different geometries of growth zoning, can be evaluated. This in turn allows visualization of the relative effects of boundary and initial conditions on calculated temporal values. As such, the geospeedometer based on the diffusive relaxation of stringers appears to be the most sensitive (and thus effective) tool for deciphering the time scale of metamorphic events.

In all examples studied, the observation of very limited degrees of diffusive modification of garnet compositions indicates extremely rapid rates of subduction-zone metamorphism, with temperature and pressure changes on the order of several hundred °C/m.y. and several cm/yr, respectively.

Acknowledgments

We are indebted to L. L. Perchuk for a critical reading of an earlier version of the manuscript. Special thanks also are due to L. Ya. Aranovich and V. Yu. Gerasimov for their constructive reviews of the manuscript itself. A. Perchuk acknowledges support in the form of a seven-month tenure as an Invited Professor at Université Paris 7. This work benefited from an INSU-CNRS grant to P. Philippot (Contribution 99IT211) and Russian Foundation for Basic Research grant No. 97-05-64418 to A. Perchuk. A. Perchuk acknowledges the Organizing Committee of the Krauskopf Symposium at Stanford University for funding his participation in the meeting where this paper was presented.

REFERENCES

Anderson, D. E., and Olimpio, J. C., 1977, Progressive homogenization of metamorphic garnets, South Morar, Scotland: Evidence for volume diffusion: Can. Mineral., v. 15, p. 205–216.

Aranovich, L. Ya. and Pattison, D. R. M., 1995, Reassessment of the garnet-clinopyroxene Fe-Mg exchange thermometer: I. Evaluation of the Pattison and Newton (1989) experiments: Contrib. Mineral. Petrol., v. 119, p. 16–29.

Ayres, M., and Vance, D., 1997, A comparative study of diffusion profiles in Himalayan and Dalradian garnets: Constraints on diffusion data and the relative duration of metamorphic events: Contrib. Mineral. Petrol., v. 128, p. 66–80.

Berman, R. G., 1988, Internally consistent thermodynamic data for minerals in the system Na_2O-K_2O-CaO-MgO-FeO-Fe_2O_3-Al_2O_3-SiO_2-TiO_2-H_2O-CO_2: Jour. Petrol., v. 8, p. 445–522.

Berman, R. G., and Aranovich, L. Y., 1996, Optimized standard state and solution properties of minerals. 1. Model calibration for olivine, orthopyroxene, cordierite, garnet, and ilmenite in the system FeO-MgO-CaO-Al_2O_3-TiO_2-SiO_2: Contrib. Mineral. Petrol., v.126, p. 1–24.

Burton, K. W., Kohn, M. J., Cohen, A. S., and O'Nions, R. K., 1995, The relative diffusion of Pb, Nd, Sr, and O in garnet: Earth Planet. Sci. Lett., v. 133, p. 199–211.

Camacho, A., Compson, W., McCulloch, M.., and McDougal, I., 1997, Timing and exhumation of eclogite-facies shear zones, Musgrave Block, central Australia: Jour. Metamor. Geol., v. 15, p. 735–751.

Carswell, D. A., O'Brien, P. J., Willson, R. N., and Zhai, M., 1997, Thermobarometry of phengite-bearing eclogites in the Dabie Mountains of central China: Jour. Metamor. Geol., v. 15, p. 239–252.

Chakraborty, S., and Ganguly, J., 1991, Compositional zoning and cation diffusion in garnets: Adv. Phys. Geochem., v. 8, p. 121–175.

Chakraborty, S., and Rubie, D. C., 1996, Mg tracer diffusion in aluminosilicate garnets at 750–850°C, 1 atm., and 1300°C, 8.5 GPa: Contrib. Mineral. Petrol., v. 122, p. 406–414.

Cliff, R. A.,1985, Isotopic dating in metamorphic belts: Jour. Geol. Soc. Lond., v. 142, p. 97–110.

Cliff, R. A., Barnicoat, A. C., and Inger, S., 1998, Early Tertiary eclogite-facies metamorphism in the Monviso Ophiolite: Jour. Metamor. Geol., v. 16, p. 447–455.

Coghlan, R. A. N., 1990, Studies of diffusional transport: Grain boundary transport of oxygen in feldspars, diffusion of oxygen, strontium and REEs in garnet, and thermal histories of granitic intrusions in south-central Maine using oxygen isotopes: Unpubl. Ph.D. thesis, Brown Univ, Providence, RI.

Cohen, A. S., O'Nions, R. K., Siegenthaler, R., and Griffin, W. L., 1988, Chronology of the pressure-temperature history recorded by a granulite terrain: Contrib. Mineral. Petrol., v. 98, p. 303–311.

Crank, J., 1975, The mathematics of diffusion: Oxford, UK, Oxford Univ. Press, 414 p.

Dahl, P. S., 1996, The crystal-chemical basis for Ar retention in micas: Inferences from interlayer partitioning and implications for geochronology: Contrib. Mineral. Petrol., v. 123, p. 22–39.

Dobretsov, N. L., Shatskiy, V. S., and Sobolev, N. V., 1995, Comparison of the Kokchetav and Dabie Shan metamorphic complexes: Coesite- and diamond-bearing rocks and UHP-HP accretional-collisional events: INT. GEOL. REV., v. 37, p. 636–656.

Dodson, M. H., 1973, Closure temperature in cooling geochronological and petrological systems: Contrib. Mineral. Petrol., v. 40, p. 259–274.

Duchene, S., Albarede, F., and Lardeaux, J.-M., 1998, Mineral zoning and exhumation history in the Munchberg eclogites (Bohemia): Amer. Jour. Sci., v. 298, p. 30–59.

Duchene, S., Lardeaux, J.-M., and Albarede, F., 1997, Exhumation of eclogites: Insight from depth-time analysis: Tectonophys., 1997, v. 280, p. 125–140.

Eiler, J. M., Baumgarnter, L. P., and Valley, J. W., 1992, Intercrystalline stable isotope diffusion: A fast grain boundary model: Contrib. Mineral. Petrol., v. 112, p. 543–557.

Ehlers, K., and Powell, R., 1994, An empirical modification of Dodson's equation for closure temperature in binary system: Geochim. et Cosmochim. Acta, v. 58, p. 241–248.

Ellis, D. J., and Green, E. H., 1974, An experimental study of the effect of Ca upon garnet-clinopyroxene Fe-Mg exchange equilibria: Contrib. Mineral. Petrol., v. 71, p. 13–22.

Erambert, M., and Austrheim, H., 1993, The effect of fluid and deformation on zoning and inclusion patterns in poly-metamorphic garnets: Contrib. Mineral. Petrol., v. 115, p. 2

Erdmer, P., Ghent, E. D., Archibald, D. A., and Stout, M. Z., 1998, Paleozoic and Mesozoic high-pressure metamorphism at the margin of the ancestral North America in central Yukon: Geol. Soc. Amer. Bull., v. 110, p. 615–629.

Erdmer, P., and Helmstaedt, H., 1983, Eclogite from central Yukon: A record of subduction at the western margin of ancient North America: Can. Jour. Earth Sci., v. 20, p. 1389–1408.

Ernst, W. G., 1971, Metamorphic zonations on presumably subducted lithospheric plates from Japan, California, and the Alps: Contrib. Mineral. Petrol., v. 34, p. 43–59.

Ernst, W. G., Liou, J. G., and Hacker, B. R., 1994, Petrologic significance of high- and ultrahigh-pressure metamorphic belts: Inference for subduction-zone histories: INT. GEOL. REV., v. 36, p. 213–237.

Ernst, W. G., and Peacock, S. M., 1996, A thermotectonic model for preservation of ultrahigh-pressure phases in metamorphosed continental crust, *in* Subduction: Top to bottom: Washington, DC, Amer. Geophys. Union, p. 171–178.

Freer, R., and Edwards, A., 1999, An experimental study of Ca-(Fe,Mg) interdiffusion in silicate garnets: Contrib. Mineral. Petrol., v. 134, p. 370–379.

Fisher, J. C., 1951, Calculation of diffusion penetration curves for the surface and grain boundary diffusion: Jour. Appl. Phys., v. 22, p. 74–77.

Ganguly, J., Chakraborty, S., Sharp, T. G., and Rumble, D., 1996, Constraint on the time scale of biotite-grade metamorphism during Acadian orogeny from a natural garnet-garnet diffusion couple: Amer. Mineral., v. 81, p. 1208–1216.

Ganguly, J., Cheng, W., and Chakraborty, S., 1998, Cation diffusion in aluminosilicate garnets: Experimental determination in pyrope-almandine diffusion couples: Contrib. Mineral. Petrol., v. 131, p. 171–180.

Gebauer, D., Schertl, H.-P., and Schreyer, W., 1995, 35 Ma old ultrahigh-pressure metamorphism in the Dora-Maira massif and its geodynamic implications for the Pennine zone of the Central and Western Alps [ext. abs.]: Boch. Geol. Geotech. Arbeinten, Abs. vol., v. 4, p. 49–52.

Gerasimov, V. Yu., 1992, Temperature evolution of metamorphism and reversibility of mineral equilibria: Moscow, Nauka Press, 129 p.

Gerasimov, V. Yu., and Savko, K. A., 1995, Geospeedometry and thermal evolution of garnet-cordierite metapelites of the Voronezh massif: Petrology, v. 6, p. 563–577.

Ghent, E. D., 1988, A review of chemical zoning in eclogite garnets, *in* Smith, D. C., ed., Eclogite and eclogite-facies rocks: Amsterdam, Elsevier, p. 207–236.

Gottschalk, M., 1997, Internally consistent thermodynamic data of rock-forming minerals: Eur. Jour. Mineral., v. 9, p. 175–223.

Harrison, T. M., Lavera, O. M., and Heizler, M. T., 1991, $^{40}Ar/^{39}Ar$ results for alkali feldspars containing diffusion domains with differing activation energy: Geochim. et Cosmochim. Acta, v. 55, p. 1435–1448.

Harrison, T. M., and Wood, B. J., 1980, An experimental investigation of the partitioning of REE between garnet and liquid with reference to the role of defect equilibria: Contrib. Mineral. Petrol., v. 82, p. 145–155.

Hodges, K. V., Hames, W. E., and Bowring, S., 1994, $^{40}Ar/^{39}Ar$ age gradients in micas from high-temperature–low-pressure metamorphic terrain: Evidence for very slow cooling and implications for the interpretation of age spectra: Geology, v. 22, p. 55–58.

Holland, T. J. B., and Powell R., 1998, An internally consistent thermodymanic data set for phases of petrological interest: Jour. Metamor. Geol., v. 16, p. 309–343.

Jiang, J., and Lasaga, A. C., 1990, The effect of post-growth thermal events on growth-zoned garnet: Implications for metamorphic P-T history calculations: Contrib. Mineral. Petrol., v. 105, p. 454–459.

Joesten, R., 1991, Grain-boundary diffusion kinetics in silicate and oxide minerals: Adv. Phys. Geochem., v. 8, p. 345–395.

Jaoul, O., and Sautter, V., 1999, A new approach to geospeedometry based on the compensation law: Phys. Earth Planet. Inter., v. 110, p. 95–114.

Kirschner, D. L., Cosca, M. A., Masson, H., and Hunziker, J. C., 1996, Staircase $^{40}Ar/^{39}Ar$ spectra of fine-grained white mice: Timing and duration of deformation and empirical constrains on argon diffusion: Geology, v. 24. p. 747–750.

Korzhinkii, D. S., 1959. Physiochemical basis of analysis of the paragenesis of minerals: New York, Consultants Bureau.

Krogh, E. G., 1988, The garnet-clinopyroxene Fe-Mg thermometer—reinterpretation of existing experimental data: Contrib. Mineral. Petrol., v. 99, p. 44–48.

Lasaga, A.C., 1983, Geospeedometry: An extension of geothermometry: Adv. Phys. Geochem., v. 3, p. 81–114.

______, 1998, Kinetic theory in the Earth sciences: Princeton, NJ, Princeton Univ. Press, 811 p.

Lindstrom, R., Viitanen, M., Juhanoja, J., and Holtta, P., 1991, Geospeedometry of metamorphic rocks: Examples in the Rantasalmi-Sulvaka and Kiuruvesi areas, eastern Finland: Biotite garnet diffusion couples: Jour. Metamor. Geol., v. 9, p. 181–190.

Loomis, T. P., 1983, Compositional zoning in crystals: A record of growth and reaction history: Adv. Phys. Geochem., v. 3, p. 1–60.

Liou, J. G., Maruyama, S., and Ernst, W. G., 1997, Seeing a mountain in a grain of garnet: Science, v. 276, p. 48–49.

Medaris, L. G., Wang, H. F., Misar, Z., and Jelinek, E., 1990, Thermobarometry, diffusion modeling and cooling rates of crustal garnet peridotites: Two examples from the Moldanubian zone of the Bohemian massif: Lithos, v. 25, p. 189–202.

Mezger, K., Essene, E. J., and Holliday, A. N., 1992, Closure temperature of the Sm-Nd system in metamorphic garnets: Earth Planet. Sci. Let., v. 113, p. 397–409.

Muncill, G. E., and Chamberlain, C. P., 1988, Crustal cooling rates inferred from homogenization of metamorphic garnets: Earth. Planet. Sci. Lett., v. 87, p. 390–396.

O'Brien, P. J., and Vrana, S., 1995, Eclogites with short-lived granulite facies overprint in the Moldanubian Zone, Czech Republic: Petrology, geochemistry, and diffusion modeling of garnet zoning: Geol. Rundsch., v. 84, p. 473–488.

Ozawa, K., 1984, Olivine-spinel geospeedometry: Analysis of diffusion-controlled Mg-Fe-Mg^{2+} exchange: Geochim. et Cosmochim. Acta, v. 48, p. 3597–3611.

Peacock, S. M., 1987, Creation and preservation of subduction related inverted metamorphic gradients: Jour. Geophys. Res., v. 92, p. 12,763–12,781.

Peacock, S. M., Rushmer, T., and Thompson, A. B., 1994, Partial melting of subducting oceanic crust: Earth Planet. Sci. Lett., v. 121, p. 227–244.

Perchuk, A. L., 1993, Petrology of the kyanite eclogites from the Red Cliff area, Fore Range Zone, Great Caucasus: Petrology, v. 1, p. 88–98.

Perchuk A. L., Gerasimov, V. Yu., and Philippot, P., 1996, Theoretical modeling of eclogite uplift: Petrology, v. 4, p. 518–532.

Perchuk, A. L., and Philippot, P., 1997, Rapid cooling and exhumation of eclogitic rocks from the Great Caucasus, Russia: Jour. Metamor. Geol., v. 15, p. 299–310.

______, 2000, Onset of subduction: Record in Yukon eclogite: Petrology, v. 9, p. 3–20.

Perchuk, A. L., Philippot, P., Erdmer, P., and Fialin, M., 1999, Rates of thermal equilibration at the onset of subduction deduced from diffusion modeling of eclogitic garnets, Yukon-Tanana terrain, Canada: Geology, v. 27, p. 531–534.

Perchuk, A. L., and Varlamov, D. A., 1995, A new type of prograde heterogeneity of garnet: Geokhimiya, v. 9, p. 1296–1310.

Perchuk, A. L., Yapaskurt, V. O, and Podlesskii, S. K., 1998, Genesis and exhumation dynamics of eclogites in the Kokchetav massif near Mount Sulu-Tjube, Kazakhstan: Geochem. Int., v. 36, p. 877–885.

Perchuk, L. L., 1969, The effect of temperature and pressure on the equilibria of natural Fe-Mg minerals: INT. GEOL. REV., v. 11, p. 875–901.

______, 1991, Derivation of thermodynamically consistent set of geothermometers and geobarometers for metamorphic and magmatic rocks, *in* Perchuk, L. L., ed., Progress in metamorphic and magmatic petrology: Cambridge, UK, Cambridge Univ. Press, pp. 93–112.

Perchuk, L. L., and Lavrent'yeva, I. V., 1983, Experimental investigation of exchange equilibria in the system cordierite-garnet-biotite: Adv. Phys. Geochem., v. 3, p. 199–239.

Powell, R., 1985, Regression diagnostics and robust regression in geothermometer/geobarometer calibration: The garnet clinopyroxene geothermometer revised: Jour. Metamor. Geol., v. 3, p. 231–243.

Powell, R., and White, L., 1995, Diffusive equilibration between minerals during cooling: An analytical extension to Dodson's equation for closure in the one dimension: Geol. Jour., v. 30, p. 295–305.

Reddy, S. M., Kelley, S. P., and Wheeler, J., 1995, A $^{40}Ar/^{39}Ar$ laser probe study of micas from the Sesia zone, Italian Alps: Implications for metamorphic and deformation histories: Jour. Metamor. Geol., v. 14, p. 493–508.

Rosenfeld, J. L., 1970, Rotated garnets in metamorphic rocks: Geol. Soc. Amer. Spec. Pap., v. 129, 105 p.

Rubie, D. C., 1986, The catalysis of mineral reactions by water and restrictions on the presence of aqueous fluid during metamorphism: Mineral. Mag., v. 50, p. 399–415.

Rumble, D., III, and Finnerty, T. A., 1974, Devonian grossularite-spessartine overgrowth on Ordovician almandine from eastern Vermont: Amer. Mineral., v. 59, p. 558–562.

Scaillet, S., Féraud, G., Ballèvre, M., and Amouric, M., 1992, Mg/Fe and [(Mg,Fe)Si-Al_2] compositional control on argon behaviour in high-pressure white micas: A $^{40}Ar/^{39}Ar$ continuous laser probe study from the Dora Maira nappe of the internal western Alps, Italy: Geochim. et Cosmochim. Acta, v. 56, p. 2851–2872.

Schwandt, C. S., Cygan, R. T., and Westrich, H. R., 1996, Ca self-diffusion in grossular garnet: Amer. Mineral., v. 81, p. 448–451.

Sobolev, N. V., and Shatsky, V. S., 1990, Diamond inclusions in garnets from metamorphic rocks: A new environment for diamond formation: Nature, v. 343, p. 742–746.

Spear, F. S., 1993, Metamorphic phase equilibria and pressure-temperature-time paths: Washington, DC, Min. Soc. Amer., 799 p.

Thompson, J. B., Jr., 1959, Local equilibrium in metasomatic processes, *in* Abelson, P. H., ed., Researches in geochemistry: New York, John Wiley & Sons, p. 427–457.

Tracy, R. J., 1982, Compositional zoning and inclusions in metamorphic minerals: Mineral. Soc. Amer., Rev. Mineral., v. 10, p. 355–397.

von Blanckenburg, F., Villa, I. M., Baur, H., Morteani, G., and Steiger, R. H., 1989, Time calibration of a P-T path from the western Tauern window, Eastern Alps: The problem of closure temperature: Contrib. Mineral. Petrol., v. 101, p. 1–11.

Vance, D., Meier, M., and Oberli, F., 1998, The influence of high U-Th inclusions on the U-Th-Pb systematics of almandine-pyrope garnet: Results of a combined bulk dissolution, stepwise-leaching, and SEM study: Geochim, et Cosmochim. Acta., v. 62, p. 3527–3540.

Walker, D., Longhi, J., Lasaga, A. C., Stolper, E. M., Grove, T., and Hays, J. F., 1977, Slowly cooled microgabbros 15555 and 15065, *in* Proc. 8th Lunar Sci. Conf., p. 1521–1547.

Walther, J. V., and Wood, B. J., 1984, Raise and mechanism in prograde metamorphism: Contrib. Mineral. Petrol., v. 88, p. 246–259.

Yardley, B. W. D., 1977, An empirical study of diffusion in garnet: Amer. Mineral., v. 62, p. 793–800.

Yund, R. A., 1997, Rates of grain boundary diffusion through enstatite and forsterite reaction rims: Contrib. Mineral. Petrol., v. 126, p. 224–236.

Zeck, H. P., Monie, P., Villa, I. M., and Hansen, B. T., 1992, Very high rates of cooling and uplift in the Alpine belt of the Betic Cordilleras, southern Spain: Geology, v. 20, p. 79–82.

Metasomatic Phase Relations in the System CaO-MgO-SiO_2-H_2O-NaCl at High Temperatures and Pressures

ROBERT C. NEWTON AND CRAIG E. MANNING

Department of Earth and Space Sciences, University of California at Los Angeles, Los Angeles, California 90095-1567

Abstract

An equation of state of solute silica in NaCl brines at 500 to 900°C and 4 to 15 kbar is formulated by making use of two experimentally determined properties of quartz solubility: the silica molality decreases in direct proportion to the logarithm of the NaCl mole fraction (X(NaCl)) at pressures approaching 10 kbar, and the relative silica molality (molality at a given NaCl mole fraction, m_x, divided by the molality in pure H_2O at the same P and T, m_o) is independent of temperature in the evaluated range. These two properties are expressed in the relation:

$$\log(m_x/m_o)^* = A + BX(\mathrm{NaCl}),$$

where $\log(m_x/m_o)^*$ denotes the logarithm of the ideal molality ratio, and A and B are functions of pressure, but not temperature or salinity, such that $B = -1.730 - 1.431 \times 10^{-3}P + 5.923 \times 10^{-4}P^2 - 9.243 \times 10^{-5}P^3$, and $A = 0$ at $P>10$ kbar, whereas $A = 0.6131 - 0.1256P + 6.431 \times 10^{-3}P^2$ at $P \leq 10$ kbar, as derived from fits to experimental data (Newton and Manning, 1999). The parameter A decreases from 0.214 to 0 from 4 to 9.5 kbar, and remains zero to 15 kbar; B decreases from –1.373 to –1.571 from 4 to 15 kbar. With the above relationship defining a variable X(NaCl)-T-P standard-state of solute silica, the activity of SiO_2 can be replaced by its molality for calculations of mineral-fluid equilibria over most of the conditions for metasomatism in the deep crust and upper mantle. Significant departures from ideality occur only at the lowest pressures, and low salinities.

Calculations on peridotite mineral stability in the simple system CaO-MgO-SiO_2-H_2O-NaCl at high T and P show that antigorite, brucite, and diopside are stable at 500°C and pressures of 5 to 15 kbar in the presence of concentrated NaCl solutions at low SiO_2 activities. At 700°C, anthophyllite is stable over a wide range of salinities at 5 kbar with tremolite but not with diopside. The presence of anthophyllite buffers silica solubility at a high, salinity-independent value close to quartz saturation. At 10 and 15 kbar and 700°C, talc replaces anthophyllite as the stable hydrate, and talc-tremolite assemblages buffer SiO_2 fluid concentrations at high values nearly independent of salinity. At 900°C hydrates are unstable and diopside again becomes stable and coexists with enstatite in peridotites. These stability calculations correspond well to the observed progressive metamorphic sequence in peridotite bodies in the Central Alps.

This method of analysis may be useful in interpretation of metamorphosed ultramafic bodies in general, including the basal portions of obducted ophiolitic mantle lithosphere and the mantle wedge above subduction zones. More detailed calculations, including rocks containing feldspars, must take into account the more soluble major components of rocks, especially alkalis, as these will affect the activity coefficient of SiO_2 in NaCl solutions. The solubility of silica in the presence of minerals containing these components must be determined by additional measurements.

Introduction

MODELING OF MINERAL STABILITY and metasomatism in the Earth's crust and mantle requires detailed knowledge of the thermodynamic properties of natural fluids and dissolved species. A principal source of information is measurements of solubilities of common rock-forming minerals in aqueous solutions, including those containing major amounts of other components such as CO_2 and soluble salts. The most comprehensive solubility data available are for quartz in pure H_2O. Since SiO_2 is one of the most soluble major components of common rocks at elevated T and P, these data may be used for stability calculations involving common assemblages, especially those low in alkalis, such as ultramafic systems of the upper mantle (Manning, 1995).

An important class of natural fluids is alkali chloride brine. Fluid-inclusion studies have shown that this type of fluid was involved in the crystallization and subsequent metamorphism of the Duluth Gabbro (Pasteris et al., 1995) and the Zabargad Island (Red Sea) peridotite complex (Sciuto and Ottonello, 1995), as well as many other kinds of igneous and metamorphic rocks. Experimental studies of dilute and concentrated NaCl-KCl solutiors (Aranovich and Newton, 1996, 1997) have shown that H_2O activity decreases rapidly with pressure at temperatures in the range from 500 to 900°C. Above ~2 kbar, a single-phase supercritical solution exists at these temperatures in the system NaCl-H_2O. The H_2O activity approaches the square of the mole fraction at pressures above ~5 kbar. The lowered H_2O activity in concentrated brines has a strong influence in suppressing hydrous minerals in mafic and ultramafic assemblages at high pressures. This property, in addition to changes of SiO_2 solubility with concentration of ionized solutes, could have important consequences for fluid-driven processes at depths greater than ~20 km.

The solubility of quartz in supercritical NaCl solutions has been measured recently over wide ranges in temperature, pressure, and NaCl concentration (Newton and Manning, 1999). These new data show that quartz solubility can be quite simply characterized in concentrated NaCl solutions at high temperatures and pressures, in contrast to its complex behavior in dilute solutions at low pressures (Xie and Walther, 1993). Moreover, quartz solubility is substantial in concentrated NaCl solutions of very low H_2O activity, in contrast to H_2O-CO_2 solutions, where it is as much as an order of magnitude lower at comparable H_2O activities. These properties make possible a simple but comprehensive reconstruction of the thermodynamics of silica solution applicable to the study of brine-mineral interactions in rock systems where silica is the most soluble major rock component.

Previous Work

The system H_2O-NaCl

Aranovich and Newton (1996) determined the H_2O activity in concentrated NaCl solutions at 2 to 15 kbar and 500 to 900°C by measuring the isobaric temperature depression of the brucite dehydration reaction to periclase:

$$\underset{\text{brucite}}{Mg(OH)_2} = \underset{\text{periclase}}{MgO} + \underset{\text{fluid}}{H_2O}. \quad (A)$$

They found a much greater salinity dependence of the dehydration temperatures at 10 to 15 kbar than at 2 to 4 kbar, indicating much faster decrease of the H_2O activity with NaCl concentration at high pressures. A fortunate property of the brucite-periclase H_2O activity monitor is negligibly low solubility of MgO in the H_2O-NaCl fluids. The dependence of H_2O activity on NaCl mole fraction, X (NaCl), is well described by the expression:

$$a(H_2O) = (1 - X[NaCl]/(1 + \alpha X[NaCl]), \quad (1)$$

where α is the dissociation parameter, which varies from near 0 at low pressure to nearly 1 at high pressure as aqueous NaCl progressively ionizes. Aranovich and Newton (1996) showed that from 2 to 15 kbar and 600 to 900°C α is described by

$$\alpha = \exp(4.049 - 2.925/p_o) - 134.2\ P/T, \quad (2)$$

with T in K, P in kbar, and p_o, the density of pure H_2O at T and P, in gm/cm^3. Equation 2 illustrates that α depends on pressure and temperature, but not NaCl concentration. Together, Equations 1 and 2 indicate that H_2O activity in NaCl solutions decreases rapidly with pressure above ~4 kbar. At 10 kbar and above, H_2O-NaCl solutions are very nearly ideal mixtures of completely ionized NaCl (Na^+ and Cl^-) and slightly ionized H_2O.

The system SiO_2-H_2O

The pioneering studies of quartz solubility by Kennedy (1950) and Morey and Hesselgesser (1951) demonstrated that quartz solubility increases greatly with increasing solution density at temperatures and pressures above the critical point of H_2O. At temperatures above 1000°C and pressures approaching 10 kbar, the aqueous fluid is so rich in silica that it has almost the constitution of a very hydrous siliceous liquid; indeed, the vapor and liquid phases in the system SiO_2-H_2O along the hydrous melting curve of quartz merge and become indistinguishable at a critical end point at 9.6 kbar and 1060°C (Kennedy et al., 1962).

Anderson and Burnham (1965) made the first comprehensive measurements of quartz solubility over extended ranges of P and T: 0.5 to 9.5 kbar and 500 to 900°C. These measurements were made by the weight-loss method in an internally heated gas-pressure vessel. The slow quenching rate of this apparatus limited precision at the highest temperatures at pressures approaching the critical end point, where silica solubility becomes substantial

TABLE 1. Log Molality of $SiO_{2,aq}$ in Equilibrium with Quartz

Pressure, kbar	Temperature, °C								
	500	550	600	650	700	750	800	850	900
4	–0.931	–0.779	–0.641	–0.517	–0.404	–0.303	–0.209	–0.126	–0.049
4.5	–0.904	–0.749	–0.607	–0.479	–0.362	–0.256	–0.159	–0.070	0.011
5	–0.882	–0.723	–0.578	–0.446	–0.326	–0.216	–0.116	–0.024	0.061
5.5	–0.861	–0.700	–0.552	–0.418	–0.295	–0.182	–0.078	0.017	0.103
6	–0.842	–0.679	–0.530	–0.392	–0.267	–0.152	–0.047	0.050	0.140
6.5	–0.826	–0.660	–0.509	–0.370	–0.243	–0.126	–0.018	0.082	0.174
7	–0.810	–0.644	–0.490	–0.350	–0.220	–0.101	0.008	0.110	0.203
7.5	–0.796	–0.629	–0.473	–0.331	–0.200	–0.079	0.032	0.134	0.229
8	–0.783	–0.614	–0.458	–0.314	–0.182	–0.060	0.053	0.157	0.253
8.5	–0.771	–0.600	–0.443	–0.298	–0.166	–0.041	0.072	0.177	0.275
9	–0.759	–0.589	–0.430	–0.284	–0.150	–0.025	0.090	0.197	0.295
9.5	–0.748	–0.576	–0.418	–0.270	–0.134	–0.009	0.107	0.215	0.314
10	–0.738	–0.566	–0.406	–0.257	–0.121	0.006	0.123	0.231	0.332
10.5	–0.729	–0.555	–0.395	–0.246	–0.108	0.019	0.138	0.247	0.348
11	–0.719	–0.546	–0.384	–0.234	–0.096	0.033	0.151	0.261	0.364
11.5	–0.711	–0.536	–0.374	–0.223	–0.084	0.045	0.164	0.275	0.378
12	–0.703	–0.527	–0.364	–0.213	–0.074	0.056	0.176	0.287	0.392
12.5	–0.695	–0.519	–0.356	–0.204	–0.063	0.066	0.188	0.300	0.404
13	–0.687	–0.511	–0.347	–0.194	–0.053	0.077	0.199	0.312	0.417
13.5	–0.680	–0.503	–0.338	–0.185	–0.044	0.088	0.209	0.323	0.428
14	–0.672	–0.495	–0.330	–0.177	–0.035	0.097	0.220	0.333	0.440
14.5	–0.666	–0.488	–0.323	–0.169	–0.027	0.106	0.229	0.343	0.450
15	–0.659	–0.480	–0.315	–0.161	–0.018	0.115	0.238	0.353	0.460

Source: Manning, 1994.

(several weight percent of the fluid). Manning (1994) determined the solubility of quartz over a broad P-T range, 5 to 20 kbar and 500 to 900°C, in a well-calibrated piston-cylinder apparatus with NaCl pressure medium. The rapid quench of this apparatus allows precise measurements by the method of weight loss of small single quartz crystals. Manning (1994) showed that the silica molality of aqueous fluid in equilibrium with quartz at a given temperature increases linearly with the logarithm of the H_2O density. He summarized his data and those of earlier workers in a single solubility equation based on this principle. Table 1 presents his solubility prediction in units of log molality in the ranges 4 to 15 kbar and 500 to 900°C. H_2O densities were taken from the equation of state of Halbach and Chatterjee (1982).

The system H_2O-SiO_2-NaCl

Prior to the recent study of Newton and Manning (1999), the only measurements of quartz solubility in NaCl brines at elevated P-T conditions were those of Xie and Walther (1993). The maximum conditions of their study were 2 kbar, 573°C, and NaCl mole fraction of 0.015. Silica concentrations of the fluid were measured by spectroscopic analysis of fluids rapidly extracted from equilibration runs at pressure and temperature (Walther and Orville, 1983). They found that NaCl in solution at 1 kbar and temperatures above 400°C actually enhances the solubility of quartz relative to pure H_2O. Such a "salting-in" effect also was found for KCl solutions at 3 kbar and 600°C by Anderson and Burnham (1967). Xie and Walther (1993) demonstrated that there is a threshold temperature, ~300°C at 0.5 kbar and above 400°C at 1 kbar, where salting-in behavior commences.

Newton and Manning (1999) measured quartz solubility in NaCl solutions over broad ranges of P-T-X: 2 to 15 kbar, 500 to 900°C, and salinities up to halite saturation. They used the techniques of Manning (1994), except for fluid analysis by puncture-and-dry weight loss measurements on quenched capsules. At 2 kbar and 700°C, there is a substantial salting-in effect similar to that found by Xie and Walther (1993) at 1 kbar. Quartz solubility at those conditions rises to a maximum at X(NaCl) ~0.1, which is 50% higher than in pure H_2O. The solubility does not become as low as in pure H_2O until the NaCl concentration reaches 40 mole%, or 68 wt%. At 10 and 15 kbar, however, there is an exponential decrease of SiO_2 solubility with NaCl mole fraction. At 4.35 kbar, the behavior is intermediate between these extremes. Although an absolute salting-in does not occur at this pressure, a tendency toward solubility enhancement at low X(NaCl) persists.

The combined results of Anderson and Burnham (1967), Xie and Walther (1993), and Newton and Manning (1999) imply that the change from salting-in to salting-out behavior correlates with the pressure-induced decrease of H_2O activity in NaCl solution, and attendant, progressive ionization of solute NaCl° (Newton and Manning, 1999). The simple exponential dependence of quartz solubility on NaCl mole fraction at high pressure is explainable in straightforward terms if the SiO_2 aqueous species, the H_2O molecules, and the Na^+ and Cl^- ions together form an ideal mixture—that is, if the ions and molecules mix in the solution without excess enthalpy and excess entropy. By contrast, neutral NaCl° dominates at low pressures, evidently leading to the more complex salting-in behavior. The transition from salting in to salting out occurs at densities of pure H_2O of 0.7 to 0.8 g/cm3, or α ~0.6 to 0.65. The precise mechanism of SiO_2-solute interaction cannot be elucidated at present.

Another useful property of silica solubility in high-pressure, high-temperature NaCl solutions is that the silica molality at a given NaCl concentration, m_x, divided by the molality in pure H_2O at the same P and T, m_o, is effectively temperature-independent at pressures of 4.35 kbar and above. Figure 1 shows the measurements of Newton and Manning (1999) of relative molality at 4.35 and 10 kbar. The corresponding data for 15 kbar lie very close to the 10 kbar data and are not plotted. Least-squares fits to these data at the investigated pressures, combined with interpolation between these pressures, yielded the curves for 4 to15 kbar in Fig. 1 (Newton and Manning, 1999). The near-ideal behavior of NaCl-H_2O solutions, described by equation 1, together with the near temperature-independence of the relative SiO_2 molality, makes possible a simple equation of state of solute silica that is useful for practical calculations of mineral stability in the presence of concentrated NaCl solutions.

Formulation of Activity-Concentration Relations

Figure 2 shows the quartz solubility data of Newton and Manning (1999), plotted as the logarithm of relative silica molality versus NaCl mole fraction. The experimental data at 10 and 15 kbar and all temperatures project through the origin as straight lines and are thus temperature-independent exponential arrays. At very high X(NaCl), realized at 10 kbar in the highest-temperature data, the measurements depart from the linear trend toward lower solubilities. Newton and Manning (1999) proposed that this results from the progressive destabilization of a hydrated solute species, probably H_4SiO_4, which would occur at very low H_2O activities. The 4.35 kbar data depart from linearity at low NaCl concentrations. This behavior reflects the persistence of low-pressure salting-in behavior. At X(NaCl) greater than about 0.2, the logarithm of the solubility ratio decreases linearly as at 10 and 15 kbar, suggesting that solutions of high salinity are simple mixtures even at pressures below 5 kbar.

The data arrays in Figure 2 follow the trend

$$\log (m_x/m_o)^* = A + B\, X(\text{NaCl}) \qquad (3)$$

over their linear ranges. In equation 3, $\log (m_x/m_o)^*$ is the logarithm of the molality ratio assuming linear behavior predicted by the right-hand side of the equation. This relationship can thus be used to define a standard state for solute silica of unit molality and unit activity in a hypothetical solution that obeys equation 3 and Henry's Law. The activity of SiO_2 thus defined is applicable only at 4–15 kbar, 500–900°C, and X(NaCl) to halite saturation, which

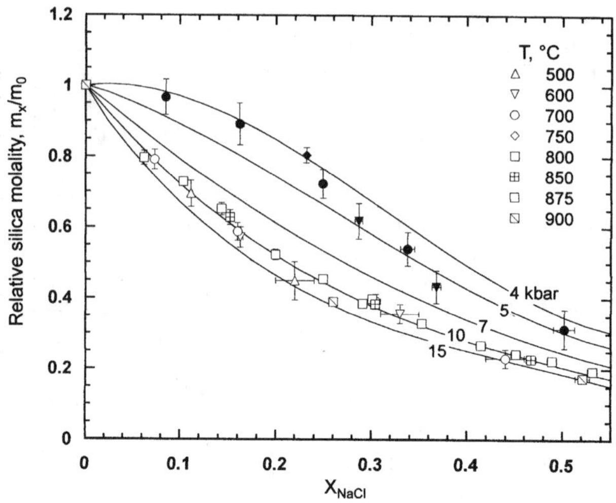

FIG. 1. Relative silica molality in NaCl solutions at pressures of 4 to 15 kbar and 500 to 900°C in equilibrium with quartz. Experimental data of Newton and Manning (1999) at 4.35 kbar (filled symbols) and 10 kbar (open symbols) and the indicated temperatures. The temperature-independent isobars shown are those modeled from the data at all pressures and temperatures by Newton and Manning (1999).

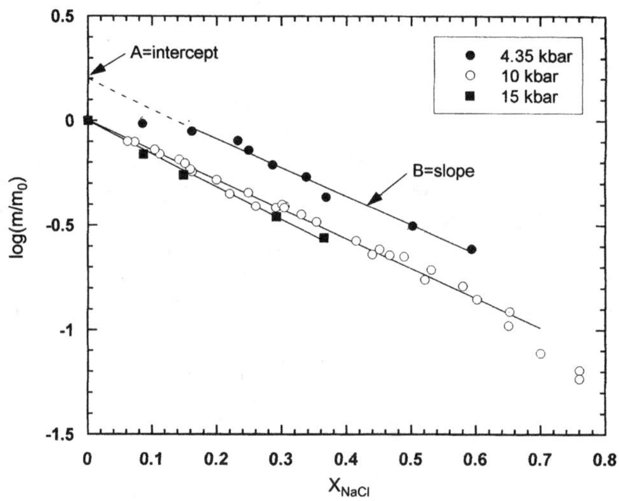

FIG. 2. Data of Figure 1 plotted on a logarithmic scale versus NaCl mole fraction. The data at all temperatures are plotted. The figure illustrates the near temperature-independence of the relative molality. Departure of the 4.35 kbar data from a linear trend occurs at NaCl mole fraction less than 0.2; this is an effect of the tendency toward "salting-in" at low pressures and low salinities (see text). Extrapolation of the linear portion of the 4.35 kbar data to X(NaCl) = 0 yields the A parameter of equation 2. The B parameter is the slope of the isobar.

reflects the range of the experimental data upon which it is based. At most of these conditions, silica activity will be equal to its molality. Significant departures from ideality relative to this standard state correspond to departures from linear behavior. Such departures are encountered only at the lowest pressures and NaCl concentrations, and at very high NaCl concentrations at high temperatures.

As illustrated in Figure 2, the A and B parameters correspond to the intercepts and slopes of each isobaric array of solubility data. Expressions for the variations in the slopes and intercepts were derived

TABLE 2. Values of A and B Parameters Used to Calculate Activity Coefficients at Discrete Pressures

Pressure, kbar	A	B
4.0	0.214	–1.373
4.5	0.178	–1.373
5.0	0.146	–1.374
5.5	0.117	–1.376
6.0	0.091	–1.378
6.5	0.068	–1.380
7.0	0.049	–1.383
7.5	0.033	–1.387
8.0	0.020	–1.391
8.5	0.010	–1.397
9.0	0.004	–1.403
9.5	0.000	–1.410
10.0	0.000	–1.418
10.5	0.000	–1.427
11.0	0.000	–1.437
11.5	0.000	–1.449
12.0	0.000	–1.462
12.5	0.000	–1.476
13.0	0.000	–1.492
13.5	0.000	–1.509
14.0	0.000	–1.528
14.5	0.000	–1.548
15.0	0.000	–1.571

by linear least-squares fits to the experimental data at 4.35, 10, and 15 kbar (Fig. 2), combined with interpolation between these pressures. The resulting equations for A and B are:

$$A = 0.6131 - 0.1256P + 6.431 \times 10^{-3}P^2 \quad P \le 10 \text{ kbar}$$
$$A = 0 \quad P > 10 \text{ kbar};$$

and

$$B = -1.730 - 1.431 \times 10^{-3}P + 5.923 \times 10^{-4}P^2 - 9.243 \times 10^{-5}P^3 \quad P = 4\text{–}15 \text{ kbar}$$

The two fit regions for A arise from the observation that nonzero intercepts occur only at P<10 kbar (Fig. 2). Table 2 gives A and B values at discrete pressures over the P and T ranges of interest. Figure 3 illustrates that A has small positive values that decrease to zero with increasing pressure, whereas B becomes progressively more negative as P increases, consistent with the changes in slope and intercept illustrated in Figure 2.

Deviations from ideality for aqueous silica in equilibrium with quartz exist where equation 3 does not describe accurately the experimentally determined values of $\log(m_x/m_o)$. Thus, the activity coefficient of solute silica is given by:

$$\log \gamma (SiO_2) = A + B\,X(NaCl) - \log(m_x/m_o). \quad (4)$$

In equation 4, $\log(m_x/m_o)$ is the value consistent with experimental data (Manning, 1994; Newton and Manning, 1999) as opposed to the ideal value, $\log(m_x/m_o)^*$. Comparison of equations 3 and 4 illustrates that the logarithm of the activity coefficient for aqueous silica is simply the difference between $\log(m_x/m_o)^*$ and $\log(m_x/m_o)$. Table 3 gives values of $\log \gamma (SiO_2)$ at the range of P and X(NaCl) where nonideality occurs. (We omit consideration of departures from ideality at very high salinity (X(NaCl))>0.5, because such solutions are likely to be geologically rare.)

The quartz-solution equilibrium is expressed formally by:

$$SiO_2(\text{quartz}) = SiO_2(\text{aq}), \quad (B)$$

for which the standard Gibbs free energy change is a function of temperature, pressure, and salinity, given by the relation:

$$\Delta G_r^o/2.303RT = \log a_{SiO_2(aq)} = A + BX(NaCl) + \log m_o. \quad (5)$$

The standard state for quartz is unit activity of the pure phase at any pressure and temperature. ΔG_r^o is interpreted as the Gibbs free energy change when one mole of quartz is dissolved at fixed T and P into a very large quantity of solvent of concentration X(NaCl). For 5 kbar and 700°C, equation 5 and Tables 1 to 3 indicate that the solution reaction in pure H_2O is attended by a Gibbs free energy change of 3.33 kJ. The corresponding Gibbs free energy change for solution into pure (metastable) molten NaCl is 28.93 kJ. The latter quantity is purely hypothetical in that it implies the existence of a hydrous solute silica species in a completely anhydrous solvent. Nevertheless, the changes in ΔG_r^o with X(NaCl) illustrate the magnitude of the increased

TABLE 3. Values of $\log\gamma_{SiO_2,aq}$ at Discrete Values of Pressure and X_{NaCl}

	XNaCl								
Pressure, kbar	0.000	0.025	0.050	0.075	0.100	0.125	0.150	0.175	0.200
4.0	0.214	0.179	0.145	0.111	0.076	0.042	0.008	0.000	0.000
4.5	0.178	0.144	0.109	0.075	0.041	0.006	0.000	0.000	0.000
5.0	0.146	0.112	0.077	0.043	0.008	0.000	0.000	0.000	0.000
5.5	0.117	0.082	0.048	0.014	0.000	0.000	0.000	0.000	0.000
6.0	0.091	0.057	0.022	0.000	0.000	0.000	0.000	0.000	0.000
6.5	0.068	0.034	0.000	0.000	0.000	0.000	0.000	0.000	0.000
7.0	0.049	0.014	0.000	0.000	0.000	0.000	0.000	0.000	0.000
7.5	0.033	0.000	0.000	0.000	0.000	0.000	0.000	0.000	0.000
8.0	0.020	0.000	0.000	0.000	0.000	0.000	0.000	0.000	0.000
8.5	0.010	0.000	0.000	0.000	0.000	0.000	0.000	0.000	0.000
9.0	0.004	0.000	0.000	0.000	0.000	0.000	0.000	0.000	0.000
9.5	0.000	0.000	0.000	0.000	0.000	0.000	0.000	0.000	0.000
10.0	0.000	0.000	0.000	0.000	0.000	0.000	0.000	0.000	0.000

free-energy change associated with the dissolution of quartz with increasing salinity.

Calculation of ΔG^o for quartz at the P and T of interest allows equation 5 to be rewritten in terms of the Gibbs free energy of aqueous silica NaCl-bearing solutions:

$$\Delta G^o_{SiO_2(aq)} = -2.303RT(A + B\,X(NaCl) + \log m_o) + \Delta G^o_{quartz}. \quad (6)$$

Equation 6 permits calculation of metasomatic phase relations as a function of X(NaCl) and silica concentration.

Application to the System CaO-MgO-SiO_2-H_2O-NaCl

The above formulation of the thermodynamics of quartz solubility may be applied to a simple ultramafic system that models the mineralogy of the upper mantle, the basal sections of ophiolites, and ultramafic intrusions in the crust with saline pore fluids, at elevated P-T conditions (T>500°C; P>5 kbar). As an example of calculation, diagrams of log $m(SiO_2)$ versus X(NaCl) are constructed at 500, 700, and 900°C, and 5, 10, and 15 kbar (Fig. 4). The minerals considered are, in addition to quartz (qz), diopside (di), tremolite (tr), enstatite (en), forsterite (fo), talc (ta), anthophyllite (an), antigorite (at), brucite (br), and periclase (pe). The non-peridotite CMSH minerals wollastonite, monticellite, larnite, portlandite, and lime were not considered. Fifteen isobaric, isothermal univariant equilibria, including reactions A and B, were found to have stable portions in the ranges of 5 to15 kbar and 500 to 900°C:

fo + SiO_2(aq) = 2 en (C)
H_2O + 3 en + SiO_2(aq) = ta (D)
H_2O + 7 en + SiO_2(aq) = an (E)
H_2O + 3/4 an + SiO_2(aq) = 7/4 ta (F)
2/9 H_2O + 7/9 fo + SiO_2(aq) = 2/9 an (G)
31/10 H_2O + 12/5 fo + SiO_2(aq) = 1/10 at (H)
1/30 at + SiO_2(aq) = 8/15 ta +1/2 H_2O (I)
24/17 br + SiO_2(aq) = 1/34 at + 17/34 H_2O (J)
2 br + SiO_2(aq) = fo +2 H_2O (K)
2 pe + SiO_2(aq) = fo (L)
2/5 H_2O + 2/5 fo + 4/5 di + SiO_2(aq) = 2/5 tr (M)
1/30 at + 16/15 di + SiO_2(aq) = 8/15 tr + 1/2 H_2O (N)
H_2O + 3 en + 2 di + SiO_2(aq) = tr (O)

The reactions are normalized to one SiO_2 (aq) for convenience. The SiO_2-independent reaction A is a consequence of K and L. Calculations were performed with the Holland and Powell (1998) thermodynamic data set and the Halbach and Chatterjee (1982) equation of state of H_2O, for which we adopted a standard state of unit activity for the pure phase at all T and P. Activity of H_2O was calculated

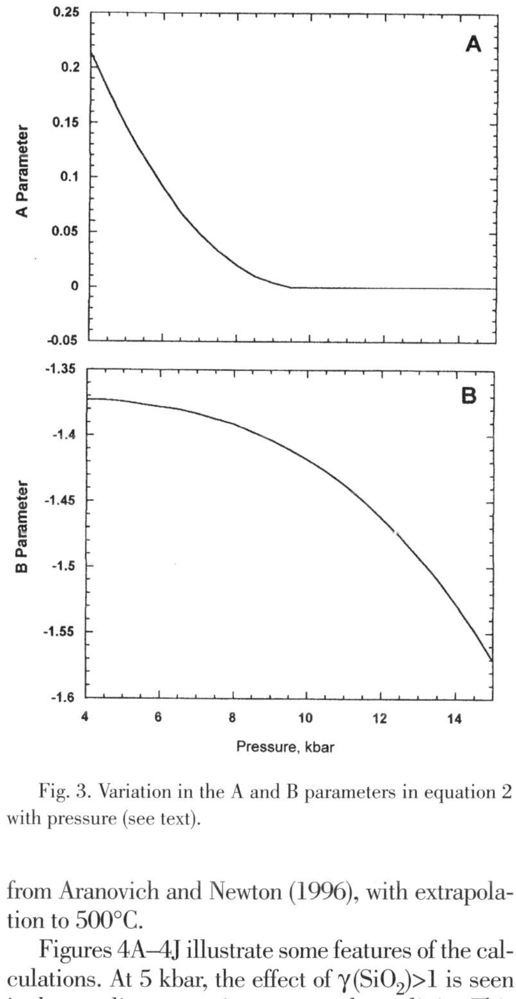

Fig. 3. Variation in the A and B parameters in equation 2 with pressure (see text).

from Aranovich and Newton (1996), with extrapolation to 500°C.

Figures 4A–4J illustrate some features of the calculations. At 5 kbar, the effect of $\gamma(SiO_2)>1$ is seen in the non-linear reaction curves at low salinity. This behavior is a carry-over from the salting-in effect at low pressure and low salinity where $NaCl^0$ is predominantly associated. At 500°C, forsterite, antigorite, talc, brucite, tremolite, and diopside are stable phases at 5 to 15 kbar, and enstatite and periclase are not stable in that range. A small range of stability of reaction J was found at low salinity at 15 kbar. Saline solutions are restrictive to antigorite stability. Maximum salinities are limited by halite saturation near a NaCl mole fraction of 0.2. At 700° C, anthophyllite is stable at 5 kbar. Its stability relative to enstatite is almost unaffected by salinity. A small range of stability of reaction G was found at low salinity. Salinity is extremely inimical to talc stability at 5 kbar, but the talc field expands strongly at 10 and 15 kbar, and anthophyllite disappears. Enstatite is stable throughout the pressure range at 700°C. Periclase is stable along with brucite. At 900°C, the only stable hydrate is brucite at 15 kbar. At this temperature diopside can coexist with enstatite or forsterite over the complete range of silica activity and NaCl concentration.

Geological Applications

The major application of the present analysis is to middle- to high-grade metamorphism of ultramafic bodies, principally alpine peridotites. Lower-grade parageneses pertinent to serpentinization were formed at temperatures below the applicability of the present work. Trommsdorff and Evans (1974) described ultramafic assemblages in a progressive metamorphic suite from the central Alps. Evans and Trommsdorf (1974) inferred pressures of metamorphism of these rocks near 7 kbar. Their lowest temperature assemblage of chrysotile-brucite would be stable at temperatures much below 500°C. Antigorite-brucite-diopside is found in higher-grade metaperidotites, but would indicate temperatures below 500°C if pore solutions are even moderately salty. Figures 4A–4C show that saline fluids are prejudicial to this assemblage, since the field boundaries of brucite and antigorite diverge strongly with increasing salinity. An antigorite-out isograd in the Alps was thought to occur between the staurolite and sillimanite isograds, i.e., between 500 and 700°C, consonant with Figure 4. Successively higher grade assemblages in the Alpine metaperidotites include talc-tremolite-enstatite, tremolite-forsterite-enstatite, and diopside-forsterite-enstatite. Figure 4 illustrates that these assemblages are all stable at 700°C over a range of pressure and NaCl concentration.

Figure 4 indicates that pore-fluid salinity and silica activity have to be taken into account in the assessment of metamorphic grade. Moreover, certain mineral assemblages are diagnostic of a restricted range of silica activity and salinity. For instance, at 500°C and 5 kbar, the assemblage forsterite + tremolite requires high X(NaCl) and low silica concentrations. Similarly, the anhydrous peridotite assemblage forsterite-enstatite-diopside can occur at 700°C and 10 kbar in the presence of aqueous pore fluids only if they are very salty (X(NaCl) < 0.3). In general, high fluid salinity and low silica activity have the same effects as increasing temperature in favoring anhydrous and low-silica assemblages. Periclase and brucite occur at such low SiO_2 activities that their presence is not likely in

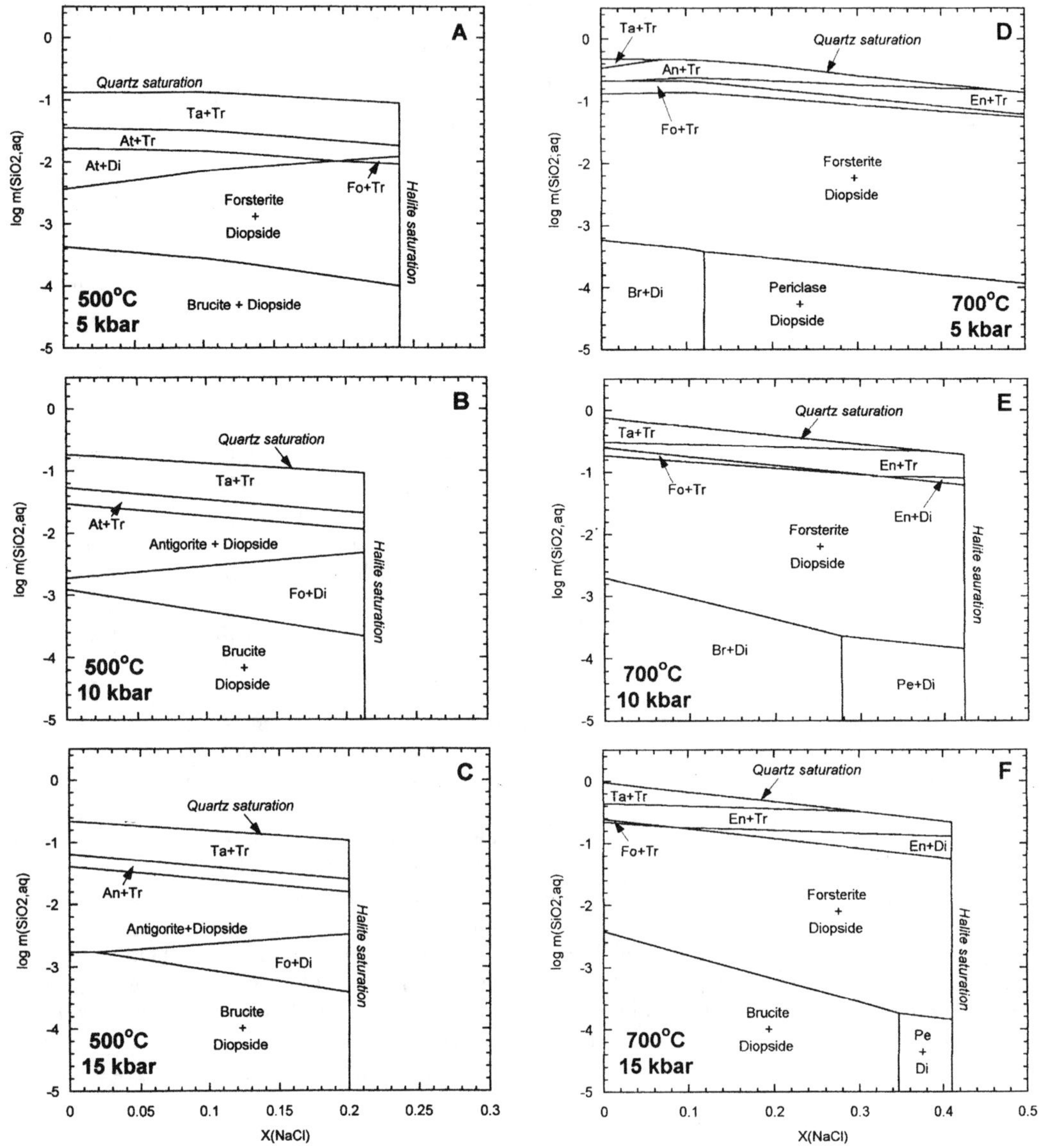

Fig. 4. A–I (G–I on facing page). Stability relations and silica solubilities of model fluid-saturated peridotite assemblages in the simple system CaO-MgO-SiO_2-H_2O-NaCl at various P-T conditions calculated with the present SiO_2 equation of state and the thermodynamic data for the minerals of Holland and Powell (1998). H_2O fugacities are from Halbach and Chatterjee (1982). H_2O activities in NaCl solutions are from Aranovich and Newton (1996). Mineral abbreviations: an = antigorite; at = anthophyllite; br = brucite; di = diopside; en = enstatite; fo = forsterite; pe = periclase; ta = talc; tr = tremolite.

high-grade peridotites of an initial forsterite-enstatite-diopside assemblage.

The phase relations depicted in Figure 4 also aid interpretation of paragenetic relations. Evans and Trommsdorff (1974) regarded anthophyllite in the Val d'Efra metaperidotite as a retrograde phase, possibly formed in decompression below 7 kbar of an original talc-enstatite assemblage. Figures 4D and 4E uphold this concept, and show that anthophyllite would be stable at 5 kbar and 700°C with fluids of a

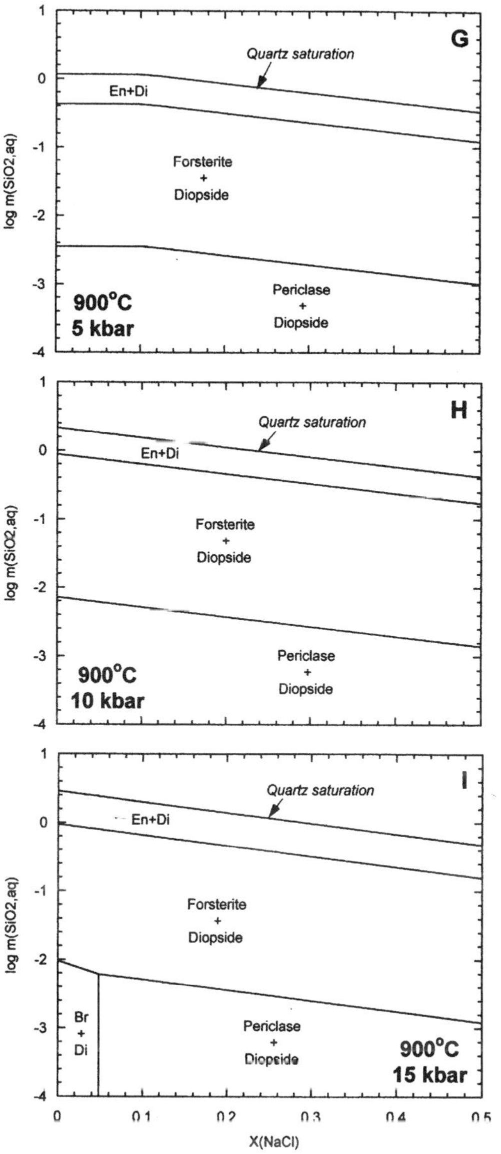

Fig. 4. Continued.

wide range of salinity. Anthophyllite indicates silica molality near 0.2, nearly independent of salinity. The high silica activity, close to quartz saturation, indicates considerable silica metasomatism of an originally ultramafic body. Another possibility suggested by Evans and Trommsdorff (1974) for anthophyllite formation at Val d'Efra is introduction of iron, which could stabilize anthophyllite relative to talc, a phase restrictive to Fe. NaCl solutions could be effective in this regard, as they readily dissolve and transport Fe oxides under high-grade metamorphic conditions (Chou and Eugster, 1977).

The major utility of the present type of analysis may be in quantifying the amounts, H_2O activities, and silica-transporting abilities of metamorphic fluids active in the metamorphism of peridotite bodies. The fact that these bodies often are profoundly metasomatic in character indicates that SiO_2 activity in fluids of various types coexisting with peridotite assemblages is one of the most important parameters determining their mineralogy. The present calculations are necessarily restricted to systems where SiO_2 is the most soluble component at elevated temperatures and pressures. The analysis may be extended to systems with other soluble components, including alkalis, when additional solubility data in feldspar-bearing assemblages become available. Hydrogen ion potential is another important parameter that will have to be characterized experimentally before a comprehensive account of the solubilities can be erected (Anderson and Burnham, 1983).

Acknowledgments

Primary funding of this project was from a National Science Foundation grant for the study of mineral-fluid interactions, #EAR-9405999. We thank D. Bird, T. Fridriksson, and P. Neuhoff for constructive review of the manuscript.

REFERENCES

Anderson, G. M., and Burnham, C. W., 1965, The solubility of quartz in supercritical water: Amer. Jour. Sci., v. 263, p. 494–511.

______, 1967, Reactions of quartz and corundum with aqueous chloride and hydroxide solutions at high temperatures and pressures: Amer. Jour. Sci., v. 265, p. 12–27.

______, 1983, Feldspar solubility and the transport of aluminum under metamorphic conditions: Amer. Jour. Sci., v. 283-A, p. 283–297.

Aranovich, L. Ya., and Newton, R. C., 1996, H_2O activity in concentrated NaCl solutions at high pressures and temperatures measured by the brucite-periclase equilibrium: Contrib. Mineral. Petrol., v. 125, p. 200–212.

______, 1997, H_2O activity in concentrated KCl and KCl-NaCl solutions at high temperatures and pressures measured by the brucite-periclase equilibrium: Contrib. Mineral. Petrol., v. 127, p. 261–271.

Chou, I-M., and Eugster, H. P., 1977, Solubility of magnetite in super-critical chloride solutions: Amer. Jour. Sci., v. 277, p. 1296–1314.

Evans, B. W., and Trommsdorff, V., 1974, Stability of enstatite + talc, and CO_2 metasomatism of metaperidotite, Val d'Efra, Lepontine Alps: Amer. Jour. Sci., v. 274, p. 274–296.

Halbach, H., and Chatterjee, N. D., 1982, An empirical Redlich-Kwong-type equation of state for water to 1000°C and 200 kbar: Contrib. Mineral. Petrol., v. 79, p. 337–345.

Holland, T. J. B., and Powell, R., 1998, An internally consistent thermodynamic data set for phases of petrologic interest: Jour. Metamor. Geol., v. 16, p. 309–344.

Kennedy, G. C., 1950, A portion of the system silica-water: Econ. Geol., v. 45, p. 629–653.

Kennedy, G. C., Wasserburg, J. G., Heard, H. C., and Newton, R. C., 1962, The upper three-phase region in the system SiO_2-H_2O: Amer. Jour. Sci., v. 260, p. 501–521.

Manning, C. E., 1994, The solubility of quartz in H_2O in the lower crust and upper mantle: Geochim. et Cosmochim. Acta, v. 58, p. 4831–4839.

______, 1995, Phase-equilibrium controls on SiO_2 metasomatism by aqueous fluids in subduction zones: Reaction at constant pressure and temperature: INT. GEOL. REV., v. 37, p. 1074–1093.

Morey, G. W., and Hesselgesser, J. M., 1951, The solubility of some minerals in supercritical steam at high pressures: Econ. Geol., v. 46, p. 821–835.

Newton, R. C.,. and Manning, C. E., 1999, Quartz solubility in concentrated NaCl solutions at deep crust–upper mantle metamorphic conditions: 2–15 kbar and 500–900°C: Geochim. et Cosmochim. Acta (submitted).

Pasteris, J. D., Harris, T. N., and Sassani, D. C., 1995, Interactions of mixed volatile-brine fluids in rocks of the southwestern footwall of the Duluth Complex, Minnesota—evidence from aqueous fluid inclusions: Amer. Jour. Sci., v. 295, p. 125–172.

Sciuto, P. F., and Ottonello, G., 1995, Water-rock interaction on Zabargad Island, Red Sea—a case study. I. Application of the concept of local equilibrium: Geochim. et Cosmochim. Acta, v. 59, p. 2187–2206.

Trommsdorff, V., and Evans, B. W., 1974, Alpine metamorphism of peridotitic rocks: Schweiz. Mineral. Mitteil, v. 54, p. 333–354.

Walther, J. V., and Orville, P. M., 1983, The extraction-quench technique for determination of the thermodynamic properties of solute complexes. Application to quartz solubility in fluid mixtures: Amer. Mineral., v. 68, 731–741.

Xie, Z., and Walther, J. V., 1993, Quartz solubilities in NaCl solutions with and without wollastonite at elevated temperatures and pressures: Geochim. et Cosmochim. Acta, v. 57, p. 1947–1955.

Petrological and Geochemical Characteristics of Ultrahigh-Pressure Metamorphic Rocks from the Dabie-Sulu Terrane, East-Central China

J. G. LIOU, R. Y. ZHANG,

Department of Geological and Environmental Sciences, Stanford University, Stanford, California 94305

AND BOR-MING JAHN

Geosciences Rennes-CNRS, Universite de Rennes 1, 35042 Rennes cedex, France

Abstract

The Dabie-Sulu ultrahigh-pressure (UHP) terrane is the largest in the world. Mafic-ultramafic rocks occur as ubiquitous minor components in it, and preserve UHP index minerals such as coesite and microdiamond. Eclogites and garnet peridotites together with their country rocks including quartzofeldspathic, pelitic, psammatic, and carbonate rocks were subjected to coeval UHP metamorphism during the Triassic collision between the Sino-Korean and Yangtze cratons. This review summarizes petrological and geochemical characteristics of eclogites, garnet peridotites, and UHP metasediments from published data and ongoing research in the Dabie-Sulu belt of east-central China.

Parageneses of minerals from coesite-bearing eclogites encased in gneiss, garnet peridotite, and marble delineate clockwise P-T paths characterized by nearly isothermal decompression. Many eclogites have been overprinted by Barrovian-zone amphibolite- to granulite-facies assemblages, whereas others preserve primary igneous minerals and textures in the cores of coesite-bearing eclogitic blocks. In addition, eclogites contain UHP hydrous and carbonate phases including talc, epidote, zoisite, magnesite, and dolomite; these together with Ti-clinohumite, phlogopite, amphibole, chlorite, and possible talc in garnet peridotites and OH-topaz in kyanite quartzites document the role of hydrous mineral transport to the deep mantle in fluid-deficient UHP metamorphic regions. Both crustal- and mantle-derived garnet peridotites from the Dabie-Sulu region were recrystallized within the diamond stability field, with the Earth's lowest recorded geothermal gradient, <5°C/km, suggesting that UHP metamorphic recrystallization took place in a previously unrecognized, forbidden P-T region.

Geochemical and isotopic data indicate that Dabie-Sulu mafic-ultramafic rocks have diverse origins; their compositions in some cases have been complicated by metamorphic recrystallization, crustal contamination, and fluid metasomatism. Nevertheless, REE geochemical and Nd isotope data clearly indicate that they have "continental" affinities and cannot represent a subducted Tethyan oceanic slab. Garnet peridotites and their enclosing eclogites display variable isotopic compositions; mantle-derived fragments preserve a mantle signature, whereas crust-hosted mafic-ultramafics display distinct crustal contamination and metasomatism. Among the many outstanding projects remaining to be investigated, geochemical and isotopic constraints of mantle-derived garnet peridotites and eclogites should provide an additional window to our understanding of mantle heterogeneity, metasomatism, slab/mantle interactions, and lithospheric evolution of the Sino-Korean craton.

Introduction

ECLOGITES AND GARNET PERIDOTITES, although volumetrically small, are widespread in all recognized ultrahigh-pressure (UHP) terranes of the world. They are the most significant components, inasmuch as they preserve the most completely preserved UHP index minerals such as coesite and microdiamond (e.g., Liou et al., 1994; Coleman and Wang, 1995). These mafic-ultramafic rocks have originated from different tectonic settings and were subjected to coeval UHP metamorphism, deformation, and retrogression with their enclosing granitic gneisses and supracrustal rocks. They may represent: (1) fragments of a subducted slab capped mainly by continental ± minor oceanic crust; (2) slices of the mantle wedge above a subduction zone that were incorporated in subducted continental crust; or (3) products of subduction-zone metamorphism previously emplaced in the crust. Because of better preservation of UHP metamorphism compared with adjacent gneissic rocks, eclogites and garnet peridotites provide important petrochemical and isotopic constraints on tectonic models of continental subduction, collision, and subsequent exhumation for a UHP terrane.

Among a dozen or more documented UHP terranes in the world, the Dabie-Sulu terrane of east-central China constitutes the most extensive exposure of UHP rocks, and hence is the target of the Chinese Deep Continental Drilling program beginning in 2000. UHP assemblages of eclogites and garnet peridotites from this terrane have been intensely overprinted by high-ranking amphibolite-facies mineral associations. Coesite and coesite pseudomorphs are present only as relict inclusions in zircon, garnet, omphacite, kyanite, zoisite, epidote, and dolomite (e.g., Liou et al., 1998); microdiamond also has been reported from both eclogites and garnet peridotites (Xu et al., 1992; Xu et al., 1998). The purpose of this article is to review the key petrochemical characteristics of eclogites and garnet peridotites from the Dabie-Sulu terrane in order to: (1) identify their tectonic settings and the processes for their formation, metamorphism, and modification; (2) classify and establish the database for various types of mafic-ultramafic rocks; and (3) suggest some outstanding petrochemical problems for future investigation.

The Dabie-Sulu UHP Terrane, the Largest in the World

The Qinling-Dabie-Sulu belt of east-central China marks the E-W–trending collisional orogenic belt between the Sino-Korean and Yangtze cratons (Fig. 1). Since the discovery of coesite in the Dabie-Sulu terrane (Xu, 1987; Okay et al., 1989; Wang et al., 1989; Hirajima et al., 1990), the Dabie-Sulu terrane has been the subject of numerous investigations. UHP metamorphic rocks were formed by N-directed subduction of the Yangtze craton or a microcontinent beneath the Sino-Korean craton (Ernst et al., 1991; Liou et al., 1996; Hacker et al., 2000). The central part of the 2000 km long Qinling-Dabie-Sulu belt consists of a series of fault-bounded, penetratively deformed HP and UHP units that are subdivided into three fault-bounded subunits. From north to south, they are: (1) the northern granulite/migmatitic complex; (2) the central UHP belt; and (3) the southern high-pressure (HP) belt. Protoliths of these subunits grade from early Proterozoic granitic gneisses through mid- to upper-Proterozoic bimodal volcanic rocks, platform-type carbonates, and turbidite sequences including Sinian peraluminous sediments, to volcanic rocks. The superjacent subunits were metamorphosed at different depths during continent subduction and collision.

Several NE-SW–trending faults subparallel to the Tan-Lu fault separate the collisional zone into distinct blocks (Fig. 1). East of the Tan-Lu fault lie the Sulu UHP (north) and Haizhou HP (south) terranes. Between the Tan-Lu and Shangcheng-Macheng faults (SMF), the Dabie block grades from a UHP belt southward through eclogite + epidote amphibolite and greenschist to minor exposures of the HP terrane at Susong. The Hongan-Tongbai block lies to the west of the SMF; there, progressively lower grade HP metamorphic zones (Eide and Liou, 2000) occur south of the coesite-bearing UHP belt in Henan Province (Zhang and Liou, 1994). Farther west is the Wudang-Qinling block, where abundant blueschist/greenschist–facies rocks occur with only minor eclogite (e.g., Ernst et al., 1991; Zhou et al., 1993; Hu et al., 1995). Because of differential unroofing of these blocks, exposure of the metamorphic units varies from block to block. The UHP belt is best exposed in the Sulu region and the eastern part of the Dabie block, whereas the HP belt crops out widely in the Hongan block.

Voluminous Cretaceous magmatism and deformation significantly modified the HP-UHP structures and lithologies in some areas (e.g., northern Dabie; Ratschbacher et al., 2000). A recent 400 km long, N-S deep seismic-refraction profile recorded across the Dabie Mountains shows no lower-crustal or upper-mantle signature of the Triassic subduction or collision (Wang et al., 2000). The fault-bounded UHP and HP units are unconformably overlain by Jurassic clastic strata and Cretaceous volcaniclastic cover, and are intruded by post-orogenic Mesozoic granites and gabbros (Fig. 1). Eclogites are common in gneiss, marble, and peridotite; blocks of garnet peridotite of meter to kilometer scale occur sporadically throughout the terrane. Petrotectonic study of this orogen has focused mainly on the Dabie Mountains because of their higher relief, where good

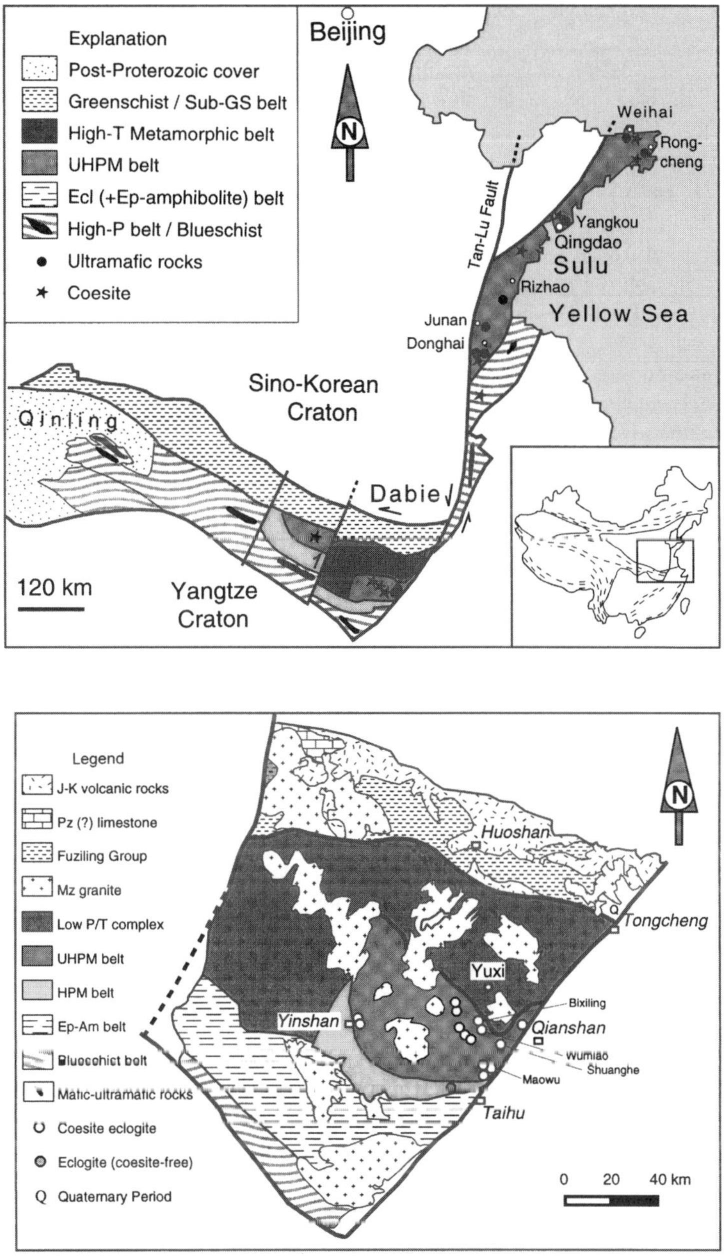

FIG. 1. Simplified regional tectonic and geologic map of the Qinling-Dabie-Sulu collision zone between the Sino-Korean and Yangtze cratons of east-central China. The major reported locations of eclogite are shown.

exposures allow geologic mapping and structural studies of UHP, HP, and other petrotectonic units. Available structural (Hacker et al., 1995, 1996, 2000; Webb et al., 1999, 2000), metamorphic (Zhang and Liou, 1994; Zhang et al., 1995a; Liou et al., 1996), and geochronological (Hacker et al., 1998) data reveal that the UHP rocks were exhumed from >125 km depth between 240 and 230 Ma by a combination of normal-sense shear from beneath the hanging-wall Sino-Korean craton, southeastward thrusting onto the footwall Yangtze craton, orogen-parallel eastward extrusion, and subvertical thinning. Similar extensive petrochemical studies of UHP rocks from Sulu have been undertaken (e.g., see Zhang and Liou, 1998b; Zhang et al., 2000); however, because of poor exposures, structural data of the UHP and HP units are limited in the Sulu region.

Figure 1 is a simplified tectonic map of the Dabie-Sulu terrane, showing most localities described in this review. The Dabie-Sulu UHP rocks are unique in several aspects: (1) the occurrence of rare but widespread coesite and hydrous phases (such as talc, zoisite/epidote, nyböite, and phengite) in eclogites (Hirajima et al., 1992; Zhang et al., 1995a; Zhang, Hirajima et al., 1995; Liou et al., 1995); (2) the world-record lowest $\delta^{18}O$ values (rutile is –15‰) for mineral separates from eclogites and metasediments (Yui et al., 1995, 1997; Zheng et al., 1996, 1998; Rumble and Yui, 1998); (3) the world-record highest Nd values of ca. +170 to 260 for Weihai eclogites (Jahn et al., 1996); (4) abundant garnet peridotites of mantle origin (Zhang et al., 1994, 2000; Zhang and Liou, 1998a, 1998b); and (5) abundant exsolution textures in UHP minerals from garnet peridotite and eclogite (Hacker et al., 1997; Zhang and Liou, 1998b, 1999; Zhang et al., 2000). The preservation of extraordinary $\delta^{18}O$ values and extremely high εNd values in Dabie-Sulu UHP rocks indicate that an open-system fluid phase was not present to promote re-equilibration of oxygen isotopes, and the temperature was not sufficiently high to reset Nd isotopes during the entire process of subduction and exhumation of supracrustal rocks (see Liou, Zhang, and Ernst, 1997 for other evidence for lack of a fluid phase during the Dabie-Sulu UHP metamorphism).

Extent of UHP Metamorphism in the Dabie-Sulu Terrane

Quartzofeldspathic gneisses are the predominant rock type in the Dabie-Sulu terrane, and they do not show clear UHP metamorphism. This lack of evidence for UHP metamorphism led to an early hypothesis that eclogites and their country rocks (gneisses, quartzite, and marble) were metamorphosed under different P-T regimes, and were subsequently juxtaposed by faulting (e.g., Smith, 1988). According to this hypothesis, the extent of UHP metamorphism was no greater than eclogite boudins and lenses now exposed, which usually measure no more than a few meters in length and thickness.

New observations, however, indicate that eclogites and country gneisses are not all fault bounded; many lithological contacts retained structural coherence throughout subduction, metamorphism, and exhumation. Mineralogical indicators of UHP metamorphism have now been found in a number of country rocks including gneisses, quartzites, and marbles. Tiny coesite inclusions have been reported: (1) in zircons from felsic gneisses (Sobolev et al., 1994; Tabata et al., 1998; Ye et al., 2000); (2) in dolomite and garnet from calc-silicate rocks and dolomite-bearing eclogite (Schertl and Okay, 1994; Zhang and Liou, 1996); and (3) in garnet and jadeite from jadeite-bearing quartzite (Zhang et al., 1995b; Liou, Zhang, and Jahn, 1997b). Detailed study of mineral compositions in quartzofeldspathic gneisses and UHP schists from Dabie (Okay, 1993; Carswell et al., 1997, 2000) shows that they were metamorphosed under P-T conditions similar to intercalated coesite eclogites and garnet peridotites.

Recent zircon ages and oxygen isotope data also support the hypothesis that eclogites and their country rocks were subjected to the same UHP metamorphism. Metamorphic overgrowths on zircons from eclogites and country-rock gneisses and schists give virtually identical U/Pb Triassic ages (Rowley et al., 1997; Hacker et al., 1998), demonstrating that both were metamorphosed at the same time. The presence of anomalously low $\delta^{18}O$ in both coesite eclogites and their country rocks suggests that both rock types remained in contact throughout subduction and both experienced UHP metamorphism (Yui et al., 1995; Zheng et al., 1996; Baker et al., 1997; Rumble, 1998; Rumble and Yui, 1998; Zheng et al., 1998). Mapping the distribution of low-$\delta^{18}O$ and low-deuterium rocks near Qinglongshan shows that a complete geothermal system measuring tens of km in outcrop length remained intact despite subduction, UHP metamorphism, and exhumation (Rumble and Yui, 1998).

In conclusion, field and petrochemical data demonstrate that the country rocks of the UHP eclogites and garnet peridotites also recorded UHP metamorphism (e.g., Carswell et al., 1997; Zhang and Liou, 1998b; Tabata et al., 1998; Carswell and Zhang, 1999; Ye et al., 2000). Consequently, a regional UHP metamorphic terrane has been proposed for the Dabie-Sulu orogen (Wang and Liou, 1991, 1993; Zhang et al., 1995b; Zhang, Hirajima et al., 1995c; Zhang and Liou, 1998b).

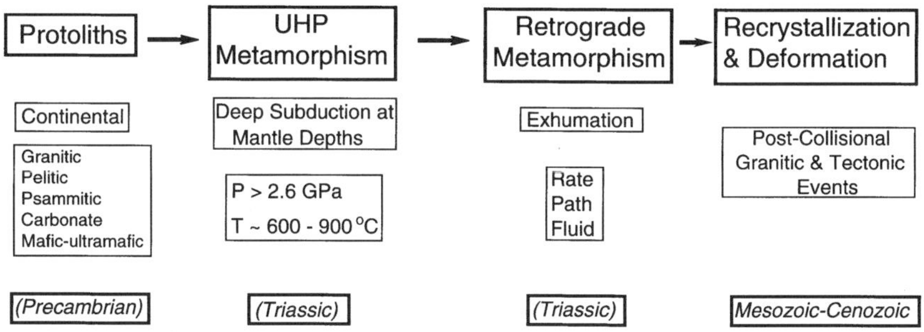

FIG. 2. Flow chart showing the process of formation and preservation of Dabie-Sulu UHP assemblages (after Liou et al., 1996).

Occurrence and Petrologic Characteristics of UHP Metamorphic Rocks

UHP rocks including eclogite, garnet peridotite, and some metasediments in the Dabie-Sulu terrane have been collected from outcrops and quarries, as well as drilled holes for rutile exploration as deep as 600 m. Figure 2 summarizes the processes, timing, nature of the protoliths, metamorphism, and recrystallization of the Dabie-Sulu UHP rocks. Precambrian protoliths include granitic, pelitic, psammitic, carbonate, and minor mafic-ultramafic rocks, presumably of Yangtze affinity. These rocks were subjected to UHP metamorphism at mantle depths prior to and during the Triassic collision between the Yangtze and Sino-Korean cratons (e.g., Li et al., 1993, 1994, 1999). Exhumation and preservation of certain UHP assemblages depend on several factors, including rate and path of returned flow, fluid infiltration, intensities of post-collisional deformation, recrystallization, and migmatization (Rubie, 1986; Mosenfield and Bohlen, 1997; Ernst and Liou, 1999). General petrologic characteristics of these UHP rocks are summarized below.

Eclogites

Dabie-Sulu UHP eclogites occur as blocks, lenses, layers, or discontinued boudins of various dimensions in gneisses (I), ultramafic rocks (II), and marbles (III). Individual bodies range from <1m to >1 km in size. Recognized localities of coesite-bearing eclogite are shown in Figure 1.

Eclogites in granitic gneisses. This type of eclogite is most abundant in Dabie, Sulu, and Hongan; inclusions of coesite and quartz pseudomorphs after coesite are common in garnet, omphacite, kyanite, zoisite, and epidote (Okay et al., 1989; Wang et al., 1989; Hirajima et al., 1990; Wang et al., 1993; Zhang and Liou, 1994, 1996, 1997; Zhang et al., 1995a, 1995b; Zhang, Hirajima et al., 1995). Most host minerals of coesite inclusions show characteristic radiating fractures; depending on the extent of retrogression, some coesite grains are rimmed by a thin palisade layer of quartz aggregates, whereas others are totally replaced by a mosaic of coarser-grained quartz aggregates. In addition to garnet, omphacite/jadeite, coesite/quartz, kyanite, apatite, and rutile, minor hydrous phases including phengite, epidote/zoisite, talc, and nyböite also occur as UHP minerals (Liou et al., 1995). A few eclogites contain magnesite and dolomite (Zhang and Liou, 1994); microdiamond has been reported as inclusions in Dabie eclogites (Xu et al., 1992; Okay, 1994), but its occurrence has not been confirmed by others.

A unique subrounded UHP slab 30 m in diameter, showing progressive mineral transformation from gabbro to coesite eclogite, was found at the Yangkou Beach locality, Sulu region (Zhang and Liou, 1997). A relict gabbro assemblage of Pl + Aug + Opx + Qtz ± Bt + Ilm/Ti-Mag (mineral abbreviations after Kretz, 1983) and ophitic textures are preserved in the core of the eclogitic block, whereas coesite eclogite occurs along the block margins; in between, rocks with transitional assemblages occur. The preservation of relict igneous minerals and textures in gabbros, and low-P assemblages in transitional rocks, and the occurrence of intergranular coesite in eclogite indicates sluggish reactions resulting from limited fluid infiltration during prograde and retrograde P-T evolution, and rapid exhumation (Liou and Zhang, 1996). The adjacent metagranitic block also shows a similar relict igneous assemblage in the core and a coesite-bearing assemblage in the block margins (e.g., Wallis et al., 1997).

Eclogites in garnet peridotite. Most eclogites associated with ultramafic rocks occur in the Sulu region. They are basically bimineralic, consisting of garnet and omphacite, with minor rutile. A few eclogites contain additional coesite/quartz, kyanite, and phengite. Round to oval-shaped inclusions of polycrystalline quartz aggregates after coesite occur in garnet and omphacite with well-developed radial fractures in eclogites from Chijiadian (Rongcheng), Xugou (Donghai), and Houshuichegou (Rizhao). Minute K-feldspar inclusions were identified in eclogitic garnet (Zhang et al., 1994; Yang et al., 1998). Rare corundum-bearing garnetite and eclogite occur as lenses in the Zhimafang garnet lherzolite (Enami and Zang, 1988).

Eclogites in marble. Eclogites occur mainly as small boudins and have sharp contacts with the marble; some are concordant layers within banded marble and calc-silicate rocks (e.g., Omori et al., 1998). A representative assemblage is Grt + Omp + Rt ± Zo ± high-Al Ttn. Some eclogites in marble from Dabie contain magnesite and dolomite. Coesite inclusions are present not only in garnet and omphacite, but also in dolomite (Schertl and Okay, 1994; Zhang and Liou, 1996).

Regardless of their occurrence, these three types of eclogite exhibit various degrees of retrograde metamorphism under amphibolite- to granulite-facies conditions. Thin, dark retrograde bands commonly developed at the contacts between the eclogite and its host rock. One typical example of granulite-facies overprinting is the Weihai coesite eclogite, which shows corona replacement of garnet by Opx, Cpx, and Pl, and intergrowths of diopside and plagioclase after omphacite (e.g., Wang et al., 1993; Zhang et al., 1995b). In contrast, amphibolite facies–overprinted eclogites show symplectic intergrowths of plagioclase + amphibole after omphacite and tarametic amphibole after garnet (e.g., Wang et al., 1995). Moreover, exsolution textures are common in some UHP minerals from various eclogites; these include quartz rods in omphacite, rutile lamellae in garnet, and lamellae of an unknown phase in apatite.

Garnet peridotites

Dabie-Sulu garnet peridotites, including garnet-bearing lherzolite, harzburgite, wehrlite, websterite, and pyroxenites, occur as blocks meters up to 1 kilometer in size, as well as thin layers of mafic-ultramafic complexes within quartzofeldspathic gneisses. They contain eclogite lenses (or nodules) and layers with minor inclusions of coesite and coesite pseudomorphs in garnet and omphacite. The garnet peridotites are partially to completely serpentinized. The mineral constituents of garnet peridotites are: olivine, enstatite, diopside and garnet, minor phlogopite, amphibole, Ti-clinohumite (Ti-Chu), chlorite, magnesite, and Fe-Ti oxides.

Two distinct types of garnet peridotite were identified based on mode of occurrence and petrochemical characteristics in the Dabie-Sulu UHP terrane (Zhang and Liou, 1998a, 1998b; Zhang et al., 2000). Type A mantle-derived peridotites originated from: (1) the mantle wedge above a subduction zone; (2) the footwall mantle of the subducted slab; or (3) represent ancient mantle fragments emplaced at crustal depths. Type B crustal peridotite and pyroxenite as parts of mafic-ultramafic layered intrusions were emplaced earlier into the continental crust prior to subduction. Both types of garnet peridotites have been subjected to UHP subduction-zone metamorphism under extremely low geothermal gradients (Zhang et al., 2000). Mineral exsolution textures are common; this includes ilmenite rods and magnetite lamellae in olivine, Mg-Al-Cr titanomagnetite and/or ilmenite and garnet rods in clinopyroxene, and rutile lamellae in garnet. Although the mechanisms for exsolution in UHP minerals are poorly known, some exsolution microstructures may have resulted from nearly isothermal decomposition of very high pressure precursor phases at mantle depths (Zhang and Liou, 1999).

Metasedimentary rocks

Metasedimentary rocks that carry a UHP metamorphic signature include kyanite quartzite, jadeite quartzite, pelite, paragneiss, and marble. These metasediments occur as interlayers with eclogite or as the country rock enclosing eclogite. Kyanite quartzite consists of quartz (>80 vol%), kyanite, pyrite, rutile, and topaz (Zhang and Liou, 1999); other quartzites interlayered with eclogite consist of quartz (>60–70 vol%) and kyanite, subordinate zoisite ± phengite ± omphacite ± epidote, and minor garnet and rutile (Zhang, Hirajima et al., 1995). Coesite-bearing jadeite quartzite from Shuanghe occurs as intercalated layers with marble and mafic eclogite, and consists of 35 to 40 vol% jadeite, 45 to 60% quartz, 5% garnet, and a trace of rutile (Liou et al., 1997b). Inclusions of coesite and of quartz pseudomorphs after coesite are present in garnet,

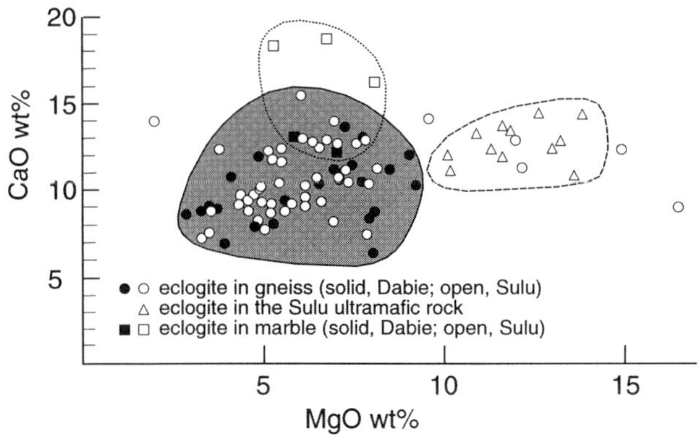

FIG. 3. Bulk-rock MgO versus CaO for eclogites from the Dabie-Sulu UHP terrane (modified after Zhang and Liou, 1998b).

jadeite, epidote, and kyanite in jadeite quartzite and in kyanite quartzite.

Eclogite-bearing UHP marbles from Xinyan in the Dabie Mountains (Omori et al., 1998) and Kongjiadian in the Sulu region (Ogasawara et al., 1998) have been particularly well studied. At Xinyan, marble beds are intercalated with calc-silicate layers; both contain coesite eclogite boudins. Pure marble is composed of dolomite, calcite, and minor graphite. Impure marble consists mainly of dolomite; silicate minerals (omphacitic clinopyroxene, phengite, and garnet) occur as discrete grains in the dolomite matrix. Inclusions of magnesite, and calcite pseudomorphs after aragonite, are found in dolomite and omphacite, respectively (Omori et al., 1998). Quartz pseudomorphs after coesite were identified as inclusions in garnet from marble and as intergranular crystals in the dolomite matrix (Zhang and Liou, 1996). Additional evidence of UHP metamorphism for the Dabie marble came from Wumiao, where inclusions of coesite are present in dolomite crystals of calc-silicate rocks (Schertl and Okay, 1994), and calcite pseudomorphs after aragonite in garnet (Wang and Liou, 1993). The Kongjiadian marble from Sulu exhibits several parageneses: Mg-Cal + Ti-clinohumite ± Dol ± Di; Mg-Cal + Dol ± Ol ± Di; and Mg-Cal + Dol + Ol + Di + Ti-clinohumite (Ogasawara et al., 1998). Although inclusions of quartz pseudomorphs after coesite have not been found in the marble, P-T estimates of 610 to 660°C and 2.5 to 3.5 GPa for X_{CO2} = 0.001 using phase equilibria and thermodynamic calculations suggest that the Kongjiadian marble also experienced UHP metamorphism (Kato et al., 1997; Ogasawara et al., 1998).

Geochemical Characteristics of UHP Rocks

Eclogites

Major, trace, and rare-earth elements. Many representative eclogites (>100 samples) from Dabie-Sulu have been analyzed for major, trace, and rare-earth elements (e.g., Zhang, 1992; Zhang et al., 1994; Zhang and Liou, 1998b; Jahn, 1998). The results indicate a wide range of bulk compositions. Such chemical variations may be attributed to the heterogeneity of the protoliths and possible modification of chemical compositions before, during, and after UHP metamorphism (Jahn, 1998). The SiO_2 contents of eclogites range from 40 to 60 wt%, but most are 45 to 50 wt%. The protoliths of most eclogites in both gneisses and ultramafic rocks are tholeiitic basalts based on the conventional AFM [(Na_2O+K_2O)-(FeOtotal)-(MgO)] diagram and the classification of Irvine and Baragar (1971) (Zhang, 1992; Zhai and Cong, 1996). However, such a classification must be used with great care because the alkali contents could have been subjected to metamorphic remobilization. Eclogites in marble could have been produced from marls or calcareous silicate rocks (Jahn, 1998, 1999) or from basaltic rocks contaminated with calcic sediments (Zhang, 1992). Although no clear distinction is apparent in the plots of SiO_2 versus TiO_2, Al_2O_3, FeO, MgO, CaO, Na_2O, K_2O, and P_2O_5 (Fig. 5 in Jahn, 1998), a significant difference is shown in the MgO versus CaO plot (Fig. 3). Eclogites in gneiss have lower MgO and CaO compared with those in ultramafic rocks and in marble, respectively; the MgO content of eclogites in gneiss is less than 10 wt%, whereas that of eclogites in ultramafic rocks is > 10 wt%.

Representative trace-element abundances of eclogites exhibit significant variation. The eclogites in gneiss have higher Ti concentrations than the two other types of eclogite; the eclogites in ultramafic rocks contain much higher Cr and Ni than the others. Eclogites also have a wide range in the Ti/V, Cr/Ni, and Ba/Nb ratios. The Ti/V ratios range from 5 to 293, covering all basaltic fields; most have Ti/V ratios of 10 to 50, suggesting an affinity with island-arc tholeiite, mid-ocean basalt (MORB), and back-arc basin basalt (Shervais et al., 1982). However, in view of other trace-element constraints, Jahn (1998) concluded that typical MORB has not been identified as protolith of the Dabie-Sulu eclogites studied so far. The Ba/Nb ratios vary from 2 to 93 for Sulu eclogites and from 0.7 to 683 for Dabie eclogites; most Ba/Nb ratios are >20, which lie outside the fields for most MORB and ocean-island basalts (OIB) (≤20).

In summary, trace-element abundance and Ti/V, Cr/Ni, and Ba/Nb ratios of analyzed eclogites span the entire field for basaltic bulk-rock compositions, and exceed the total range of melt compositions produced by magmatic differentiation alone. This suggests that multiple sources of protoliths are more probable, and metasomatism has imprinted some eclogites (Jahn, 1998). Consequently, it is difficult to apply trace-element abundances and Ti/V, Cr/Ni, and Ba/Nb ratios to identify the initial tectonic setting of these eclogites.

Eclogites in gneisses exhibit large variations in REE abundances, but they are generally LREE-enriched, as shown in Figure 4. These patterns are similar to those commonly observed in Precambrian continental basalts and amphibolites (Hall and Hughes, 1990). Eclogites from Sulu, such as Datan (Rongcheng), Yakou (Qiandao), and Majiakou (Rizhao), show convex patterns and positive Eu anomalies but depletion of La related to Ce; this may result from the presence of plagioclase and clinopyroxene-enriched cumulates. One sample from Maobei (Donghai) shows a LREE-depleted MORB-type pattern. The geochemical characteristics indicate that most eclogite protoliths in gneiss are continental basalts (including gabbro and tuff) rather than ocean-floor basalts, and form a minor component of the ancient continental crust (Zhang, 1992; Jahn, 1998; Zhang and Liou, 1998b).

The REE patterns for eclogites associated with ultramafic rocks have even a greater variability. They show highly fractionated patterns with LREE enrichment to nearly flat REE patterns. A few exhibit slight HREE depletion (Zhang et al., 1994). These features probably are related to: (1) the variable sources of the protoliths; and (2) biased sampling of coarse-grained eclogites. The latter warrants an explanation. In many cases, eclogites are coarsely crystallized and show clear metamorphic banding (e.g., Qinglongshan). It is very difficult in this case to obtain a truly representative whole-rock sample from the hand specimen. Consequently, some of the REE variability could simply result from metamorphic banding and biased sampling. Besides, a few eclogites showing slightly convex REE patterns with small positive Eu anomalies may indicate their origin by accumulation of garnet and clinopyroxene + variable trapped melt (5 to 50%) in the upper mantle (Zhang et al., 1994; Zhang and Liou, 1998b).

O-isotopic characteristics. Reported $\delta^{18}O$ values for eclogites worldwide range from +1.5 to +12.0‰ (see summary by Yui et al., 1995). Analyses of $\delta^{18}O$ values for mineral separates from Dabie-Sulu eclogites within gneiss yielded a very significant discovery. Garnet, omphacite, phengite, kyanite, quartz, and rutile from Qinglongshan eclogites and interlayered Grt/omp–bearing kyanite quartzite in Donghai possess world-record negative $\delta^{18}O$ values—such as –4 to –11‰ for garnet, –2.2 to –11.2‰ for omphacite, –1.2 to –10.7‰ for phengite, –2.1 to –11.1‰ for epidote, 0.8 to –7.7‰ for quartz, and –5.3 to –14.8 ‰ for rutile (Yui et al., 1995, 1997; Zheng et al., 1996, 1999; Rumble and Yui, 1998). In addition, Qinglongshan rocks preserve pre-metamorphic differences in $\delta^{18}O$ at the outcrop scale, suggesting that closely spaced rocks have not achieved isotopic re-equilibration during the entire metamorphic process (Rumble and Yui, 1998). Together with low δD values (from –101 to –127‰), these characteristics are interpreted as follows: (1) The protoliths were once at or near Earth's surface and were subjected to alteration in a Neoproterozoic geothermal area charged with meteoric water from high paleolatitude or high paleoaltitude (Yui et al., 1995, Zheng et al., 1996; Rumble and Yui, 1998); a similar conclusion was obtained from low $\delta^{18}O$ values of minerals from eclogites and their country rock gneiss in the Dabie Mountains (Baker et al., 1997). (2) The persistence of pre-metamorphic differences in oxygen isotopic composition between different rocks and the preservation of high-temperature isotope fractionation between UHP minerals indicate the lack of flowing fluid to mediate isotope exchange between rocks during UHP and retrograde metamor-

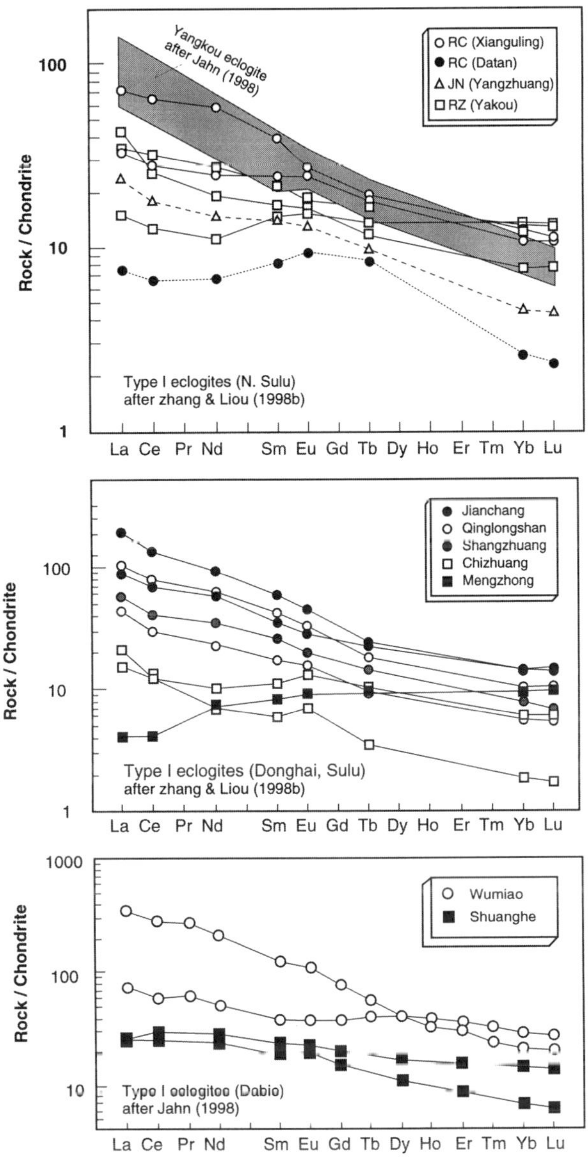

FIG. 4. REE distribution patterns of type I eclogites (in gneiss) from the Dabie-Sulu UHP terrane; they are characterized by LREE enrichment. The chondrite-normalized REE patterns of eclogites from Dabie and Sulu are modified after Jahn (1998) and Zhang and Liou (1998b), respectively. Chondrite values are from Masuda et al. (1973) for the Dabie eclogite and from Nakamura (1974) for the Sulu eclogite.

phism (Rumble and Yui, 1998). In other words, the environment for subduction of cold, dry, old supracrustal rocks constitutes a typical example of the fluid-deficient region of metamorphism (Yoder, 1955). However, other recent studies indicated that the O-isotope disequilibrium between eclogitic minerals may have been caused by retrograde hydration (Yui et al., 1997; Zheng et al., 1999) and the difference in isotopic composition of eclogite protoliths in various areas was not only controlled by different degrees of water-rock interaction before metamorphism, but also by the channelized flow of fluids during progressive metamorphism resulting from rapid plate subduction (Zheng et al., 1999).

Nd-Sr isotope compositions. The Rb-Sr isotopic compositions of eclogites in gneisses and marbles from Dabie-Sulu are highly dispersed in a conventional isochron diagram, suggesting that many Rb-Sr systems have been disturbed (Jahn, 1998). A significant number of data show relatively high $^{87}Sr/^{86}Sr$ (0.705 to 0.710) but very low $^{87}Rb/^{86}Sr$ (≤ 0.02) ratios. These high $^{87}Sr/^{86}Sr$ ratios are unsupported by their Rb/Sr ratios, indicating preferential Rb depletion. Because of the highly mobile nature of Rb relative to Sr, no unique interpretation is possible for the scatter of these data.

Eclogites and associated ultramafic rocks of the Bixiling massif (Dabie) and garnet pyroxenites of the Rizhao area (Sulu) have the lowest $^{87}Sr/^{86}Sr$ ratios (0.7036 to 0.7042). Although the intrusive age of the Bixiling layered massif is not precisely known, the very low Rb/Sr ratios readily suggest that the initial $^{87}Sr/^{86}Sr$ ratio of their formation must be close to 0.7036. Similarly, for the garnet pyroxenites of Rizhao, the initial $^{87}Sr/^{86}Sr$ ratio could be estimated at 0.7032 to 0.7036, regardless of their ages. Because the Bixiling layered massif represents a mantle-derived melt-cumulate association, and the Rizhao garnet pyroxenites are probably of mantle origin in view of their intimate association with serpentinite and harzburgite, the $^{87}Sr/^{86}Sr$ ratio of about 0.7036 may be regarded as the Sr isotopic composition of the subcontinental lithosphere beneath the Dabie Mountains and the Sulu region. Other eclogites in association with ultramafic rocks showing unsupported high $^{87}Sr/^{86}Sr$ ratios, such as those of the Maowu layered complex, the Junan eclogite, and the Rongcheng (Chijiadian) peridotite, could have been produced by crustal contamination.

REE are generally considered to be less mobile than Rb and Sr in metamorphic processes. Thus, Sm-Nd isotope tracers can provide more accurate information about the nature of the eclogite protoliths (Fig. 5). Eclogites from the European Hercynian and Alpine chains and Pan-African orogenic belts are characterized by positive εNd(T) values, suggesting an origin from subducted oceanic crust (Jahn, 1998, 1999). By contrast, eclogites from the Dabie-Sulu terranes have unusually low εNd(T) values (–6 to –20) at the time of peak metamorphism or continental collision (~220 Ma). These values are among the lowest known for eclogites, and are significantly different from those of the Hercynian and Alpine chains in western Europe. To some extent, only eclogites from the Caledonian belt (e.g., northwestern Scotland and western Norway) have Nd isotopic compositions similar to those of the Dabie orogen. The isotopic data and geochemical characteristics clearly indicate that the eclogite protoliths of the Dabie orogen have "continental" affinities, and cannot represent a subducted Triassic Tethyan oceanic slab. The protoliths are most likely the mafic components (e.g., amphibolites) of Precambrian continental crust, and their very low εNd(T) values require that they formed long before the Triassic collision. It appears that, in addition to the UHP metasedimentary rocks of the Dora-Maira, the Dabieshan, and the Western Gneiss Region are two regions where coesite- and diamond-bearing eclogites and ultramafic rocks show clear evidence of subduction of ancient, cold continental crustal blocks to mantle depths.

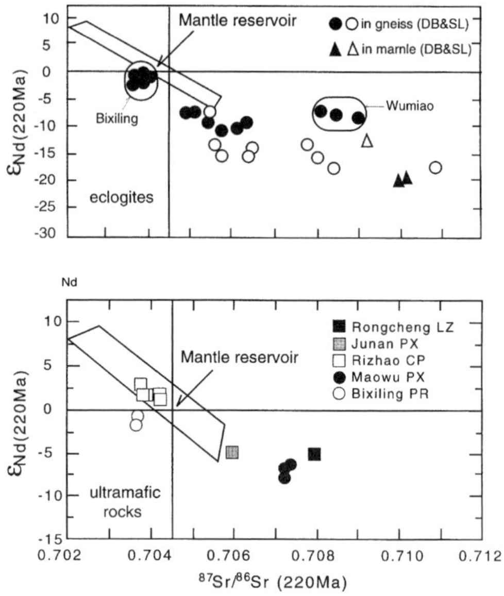

FIG. 5. εNd(220 Ma) versus $^{87}Sr/^{86}Sr$ (220 Ma) for eclogites (upper) and garnet peridotites from the Dabie-Sulu terrane. Time (t) is assumed to be 220 Ma (UHP metamorphism age), but does not represent the emplacement time of the protoliths (modified after Jahn, 1998). Solid and open circles and triangles for Dabie (DB) and Sulu (SL) eclogites. Abbreviations: CP = clinopyroxenite; LZ = lherzolite; PX = pyroxenite, PR = peridotite.

Eclogites in association with ultramafic rocks also show variable Nd isotopic compositions, but they are generally similar to the associated ultramafic rocks (Fig. 5). Some of them are close to the mantle isotopic value, as exemplified by garnet clinopyroxenites from the Rizhao area. Such a feature is consistent with their being mantle xenoliths of tectonic origin. Both the Sr and Nd isotopic compositions may be characteristic of the lithospheric mantle beneath the Dabie-Sulu terrane. The Bixiling eclogites and ultramafic rocks (garnet peridotites and garnet pyroxenites) have a narrow range of εNd(T) values (–0.1 to –2.5) and $^{147}Sm/^{144}Nd$ ratios (0.16 to 0.19). In a εNd versus $^{87}Sr/^{86}Sr$ diagram, these rocks lie just beneath the "mantle array," which suggests involvement of depleted lower-crustal material in the petrogenesis and evolution of this layered complex (Chavagnac and Jahn, 1996).

In the Sulu region, eclogite and associated garnet lherzolite from Rongcheng, and eclogite and associated garnet pyroxenite from Junan, have εNd(T) values of about –5 (Jahn, 1998). In the Dabie terrane, eclogite and associated ultramafic rocks from Maowu have εNd(220 Ma) values of –5 to –7 and "initial" $^{87}Sr/^{86}Sr$ (220 Ma) = 0.707 to 0.708 (Jahn, 1998 and unpubl.). These rocks were probably derived from long-term enriched mantle segments, or they have been subjected to crustal contamination. If the latter, it is still not clear whether such contamination was effected in the mantle, during magma differentiation (Maowu), or during tectonic transport of mantle slices.

Eclogite samples from Weihai show the most spectacular Nd isotopic compositions. Their protoliths were severely metasomatized prior to UHP metamorphism, and they have preserved extremely high εNd values (+260 to +174) and unusual Sm/Nd ratios ($f_{Sm/Nd}$ = +6.1 and +4.6), even though they have undergone UHP metamorphism and later granulite-facies retrogression. The Sm-Nd isotopic compositions allowed us to calculate precise TDM ages of 1.67 and 1.61 Ga, which constrain the minimum time for protolith formation to the middle Proterozoic (~1.7 Ga; Jahn et al., 1996). The metasomatism was not related to the UHP metamorphism, because the small time interval of ~220 Ma is not sufficient to develop such large isotopic growth (270 εNd units).

Ultramafic rocks

Major, trace, and rare-earth elements. Geochemical characteristics of garnet peridotites and pyroxenites have been extensively investigated (Zhang et al., 1994, 2000; Jahn, 1998). Type A garnet lherzolites from Sulu have lower "fertile-element" concentrations than those of the primitive mantle defined by Ringwood (1975), and show negative correlation for MgO versus TiO_2, Al_2O_3, and CaO. These features suggest an origin as depleted residual mantle rocks. In contrast, type B peridotites and pyroxenite from Dabie show large variation in MgO content and "fertile-element" concentrations as a result of variations in mineral proportions that are attributed to crystal cumulation. Type A peridotites have higher MgO (25 to 46 wt%) but lower "fertile-element" contents than type B pyroxenites (14 to 36 wt%); a negative correlation exists between MgO versus CaO and TiO_2 contents for pyroxenites (Zhang et al., 2000).

Trace-element versus MgO plots for type A peridotites show a negative correlation for V and Sc and a random Cr and Ni distribution. These relationships are consistent with peridotites derived from a depleted mantle with variation in concentrations of major and some trace elements chiefly controlled by extraction of partial melts, and to some extent by metasomatism or by reaction with melts. For example, a few lherzolites from Rongcheng have higher FeO than primitive mantle, suggesting that they may represent slightly depleted mantle that was refertilized by interaction with basaltic melt (Zhang et al., 2000).

REE contents of type A garnet lherzolites are low; some lherzolites show slight negative Eu anomalies. Two Rongcheng lherzolites and a Donghai lherzolite possess flat REE patterns with REE abundances less than the primitive mantle values of Sun and McDonough (1989), but more than or close to the chondrite values defined by Nakamura (1974), similar to primitive mantle patterns (Fig. 6). Most other peridotites have slightly to moderately fractionated REE patterns with $(La/Yb)_N$ ratios of 2 to 15; a few lherzolites show elevated middle REE values. Some garnet-free peridotites with U-shaped patterns reflect a complicated evolution that requires mantle enrichment and crustal contamination (Song and Frey, 1989; Gruau et al., 1998). The Rizhao pyroxenites (type A) show convex REE patterns (Li et al., 1993; Jahn, 1998), which may be interpreted as a result of clinopyroxene cumulation during differentiation. In fact, some pyroxenites consist dominantly of diopside (>93 vol%), with minor metamorphic garnet as irregular interstitial grains (Zhang et al., 2000).

The REE patterns of type B peridotite and pyroxenite are shown in Figure 6. The Bixiling peridotites have low REE contents and show unusual REE patterns with Eu positive anomalies. These patterns were interpreted to represent more advanced crystal accumulation, or as a residual solid after the separation of liquid (Chavagnac and Jahn, 1996). Most

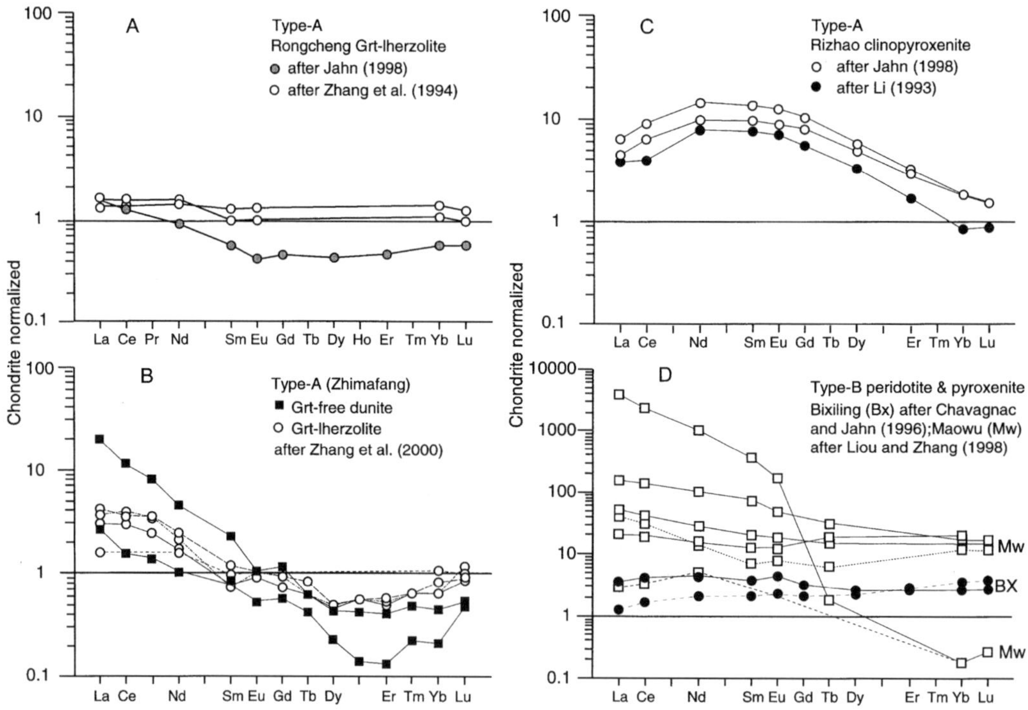

FIG. 6. Chondrite-normalized (Nakamura, 1974) REE patterns for the Dabie-Sulu UHP garnet peridotites (after Zhang et al., 2000).

Maowu pyroxenites are characterized by extremely high total REE contents and LREE enrichment with two distinct REE patterns. Monazite-bearing clinopyroxenites possess an extremely to moderately fractionated and LREE-enriched REE patterns, whereas orthopyroxenites have concave REE patterns (Liou and Zhang, 1998). Most samples do not show Eu positive anomalies except for one orthopyroxenite; the latter may be attributed to LREE metasomatism.

O-isotopic and Nd-Sr isotopic characteristics. Type A and type B garnet peridotites exhibit significantly different $\delta^{18}O$ values (Zhang et al., 1998, 2000). Type A garnet lherzolites and pyroxenites preserve mantle isotopic values; $\delta^{18}O$ values range from +4.8 to +5.6, +4.7, and +4.5 to 5.6 ‰ for garnet, olivine, and clinopyroxene, respectively. These $\delta^{18}O$ values are very close to the range (+5.1 to +6.6‰) of mantle-derived ultramafic rocks (Taylor, 1967). On the other hand, $\delta^{18}O$ values of garnet, pyroxene, and olivine of type B peridotites and pyroxenites from Maowu and Bixiling range from +2.9 to +4.5, +3.0 to +4.3, and +3.1 to +3.8‰, respectively. These values are lower than those of mantle rocks and may have resulted from rock/water interaction at crustal levels prior to Triassic UHP metamorphism (Zhang et al., 1998, 2000).

Oxygen isotope fractionation between garnet and clinopyroxene for both type A and type B ultramafic rocks, $\Delta^{18}O_{(Grt-Cpx)}$ values are small (<–0.16‰), suggesting oxygen isotopic equilibrium for most coexisting pyroxene and garnet. Some high fractionation values of about –1.0‰ for olivine-garnet and diopside-garnet pairs may result from retrogression of olivine and diopside involving O-isotopic exchange during exhumation.

The limited Rb-Sr and Sm-Nd isotopic data of whole rocks and minerals for garnet peridotites and pyroxenites from the Dabie-Sulu terrane (Chavagnac and Jahn, 1996; Jahn, 1998; Yang and Jahn, 2000) have been used to construct the history of these eclogites and garnet peridotites. Type A lherzolites from the Rongcheng (one sample) and Donghai (2 samples) bodies show unusually high initial $^{87}Sr/^{86}Sr$ ratios of ~0.708 and low εNd(220) values of ca. –5.0. In contrast, garnet clinopyroxenites from the Type A Rizhao body have the lowest $^{87}Sr/^{86}Sr$ (0.7038 to 0.7044) and the highest εNd (220) values

(+1.2 to +1.6) among all analyzed peridotites and pyroxenites. In a plot of ε_{Nd} (220 Ma) versus I_{Sr} (220 Ma) (Fig. 5), Rizhao clinopyroxenites fall within mantle values and may have crystallized in the upper mantle, consistent with our interpretation of the Rizhao pyroxenite. However, the Rongcheng and Donghai lherzolites (type A) show crustal values, indicating that they have undergone crustal contamination (e.g., Jahn, 1998; Yang and Jahn, 2000). The Rongcheng lherzolite has a low Sm/Nd ratio (0.2), a low εNd (220 Ma) value (about –5), and a high $^{87}Sr/^{86}Sr$ ratio (0.708), suggesting derivation from a long-term enriched mantle source or subjection to crustal contamination (Jahn, 1998). In the second case, there are two possible interpretations. (1) The lherzolite was derived from the mantle wedge above a subduction zone and was metasomatized by crustal fluids derived from the descending slab; the metasomatic process could have occurred in the mantle wedge, or within the subducting slab after the peridotite fragment was entrained in it. Or (2) the lherzolite was derived from an ancient mantle intruded into the crust long before the Triassic, experienced a long resident history in continental crust, and was contaminated by supracrustal materials. These modified mantle rocks were then subducted together with continental crust and subjected to UHP metamorphism (Zhang et al., 2000). The Zhimafang lherzolite in the Donghai area has been proposed to be an older mantle-derived slice that was emplaced into a low P-T environment prior to Triassic subduction (Yang and Jahn, 2000).

Type B peridotites and pyroxenites from Dabie also have a wide range of $^{87}Sr/^{86}Sr$ and $^{143}Nd/^{144}Nd$ ratios (Fig. 5). Two Bixiling peridotites have $^{87}Sr/^{86}Sr$ values of ~0.7037 and εNd(220) values of –1.8 and –0.7, but three Maowu garnet pyroxenites have high $^{87}Sr/^{86}Sr$ ratios of ~0.707, and low εNd(220) of –6.2 to –7.9 (Jahn, 1998). Detailed petrological and geochemical studies (Zhang et al., 1995a, 1998; Chavagnac and Jahn, 1996; Rowley et al., 1997; Liou and Zhang, 1998) indicate that the Maowu and Bixiling mafic-ultramafic complexes were emplaced into continental crust prior to Triassic subduction and UHP metamorphism. It seems likely that both bodies would experience crustal contamination and/or metasomatism during or after intrusion, or perhaps during metamorphism. However, in the plot of εNd (220 Ma) versus I_{Sr} (220 Ma), the Bixiling complex lies near the mantle field, and clearly is less contaminated with lower $^{87}Sr/^{86}Sr$ and higher $^{143}Nd/^{144}Nd$ than the Maowu layered pyroxenite, which plots within the crustal field.

Metasediments

Geochemical and isotopic data of UHP metasediments from Dabie-Sulu are very limited, except for some oxygen isotope measurement on marbles (Zheng et al., 1999; Rumble et al., 2000), gneissic rocks, and interlayered kyanite quartzite (e.g., Rumble and Yui, 1998). The Shuanghe jadeite quartzite is the only rock that has been studied in some detail (Liou, Zhang, and Jahn, 1997). Geochemical features of the coesite-bearing jadeite quartzite are described below, inasmuch as this lithology is unusual in terms of high jadeite content and its rare occurrence.

Two major-element bulk-rock analyses of the jadeite quartzite show an overall silicic composition (~72% SiO_2), low CaO (1.4%), very high Na_2O (6.35%), and very low K_2O (0.14%) (Cong et al., 1995; Liou, Zhang, and Jahn, 1997); these data suggest that the protolith could be either a keratophyre or an albitized fine-grained sediment. The very low contents of K, Rb, Ba, and Sr, the REE patterns of the upper continental type, the extremely high Na_2O/K_2O ratio of 45 to 137, the high $^{87}Sr/^{86}Sr$ ratio of 0.7113 (at 220 Ma), and very low εNd (220) of –22, and the very old Sm-Nd model age (TDM) of 2.58 Ga for the analyzed jadeite quartzite suggest that the protolith was ancient Precambrian clastic sediments that underwent Na metasomatism prior to in-situ UHP metamorphism with the adjacent mafic, carbonate, and gneissic rocks (see Liou, Zhang, and Jahn, 1997 for details).

Some UHP marbles from Dabie-Sulu have $\delta^{18}O$ (SMOW) from +7 to +25 and $\delta^{13}C$ (PDB) from 0 to +6 $\delta^{13}C$ (VPDB) (Zheng et al., 1999; Rumble et al., 2000). High $\delta^{13}C$ values are representative of unmodified protoliths and support stratigraphic correlation with Neoproterozoic limestones enriched in ^{13}C. High oxygen isotope ratios reflect pristine protoliths, but the low values are caused by very localized equilibration on a centimeter scale with silicate minerals of the adjacent rocks. No evidence of regional isotopic transport during UHP metamorphism and no internal isotopic equilibration on scales of 1 to 100 m were found. Only samples separated by less than 10 cm show homogeneous oxygen and carbon isotope compositions. Deposition of protolith limestones took place in a cold climate, either preceding or following—but not coincident

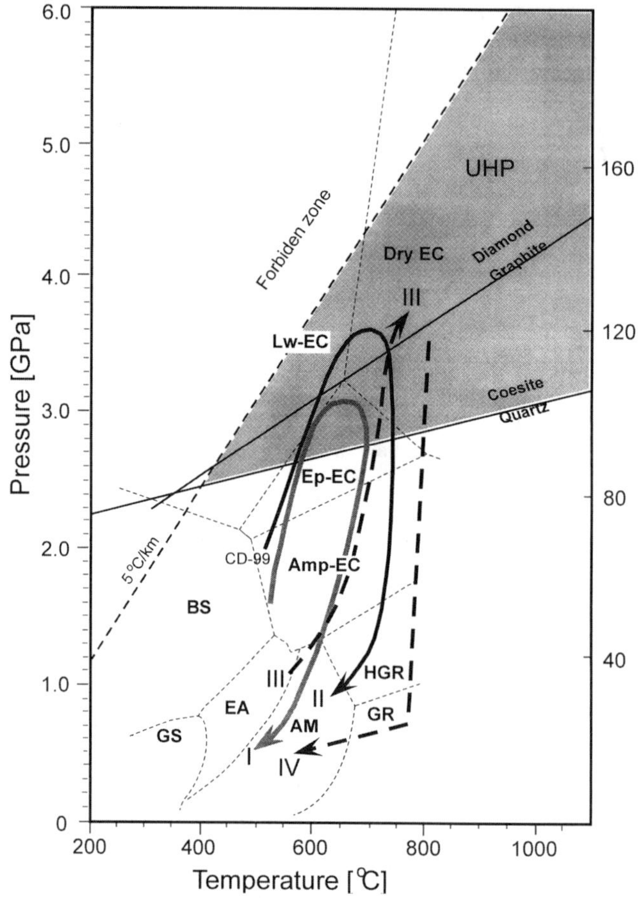

FIG. 7. Comparisons between previously published P-T paths for Dabie-Sulu UHP eclogites. I = for eastern Dabie (Zhang and Liou, 1994); II = for central Dabie (Carswell and Zhang, 1999); III = for Yangkou Beach, Sulu region (Zhang and Liou, 1997); IV = for Weihei, Sulu region (Zhang et al., 1995b); III and IV show only prograde and retrograde paths, respectively. The simplified P-T facies diagram is based on that of Okamoto and Maruyama (1999).

with—Neoproterozoic glaciation (for explanation, see Rumble et al., 2000).

Metamorphic Evolution and P-T Paths of UHP Rocks

Previous studies indicated that Dabie-Sulu UHP rocks experienced a complex evolution, with three or four stages of metamorphic recrystallization—pre-eclogite, peak (UHP) metamorphic, early decompression, and retrograde (amphibolite-facies) stages (Yang et al., 1993; Okay, 1994; Zhang et al., 1994, 1995a, 1995b; Zhang and Liou, 1994, 1997; Zhang, Hirajima et al., 1995; Wang et al., 1995; Carswell et al., 1996, 1997; Cong, 1996; Liou et al., 1996; Carswell and Zhang, 1999; Yang and Jahn, 2000). These P-T paths were delineated through detailed investigations of inclusion assemblages in UHP minerals, compositions of zoned minerals, appropriate mineral reaction equilibria, and microstructure + texture.

Eclogitic rocks

Eclogites in gneisses were selected here to discuss metamorphic evolution and deduced P-T paths (see Fig. 7). The pre-eclogite stage is mainly defined by relict mineral inclusions in garnet and prograde mineral zoning, with the assumption that the identified inclusions are representative of mineral parageneses that are stable at various stages during prolonged prograde garnet growth. The observed inclusions include discrete Phn, Kfs, Ep, Pg, Rt, Par, and Hbl, and multimineralic Bar-Pg-Zo-Mag and Grt-Zo-Hbl-Chl in garnets from Donghai eclogites (Zhang, Hirajima et al., 1995) and discrete Grt + katophorite/taramite + Qtz + Pg in the cores of garnets from the western Dabie eclogites (Zhang and Liou, 1994). These assemblages record the earliest stage of epidote amphibolite-facies metamorphism at ~500°C based on Fe-Mg partitioning between garnet and hornblende (Graham and Powell, 1984). Similarly, a well-documented record of the pre-eclogite stage was preserved in the Yangkou UHP slab; the initial metamorphism of the gabbroic protolith took place at 540 ± 50°C and ~1 GPa (Zhang and Liou, 1997). Preservation of metastable, low-P assemblages in garnet suggests that the UHP metamorphism may have a short duration, insufficient to upgrade completely the low-grade assemblage to the coesite-eclogite facies. For most localities such as that of the Weihai eclogite, however, only an exhumation P-T path from peak to retrograde stage has been deduced.

Peak-stage assemblages were identified by coexistence of garnet, jadeite/omphacite, and coesite together with minor kyanite, phengite, and some hydrous phases including epidote/zoisite, nyböite, talc, and carbonate (magnesite and dolomite). Peak temperatures were estimated based on Fe-Mg partitioning between coexisting garnet and clinopyroxene using the calibration of Powell (1985). They range from 650 to 850°C at 3 GPa. The peak metamorphic temperature appears to decrease roughly westward from Sulu through the Dabie to the Hongan block (Fig. 7). However, the Fe^{3+} estimate of omphacite strongly affects the T calculation, because of a strong negative correlation between the calculated Fe^{3+}/Fe^{total} ratio of omphacite and the calculated temperature from garnet-omphacite Fe/Mg partitioning (Carswell et al., 1997). Omphacite separates from two Dabie UHP eclogite samples contain roughly 50% of the total Fe as ferric, based on Mössbauer spectroscopy (P. O'Brien, pers. commun., 1998). In contrast, the coexisting garnets contain only around 6% Fe^{3+}. Using an assumed Fe^{3+}/Fe^{total} = 0.5 for omphacite and 0.06 for coexisting garnet, most Dabie eclogites yield garnet-omphacite temperatures just over 700°C (e.g., Fig. 7 of Carswell and Zhang, 1999). Most published P estimates for coesite eclogite gave minimum pressures based on the presence of coesite; other P-calculations of 3–4 GPa for some Dabie eclogites used the Grt-Omp-Phn barometer (Carswell et al., 1997), and/or geochemical calculations (Zhang and Liou, 1994).

The early decompression stage is identified by continuous growth of porphyroblastic barroisitic amphibole, which contains abundant inclusions of eclogitic minerals except for the absence of inclusions of coesite and quartz pseudomorphs after coesite. This stage occurred under eclogite-facies conditions at pressures below the coesite-quartz transition. The retrograde stage is characterized by granulite/amphibolite–facies overprinting. The coexisting neoblastic Grt-Cpx pairs of granulite-facies recrystallization yield a P-T estimate of ~760°C, 0.9 GPa for the Weihai eclogite (Zhang et al., 1995b). Amphibolite-facies overprints producing the Pl + Hbl assemblage occurred at 500 to 600°C, <1 GPa (Wang et al., 1995; Zhang et al., 1995a, 1995b, 1998; Zhang, Hirajima et al., 1995).

Several representative P-T paths, including the Sulu, Dabie, and two single localities—Weihai and Xinxian—are shown in Figure 7. Except for the Sulu block and Xinxian, the prograde segments are not well constrained. In many previous studies, the P-T-t paths for UHP eclogites were constructed with an assumption that P_{max} and T_{max} conditions were synchronously achieved. However, unless the uplift rate is extremely fast, deeply subducted rocks may well continue to experience heating during the initial stages of exhumation (England and Thompson, 1984). In that case, it may be anticipated that the maximum subduction depth (P_{max}) will be reached well before the rocks attain their T_{max} condition, and the latter may only be achieved at much shallower depths along the exhumation path (Carswell and Zhang, 1999). A good example, that of the P-T path deduced for Dabie UHP eclogite sample CD99 in Figure 7, indicates that maximum temperatures were probably attained at pressures well below the field of coesite stability and that a substantial part of the exhumation history was probably accomplished under near-isothermal conditions. If UHP eclogites customarily follow the P-T-time path of sample CD99, then there are important implications for the interpretation of radiometric age dates for UHP rocks. The closure of the U-Pb and Sm-Nd isotopic systems used to date the timing for formation of UHP minerals would be expected to occur some time after the attainment of maximum subduction depth (Carswell and Zhang, 1999); consequently, reconsideration and significant reduction in deduced exhumation rates may be required.

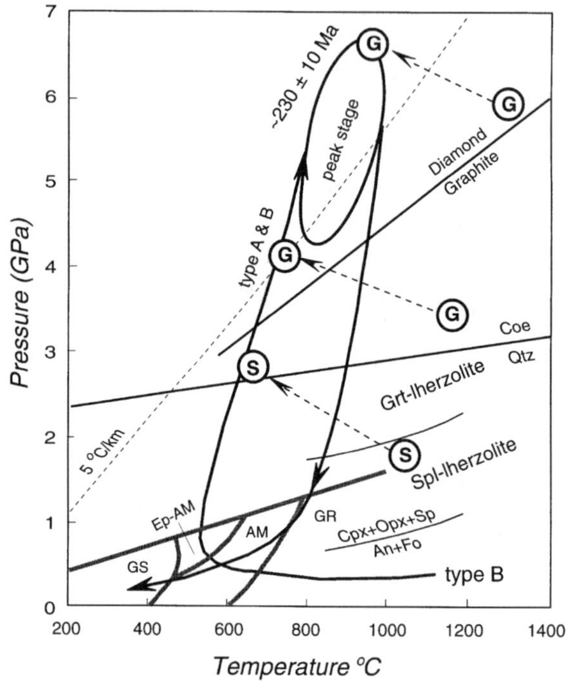

FIG. 8. P-T paths of type A and B garnet peridotites in the Dabie-Sulu UHP terrane, modified from Zhang et al. (2000). The arrows joining G-G and S-S show P-T paths for the proposed protoliths of garnet and spinel peridotites from the mantle wedge inserted into the subducted slab, respectively. Equilibrium lines for the transformations Fo + An (low-P) = Cpx + Opx + Spl (high-P) in the system CMAS (Obata, 1976), Grt- and Spl-lherzolite (Perkins et al., 1981), coesite-quartz (Bohlen and Boettcher, 1982), and diamond-graphite (Bundy, 1980) are shown.

Ultramafic rocks

P-T conditions of various metamorphic stages have been estimated for most peridotites from the Dabie-Sulu terrane (Zhang et al., 1994, 1995a, 2000; Hiramatsu and Hirajima, 1995; Hiramatsu et al., 1995; Liou and Zhang, 1998; Yang et al., 1993; Yang and Jahn, 2000). In this review, only those estimates for pre-UHP and peak UHP metamorphism are summarized below. The retrograde history is similar to that of UHP eclogites described above.

Type A peridotites. Protoliths of mantle-derived type A peridotite are proposed to be from different depths. The preservation of rare Cr-spinel inclusions in garnet from some Sulu lherzolites suggests that some protolith could have formed in the spinel-peridotite facies. However, garnet peridotites are probably the most important protolith for most type A peridotites from Sulu (Zhang et al., 2000).

The original assemblage and composition of pre-UHP minerals for garnet peridotites are seldom preserved. Some inclusion associations of Spl + Cpx ± Opx + Ilm ± Amp in porphyroblastic garnets from the Rizhao clinopyroxenite (type A) were suggested to be a possible protolith assemblage, recrystallized at a low-P condition in the spinel-peridotite facies <1.5 GPa and 900 to 1000°C (Zhang et al., 1994; Hiramatsu and Hirajima, 1995). Similarly coarse-grained garnets of some Donghai garnet peridotites contain amphibole + spinel inclusions; the ultramafics were suggested to have been emplaced in continental crust and subjected to low-P metamorphism prior to Triassic UHP recrystallization (Yang and Jahn, 2000). The lack of P-T constraints on the earliest history of the garnet peridotites has led to ambiguous assignments of their original tectonic setting.

P-T conditions of peak UHP metamorphism were estimated by a combination of the Grt-Opx barometers based on Al partitioning (Brey and Köhler, 1990) and the Fe-Mg exchange thermometers between Grt-Cpx (Powell, 1985) and/or Grt-Opx (Harley, 1984) as well as two-pyroxene thermometers (Brey and Köhler, 1990). Total Fe of garnet, diopside, and enstatite are expressed as FeO. As shown in Figure 8, metamorphic temperatures of the UHP stage range from 760 to 970°C, and pressures from 4.0 to 6.5 ± 0.2 GPa. Medaris (1999) recalculated P-T conditions for Dabie-Sulu garnet peridotites including type A and B from published compositions of coexisting garnet, orthopyroxene, and olivine (Yang et al., 1993; Zhang et al., 1994, 1995b; Liou and Zhang, 1998) using the olivine-garnet Fe-Mg exchange thermometer (O'Neill and Wood, 1979; O'Neill, 1980) and Al-in-Opx barometer. His results yield a similar pressure range, but slightly different temperatures.

Type B peridotites. The protoliths of type B garnet peridotite-pyroxenite-eclogite complexes were crystallized at high temperature (~1000°C) and low pressure (<1 GPa) at crustal levels. Prior to continental subduction, some of these rocks, such as the Maowu pyroxenite, were subjected to low P/T (730 ± 30°C, and 0.4 ± 0.2 GPa) metamorphism and crustal metasomatism, indicated by the inclusions of talc, amphibole, clinochlore, enstatite, gedrite, hornblende, mica, and corundum in garnet (Okay, 1994) and by unusual geochemical characteristics (Zhang et al., 1998; Liou and Zhang, 1998). The peak assemblages are Grt + Ol ± Di ± En ± Mgs ± Ti-clinohumite for peridotite and Grt ± Opx ± Cpx ± Mgs ± Ti-clinohumite for pyroxenite; peak pressures of 4.2 to 6.7 GPa and temperatures of 710 to 970°C were estimated for garnet peridotites and pyroxenites from the Bixiling and Maowu complexes (Okay, 1994; Liou and Zhang, 1998; Medaris, 1999) (Fig. 8).

Discussion

Triassic subduction of a cold continental slab of the Yangtze craton: Lowest geothermal gradient for the Dabie-Sulu UHP rocks

Figure 8 shows P-T estimates for garnet peridotites from the Dabie-Sulu terrane, and for several other UHP terranes compiled by Medaris (1999). Among all reported Eurasian garnet peridotites, those in the Dabie-Sulu UHP terrane yield the highest P/T ratios (Yang et al., 1993; Hiramatsu et al., 1995; Zhang et al., 1995a, 1999, 2000; Zhang and Liou, 1998b; Medaris, 1999; Liou et al., 2000). They plot well within the diamond stability field at temperatures compatible with the presence of Ti-clinohumite and the stability of magnesite at low X_{CO2}. The exceptionally high pressures of the Dabie-Sulu suite reflect the consistently low Al_2O_3 contents of the associated orthopyroxenes. Sulu samples range from 670°C/4.1 GPa to 925°C/7.0 GPa, and Dabie samples from 705°C/4.2 GPa to 885°C/6.7 GPa. P-T results for the Dabie-Sulu garnet peridotites are consistent with steady-state P-T paths calculated by Peacock (1995) for the top of subducting oceanic crust at different shear stresses and convergence rates.

Conditions within the P-T field characterized by geothermal gradients <5°C/km have been considered as a forbidden zone for metamorphic recrystallization (Schreyer, 1988), and indeed, most P-T estimates of global HP-UHP rocks define slightly higher geothermal environments (5 to 10°C/km) (e.g., Maruyama et al., 1996). The P-T estimates (750 to 970°C and ≥ 4.0 to 6.7 GPa) for Dabie-Sulu UHP garnet peridotites shown in Figure 8 require subduction of a cold, old, dry continent slab under conditions of a ≤5°C/km geothermal gradient. Inasmuch as most eclogites contain rather simple assemblages, they do not provide a well-constrained pressure for the peak metamorphism; previous estimates only yield a minimum pressure based on the stability of coesite or diamond (e.g., Liou et al., 1996). Whether the Dabie-Sulu eclogitic rocks possess similar P-T trajectories within the extremely low geothermal gradient cited here remains to be proven. Recent P-T estimates of the marble-eclogite association from the Dabie Mountains yield P > 4.5 GPa and T = 700 ± 50°C (R. Oberhansli, pers. commun., 1999). If UHP metamorphism of the Dabie-Sulu terrane indeed occurred within such a low geothermal gradient, 3D modeling of the thermal structure of continental subduction, similar to thermal models for various depths of subducting oceanic slabs and various subduction rates (Peacock et al., 1994; Peacock, 1995; Peacock and Wang, 1999), is needed to explain this unexpected finding.

Sinian magmatic-hydrothermal system for protoliths of Donghai UHP rocks

Determination of the physical conditions of UHP metamorphism places constraints on the process of active subduction. Age-dating of key episodes of metamorphism, cooling, and decompression sets limits on exhumation as well. Of equal importance in deducing the controls on UHP metamorphism may be the pre-collisional plate motions. Collision between a straight-edged plate and the promontory of an irregularly shaped plate, for example, would lead to complex movements that could influence the course of subduction and exhumation. An important aspect of the plate tectonic history of the Dabie-Sulu terrane is to verify the geometry of the parent plate from which the UHP terranes were derived. Accordingly, we seek to establish geologic ages and depositional conditions of UHP rocks in order to understand processes leading up to the collision.

Our working hypothesis is that UHP protoliths in the Dabie-Sulu terrane correlate stratigraphically with unmetamorphosed upper Proterozoic rocks of the Yangtze craton. The correlation is based on two independent geochemical criteria: (1) measured U-Pb age dates of zircons from eclogites and metagranites, and (2) oxygen and hydrogen isotope analyses of coesite eclogites and their country rocks. The U-Pb dates of zircon cores suggest that protoliths of the UHP rocks are of the same age (late Proterozoic) as the Yangtze craton. Stable isotope studies indicate that UHP protoliths were exposed to a cold climate that may correlate with Sinian glacial deposits on the Yangtze platform. The Qinglongshan rocks are unusually depleted in ^{18}O and deuterium (Rumble and Yui, 1998). This depletion must reflect isotopic exchange between protoliths and ground waters under cold climate conditions. Such depleted isotopic values have been confirmed independently in both Dabie and Sulu by Yui et al. (1995, 1997), Baker et al. (1997), and Zheng et al. (1996, 1998). Recent mapping near Donghai shows that a geothermal system complete with magmatic heat source and charged with groundwater from a cold climate was present prior to UHP metamorphism (Rumble and Yui, 1998). The heat engine that drove groundwater convection was an intrusive granite whose quartz has $\delta^{18}O_{VSMOW}$ of –1.5 to –2.5‰ (Rumble and Yui, 1998) and whose zircons have 223 ± 4 Ma concordant rims and ~ 800 Ma cores, with some discordant core ages at 400 to 500 Ma (unpubl. SHRIMP analyses). The country rocks are coesite eclogite, phengite schist, amphibole gneiss, and kyanite-topaz quartzite, whose garnet $\delta^{18}O_{VSMOW}$ values are –1.6 to –11.1‰ (Rumble and Yui, 1998). Conventional U-Pb analysis of zircons from coesite eclogite yielded a lower intercept age of 217 ± 9 Ma and an upper intercept age of 762 ± 28 Ma (Ames et al., 1996). The lower intercept gives a metamorphic age, whereas the upper is interpreted as the protolith age. The Qinglongshan rocks are tentatively correlated with the Sinian (late Proterozoic) succession of the Yangtze craton on the basis of zircon dates; the cold-climate oxygen and hydrogen isotope values are consistent with stratigraphic evidence of Sinian glaciation.

Thus, stable isotope geochemistry supports the hypothesis that UHP rocks from Qinglongshan once belonged to a passive margin along what was once the northern edge of the Yangtze craton. The correlation is reinforced by the occurrence of UHP marbles in Dabieshan with high $^{13}C/^{12}C$, similar to those of unmetamorphosed carbonates on the Yangtze platform (Zheng et al., 1998; Rumble et al., 2000), and by the occurrence of phosphate-rich pelites and quartzites. It is these rocks that were overridden by the Sino-Korean craton and subducted into the upper mantle.

Original setting of mafic-ultramafic rocks

All the evidence described above points to the fact that the Dabie-Sulu eclogites and garnet peridotites possess significantly different petrochemical and isotopic characteristics, despite their similar metamorphic histories. Compositional heterogeneity of the UHP eclogitic rocks is further complicated by post-magmatic metamorphic recrystallization, crustal contamination, and fluid metasomatism. Hence, identification of the tectonic setting for the protoliths is not straightforward. Nevertheless, as concluded by Jahn (1998), the eclogites enclosed in granitic gneisses may represent amphibolites and granulites of the Precambrian gneiss terrane, whereas those in the marbles and metasediments could have been derived from marls or calcareous silicate layers. Their protoliths have a clear continental affinity, and thus cannot be fragments of the subducted Tethyan oceanic slab.

Zhang and Liou (1998a, 1998b) divided such mafic-ultramafic associations into at least two types—mantle-derived (type A) and crustal-hosted (type B)—based on their modes of occurrence and geochemical/isotopic characteristics described above. Type A peridotites (mainly lherzolites) were derived from the mantle wedge above the subduction zone. They were never at the surface prior to UHPM. Fragments of mantle-wedge ultramafics were tectonically inserted into subducting lithosphere, and subsequently recrystallized together with supracrustal materials at depths greater than 100 to 180 km (Fig. 9A). Depending on the thickness of the mantle wedge above the subduction zone, both spinel and garnet peridotites could have been sequestered into subducting materials, and subsequently subjected to coeval UHP metamorphism. Bruckner (1998) recently proposed a "sinking-intrusion model" to explain the occurrence of garnet peridotites in the Western Gneiss Region of the Nor-

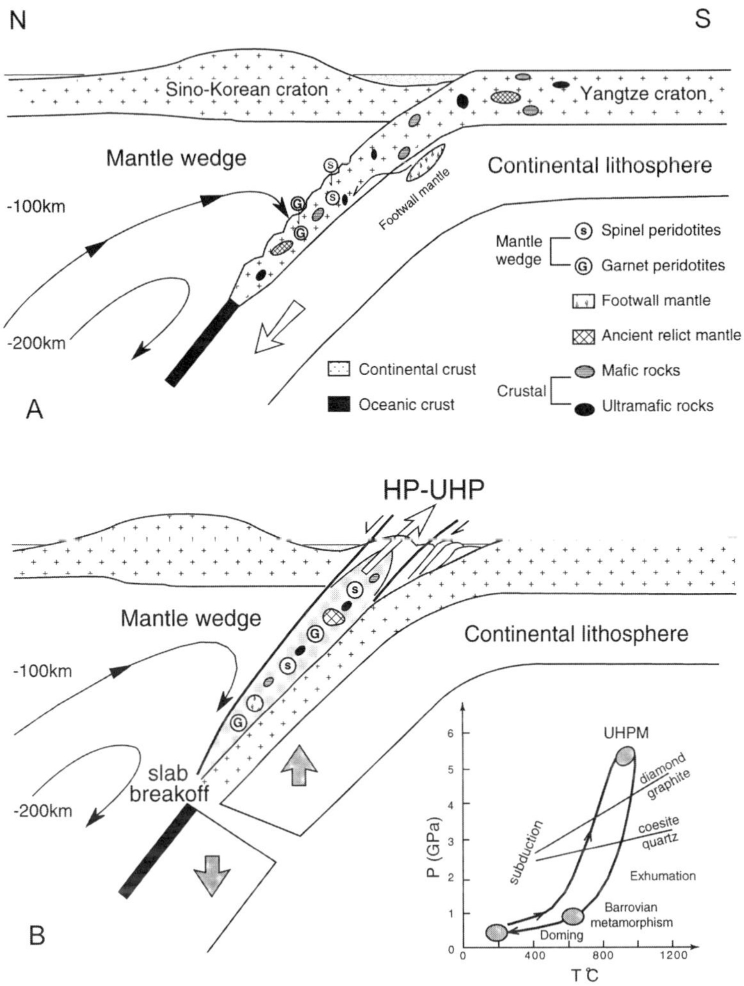

FIG. 9. Interpretative model for the origin and emplacement of Dabie-Sulu eclogites and garnet peridotites (modified after Zhang and Liou, 1998a, Zhang et al., 2000).

wegian Caledonides. He noted that a denser mantle wedge above less-dense subducting supracrustal material is gravitationally unstable; mixing of mantle ultramafics in the deep subduction zone evidently occurred at the interface between crust and mantle.

Type B peridotites, which contain considerable pyroxenite and more hydrous and carbonate phases, have low $\delta^{18}O$ values (+1.5 to +4.1‰) and high $^{87}Sr/^{86}Sr$ ratios (such as 0.707 to 0.708 for the Maowu pyroxenite), suggesting distinct crustal contamination and metasomatism. The protoliths of these mafic-ultramafic complexes crystallized within the continental crust before Triassic subduction and UHP metamorphism; hence they share the same prograde and retrograde P-T-t path with the adjacent eclogites and enclosing quartzofeldspathic gneisses.

A schematic model for the origin of these two types of garnet peridotites from the Dabie-Sulu terrane is shown in Figure 9. Figure 9A illustrates the original settings of type A and type B peridotites before and during Triassic subduction of the Yangtze craton beneath the Sino-Korean craton, and the coeval nature of UHP and retrograde events with the surrounding gneissic rocks. Because of their different settings, type A and type B garnet peridotites possess different isotopic and metasomatic signatures. Figure 9B shows exhumation of the UHP slab, which initially could be mainly the result of buoyant return after slab breakoff and later by a doming processes (e.g., Ernst et al., 1997). These peridotites share similar retrograde overprints by granulite- through amphibolite- to greenschist-facies assemblages.

A multiple origin of Alpine-type garnet peridotites has long been described (for a review, see Medaris, 1999) and such rocks are common in most HP and UHP subduction complexes. Those from the Dabie-Sulu terrane are unique in that they contain inclusions of coesite in the enclosing eclogite, or omphacitite and ilmenite rods and magnetite plates in olivine. Moreover, because of the limited fluid/rock interaction during the UHP event, type A and B garnet peridotites preserve their pre-UHP metamorphic geochemical and isotopic signatures. Systematic investigation of these garnet peridotites from this terrane provides a window into geochemical characteristics of the mantle wedge and deep subduction processes involving the role of different fluids (surfacial versus juvenile) sequestered within the subducting continental lithospheric plate.

Unresolved Problems

Despite extensive investigation of mafic-ultramafic rocks from the Dabie-Sulu terrane, outstanding questions remain to be answered. They include the following.

(1) What is the complete spectrum of the primary lithologic types affected by UHP metamorphism? Are all of them of supracrustal origin? Can they be correlated with existing lithologies found in the non-subducted part of the Yangtze craton?

(2) When did fluid/rock interactions take place? Is it possible to date these events? What is the nature of crust-mantle interactions that take place when continental material is subducted to great depths? How do such processes affect crustal growth and global geochemical recycling?

(3) What is the maximum pressure for the stability of the hydrous-carbonate phases contained in the Dabie-Sulu UHP rocks? What role did they play in the fluid/rock interaction and metasomatism at mantle depths? What was the role of these phases as major fluid carriers for cold, mantle-penetrating subducting slabs?

(4) What new constraints on petrotectonic processes do UHP garnet peridotites of the mantle hanging wall (Sino-Korean plate) provide? How do these rocks differ from mantle xenoliths in Cenozoic basalts, and from ultramafics in the subducting slab (the Yangtze plate)? What do these imply about the lithospheric evolution of eastern China?

(5) What exhumation mechanisms and rates of ascent prevent UHP mineral assemblages from being completely obliterated by Cretaceous metamorphic overprinting and/or partial melting?

Among these questions, petrochemical studies of garnet peridotites need to be emphasized. Since 1987, studies of UHP rocks from Dabie and Sulu have established a large mineralogical-petrological-geochemical database for further investigation of mantle heterogeneity and metasomatism of garnet peridotites from the mantle wedge. The preservation of the very low $\delta^{18}O$ values and extremely high εNd in eclogites of low Nd concentrations suggests limited fluid/rock interactions during UHP metamorphism and subsequent exhumation. Major directions of future study should include: (1) petrological characteristics and metamorphic evolution of garnet peridotites from drill core samples recovered from the Chinese Deep Continental Scientific Drilling program in Donghai; (2) geochemical characterization of their source and tectonic setting; (3) radiogenic and stable isotopic studies of crust/mantle and rock/fluid interactions; (4) identification of additional UHP phases, such as exsolution lamellae and other microphases; and (5) precise age determinations using various methods to establish the second-order events of the UHP rocks. Such comprehensive petrochemical and geochronological constraints are prerequisites for a better evaluation of the entire process of continental collision and subduction.

Acknowledgments

This paper is dedicated to Prof. Konrad B. Krauskopf for his contributions in geochemistry and petrology over the past four decades. The senior author appreciates his continuous guidance and friendship during the last 28 years as a colleague at Stanford University. This research was supported by a U.S.-China scientific cooperative project (NSF EAR-89-017319, 92-04563, 95-064688, and 98-14468) and by the Stanford-China Industrial Affiliates program to JGL; and by a French INSU (CNRS) research Programme "Intérieur de la Terre" to BmJ. We thank many Chinese colleagues, particularly Prof. Cong B. L. and his associates at the Chinese Academy of Sciences (1993–1998) and Profs. Xu Zhi-qin and Yang Jingsui and their associates at the

Chinese Academy of Geological Sciences for their cooperation and field assistance (1998–present). We also thank many UHP coworkers, particularly W. G. Ernst, Bradley Hacker, Shige Maruyama, and Doug Rumble, for many discussions and fruitful cooperation during the past decade. This paper was critically reviewed and materially improved by W. G. Ernst.

REFERENCES

Ames, L., Zhou, G., and Xiong, B., 1996, Geochronology and geochemistry of ultrahigh-pressure metamorphism with implications for collision of the Sino–Korean and Yangtze cratons, central China: Tectonics, v. 15, p. 472–489.

Baker, J., Matthews, A., Mattey, D., Rowley, D. B., and Xue, F., 1997, Fluid-rock interaction during ultra-high pressure metamorphism, Dabie Shan, China: Geochim. et Cosmochim. Acta, v. 61, p. 1685–1696.

Bohlen, S. R., and Boettcher, A. L., 1982, The quartz-coesite transformation: A pressure determination and the effects of other composition: Jour. Geophys. Res., v. 87, p. 7073–7078.

Brey, G. P., and Köhler, T., 1990, Geothermobarometery in four-phase lherzolite II. New thermobarometers, and practical assessment of existing thermobarometers: Jour. Petrol., v. 31, p. 1353–1378.

Bruckner, H. K., 1998, Sinking intrusion model for the introduction of garnet-bearing peridotites into continent collision orogens: Geology, v. 26, p. 631–634.

Bundy, F. R., 1980, The P, T phase and reaction diagram for elemental carbon, 1979: Jour. Geophys. Res., v. 85, p. 6930–6936.

Carswell, D. A., O'Brien, P. J., Wilson, R. N., and Zhai, M., 1997, Thermobarometry of phengite-bearing eclogites in the Dabie Mountains of central China: Jour. Metamor. Geol., v. 15, p. 239–252.

Carswell, D. A., Wilson, R. N., and Zhai, M., 1996, Ultra-high pressure aluminous titanites in carbonate-bearing eclogites at Shuanghe in Dabieshan, central China: Mineral. Mag., v. 60, p. 461–471.

______, 2000, Metamorphic evolution, mineral chemistry and thermobarometry of schists and orthogneisses hosting ultra-high pressure eclogites in the Dabieshan of central China: Lithos, forthcoming.

Carswell, D. A., and Zhang, R. Y., 1999, Petrographic characteristics and metamorphic evolution of ultra-high-pressure eclogites in plate collision belts: INT. GEOL. REV., v. 41, p. 781–798.

Chavagnac, V., and Jahn, B. M., 1996, Coesite-bearing eclogites from the Bixiling Complex, Dabie Mountains, China: Sm-Nd ages, geochemical characteristics, and tectonic implications: Chem. Geol., v. 133, p. 29–51.

Coleman, R. G., and Wang, X., eds., 1995, Ultrahigh-pressure metamorphism: New York, Cambridge Univ. Press, 528 p.

Cong, B., ed., 1996, Ultrahigh-pressure metamorphic rocks in the Dabieshan-Sulu region of China: Beijing and Dordrecht, Netherlands, Science Press and Kluwer Acad. Publ., 224 p.

Cong, B., Zhai, M., Carswell, D. A., Wilson, R. N., Wang, Q., Zhao, Z., and Windley, B. F., 1995, Petrogenesis of the ultrahigh-pressure rocks and their country rocks at Shuanghe in Dabieshan, central China: Eur. Jour. Mineral., v. 7, p. 119–138.

Eide, E. A., and Liou, J. G., 2000, High-pressure blueschists and eclogites in Hong'an: A framework for addressing the evolution of ultrahigh-pressure rocks in central China: Lithos, forthcoming.

Enami, M., and Zang, Q., 1988, Magnesian staurolite in garnet-corundum rocks and eclogite from the Donghai district, Jiangsu province, east China: Amer. Mineral., v. 73, p. 48–56.

England, P. C., and Thompson, A. B., 1984, Pressure-temperature-time paths of regional metamorphism I. Heat transfer during the evolution of regions of thickened continental crust: Jour. Petrol., v. 25, p. 894–928.

Ernst, W. G., and Liou, J. G., 1995, Contrasting plate-tectonic styles of the Qinling-Dabie-Sulu, Alpine-type, and Franciscan, Pacific-type, metamorphic belts: Geology, v. 23, p. 353–356.

______, 1999, Overview of UHP metamorphism and tectonics in well-studied collisional orogens: INT. GEOL. REV., v. 41, p. 477–493.

Ernst, W. G., Maruyama, S., and Countryis, S., 1997, Buoyancy-driven, rapid exhumation of ultrahigh-pressure metamorphosed continental crust: Proc. Nat. Acad. Sci., v. 94, p. 9532–9537.

Ernst, W. G., Zhou, G., Liou, J. G., Eide, E., and Wang, X., 1991, High-pressure and superhigh-pressure metamorphic terranes in the Qinling-Dabie mountain belt, central China: Early- to mid-Phanerozoic accretion of the western paleo-Pacific Rim: Pacific Sci. Assoc., Inf. Bull., v. 43, p. 6–15.

Graham, C. M., and Powell, R., 1984, A garnet-hornblende geothermometer: Calibration, testing and application to the Pelona Schist, Southern California: Jour. Metamor. Geol. v. 2, p. 13–31.

Gruau, G., Bernard Griffiths, J., and Lecuyer, C., 1998, The origin of U-shaped rare earth patterns in ophiolite peridotites: Assessing the role of secondary alteration and melt/rock reaction: Geochim. et Cosmochim. Acta, v. 62, p. 3545–3560.

Hacker, B. R., Ratschbacher, L., Webb, L., and Dong, Shuwen, 1995, What brought them up? Exhumation of the Dabie Shan ultrahigh-pressure rocks: Geology, v. 23, p. 743–746.

Hacker, B. R., Ratschbacher, L., Webb, L. E., Ireland, T. R., Calvert, A., Gans, P., and Dong, S., 2000, Exhumation of ultrahigh-pressure continental crust in east–

central China: Late Triassic–Early Jurassic tectonic unroofing: Jour. Geophys. Res., forthcoming.

Hacker, B. R., Ratschbacher, L., Webb, L., Ireland, T., Walker, D., and Dong, S., 1998, U/Pb zircon ages constrain the architecture of the ultrahigh-pressure Qinling–Dabie Orogen, China: Earth Planet. Sci. Lett., v. 161, p. 215–230.

Hacker, B.R., Sharp, T., Zhang, R. Y., Liou, J. G., and Hervig, R. L., 1997, Determining the origin of ultrahigh-pressure lherzolites: Science, v. 258, p. 702–704.

Hacker, B. R., Wang, X., Eide, E. A., and Ratschbacher, L., 1996, The Qinling-Dabie ultrahigh-pressure collisional orogen, *in* Yin, A., and Harrison, T. M., eds., The tectonic development of Asia: New York, Cambridge Univ. Press, p. 345–370.

Hall, R. P., and Hughes, D. J., 1990, Early Precambrian basic magmatism: Glasgow and London, Blackie.

Harley, S. L., 1984, An experimental study of the partitioning of Fe and Mg between garnet and orthopyroxene: Contrib. Mineral. Petrol., v. 86, p. 359–373.

Hirajima, T., Ishiwatari, A., Cong, B., Zhang, R., Banno, S., and Nozaka, T., 1990, Coesite from Mengzhong eclogite at Donghai county, northeastern Jiangsu province, China: Mineral. Mag., v. 54, p. 579–583.

Hirajima, T., Zhang, R. Y., Li, J. J., and Cong, B. L., 1992, Nyböite from the Donghai area, Jiangsu Province, eastern China: Mineral. Mag., v. 56, p. 37–46.

Hiramatsu, N., Banno, S., Hirajama, T., and Cong, B., 1995, Ultrahigh-pressure garnet lherzolite from Chijiadian, Rongcheng County, in the Sulu region of eastern China: The Island Arc, v. 4, p. 324–333.

Hiramatsu, N., and Hirajima, T., 1995, Petrology of the Hujialin garnet clinopyroxenite in the Su-Lu ultrahigh-pressure province, eastern China: The Island Arc, v. 4, p. 310–323.

Hu, N. G., Zhao, D. L., Xu, B. Q., and Wang, T., 1995, Petrography and metamorphism study on high-ultrahigh pressure eclogite from Guanpo area, northern Qinling Mountains: Jour. Mineral Petrol., v. 15, p. 1–9 (in Chinese with English abstract).

Irvine, T. N., and Baragar, W. R. A., 1971, A guide to the chemical classification of the common volcanic rocks: Can. Jour. Earth Sci., v. 8, p. 523–548.

Jahn, B-M., 1998, Geochemical and isotopic characteristics of UHP eclogites of the Dabie orogen: Implications for continental subduction and collisional tectonics, *in* Hacker, B., and Liou, J. G., eds., When continents collide: Geodynamics and geochemistry of ultrahigh-pressure rocks: Dordrecht, Kluwer Acad. Publ., p. 203–240.

______, 1999, Sm-Nd isotope tracer study of UHP metamorphic rocks: Implications for continental subduction and collisional tectonics: INT. GEOL. REV., v. 41, p. 859–885.

Jahn, B.-M., Cornichet, J., Cong, B., and Yui, T.-F., 1996, Ultrahigh-εNd eclogites from an ultrahigh-pressure metamorphic terrane of China: Chem. Geol., v. 127, p. 61–79.

Kato, T., Enami, A., and Zhai, M., 1997, Ultrahigh-pressure marble and eclogite in the Su-Lu UHP terrane, eastern China: Jour. Metamor. Geol., v. 15, p. 169–182.

Kretz, R., 1983, Symbols for rock-forming minerals: Amer. Mineral., v. 68, p. 277–279.

Li, S., Chen, Y., Cong, B., Zhang, Z., Zhang, R. Y., Liou, D., Hart, S. R., and Ge, N., 1993, Collision of the North China and Yangtze blocks and formation of coesite-bearing eclogites: Timing and processes: Chem. Geol., v. 109, p. 70–89.

Li, S., Chen, Y., Song, M., Zhang, Z., Yang, C., and Zhao, D., 1994, U-Pb zircon ages of amphibolite from the Haiyangsuo area, eastern Shandong province — an example for influence of multimetamorphism to lower and upper intercepts of zircon discordia line at the concordia curve: Acta Geosci. Sinica, v. 1, p. 37–42.

Li, S., Jagoutz, E., Lo, C. H., Li, Q., and Xiao, Y., 1999, Sm/Nd, Rb/Sr, and $^{40}Ar/^{39}Ar$ isotopic systematics of ultrahigh-pressure metamorphic rocks in the Dabie-Sulu belt, central China: A retrospective view: INT. GEOL. REV., v. 41, p. 1114–1124.

Liou, J. G., Hacker, B. R., and Zhang, R. Y., 2000, Ultrahigh-pressure (UHP) metamorphism in the forbidden zone: Science, forthcoming.

______, 1996, Occurrence of intergranular coesite in Sulu ultrahigh-P rocks from China: Implications for fluid activity during exhumation: Amer. Mineral., v. 81, p. 1217–1221.

______, 1998, Petrogenesis of ultrahigh-P garnet-bearing ultramafic body from Maowu, the Dabie Mountains, Central China: The Island Arc, v. 7, p. 115–134.

Liou, J. G., Zhang, R. Y., Eide, E. A., Maruyama, S., Wang, X., and Ernst, W. G., 1996, Metamorphism and tectonics of high-P and ultrahigh-P belts in the Dabie-Sulu regions, eastern central China, *in* Yin, A., and Harrison, T. M., eds., The tectonic evolution of Asia: New York, Cambridge Univ. Press, p. 300–343.

Liou, J. G., Zhang, R. Y., and Ernst, W. G., 1994, An introduction to ultrahigh-P metamorphism: The Island Arc, v. 3, p. 1–24.

______, 1995, Occurrence of some hydrous and carbonate phases in ultrahigh-P rocks from east-central China—implications for the role of fluids in deep subduction zones: The Island Arc, v. 4, p. 362–375.

______, 1997, Lack of fluid during ultrahigh-P metamorphism in the Dabie-Sulu region, eastern China: Proc. 30th Int. Geol. Congr., v. 17, p. 141–156.

Liou, J. G., Zhang, R. Y., Ernst, W. G., Rumble, D., and Maruyama, S., 1998, High-pressure minerals from deeply subducted metamorphic rocks: Rev. Mineral., v. 37, p. 33–96.

Liou, J. G., Zhang, R. Y., and Jahn, B. M., 1997, Petrogenesis of ultrahigh-pressure jadeite quartzite from the

Dabie region, East-central China: Lithos, v. 41, p. 59–78.

Maruyama, S., Liou, J. G., and Terabayashi, M., 1996, Blueschists and eclogites of the world, and their exhumation: INT. GEOL. REV., v. 38, p. 485–594.

Masuda, A., Nakamura, N., and Tanaka, T., 1973, Fine structure of mutually normalized rare-earth patterns of chondrites: Geochim. et Cosmochim. Acta, v. 37, p. 239–244.

Medaris, L. G., 1999, Garnet peridotite in Eurasian HP and UHP terranes: A diversity of origins and thermal histories: INT. GEOL. REV., v. 41, p. 799–815.

Mosenfelder, J. L., and Bohlen, S. R., 1997, Kinetics of the coesite to quartz transformation: Earth Planet Sci. Lett., v. 153, p. 133–147.

Nakamura, N., 1974, Determination of REE, Ba, Fe, Mg, Na, and K in carbonaceous and ordinary chondrites: Geochim. et Cosmochim. Acta, v. 38, p. 757–775.

Obata, M., 1976, The solubility of Al_2O_3 in orthopyroxene in spinel and plagioclase peridotites and spinel pyroxenite: Amer. Mineral., v. 61, p. 804–816.

Ogasawara, Y., Zhang, R. Y., and Liou, J. G., 1998, Petrogenesis of dolomitic marbles from Rongcheng in the Sulu UHP metamorphic terrane, eastern China: The Island Arc, v. 7, p. 82–97.

Okamoto, K., and Maruyama, S., 1999, The high pressure synthesis of lawsonite in the MORB + H_2O system: Amer. Mineral., v. 84, p. 362–373.

Okay, A. I., 1993, Petrology of a diamond and coesite-bearing metamorphic terrain, Dabie Shan, China: Eur. Jour. Mineral., v. 5, p. 659–675.

______, 1994, Sapphirine and Ti-clinohumite in ultra-high-pressure garnet pyroxenite and eclogite from Dabie Shan, China: Contrib. Mineral. Petrol., v. 116, p. 145–155.

Okay, A. I., Xu, S., and Sengör, A. M. C., 1989, Coesite from the Dabie Shan eclogites, central China: Eur. Jour. Mineral., v. 1, p. 595–598.

Omori, S., Liou, J. G., Zhang, R. Y., and Ogasawara, Y., 1998, Petrogenesis of impure dolomitic marble from the Dabie Mountains, central China: The Island Arc, v. 7, p. 98–114.

O'Neill, H. St. C., 1980, An experimental study of Fe-Mg partioning between garnet and olivine and its calibrations as a geothermometer: Contrib. Mineral. Petrol., v. 72, p. 337.

O'Neill, H. St. C., and Wood, B. J. 1979, An experimental study of Fe-Mg partioning between garnet and olivine and its calibrations as a geothermometer: Contrib. Mineral. Petrol., v. 70, p. 59–70.

Perkins, D., III, Holland, T. J. B., and Newton, R. C., 1981, The Al_2O_3 contents of enstatite in equilibrium with garnet in the system $MgO-Al_2O_3-SiO_2$ at 15–40 kbar and 900°–1600° C: Contrib. Mineral. Petrol., v. 78, p. 99–109.

Peacock, S. M., 1995, Ultrahigh-pressure metamorphic rocks and the thermal evolution of continent collision belts: The Island Arc, v. 4, p. 376–383.

Peacock, S. M., Rushmer, T., and Thompson, A. B., 1994, Partial melting of subducting oceanic crust: Earth Planet. Sci. Lett., v. 121, p. 227–244.

Peacock, S. M., and Wang, K., 1999, Seismic consequences of warm versus cool subduction metamorphism: Examples from southwest and northeast Japan: Science, v. 266, p. 937–939.

Powell, R., 1985, Regression diagnostic and robust regression in geothermometer/geobarometer calibration: The garnet-clinopyroxene geothermometer revisited: Jour. Metamor. Geol., v. 3, p. 231–243.

Ratschbacher, L., Hacker, B. R., Webb, L. E., Ireland, T. R., Calvert, A., Gans, P., and Dong, S., 2000, Exhumation of ultrahigh-pressure rocks in the Dabie–Hong'an area: Cretaceous–Recent extension: Jour. Geophys. Res., forthcoming

Ringwood, A. E., 1975. Composition and petrology of the Earth's mantle: New York, McGraw-Hill, 618 p.

Rowley, D. B., Xue, F., Tucker, R. D., Peng, Z. X., Baker, J., and Davis, A., 1997, Ages of ultrahigh pressure metamorphism and protolith orthogneisses from the eastern Dabie Shan: U/Pb zircon geochronology: Earth Planet. Sci. Lett., v. 151, p. 21–36.

Rubie, D. C., 1986, The catalysis of mineral reactions by water and restrictions on the presence of aqueous fluid during metamorphism: Mineral. Mag., v. 50, p. 399–415.

Rumble, D., 1998, Stable isotope geochemistry of ultra-high pressure rocks, *in* Hacker B., and Liou, J. G., eds., When continents collide: Geodynamics and geochemistry of ultrahigh-pressure rocks: Dordrecht, Kluwer Acad Publ., p. 241–260.

Rumble, D., and Yui, T. F., 1998, The Qinglongshan oxygen and hydrogen isotope anomaly near Donghai in Jiangsu Province, China: Geochim. et Cosmochim. Acta, v. 62, p. 3307–3321.

Rumble, D., Wang, Q., and Zhang, R. 2000, Stable isotope geochemistry of marbles from the coesite UHP terrains of Dabieshan and Sulu, China: Lithos, forthcoming.

Schreyer, W., 1988, Experimental studies on metamorphism of crustal rocks under mantle pressures: Mineral. Mag., v. 51, p. 1–26.

Schertl, H. P., and Okay, A. I., 1994, A coesite inclusion in dolomite in Dabie Shan, China: Petrological and rheological significance: Eur. Jour. Mineral., v. 6, p. 995–1000.

Shervais, J. W., 1982, Ti-V plots and the petrogenesis of modern and ophiolitic lavas: Earth Planet. Sci. Lett., v. 57, p. 101–118.

Smith, D. C., 1988, A review of the peculiar mineralogy of the "Norwegian coesite-eclogite province" with crystal-chemical, petrological, geochemical, and geodynamical notes and an extensive bibliography, *in* Smith,

D. C., ed., Eclogites and eclogite-facies rocks: Amsterdam, Elsevier, p. 1–206.

Sobolev, N. V., Shatsky, V. S., Vavilov, M. A., and Goryainov, S. V., 1994, Zircon in high pressure metamorphic rocks of folded regions as a unique container of inclusions of diamond, coesite, and coexisting minerals: Doklady Russ. Akad. Nauk, v. 334, p. 488–492 (in Russian).

Song, Y., and Frey, F. A., 1989, Geochemistry of peridotite xenoliths in basalts from Hannuoba, eastern China: Implications for subcontinental mantle heterogeneity: Earth Planet. Sci. Lett., v. 53, p. 97–113.

Sun, S. S., and McDonough, W. F., 1989, Chemical and isotopic systematics of oceanic basalts: Implications for mantle composition and processes, *in* Sanders, A. D., and Norry, M. J., eds., Magmatism in the ocean basins: Geol. Soc. Lond. Spec. Publ., v. 42, p. 313–346.

Tabata, H., Yamauchi, K., Maruyama, S., and Liou, J. G., 1998, Tracing the extent of a UHP metamorphic terrane: Mineral-inclusion study of zircons in gneisses from the Dabie Shan, *in* Hacker, B., and Liou, J. G., eds., When continents collide: Geodynamics and geochemistry of ultrahigh-pressure rocks: Dordrecht, Kluwer Acad. Publ., p. 261–274.

Taylor, H. P., Jr., 1967, Stable isotope studies of ultramafic rocks and meteorites, *in* Wyllie, P. J., ed., Ultramafic and related rocks: New York, John Wiley and Sons, Inc., p. 362–380.

Wallis, S. R., Ishiwatari, A., Hirajima, T., Ye, K., Guo, J., Nakamura, D., Kato, K., Zhai, M., Enami, M., Cong, B., and Banno, S., 1997, Occurrence and field relationships of ultrahigh-pressure metagranitoid and coesite eclogite in the Sulu terrane, eastern China: Jour. Geol. Soc. Lond., v. 154, p. 45–54.

Wang, C. Y., Zeng, R. S., Mooney, W. D., and Hacker, B. R., 2000, Ultrahigh-pressure Dabie Shan orogenic belt: Support for an Alpine subduction model from deep seismic-refraction profiling: Jour. Geophys. Res., forthcoming.

Wang, Q., Ishiwatari, A., Zhao, Z., Hirajima, T., Enami, M., Zhai, M., Li, J., and Cong, B., 1993, Coesite-bearing granulite retrograded from eclogite in Weihai, eastern China: A preliminary study: Eur. Jour. Mineral., v. 5, p. 141–152.

Wang, X., and Liou, J. G., 1991, Regional ultrahigh-pressure metamorphic terrane in central China: Evidence from coesite-bearing eclogite, marble, and metapelite: Geology, v. 19, p. 933–936.

______, 1993, Ultrahigh-pressure metamorphism of carbonate rocks in the Dabie Mountains, central China: Jour. Metamor. Geol., v. 11, p. 575–588.

Wang, X., Liou, J., and Mao, H. K., 1989, Coesite-bearing eclogite from the Dabie mountains in central China: Geology, v. 17, p. 1085–1088.

Wang, X., Zhang, R., and Liou, J. G., 1995, Ultrahigh pressure metamorphic terrane in eastern central China, *in* Coleman, R. G., and Wang, X., eds., Ultrahigh-pressure metamorphism: Cambridge Univ. Press, p. 356–390.

Webb, L. E., Hacker, B. R., Ratschbacher, L., and Dong, S., 2000, Mesozoic tectonics in the Hong'an and Tongbai Shan of the Qinling-Dabie collisional orogen, China: Implications for the exhumation of ultrahigh-pressure rocks, *in* Hendrix, M. S., and Davis, G. A., eds., Paleozoic and Mesozoic tectonic evolution of central Asia—from continental assembly to intracontinental deformation: Boulder, CO, Geol. Soc. Amer. Spec. Pap., forthcoming.

Webb, L. E., Hacker, B. R., Ratschbacher, L., McWilliams, M. O., and Dong, S., 1999, $^{40}Ar/^{39}Ar$ thermochronological constraints on deformation and cooling history of high and ultrahigh-pressure rocks in the Qinling-Dabie orogen: Tectonics, v. 18, p. 621–638.

Xu, S., Okay, A.I., Ji, S., Sengor, A. M. C., Su, W., Liu, Y., and Jiang, L., 1992, Diamond from the Dabie Shan metamorphic rocks and its implication for tectonic setting: Science, v. 256, p. 80–82.

Xu, Z. Q., 1987, Etude tectonique et microtectonique de la chaine Paleozoique et Triasique de Qinglings: Unpubl. Ph.D. dissertation, L'Universite des Sciences et Technol. du Languedoc.

Xu, Z. Q., Yang, W. C., Zhang, Z. M., and Yang, T. M., 1998, Scientific significance and site-selection researches of the first Chinese continental scientific deep drillhole: Continental Dynamics, v. 3, p. 1–13.

Yang, J. J., Godard, G., Kienast, J. R., Lu, Y., and Sun, J., 1993, Ultrahigh-pressure (60 kbar) magnesite-bearing garnet peridotites from northeastern Jiangsu, China: Jour. Geol., v. 101, p. 541–554.

Yang, J. J., Godard, G., and Smith, D. C., 1998, K-feldspar in the coesite pseudomorphs in an eclogite from Lanshantou (Eastern China): Eur. Jour. Mineral., v. 10, p. 969–985.

Yang, J. J., and Jahn, B-M., 2000, Deep subduction of mantle-derived garnet peridotites from the Su-Lu UHPM terrane in China: Jour. Metamor. Geol., v. 18, forthcoming.

Ye, K., Yao, Y., Liu, J., Katayama, I., Cong, B., Wang, Q., and Maruyama, S., 2000, Areal extent of UHPM: Implications from a coesite inclusion in zircon of granitic gneiss in the Sulu UHP terrane, eastern China: Lithos, forthcoming.

Yoder, H. S., 1955, Role of water in metamorphism: Geol. Soc. Amer., Spec. Pap., v. 62, p. 505–523.

Yui, T. F., Rumble, D., Chen, C. H., and Lo, C. H., 1997, Stable isotope characteristics of eclogites from the ultrahigh-pressure metamorphic terrain, east-central China: Chem. Geol., v. 137, p. 135–147.

Yui, T. F., Rumble, D., and Lo, C. H., 1995, Unusually low $\delta^{18}O$ ultrahigh-pressure metamorphic rocks from Sulu terrane, China: Geochim. et Cosmochim. Acta, v. 59, p. 2859–2864.

Zhai, M., and Cong, B., 1996, Major and trace element geochemistry of eclogites and related rocks, *in* Cong, B., ed., Ultrahigh-pressure metamorphic rocks in the Dabieshan-Sulu Region of China: Beijing, Sci. Press, p. 69–89.

Zhang, R. Y., 1992, Petrogenesis of high pressure metamorphic rocks in the Sulu and Dianxi regions, China: Unpubl. Ph.D. thesis, Kyoto University.

Zhang, R. Y., Hirajima, T., Banno, S., Cong, B., and Liou, J. G., 1995, Petrology of ultrahigh-pressure rocks from the southern Sulu region, eastern China: Jour. Metamor. Geol., v. 13, p. 659–675.

Zhang, R. Y., and Liou, J. G., 1994, Coesite-bearing eclogite in Henan Province, central China: Detailed petrography, glaucophane stability, and PT path: Eur. Jour. Mineral., v. 6, p. 217–233.

______,1996, Significance of coesite inclusions in dolomite from eclogite in the southern Dabie mountains, China: Amer. Mineral., v. 80, p. 181–186.

______, 1997, Partial transformation of gabbro to coesite-bearing eclogites from Yangkou, the Sulu Terrane, eastern China: Jour. Metamor. Geol., v. 15, p. 183–202.

______, 1998a, Dual origin of garnet peridotites of Dabie-Sulu UHP terrane, eastern-central China: Episodes, v. 21, p. 229–234.

______, 1998b, Ultrahigh-pressure metamorphism of the Sulu terrane, eastern China: A prospective view: Continental Dynamics, v. 3, p. 32–53.

______, 1999, Exsolution lamellae in minerals from ultrahigh-P rocks: INT. GEOL. REV., v. 41, p. 981–993.

Zhang, R. Y., Liou, J. G., and Cong, B., 1994, Petrogenesis of garnet-bearing ultramafic rocks and associated eclogites in the Sulu ultrahigh-P metamorphic terrane, eastern China: Jour. Metamor. Geol., v. 12, p. 169–186.

______, 1995a, Talc-, magnesite-, and Ti-clinohumite-bearing ultrahigh-pressure mafic and ultramafic complex in the Dabie Mountains, China: Jour. Petrol., v. 36, p. 1011–1037.

Zhang, R. Y., Liou, J. G., and Ernst, W. G., 1995b, Ultrahigh-pressure metamorphism and decompressional P-T path of eclogites and country rocks from Weihai, eastern China: The Island Arc, v. 4, p. 293–309.

Zhang, R. Y., Liou, J. G., Yang, J. S., and Yui, T-F., 2000, Petrochemical constraints for dual origin of garnet peridotites from the Dabie-Sulu UHP terrane, eastern-central China: Jour. Metamor. Geol., v.18, forthcoming.

Zhang, R. Y., Rumble, D., Liou, J. G., and Wang, Q. C., 1998, Low delta ^{18}O ultrahigh-P garnet mafic-ultramafic rocks from Dabieshan, China by in-situ UV laser probe: Chem. Geol., v. 150, p. 161–170.

Zhang, R. Y., Shu, J. F., Mao, H. K., and Liou, J. G., 1999, Magnetite lamellae in olivine and clinohumite from Dabie UHP ultramafic rocks, central China: Amer. Mineral., v. 84, p. 564–569.

Zheng, Y. F., Fue, B., Gong, B., and Li, S. G., 1996, Extreme ^{18}O depletion in eclogite from the Sulu terrane in East China: Eur. Jour. Mineral., v. 8, p. 317–323.

Zheng, Y. F., Fu, B., Gong, B., and Wang, Z. R., 1998, Carbon isotope anomaly in marbles associated with eclogites from the Dabie Mountains in China: Jour. Geol., v. 106, p. 97–104.

Zheng, Y. F., Fu, B., Li, Y., Ziao, Y., and Li, S., 1998, Oxygen and hydrogen isotope geochemistry of ultrahigh-pressure eclogites from the Dabie Mountains and the Sulu terrane: Earth Planet. Sci. Lett., v. 155, p. 113–129.

Zhou, G., Liu, Y. J., Eide, E. A., Liou, J. G., and Ernst, W. G., 1993, Progressive, high pressure–low temperature metamorphism in northern Hubei Province, central China: Jour. Metamor. Geol., v. 11, p. 561–574.

Fluid-Rock Interactions during High-Pressure and Ultrahigh-Pressure Metamorphism

P. PHILIPPOT

CNRS-ESA 7058, Laboratoire de Pétrologie, Université Paris-Jussieu, Case 110, T26-E3, 4 place Jussieu, 75252 Paris cedex 05, France

AND D. RUMBLE, III

Geophysical Laboratory, 5251 Broad Branch Road NW, Washington, DC 20015-1305

Abstract

Fluid inclusions and stable-isotope data of HP and UHP terrains from China and the European Alps are used to evaluate fluid-rock interactions during subduction and continental collision. The choice of the European Alps and China for a comparative study is compelling, because the rocks of the Alps exemplify subduction of a cold (mature) oceanic lithosphere, whereas those of China illustrate the responses of continental crust to subduction under relatively higher temperatures representative of the base of a thickened continental wedge. Our study demonstrates that there is no difference between the fluid-rock behavior of continental and oceanic crust during subduction and metamorphism. In both regions, there is evidence for the presence of a grain-boundary fluid phase during HP metamorphism. Recognition of grain- to hand specimen–scale chemical and isotopic heterogeneity attests to local buffering by mineral assemblages of fluid compositions. Assuming a one-dimensional fluid flow perpendicular to rock-layer contacts, we estimated fluid fluxes of 15 to 465 cm^3/cm^2 to as low as 1.6 cm^3/cm^2 in China and Alpine rocks, respectively. The values from China are to be considered as maxima because, in most cases, sampling was at a reconnaissance scale and does not permit evaluation of disequilibrium on a scale of one centimeter.

An important finding, characteristic of the entire range of conditions represented by these rocks, is the retention of recognizable vestiges of pre-subduction fluid-rock interactions, be they a geothermal system with a surface-exposure length scale on the order of 100 kilometers in continental rocks of China, or relics of high- and low-temperature hydrothermal alteration features in oceanic material of the Western Alps. A key issue arising from these results concerns the plausibility of transferring a slab component to the source of arc-magma genesis in a low fluid/rock regime. Possible scenarios are discussed in light of several recent innovative experiments that have dispelled some of the speculation obscuring our understanding of fluid-melt-rock interactions during HP and UHP metamorphism.

Introduction

FLUIDS IN SUBDUCTING slabs have been implicated in a number of plate tectonic processes. Water released by metamorphic dehydration reactions transiently increases pore pressure and may trigger slab earthquakes (Kirby et al., 1996). The same water, rising into the overlying mantle wedge, may play a decisive role in arc magmatism by fluxing melting (Tatsumi, 1989). If all of the effects of slab metamorphism are considered, including the transport of water chemically bound in hydrous minerals into the mantle, mineral dehydration, rock densification, and the impact of these fluid events on mechanical properties, then one is forced to conclude that fluid-rock interactions are a controlling factor in processes at convergent margins. It is manifest that these events are taking place at depths that will never be accessible to direct observation. The discovery of a number of ultrahigh-pressure (UHP) metamorphic terrains in recent years, however, makes available for study rocks that experienced subduction to depths greater than 100 to 185 km (Chopin, 1984; Sobolev and Shatsky, 1990; van Roermund and Drury, 1998). Study of exhumed slabs is now providing constraints for the necessarily speculative hypotheses that have arisen. In what follows, we show that subducted rocks carry a remarkably complete record of fluid interactions. We compare two different UHP and HP terrains featuring subducted continental and oceanic crust, respectively, to establish generally prevalent characteristics of fluids in subducted slabs.

We propose to use abundant new and high-quality data on fluid inclusions and stable-isotope compositions in HP and UHP terrains in China and the European Alps to evaluate fluid-rock interactions during subduction and continental collision. The choice of the European Alps and China for a comparative study is compelling, because the rocks of China exemplify the responses of continental crust to subduction, whereas those of the Alps afford representative oceanic crust. Another reason concerns the temperature conditions of rock equilibration: Alpine rocks have experienced a prograde P-T evolution characteristic of a cold (mature) subduction-zone environment, whereas China eclogites equilibrated under relatively higher temperatures representative of the base of a thickened continental wedge. Contrasting rocks from two such distinct provenances provide a certain geologic control, so that one may discern effects dictated by specific local rock types from those imposed regionally by plate tectonics. In particular, we wish to focus attention on the related questions: (1) Is HP and UHP metamorphism characterized by a pervasive fluid phase? (2) What is the nature of slab fluids? (3) Are slab fluids heterogeneous? (4) Are fluid fluxes low during metamorphism, as indicated by rock-buffered systems? These questions are crucial in evaluating the role played by slab volatiles in seismicity and magmatism at convergent margins.

Of particular interest is the relationship between slab fluids and arc magmatism. In this connection, the choice of the Chinese Triassic collision zone and the Tertiary Alps for study may seem inappropriate because neither are known for profuse syn-collisional magmatism (Jahn, 1998). Of course, the lack of outcrops of magmatic rock could simply be an accident of preservation. A hypothetical Chinese arc may have been eroded away; Alpine volcanics may now be buried in the subducted Apulian plate or overlain by sediments of the Po basin. But there may be more profoundly fundamental explanations for disparities in the prevalence of arc magmatism at convergent margins (Mibe et al., 1999; Peacock and Wang, 1999). We do not offer a comprehensive scenario: important data such as rate of subduction are difficult to estimate for ancient convergent margins whose magnetic stripes have been erased. We propose, however, a significant constraint to assist in the formulation of improved models. Our study demonstrates that there is no difference between the fluid-rock behavior of continental and oceanic crust during subduction and metamorphism.

Continental Collision and Subduction of Continental Crust

The two UHP terrains of China—i.e., Dabieshan and Sulu—occur in distinct outcrop areas separated by approximately 500 km of left-lateral strike-slip displacement along the postmetamorphic Tanlu fault (Xu, 1993; Wang et al., 1995) (Fig. 1). The conditions for UHP metamorphism were created by Triassic continental collision, during which the passive margin of the Yangtze plate was overridden by the Sino-Korean craton. Protoliths extend from mid- to late Proterozoic in age (Jahn, 1998). Peak metamorphic conditions are defined by coesite-bearing eclogites with temperatures of 650 to 850°C and pressures of 2.5 to 4.0 GPa (Liou et al., 1998). The principle available data on fluid-rock interactions are stable-isotope analyses (Rumble, 1998).

Outcrop scale

Let us first consider stable-isotope data on the scale of hand specimens, single outcrops, and small outcrop areas with length scales of no more than 100 meters. Analysis of local data demonstrates to what extent mineral–mineral isotope exchange equilibrium was achieved and also shows the success or failure of adjacent rocks to equilibrate. The two chief characteristics of the oxygen-isotope data are as follows. (1) In rocks free from retrograde effects, minerals separated from centimeter-scale hand specimens show a tendency to approach ^{18}O-^{16}O exchange equilibrium (Fig. 2). Minerals in oxygen-isotope exchange equilibrium include members of the eclogite-facies assemblage pyrope, omphacite, kyanite, phengite, and epidote. In many samples, oxygen-isotope geothermometry agrees with cation exchange (Fe-Mg) mineral thermometry within ± 50°C (Rumble and Yui, 1998; Zheng, Fu, Li et al., 1998, 1999; Fu et al., 1999). (2) Within a single outcrop area measuring no more than 100 meters, a given mineral shows heterogeneity of as much as 10‰ in $\delta^{18}O$. Specific outcrop areas (Fig. 1) in Dabieshan (Anhui Province) that exhibit extreme mineral-by-mineral oxygen-isotope heterogeneity are Shima (Baker et al.,1997) and Shuanghe (Zheng, Fu, Li et al., 1998; Fu et al. 1999), where garnets range over 10‰, and at Huangzhen and Hualiangting (Anhui Province), where garnets are heter-

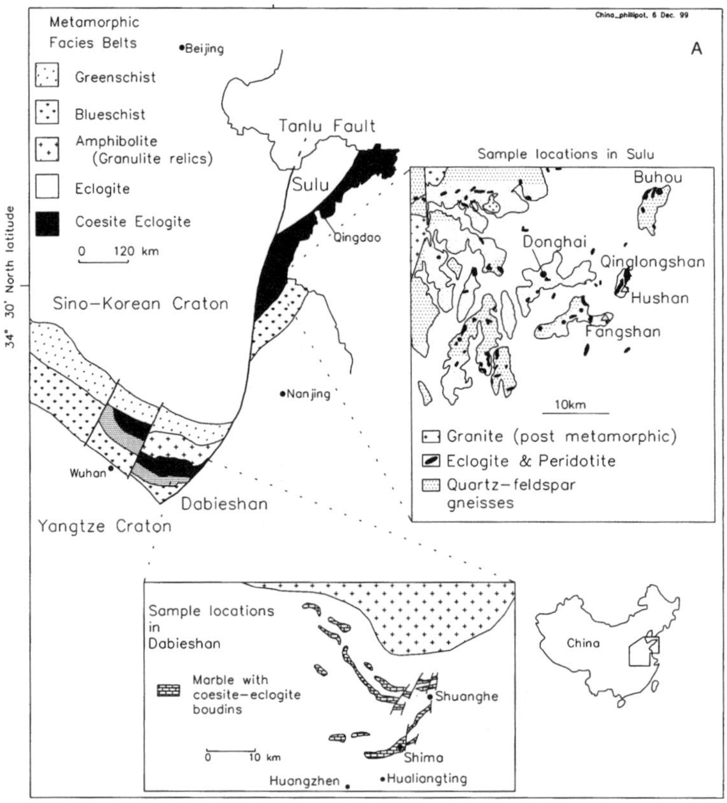

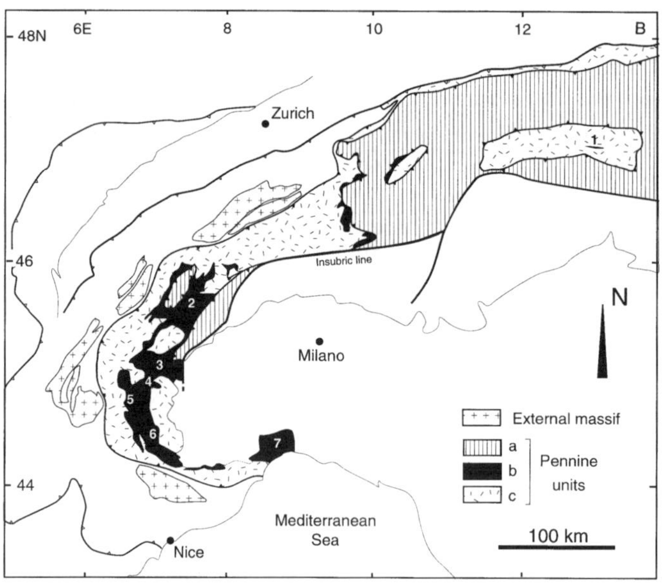

FIG. 1. Maps of China (A) and Alps (B) showing localities mentioned in text. Symbols in Fig 1B: 1 = Tauern window; 3 = Zermat Zone; 3 = Viu area; 4 = Rocciavrè massif; 5 = Chenaillet massif; 6 = Monviso massif; 7 = Voltri aind Erro Tobbio groups.

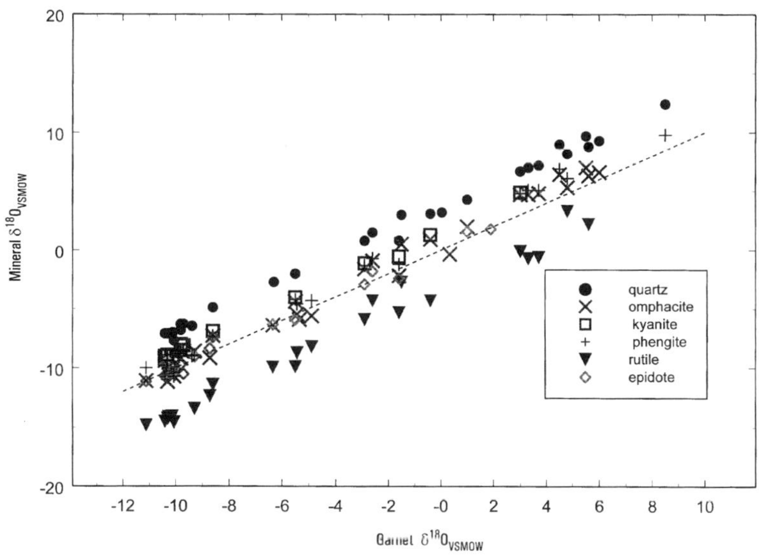

FIG. 2. Plot of $\delta^{18}O$ mineral versus $\delta^{18}O$ of garnet for Dabieshan and Sulu coesite-eclogite facies rocks. Data points lying along linear trends (kyanite, epidote) denote garnet-mineral pairs with systematic attainment of oxygen-isotope exchange equilibrium at high temperature. Scattered data points (omphacite, phengite) show retrograde alteration. Quartz and rutile mineral separates are both vulnerable to physical dispersal under a CO_2 laser beam. The analyzed separates were of sufficiently small grain size so that analytical precision may have been compromised by such dispersal. Data are from Baker et al. (1997), Rumble and Yui (1998), Zheng, Fu, Gong et al. (1998), Zheng, Fu, Li et al. (1998), and Fu et al. (1999).

ogeneous by 6 to 8‰ in $\delta^{18}O$ (Baker et al., 1997; Zheng et al. 1999). Carbonate minerals at Shuanghe in Dabieshan (Anhui Province) have a range of 10‰ in $\delta^{18}O$ and 6‰ in $\delta^{13}C$ (Zheng, Fu, Gong et al., 1998b; Rumble et al., in press). Qinglongshan, in the Sulu UHP terrain, Jiangsu Province, has garnets that differ by more than 10‰ in $\delta^{18}O$ and hydrogen isotopes in phengite that differ by 25‰ (Rumble and Yui, 1998; Zheng, Fu, Li et al. 1998).

The foregoing isotopic behaviors signify localized fluid-rock interaction effects during UHP metamorphism. The tendency toward millimeter- to centimeter-scale equilibration, in the first instance, demonstrates that oxygen isotopes were free to exchange between adjacent mineral grains during UHP metamorphism. Experiments on intermineral isotope exchange in the laboratory suggest that a grain boundary film of H_2O greatly facilitates equilibration. Moreover, persistence of grain-boundary H_2O would lead to rapid down-temperature re-equilibration and re-setting of oxygen-isotope mineral thermometers (Giletti, 1986; Eiler et al., 1992; Jenkin et al., 1994; Farquhar et al., 1996). Thus, one is led to the somewhat imprecise conclusion that grain boundaries had a film of H_2O to facilitate up-temperature equilibration between mineral grains but lacked sufficient H_2O to promote down-temperature retrograde exchange. The finding of meter-scale isotopic heterogeneity, in the second instance, proves that grain boundary fluids were not freely flowing between rock types and were unable to impose infiltration metasomatism. If flowing fluid had been present, then a tendency toward homogenization of mineral compositions between different rock types, as at Beaver Brook, New Hampshire, USA (Rumble et al., 1982), would have been observed. The conclusion that isotopic compositions were dominated by local fluid-rock effects on a scale of meters to centimeters (perhaps millimeters) has important consequences, for it suggests that rocks may retain identifying signatures of their protoliths and conditions of origin (see discussion of Nadeau et al., 1993; Ganor et al., 1996; Matthews et al., 1996).

An important discovery is that not merely eclogites, as at Qinglongshan, but also eclogite wall rocks such as biotite gneisses, quartzites, and metagranites, although themselves heterogeneous, are unusually low in $\delta^{18}O$ (Yui et al., 1995; Zheng et al. 1996; Baker et al., 1997). Quartz in metagranites at Hushan (Sulu, near Qinglongshan) and at Shuanghe (Dabieshan) has values as low as –2.5‰ in $\delta^{18}O$ (Rumble and Yui, 1998; Fu et al., 1999). Given the observation that oxygen-isotope exchange equilibrium was approached between UHP eclogite minerals and in the mineral assemblages of their wall rocks, it is concluded that oxygen-isotope compositions depleted in $\delta^{18}O$ were imposed on all the rock types at a time prior to UHP metamorphism. The implications are that both eclogites and their wall rocks experienced the same fluid-rock alteration history prior to subduction and metamorphism.

Isotopic heterogeneity

Before consideration of the China data on a regional scale (with lengths ranging from a 1 to tens of kilometers), it is necessary first to consider the significance of measured isotopic heterogeneity. If the presence of rock-dominated, UHP metamorphic fluid-rock systems raises the possibility that relict protolith features persist, then we should ask whether the observed fact of heterogeneity carries with it protolith information. Heterogeneity in oxygen-isotope compositions may arise in a number of ways, both primary and secondary. A leading primary cause of heterogeneity is sedimentary superposition of different rock types such as sandstones, shales, and limestones, juxtaposing rocks with a range in $\delta^{18}O$ of as much as 20‰, depending on detrital provenance. It has been found that these protolith differences in $\delta^{18}O$ persist through the amphibolite and into the granulite metamorphic facies (Anderson, 1967; Rumble, 1978; Rumble and Spear, 1983; Valley and O'Neil, 1984).

Oxygen- and hydrogen-isotope heterogeneity may be imposed secondarily by alteration, as is observed in near-surface geothermal and hydrothermal systems (Sheppard and Taylor, 1974). In these systems, $\delta^{18}O$ and δD rock and mineral values are controlled by fluid-rock ratios—that is, they are controlled by the permeability dictated by such structures as sedimentary contacts, fractures, faults, and bedding. Lateral gradients in isotopic composition from vein fillings into wall rocks and between rock types of different permeability are a common occurrence. Isotopic evidence of secondary alteration in geothermal systems is most easily recognized in localities experiencing a cold climate. Cold-climate (high-latitude, high-altitude) meteoric waters are strongly out of isotopic equilibrium with typical igneous and sedimentary rocks; thus they leave an unmistakable record in the unusually low $\delta^{18}O$ and δD values of altered rocks from geothermal systems (Blattner et al., 1997).

In applying the foregoing characteristics of primary and secondary isotope heterogeneity to the present investigation, one finds in the low $\delta^{18}O$ and δD rocks of Qinglongshan a likelihood that heterogeneity was imposed by a near-surface geothermal system (Rumble and Yui, 1998; Zheng, Fu, Li et al, 1998). In contrast, the heterogeneity in oxygen isotopes at Shuanghe, with its interlayered marbles, schists, gneisses, and eclogites, may have arisen at least in part from differences in composition of sedimentary precursors (Rumble et al., in press; Fu et al., 1999). Thus, we accept as a working hypothesis that the isotopic heterogeneity of the Chinese UHP rocks is a protolith characteristic that has survived subduction, metamorphism, and exhumation.

Regional scale

One of the most exciting discoveries of the new research on Chinese UHP rocks is the widespread regional occurrence of unusually low $\delta^{18}O$ values in both the Sulu and Dabieshan terrains. Not restricted to the type locality at Qinglongshan, values as low as 0 to –5.5‰ are found at Fangshan, Hushan, Hongtoushan, Mengzhong, and Yanshuifang, across a region in southern Sulu measuring some 20 by 30 km (Yui et al., 1995; Zheng et al., 1996; Rumble, 1998; Rumble and Yui, 1998; Zheng, Fu, Li et al., 1998). Not only are low values widely distributed, but, heterogeneity is a regional feature. Eclogites with high garnet $\delta^{18}O$ of +3.3 to +8.2‰ are found at Buhou, Dahling, and Duchanggo in the same area including Qinglongshan (Rumble and Yui, 1998; Zheng Fu, Li et al., 1998) (Fig. 1).

Dabieshan likewise has low and heterogeneous $\delta^{18}O$ values distributed over tens of kilometers. In coesite-eclogite facies rocks, Hualiangting, Maowu, Shima, and Shuanghe all have heterogeneous garnet $\delta^{18}O$ ranging from –5.5 to +8.5‰ (Baker et al., 1997; Yui et al., 1997; Zhang et al., 1998; Fu et al., 1999). The Huangzhen outcrops from southern Dabieshan in the quartz-eclogite facies have garnets from –5.5 to +4.2‰ (Baker et al., 1997; Yui et al., 1997; Zheng et al., 1999) (Fig. 1).

Comparison of the Dabieshan and Sulu outcrops is facilitated if 500 km of post-UHP metamorphism, left-lateral strike slip on the Tanlu fault is restored (Xu, 1993). By resetting to a pre-fault configuration, not only are the coesite-eclogite facies outcrop areas of Dabieshan and Sulu juxtaposed, but the areas of low $\delta^{18}O$ are made contiguous (Rumble, 1998). The resulting area of low and heterogeneous $\delta^{18}O$ values is comparable in areal extent to hydrothermal systems such as those studied by Criss and Taylor (1986). Thus we find that in their regional extent, low $\delta^{18}O$ values of Dabieshan and Sulu map a pre-subduction, pre-UHP metamorphism geothermal system with a surface-exposure length scale on the order of 100 kilometers.

Continental Collision and Subduction of Oceanic Crust

Meta-ophiolites from the European Alps, representing the Jurassic crust of the Piemonte ocean destroyed in the Alpine orogeny, are preserved in an arcuate zone stretching from eastern Switzerland to the Mediterranean coast (Ligurian Alps; Fig. 1). Small remnants of oceanic material also are present in the Tauern window farther east. Three major nappes piled one above the other and consisting of oceanic metasediments (calc-schists of the so-called "schistes lustrés" nappe) with large masses of more or less coherent ophiolitic sequences have been recognized. The structurally lowest unit (Tauern window eclogites, Zermatt-Saas zone, Viu area, Rocciavrè and Monviso massifs, Voltri Group, Erro Tobbio) comprises eclogitic to high-grade blueschist-facies rocks (450 to 750°C and 1.8 to 3.0 GPa). It is overlain by a lower-grade blueschist- to greenschist-facies unit (Queyras-Ubaye region, Combin zone, Platta unit; 350 to 500°C and 0.5 to 1.8 GPa). HP metamorphism in these two nappes pre-dates continental collision and resulted from southeastward subduction of Piemonte oceanic crust under the margin of the Apulian continent (a small plate acting as a promontory of Africa; Dal Piaz, 1974). In the following, we will refer to these as eclogitic and blueschist-facies units, respectively. The structurally highest unit (Chenaillet Massif) escaped subduction-zone metamorphism and retains oceanic amphibolite-facies metamorphism (Mével et al., 1978). This unit is of fundamental importance for unraveling fluid-rock interaction processes during subduction-zone metamorphism because it represents the hydrothermally altered equivalent of the blueschist and eclogitic units. Most fluid-inclusion and stable-isotope studies from the Alps were performed in the structurally lower, eclogite-facies unit. A review of the results obtained in HP and UHP rocks of oceanic affinities is presented below. For complementary information on fluid and stable-isotopic data on HP rocks of continental affinity, the reader is referred to the work of Desmons and ONeil (1978), Robert et al., (1985), Philippot (1993), Sharp et al., (1993), Fruh-Green (1994), Philippot et al., (1995), and Scambelluri et al. (1998).

HP-fluid occurrence and composition

With the exception of one study performed in the blueschist-facies metasediments of the Ubaye-Queyras region (Agard, 1999), all fluid-inclusion studies were performed in the structurally lower, eclogite-facies Piemonte unit. Unequivocal evidence for the presence of a free fluid phase during eclogitic metamorphism is the occurrence of HP segregation vein systems containing primary fluid inclusions in all the different protoliths forming a representative section of the oceanic crust (meta-peridotites, layered and massive metagabbros, layered metabasalts, and their sedimentary cover; Philippot, 1987; Philippot and Selverstone, 1991; Selverstone et al., 1992; Scambelluri et al., 1997). Segregation veins contain the same mineral assemblage as the wall rock, indicating that the material filling-vein was locally derived. Several textural features such as fluid-inclusion clusters not bound to healed fractures, inclusion bands lining oscillatory-zoned omphacite (Philippot and Selverstone, 1991; Scambelluri et al., 1997), and crack-seal microstructures (Philippot, 1987) indicate that the fluid inclusions are primary with respect to the veining process. Vein inclusion fluids belong essentially to the system H_2O-salts (NaCl ± KCl ± $CaCl_2$ ± $FeCl_2$ ± $MgCl_2$; Monviso, Rocciavrè; Tauern window; Erro Tobbio). Two other systems were identified in the Tauern eclogites. These are: H_2O-CO_2-salts ± N_2 and CO_2-N_2 (Luckscheiter and Morteani, 1980; Thomas, 1991; Selverstone et al., 1992). A large variety of daughter minerals were recognized. These include: in Monviso—halite, sylvite, calcite, dolomite, albite, anhydrite/gypsum, barite, baddeleyite, rutile, sphene, Fe-oxides, pyrite, and monazite (Philippot and Selverstone, 1991); in the Tauern window—chlorides, carbonates, Fe-Ti oxides, rutile, apatite, mica, zoisite, kyanite, Ca-sulfate, and omphacite (Thomas, 1991; Selverstone et al., 1992); and in Erro Tobbio metaperidotites, Scambelluri et al. (1997) reported the presence of chlorides, ilmenite, and magnetite. These minerals should not all be considered as true daughter crystals, however. Indeed, as discussed by Philippot and Selverstone (1991), despite the presence of carbonate, sulfate, and sulfide, neither CO_2, N_2, CH_4, nor sulfur species were detectable by microthermometry or Raman techniques. This implies that some of the trapped solid precipitated as a result of interaction between inclusion fluid and host mineral. Although not distinguished genetically at that time, we follow the interpretation of Heinrich and Gottschalk (1995) and Svensen et al. (1999) to term carbonate and sulfate as step-daughter crystals since they likely contain components from both the fluid and the host. In contrast to the moderately to highly saline brine recorded in the eclogite-facies rocks, fluid inclusions present in blueschist-facies metasediments contain low-salinity aqueous solutions (0 to 9 wt% eq. NaCl; Agard, 1999).

A common feature of undeformed eclogitic metatgabbros is the presence of clusters of primary aqueous brine inclusions located within the core of omphacite porphyroblasts pseudomorphing precursor magmatic augite (Philippot and Selverstone, 1991; Nadeau et al., 1993; Vallis and Scambelluri, 1996; Philippot et al., 1998). Generally two-phase (liquid + vapor) aqueous inclusions at room temperature, and additional halite or calcite and pyrite solid phases are present in Rocciavrè and Monviso eclogites, respectively. These inclusions are distinctly different from the vein and wall-rock inclusions both in terms of their salinity and chemical and hydrogen isotopic compositions. As shown by Nadeau et al., (1993) the hydrogen isotopic signature of inclusion water (–72 ‰) in omphacite is similar to that of hydrothermal chlorite and amphibole-altering augite in Fe-metagabbros from the Chenaillet Massif, suggesting that the inclusion fluid was derived from the breakdown of amphibole and chlorite during prograde metamorphism (i.e., dehydration fluid trapped along the burial path). Similar conclusions were reached by Scambelluri et al. (1997), who showed that the salt-saturated brines preserved in the olivine + titanian clinohumite + antigorite UHP mineral assemblage from the Erro Tobbio metaperidotite were likely derived from the breakdown of ocean-floor Cl-bearing serpentine and brucite pseudomorphing relict mantle minerals in the least-deformed adjacent metaperidotites.

Fluid heterogeneities

On the basis of the two compositionaly distinct brines present in Monviso undeformed rocks and veins, respectively, Philippot and Selverstone (1991) showed that fluid-inclusion heterogeneities also are preserved between and within adjacent omphacite grains in a single eclogitic vein. An unequivocal observation is the presence of adjacent fluid-inclusion bands decorating different concentric zones of oscillatory-zoned omphacite, and containing different daughter-mineral assemblages, hence suggesting that the fluid composition varied between adjacent fluid-inclusion bands. Location of fluid-inclusion bands exclusively within diopside-rich zones indicates that there is a close relationship between the presence of fluid inclusions and omphacite compositional zoning. This observation suggests that chemical variations in omphacite are associated with local variations in fluid composition that are manifested in crystal-fluid boundary kinetics.

Using a different approach, Selverstone et al. (1992) calculated fluid activity ratios between millimeter- to centimeter-scale layers in banded mafic eclogite from the Tauern window. Results of the calculation indicate that CO_2 was constant between layers but that H_2O activity varied dramatically from 0.9 down to 0.2 between the same layers. An important observation made by Getty and Selverstone (1994) in the same material is that major elements are well equilibrated among garnet, omphacite, and clinozoisite rims but that, on the contrary, Ti, V, Cr, Y, Sr, or Zr show little systematic partitioning behavior among these same minerals. Most important is that despite the apparent lack of mineral–mineral trace element equilibration, the trace elements vary systematically with water activity calculated in each layer. As in the Monviso samples, trace-element behavior during eclogite metamorphism thus appears to have been controlled largely by mineral-fluid interactions along grain boundaries.

Isotopic heterogeneities

Variations in oxygen-isotope ratios of omphacite between veins and wall rocks of +3.0 to +5.3‰ and within the same microstructural domains (variations of 2.3‰ and of 1‰ within the different veins and mylonites, respectively) indicate that the eclogitic rocks from Monviso did not equilibrate during UHP metamorphism at about 2.4 GPa and 650°C (Nadeau et al., 1993). Recognition of $\delta^{18}O$ variations between vein–wall rock pairs sampled 5 cm apart indicates centimeter-scale equilibration. The δD of primary-inclusion fluid trapped in omphacite from veins and wall rocks show similar variation as that described for oxygen. Values of δD from –34‰ in the vein to –84‰ in the adjacent mylonitic wall rock have been attributed to kinetic fractionation (diffusion related) during partial leakage of the fluids attending dynamic recrystallization of omphacite (see Nadeau et al., 1993, p. 443).

Carbon isotopes behave in a similar way. On the basis of step-heating analysis, Nadeau et al. (1993) showed that the fraction of carbon extracted at high temperature (T >1200°C) in the eclogitic veins was high in ^{13}C ($\delta^{13}C$ values between –3.4 and –7.6‰), compared to the high-T fraction in mylonites and low-strain rocks and the low-T fraction in all analyzed samples (mean $\delta^{13}C$ value of –25‰). The observation that the hydrothermally altered gabbro from Chenaillet shows a similar low $\delta^{13}C$ value of ca. –24‰ has led to the interpretation that the enrichment in $\delta^{13}C$ of some of the eclogitic vein carbonate (daughter phase trapped in fluid inclusions) resulted from complex adsorption-desorption and oxidation reaction controlled by kinetic isotopic fractionation on fluid-mineral interfaces rather than equilibrium isotopic reactions.

Getty and Selverstone (1994) showed that, in addition to the mineral trace-element contents, the millimeter-scale layered tuffaceous material from the Tauern window preserved layer-scale heterogeneities in $\delta^{18}O$ values of dolomite that were generated during eclogitic metamorphism. These heterogeneities were correlated with a_{H2O} and dolomite modal abundances calculated in different layers, indicating that equilibration was achieved between fluid composition, oxygen isotopes, and trace-element composition of the eclogitic minerals on a layer-scale.

Vestiges of sea-floor hydrothermal alteration

In addition to the occurrence of Chenaillet hydrothermally altered ophiolites, an important feature of the Alpine HP rocks is the preservation of relics of pre-eclogitic metamorphism. Such relics were first recognized on the basis of detailed petrological investigations and major- and trace-element analysis of undeformed eclogitic metagabbros preserving corona textures (Pognante, 1981, 1989; Messiga, 1987; Messiga and Scambelluri, 1988; Barnicoat and Cartwright, 1997; Messiga et al., 1995, 1999; Scambelluri et al., 1995). This observation was supported further by fluid-inclusion and hydrogen stable-isotope data for low-strain metagabbros from Monviso and metaperidotites from Erro Tobbio (see above; Nadeau et al., 1993; Scambelluri et al., 1997). Oxygen- and carbon-isotope studies have provided unequivocal arguments in favor of this interpretation.

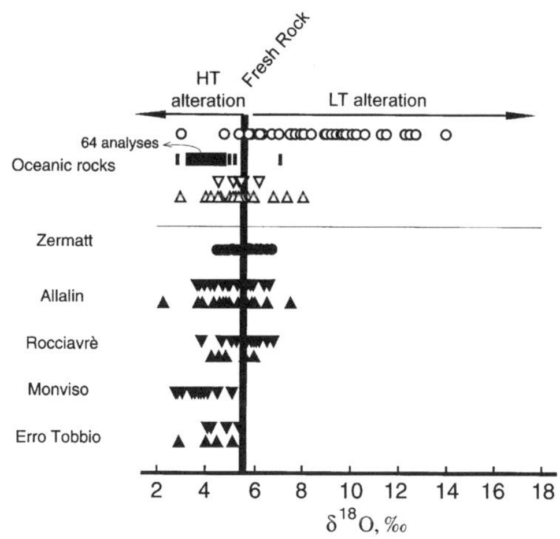

FIG. 3. Oxygen-isotope compositions of Alpine high-pressure rocks compared with their unmetamorphosed or hydrothermally altered equivalent. The data are from modern oceanic environments (Muehlenbachs, 1986; Marion, 1992; Agrinier, Hékinian et al., 1995; Agrinier, Laverne et al.,1995) and Alpine meta-ophiolites (Nadeau et al., 1993; Barnicoat and Cartwright, 1997; Philippot et al., 1998; Vallis, 1997). The different symbols refer to differentiated gabbros (triangles), sheeted dike complex (vertical bars), and basalts (circles). These symbols correspond to whole-rock analysis, whereas reversed triangles refer to analysis performed on pyroxene separates (augite in ocean gabbro and omphacite in high-pressure metagabbro). Oceanic rocks are shown as open symbols, high-pressure rocks as filled symbols.

Figure 3 shows a compilation of all oxygen-isotope data of oceanic crustal material from the Western Alps. The data are from eclogitic and blueschist metabasalts and associated talc- and chloritoid-bearing rocks of Aosta Valley and the Zermatt-Saas zone (Barnicoat and Cartwright, 1997), and olivine-bearing and Fe-rich metagabbros from Monviso (Nadeau et al., 1993), Rocciavrè (Barnicoat and Cartwright, 1997; Philippot et al., 1998) and Erro Tobbio (Vallis, 1997; Vallis et al., 1997). Also shown is the range of $\delta^{18}O$ values of clinopyroxene separate and whole-rock from altered oceanic material and ophiolites.

There is a remarkable similarity between the isotopic composition of samples from the Alps and the hydrothermally altered ocean floor and ophiolites. All eclogitic metagabbros show a range in $\delta^{18}O$ from 2.3 to 7.6‰, with a mean $\delta^{18}O$ of 5.0 ± 0.9‰. The

$\delta^{18}O$ values of ocean-floor gabbros altered at high temperature (> 400°C; e.g., Muehlenbachs, 1986) are typically between 3.0 and 5.7‰. Values as high as 7.6‰ in the metagabbros and the high $\delta^{18}O$ compositions of the talc-bearing rocks and mineralized quartzite from the Zermatt zone can be attributed to a stage of hydrothermal alteration at low temperature (Muehlenbachs, 1986). In any case, fractionation of oxygen isotopes during devolatilization is known to be small at temperatures of 400 to 600°C, and in all cases should be less than 1‰ (Valley, 1986). Barnicoat and Cartwright (1997) calculated that the effect of removing the water from an assemblage composed of 60% amphibole, 20% plagioclase, 10% quartz, 5% chlorite, and 5% water at 500°C is to lower $\delta^{18}O$ values by less than 0.2‰ during metamorphism. Recognition that the fluid salinity-$\delta^{18}O$ systematics of HP and UHP oceanic material is similar to that documented along a typical section of the altered oceanic crust and unmetamorphosed ophiolites provides further support in favor of retention of vestiges of pre-subduction fluid-rock interactions (Philippot et al., 1998).

In the banded eclogites from the Tauern window, Selverstone and Getty (1994) showed that in the same sampled traverse consisting of 23 different dolomite-bearing layers, dolomite exhibits differences in $\delta^{13}C$ values of more than 1‰. These differences are independent of calculated $a_{(CO2)}$ and dolomite modal abundance. Because dolomite and the fluid phase are the only two carbon-bearing phases, the CO_2 component of the fluid per layer must have maintained $\delta^{13}C$ values proportional to that of the accompanying dolomite. These contrasting $\delta^{13}C$ values for the fluid most probably mirrored the original, pre-eclogitic gradient in $\delta^{13}C$ of the host, likely inherited from a stage of hydrothermal alteration ($\delta^{13}C$ values between 0 and 1‰ fall within the range of oceanic carbon of –2 to 2‰; Fruh-Green et al., 1996).

Comparisons, Contrasts, and Implications

Slab fluids from China and the Alps share many features in common. In both regions there is evidence for the presence of a grain-boundary fluid phase during HP and UHP metamorphism. That coexisting minerals achieved oxygen-isotope exchange equilibrium under UHP conditions is a permissive test for grain boundary fluid in the Chinese rocks. Fluid inclusions and veins afford a more direct and convincing proof of fluid presence in Alpine rocks. Data from Alpine fluid inclusions demonstrate that fluids were saturated solutions of their host materials and contain an impressive inventory of elements (Cl, Br, Na, K, Ca, Mg, Fe, Zr, Ti, P, Ba, Ce, La, Th, S ± CO_2 and N_2). Such inclusion fluids provide direct evidence for transport of high-field-strength, large-ion-lithophile, and light-rare-earth elements as dissolved species during subduction. Observed chemical and isotopic heterogeneity between contiguous but different rock types attests to local buffering of fluid compositions by mineral assemblages. Not only were fluid species proportions (activities of volatile components such as H_2O, CO_2, etc.) different on a centimeter scale but also the isotopic compositions of grain boundary fluids must have been different from rock to rock. The hand-specimen-scale control of fluid chemical and isotopic compositions imposes an upper limit on the magnitude of fluid flow through the subducting slab. This upper limit may be described as a failure of flowing fluid to exhaust the buffer capacity of a local mineral assemblage. To maintain isotopic heterogeneity the exchange capacity of a rock must not be exceeded. Fluids were not necessarily immobilized on grain boundaries but neither were they flowing freely across lithologic contacts.

Estimated fluid fluxes

The preservation of pre-metamorphic differences in stable-isotope composition between intercalated rock types makes it possible to estimate an upper bound on fluid fluxes during HP and UHP metamorphism. The available data consist of oxygen-, carbon-, and hydrogen-isotope analyses of minerals separated by as little as centimeters and as much as meters in outcrop. We seek to obtain fluid-flux estimates based on the measured failure to achieve isotopic equilibrium between two samples collected close to one another from the same outcrop. Equation (5) of Ferry and Gerdes (1998, p. 267) (cf. Barnicoat and Cartwright, 1997, p. 59)

$$Z_f = N_f\, q_m / \alpha V_r\,, \qquad (1)$$

gives the distance (Z_f) a reaction front is displaced by the molar, time-integrated fluid flux q_m by infiltrating a fluid with N_f moles of reactive element (O, C, or H) per mole of fluid through a rock with V_r moles of reactive element per unit volume of rock.

In the following discussion, we solve for q_m and substitute $q_v/V_f = q_m$ where q_v is the time-integrated volumetric fluid flux, V_f is the molar volume of fluid,

$N_f = 1$ for H_2O, and α, the isotope fractionation factor, is assumed = 1.0. Values of V_r for oxygen range from 0.11 for kyanite to 0.08 (moles / cm^3) for feldspars with many HP and UHP rock-forming minerals falling in the range of 0.09 to 0.105; we choose a value of 0.10 to represent a hypothetical rock. The molar volume of water (V_f) at 3.0 GPa and 675°C is 15.5 cm^3/mole (Kerrick and Jacobs, 1981). The units of q_v are cm^3/cm^2. Applying this equation to two unequilibrated samples separated in outcrop by a distance Z_f, we calculate the flux that would have been required to equilibrate them and designate this value as the upper limit of flux between them. Equation (A) was derived from the mass continuity equation (equation 2 of Ferry and Gerdes, 1998) by assuming local equilibrium with no mass transfer via diffusion or dispersion. Equation 1 is applicable only to circumstances where fluid infiltration is the dominant mode of mass transfer. In the present application, the equation affords estimates of the maximum possible time-integrated fluid flux consistent with observed differences in isotopic composition between rock types. If diffusive or dispersive mass transfer were operational, use of the equation would overestimate the amount of advection. Note that as used herein, one-dimensional fluid flow perpendicular to rock-layer contacts is assumed. The estimates we present do not represent fluid flow parallel to contacts, rock layers, or parallel to the walls of veins.

Baker et al. (1997, p. 1692) estimated the length scale of oxygen-isotope exchange between marble hosts and coesite-eclogite pods in Dabieshan to be greater than 10 cm but less than 3 m. The corresponding flux estimates (q_v) are 15.5 and 465 cm^3/cm^2, respectively. The 465 cm^3/cm^2 estimate applies, as well, to Qinglongshan (Rumble and Yui, 1998, p. 3309, Fig. 1). Getty and Selverstone (Fig. 4, p. 752, 1994) reported dolomites not equilibrated with respect to both oxygen and carbon isotopes and separated by as little as 1 cm in outcrops of banded eclogite from the Tauern window, Austria. Application of equation (5) yields an estimated q_v of 1.65 cm^3/cm^2 for oxygen (V_f = 16.48 cm^3 /mole at 2.0 GPa, 625°C).

The foregoing discussion results in estimates of time-integrated volumetric fluid flux across rock contacts no higher than 465 cm^3/cm^2. It is suspected that these values are, if anything, too high because in most cases sampling was at a reconnaissance scale and does not permit evaluation of disequilibrium on a scale of one centimeter. The values of q_v from UHP and HP rocks may be compared with those tabulated from shallower regional and contact-metamorphic terrains. Ferry and Gerdes (Table 1, 1998) give a range of q_v from 20 to 10^6 cm^3/cm^2; Manning and Ingebritsen (1999, Table 2) reported 10^4 to 10^7 cm^3/cm^2. It may be seen that HP and UHP terrains show much less flux than lower-pressure metamorphic rocks. The observation of decreasing flux with depth of metamorphism is consistent with an estimate of permeability decreasing according to the relationship log (k) = –3.0 log (z) - 14.2 (k in m^2, z in km) over the depth range from surface to 30 km (Manning and Ingebritsen, 1999). However, although the flux of fluid in the direction perpendicular to rock layers is low, fluid flow parallel to structural channels or metamorphic aquifers may be effective. For example, in the banded eclogites from the Tauern window, Selverstone et al. (1992) suggested that, although little to no transport of fluids occurred across the layers, considerable fluid flow may have taken place within individual layers. A striking example of channelized fluid flow concerns eclogitization of metastable anhydrous granulites accomplished through hydration reactions in the Norwegian Caledonides (Austhreim, 1987; Jamtveit et al., 1990). Ferry and Gerdes (1998, p. 261, Table 1), estimated fracture flow of 8 x 10^4 cm^3/cm^2 through the eclogitized veins of the Western Gneiss Region, Norway, a UHP terrain not reviewed herein (see Smith, 1984; van Roermund and Drury, 1998). Thus it may be seen that in particular circumstances, fracture or aquifer flow is a dominant transport mechanism in relation to flow across rock layers.

Implications

It may not be surprising that a correlation between limited slab fluid flux and absence of calc-alkaline magmatism can be documented in China and the Alps. A low flux slab should yield little of the H_2O needed for flux melting. Alternatively, it is possible that the exhumed UHP rocks do not represent the levels where fluids were extracted from the slab. In the basaltic and ultramafic systems, phengite, lawsonite, and antigorite are all stable above 5 GPa at 600 to 700°C (Ulmer and Trommsdorf, 1995; Schmidt and Poli, 1998). These minerals, which can contain up to 13 wt% in their structure (antigorite), are still preserved in most Alpine rock, implying that the oceanic lithosphere did not evolve most of its water while subducted to about 100 km depth. Similar reasoning can apply to continental rocks (Ernst, 1999), as phengite and biotite remain stable in quartzofeldspathic assemblages to pressures of 3.5 GPa. Alternatively, as discussed by Schreyer et al. (1987), the assemblage K-feldspar–phlogopite in the presence of excess water and in the presure-temperature range of 1.5 to 2.0 GPa and 400 to 700°C can form very MgSi-rich phengite and quartz plus a KMg-rich fluid. During continental collision, such highly reactive KMg-rich fluids when interacting with neighboring mantle rocks can cause the commonly observed mantle metasomatism, producing K-rich igneous rocks of the lamproite family. These types of spatial relations have been documented in Spain, the Colorado Plateau, the Western Alps (lamprophyre dikes of the Sesia Zone; Dal Piaz et al., 1979), Corsica, and the Karakorum. Similar types of rocks should be actively looked for in the Dabieshan and Sulu UHP terrains if absence of syncollisional magmatism is to be fully documented. Despite uncertainties regarding the fate of volatiles attending subduction of Alpine and China rocks, the foregoing data demonstrate that continental and oceanic crust do not differ in their fluid-rock behavior. Is it possible that our results are applicable to downgoing slabs beneath active volcanic arcs? What are the indicated mechanisms of water release during subduction?

Hypotheses of water release invoking a unique major dehydration event to explain volcanic fronts are not compatible with our observations in China and the Alps. There is simply no evidence for such an event in the rock record. However, a model of gradual and continuous water release down to depths of 150 to 200 km (beyond the depth limit of our study) is not inconsistent with our findings. Recent experimental determinations of the thermal stabilities of the hydrous mineral chlorite, lawsonite, epidote-zoisite, amphibole, paragonite, chloritoid, talc, and phengite in naturally occurring assemblages show that a major dehydration front located in the slab is unlikely (Schmidt and Poli, 1998). Continuous dehydration is to be expected because different bulk compositions will dehydrate under different P-T conditions. Also, dehydration is not a singular occurrence in the sense that all water is lost upon the breakdown of a single hydrous phase (such as amphibole). On the contrary, in the basaltic and ultramafic systems, there is a succession of dehydration reactions, many of which actually produce hydrous phases (less hydrous than reactants but still hydrous) so that dehydration is smeared out over a depth interval from shallow to as deep as 200 km. In HP and UHP metasediments, numerous dense hydrous silicates can form with prograde metamorphism (ellenbergerite, sudoite, yoderite, Mg-carpholite, Mg-chloritoid, Mg-staurolite, K-cymrite; see Schreyer, 1988 for a review). Most of the reactions forming these dense hydrous phases proceed toward higher pressure either by preserving the original water content or by absorbing additional free water. This has led Philippot (1993) to suggest that in subducted metasediments, part of the dehydration water could be transferred from one hydrous mineral to another during burial. Accordingly, a hypothesis of gradual, continuous dehydration is compatible with our observations of limited fluid flow in the slab and provides sufficient water to account for flux melting of the mantle.

We also wish to call attention to the implications of pressure-enhanced mineral solubilities as expressed in the concentrated metal-rich brines of Alpine UHP fluid inclusions. Most experimental works have found higher solute concentrations in fluids with increasing pressure (Ryabchikov and Boettcher, 1980; Massonne, 1992; Ryabchikov, 1993; Manning, 1994, 1998, 1999; Shen and Keppler, 1997; Domanik and Holloway, 1998; Bureau and Keppler, 1999). Massonne (1992) showed that the solubility of metasedimentary rocks (2 micas, K-feldspar, quartz) in aqueous fluids may be as high as 50 wt% at 2.0 GPa and 550°C. Comparable concentrations of silicate component could be dissolved in fluids in equilibrium with lherzolite at 150 km depth (Ryabchikov, 1993). Although not demonstrated by Massonne or Ryabchikov, both authors suggested that the transition from aqueous solution to siliceous fluid, or melt, may be supercritical (gradual) in nature. Recent experiments by Shen and Keppler (1997) and Bureau and Keppler (1999) point to the presence of complete miscibility between H_2O and silicate melt for a variety of melt compositions including albite, nepheline, jadeite, dacite, haplogranite, Ca-bearing granite, and basalt. In all systems, the critical temperature decreases with increasing pressure. At 1.5 Gpa, the critical temperature is around 800°C for jadeite. Critical temperatures at a given pressure are primarily controlled by the silica content in the system (at 2 GPa the critical temperature is 700°C for jadeite with ~60% SiO_2). Bureau and Keppler (1999) showed that, although the conditions of amphibole dehydration in a subducted slab are probably below the critical curves in an intermediate to silicic system, the decomposition of phengite and lawsonite occurs at conditions far beyond the critical curve. This implies that the progressive breakdown of phengite and lawsonite at depths >100 km should be highly enriched in silica, the actual contents of dissolved silicate depending on pressure and temperature. Domanik and Holloway (1998) showed that fluids produced during recrystallization of calcareous metapelite from the Dabieshan at 6.5 to 11 GPa and 750 to 1050°C, contain approximately 40 mol% solute at 8 GPa and 60 mol% at 11 GPa. Recognition that the fluid phase present on either side of the phengite melting reaction had similar physical characteristics has led Domanik and Holloway (1998) to propose that phengite breakdown lies beyond the second critical endpoint. Phengite and to a lesser extent lawsonite are framework-supporting minerals in basaltic and metasedimentary rocks. A gradual increase in solute content of slab fluids with increasing pressure can account for the widespread presence of omphacite-quartz and quartz-kyanite segregations in eclogite-facies rocks (Philippot, 1996; Manning, 1998), as well as precipitation of quartz daughter crystals during cooling of fluid inclusions trapped in vein omphacites (Scambelluri et al., 1998). In addition, these results provide support to models advocating the transfer of water into the mantle wedge by the migration of a silicate-rich aqueous soup having the physical properties of a hydrous melt (Philippot, 1993).

A key remaining issue concerns the plausibility of transferring a slab component to the source of arc magma genesis in a low fluid/rock regime. The observation that rocks that contain a stationary grain-boundary fluid phase may be responsible for the transfer of an appreciable amount of soluble material on a regional scale means that substantial effort has to be given to defining transport characteristics of fluids during UHP metamorphism. Several recent innovative experiments have dispelled some of the speculation obscuring our understanding of material transfer during subduction. Among these, Watson and Wark (1997) showed that aqueous SiO_2 diffusion in a stationary fluid phase at geologic P-T conditions can occur at rates comparable with that of thermal conductivity and over impressive distances (typically the km scale per Ma for PT conditions appropriate to the upper mantle). Most important is that diffusive flux through interconnected stationary fluid arises solely from the temperature gradient and is independent of the vertical distance over which the porosity is interconnected. Subduction zones, which can display a non-vertical thermal gradient up to 100°C/km, are of major interest in this respect. Although direct measurements of solute viscosity under high P-T conditions are required to fully document the effectiveness of diffusive transport properties of fluids in deep metamorphic environments, these results may explain both limited fluid circulation and mantle metasomatism in subduction zones and represent a valuable approach if uncertainties concerning fluid-melt-rock interaction processes during subduction-zone metamorphism are to be resolved.

Acknowledgments

Rumble acknowledges support of NSF grants EAR-9526700 and EAR-9725347. Philippot acknowledges support from the CNRS-INSU, Programme "Intérieur de la Terre" (Contribution no. 211). Special thanks are extended to J. M. Ferry and B. O. Mysen for their advice and constructive reviews of an earlier version of the manuscript.

REFERENCES

Agard, P., 1999, Evolution métamorphique et structurale des métapélites océaniques dans l'orogène alpin: l'exemple des schistes lustrés des Alpes occidentales: Unpubl. Ph.D. thesis, Université Jussieu, Paris, 279 p.

Agrinier, P., Hekinian, R., Bideau, D., and Javoy, M., 1995, O and H stable isotope compositions of oceanic crust and upper mantle rocks in the Hess Deep near the Galapagos Triple Junction: Earth Planet. Sci. Lett., v. 136, p. 183–196.

Agrinier, P., Laverne, C., and Tartarotti, P., 1995, Stable isotope ratios (oxygen, hydrogen) and petrology of hydrothermally altered dolerites at the bottom of the sheeted dike complex of Hole 504B, *in* Erzinger, J., Becker, K, Dick, H. J. B., and Stokking, L. B., eds., Proc. Ocean Drilling Program, Science Results. College Station, TX, p. 99–106.

Anderson, A. T., 1967, The dimensions of oxygen isotopic equilibrium attainment during prograde metamorphism: Jour. Geol., v. 75, p. 323–332.

Austrheim, H., 1987, Eclogitization of lower crustal granulites by fluid migration through shear zones: Earth Planet. Sci. Lett., v. 81, p. 221–232.

Baker, J., Matthews, A., Mattey, D., Rowley, D. B., and Xue, F., 1997, Fluid-rock interaction during ultra-high pressure metamorphism, Dabie Shan, China: Geochim. et Cosmochim. Acta, v. 61, p. 1685–1696.

Barnicoat, A. C., and Cartwright, I., 1997, Focused fluid flow during subduction: Oxygen isotope data from high pressure ophiolites of the western Alps: Earth Planet. Sci. Lett., v. 132, p. 53–61.

Blattner, P., Grindley, G. W., and Adams, C. J., 1997, Low ^{18}O terranes tracking Mesozoic polar climates in the South Pacific: Geochim. et Cosmochim. Acta, v. 61, p. 569–576.

Bureau, H., and Keppler, H., 1999, Complete miscibility between silicate melts and hydrous fluids in the upper mantle: Experimental evidence and geochemical implications: Earth Planet. Sci. Lett., v. 165, p. 187–196.

Chopin, C., 1984, Coesite and pure pyrope in high grade blueschists of the Western Alps: A first record and some consequences: Contrib. Mineral. Petrol., v. 86, p. 107–118.

Criss, R. E., and Taylor, H. P., 1986, Meteoric-hydrothermal systems, *in* Valley, J. W., Taylor, H. P., and O'Neil, J. R., ed., Stable isotopes in high temperature geological Processes: Washington, DC, Mineral. Soc. Amer., v. 16, p. 373–424.

Dal Piaz, G., 1974, Le métamorphisme alpin de haute pression et basse température dans l'évolution structurale du bassin ophiolitique alpino-apenninique: Schweiz. Mineral. Petrog. Mitteil., v. 54, p. 399–424.

Dal Piaz, G.V., Venturelli, G., and Scolari, A., 1979, Calc-alkaline to ultrapotassic postcollisional volcanic activity in the internal northwestern Alps: Mem. Sci. Geol. Padova, v. 32, pp. 16.

Desmons, J., and O'Neil, J. R., 1978, Oxygen and hydrogen isotope compositions of eclogites and associated rocks from the Eastern Sesia Zone (Western Alps, Italy): Contrib. Mineral. Petrol., v. 67, p. 79–85.

Domanik, K. J., and Holloway, J. R., 1998, Experimental determination of the stability and phase relations of phengitic muscovite in a calcareous metapelite from the Dabie Mountains, China, *in* Liou, J. G., and Ernst, W. G., eds., Int. Workshop on UHP Metamorphism and Exhumation: Stanford, Stanford Univ., p. A206–A207.

Eiler, J. M., Baumgartner, L. P., and Valley, J. W., 1992, Intercrystalline stable isotope diffusion: A fast grain boundary model: Contrib. Mineral. Petrol., v. 112, p. 543–557.

Ernst, W. G., 1999, Metamorphism, partial preservation, and exhumation of ultrahigh-pressure belts: The Island Arc, v. 8, p. 125–153.

Farquhar, J., Chacko, T., and Ellis, D. J., 1996, Preservation of oxygen isotope compositions in granulites from Northwestern Canada and Enderby Land, Antarctica: Implications for high-temperature isotopic thermometry: Contrib. Mineral. Petrol., v. 125, p. 213–224.

Ferry, J. M., and Gerdes, M. L., 1998, Chemically reactive fluid flow during metamorphism: Ann. Rev. Earth Planet. Sci., v. 26, p. 255–287.

Früh-Green, G. L., 1994, Interdependence of deformation, fluid infiltration, and reaction progress recorded in eclogitic metagranitoids (Sesia Zone, western Alps): Jour. Metamor. Geol., v. 12, p. 327–343.

Früh-Green, G. L., Plas, A., and Lecuyer, C., 1996, Petrologic and stable isotope constraints on hydrothermal alteration and serpentinization of the EPR shallow mantle at Hess Deep (site 895), *in* Mevel, C., Gillis, K. M., Allan, J. F., and Meyer, P. S., eds., Proc. of the Ocean Drilling Program, Scientific Results: College Station, TX, p. 255–287.

Fu, B., Zheng, Y.-F., Wang, Z., Xiao, Y., Gong, B., and Li, S., 1999, Oxygen and hydrogen isotope geochemistry of gneisses associated with ultrahigh pressure eclogites at Shuanghe in the Dabie Mountains: Contrib. Mineral. Petrol., v. 134, p. 52–66.

Ganor, J., Matthews, A., Schliestedt, M., and Garfunkel, Z., 1996, Oxygen isotope heterogeneities of metamorphic rocks: An original tectonostratigraphic signature, or an imprint of exotic fluids? A case study of Sifnos and Tinos islands (Greece): Eur. Jour. Mineral., v. 8, p. 719–732.

Getty, S. R., and Selverstone, J., 1994, Stable isotope and trace element evidence for restricted fluid migration in 2 GPa eclogites: Jour. Metamor. Geol., v. 12, p. 747–760.

Giletti, B. J., 1986, Diffusion effects on oxygen isotope temperatures of slowly cooled igneous and metamorphic rocks: Earth Planet. Sci. Lett., v. 77, p. 218–228.

Heinrich, W., and Gottschalk, M., 1995, Metamorphic reations between fluid inclusions and mineral host: I. Progress of the reaction calcite + quartz = wollastonite + CO_2 in natural wollastonite-hosted fluid inclusions: Contrib. Mineral. Petrol., v. 122, p. 51–61.

Jahn, B. M., 1998, Geochemical and isotopic characteristics of UHP eclogites and ultramafic rocks of the Dabie orogen: Implications for continental subduction and collisional tectonics, in Hacker, B. R., and Liou, J. G., eds., When continents collide: Geodynamics and geochemistry of ultrahigh-pressure rocks: Dordrecht, Kluwer Acad. Publ., p. 203–239.

Jamtveit, B., Bucher-Nurminen, K., and Austrheim, H., 1990, Fluid controlled eclogitization of granulites in deep crustal shear zones, Bergen arcs, Western Norway: Contrib. Mineral. Petrol., v. 104, p. 184–193.

Jenkin, G. R. T., Farrow, C. M., Fallick, A. E., and Higgins, D., 1994, Oxygen isotope exchange and closure temperatures in cooling rocks: Jour. Metamor. Geol., v. 12, p. 221–235.

Kerrick, D. M., and Jacobs, G. K., 1981, A modified Redlich-Kwong equation for H_2O, CO_2, and H_2O-CO_2 mixtures at elevated pressures and temperatures: Amer. Jour. Sci., v. 281, p. 735–767.

Kirby, S. H., Engdahl, E. R., and Denlinger, R., 1996, Intermediate-depth intraslab earthquakes and arc volcanism as physical expressions of crustal and uppermost mantle metamorphism in subducting slabs, *in* Bebout, G. E., Scholl, D. W., Kirby, S. H., and Platt, J. P., eds., Subduction top to bottom: Washington, DC, Amer. Geophys. Union, p. 195–214.

Luckscheiter, B., and Morteani, G., 1980, Microthermometrical and chemical studies of fluid inclusions from Alpine veins from the penninic rocks of the central and western Tauern Window, Austria, Italy: Lithos, v. 13, p. 61–77.

Manning, C. E., 1994, The solubility of quartz in H_2O in the lower crust and upper mantle: Geochim. et Cosmochim. Acta, v. 58, p. 4831–4839.

______, 1998, Fluid composition at the blueschist-eclogite transitions in the model system Na_2O-MgO-Al_2O_3-SiO_2-H_2O-HCl: Schweiz. Mineral. Petrog. Mitteil., v. 78, p. 225–242.

______, 1999, Solubility of corundum + kyanite in H_2O at 700 Deg C, 1 and 1.5 GPA: Evidence of high Al mobility in deep metasomatic environments [abs.]: Geol. Soc. Amer. Abs. Prog., 1999, p. A-354.

Manning, C. E., and Ingebritsen, S. E., 1999, Permeability of the continental crust: Implications of geothermal data and metamorphic systems: Rev. Geophys. Space Phys., v. 37, p. 127–150.

Marion, E., 1992, Intéractions croûte océanique profonde-eau de mer dans les gabbros de la zone M.A.R.K.: Unpubl. Ph.D. thesis, Paris, Université Paris 6, 159 p.

Matthews, A., Liata, A., Mposkos, E., and Skarpelis, N., 1996, Oxygen isotope geochemistry of the Rhodope polymetamorphic terrain in northern Greece: Evidence for preservation of pre-metamorphic isotopic compositions: Eur. Jour. Mineral., v. 8, p. 1139–1152.

Massonne, H. J., 1992, Evidence for low-temperature ultrapotassic siliceous fluids in subduction zone environment from experiments in the system K_2O-MgO-Al_2O_3-SiO_2-H_2O (KMASH): Lithos, v. 28, p. 421–434.

Messiga, B., 1987, Alpine metamorphic evolution of Ligurian Alps (North-West Italy): Chemography and petrological constraints inferred from metamorphic climax assemblages: Contrib. Mineral. Petrol., v. 95, p. 269–277.

Messiga, B., Kienast, J. R., Rebay, G., Riccardi, M. P., and Tribuzio, R., 1999, Cr-rich magnesochloritoid eclogites from the Monviso ophiolites (western Alps, Italy): Jour. Metamor. Geol., v. 17, p. 287–299.

Messiga, B., and Scambelluri, M., 1988, Comparison between two types of coronitic eclogites from the western Alps: Implications for a pre-eclogitic evolution: Schweiz. Mineral. Petrog. Mitteil., v. 68, p. 225–235.

Messiga, B., Scambelluri, M., and Piccardo, G. B., 1995, Chloritoid-bearing assemblages in mafic systems and the eclogitic hydration of Alpine Mg-Al metagabbros (Erro-Tobbio unit, Ligurian Western Alps): Eur. Jour. Mineral., v. 7, p. 1149–1167.

Mével, C., Caby, R., and Kienast, J. R., 1978, Amphibolite facies conditions in the oceanic crust: Example of amphibolitized flaser gabbros and amphibolites from the Chenaillet ophiolite massif (Hautes Alpes, France): Earth Planet. Sci. Lett., v. 39, p. 98–108.

Mibe, K., Fujii, T., and Yasuda, A., 1999, Control of the location of the volcanic front in island arcs by aqueous fluid connectivity in the mantle wedge: Nature, v. 401, p. 259–262.

Muehlenbachs, K., 1986, Alteration of the oceanic crust and the ^{18}O history of seawater, *in* Valley, J. W., Taylor, H. P. J., and O'Neil, J. R., eds., Stable isotopes in high temperature geological processes: Mineral. Soc. Amer., Rev. Mineral., v. 16, p. 425–444.

Nadeau, S., Philippot, P., and Pineau, F., 1993, Fluid inclusion and mineral isotopic compositions (H-C-O) in eclogite rocks as tracers of local fluid migration during high pressure metamorphism: Earth Planet. Sci. Lett., v. 114, p. 431–448.

Peacock, S. M., and Wang, K., 1999, Seismic consequences of warm versus cool subduction metamorphism: Examples from southwest and northeast Japan: Science, v. 286, p. 937–939.

Philippot, P., 1987, Crack-seal vein geometry in eclogitic rocks: Geodin. Acta, v. 1, p. 171–181.

______, 1993, Fluid-melt-rock interaction in mafic eclogites and coesite-bearing metasediments: Constraints on volatile recycling during subduction: Chem. Geol., v. 108, p. 93–112.

______,1996, The composition of high-pressure fluids from the Alps: Natural observations vs. experimental constraints: Earth Sci. Frontiers, v. 3, p. 39–52.

Philippot, P., Agrinier, P., and Scambelluri, M., 1998, Chlorine cycling during subduction of altered oceanic crust: Earth Planet. Sci. Lett., v. 161, p. 33–44.

Philippot, P., Chevallier, P., Chopin, C., and Dubessy, J., 1995, Fluid composition and evolution in coesite bearing rocks (Dora Maira massif, western Alps): Implications for element recycling during subduction: Contrib. Mineral. Petrol., v. 121, p. 29–44.

Philippot, P., and Selverstone, J., 1991, Trace-element brines in eclogitic veins: Implications for fluid composition and transport during subduction: Contrib. Mineral. Petrol., v. 106, p. 417–430.

Pognante, U., 1981, Magmatic and metamorphic evolution of two Fe-Ti gabbroic series from the Piemonte ophiolite nappe in the Susa valley area, Italian Western Alps: Mem. Sci. Geol. Padova, v. 35, p. 21–34.

______, 1989, Early alpine eclogitization in talc/chloritoid-bearing Mg-metagabbros and in jadeite-Fe omphacite bearing metatrondjemites from the ophiolites of the western Alps: Rend. Soc. Ital. Mineral. Petrol., v. 43, p. 687–704.

Robert, C., Javoy, M., and Kienast, J. R., 1985, Coefficients de distribution et mesures isotopiques $^{18}O/^{16}O$: comparaisons thermométriques et barométriques sur quelques éclogites et micaschists de la zone Sesia-Lanzo (Alpes italiennes): Bull. Mineral., v. 108, p. 699–711.

Rumble, D., 1978, Mineralogy, petrology, and oxygen isotope geochemistry of the Clough Formation, Black Mountain, western New Hampshire, USA: Jour. Petrol., v. 19, p. 317–340.

______, 1998, Stable isotope geochemistry of ultrahigh-pressure rocks, *in* Hacker, B. R., and Liou, J. G., eds., When continents collide: Geodynamics and geochemistry of ultrahigh-pressure rocks: Dordrecht, Kluwer Acad. Publ., p. 241–259.

Rumble, D., Ferry, J. M., Hoering, T. C., and Boucot, A. J., 1982, Fluid flow during metamorphism at the Beaver Brook fossil locality, New Hampshire: Amer. Jour. Sci., v. 282, p. 886–919.

Rumble, D., and Spear, F. S., 1983, Oxygen-isotope equilibration and permeability enhancement during regional metamorphism: Jour. Geol. Soc. Lond., v. 140, p. 619–628.

Rumble, D., Wang, Q. C., and Zhang, R. Y., Stable isotope geochemistry of marbles from UHP terrains in Dabieshan and Sulu, China : Lithos, in press.

Rumble, D., and Yui, T. F., 1998, The Qinglongshan oxygen and hydrogen isotope anomaly near Donghai in Jiangsu province, China: Geochim. et Cosmochim. Acta, v. 62, p. 3307–3321.

Ryabchikov, I. D., 1993, Fluid transport of ore metals in ultramafic mantle rocks: Proc. Eighth Quadrennial IAGOD Symp., Vol. 8, Stuggart, p. 425–433.

Ryabchikov, I. D., and Boettcher, A. L., 1980, Experimental evidence at high pressure for potassic metasomatism in the mantle of the Earth: Amer. Mineral., v. 65, p. 915–919.

Scambelluri, M., Muentener, O., Hermann, J., Piccardo, G. B., and Trommsdorff, V., 1995, Subduction of water into the mantle: History of an Alpine peridotite: Geology, v. 23, p. 459–462.

Scambelluri, M., Pennacchioni, G., and Philippot, P., 1998, Salt rich aqueous fluids formed during eclogitisation of metabasites in the alpine continental crust (Austroalpine Mt. Emilius unit, Italian Western Alps): Lithos, v. 43, p. 151–167.

Scambelluri, M., and Philippot, P., 2000, Deep cycling of fluids in subduction zones, *in* Fluid inclusion studies at the turn of the century: Eur. Mineral. Union.

Scambelluri, M., Piccardo, G., Philippot, P., Robbiano, A., and Negretti, L., 1997, High-salinity brines from recycled seawater in deeply subducted serpentinite: Earth Planet. Sci. Lett., v. 148, p. 485–499.

Schmidt, M. W., and Poli, S., 1998, Experimentally based water budgets for dehydrating slabs and consequences for arc magma generation: Earth Planet. Sci. Lett., v. 163, p. 361–379.

Schreyer, W., 1988, Experimental studies on metamorphism of crustal rocks under mantle pressure: Mineral. Mag., v. 364, p. 1–36.

Schreyer, W., Massonne, H. J., and Chopin, C., 1987, Continental crust subducted to depths near 100 km: Implications for magma and fluid genesis in collision zones, *in* Mysen, B. O., ed., Magmatic processes: Physiochemical principles: Univ. Park, PA, Geochem. Soc. Spec. Publ., v. 1, p. 155–163.

Selverstone, J., Franz, G., Thomas, S., and Getty, S. R., 1992, Fluid variability in 2 GPa eclogites as an indicator of fluid behaviour during subduction: Contrib. Mineral. Petrol., v. 112, p. 341–357.

Sharp, Z. D., Essene, E. J., and Hunziker, J. C., 1993, Stable isotope geochemistry and phase equilibria of coesite-bearing whiteschists, Dora Maira massif, western Alps: Contrib. Mineral. Petrol., v. 114, p. 1–12.

Shen, A. H., and Keppler, H., 1997, Direct observation of complete miscibility in the albite H_2O system: Nature, v. 385, p. 710–712.

Sheppard, S. M. F., and Taylor, H. P., Jr., 1974, Hydrogen and oxygen isotope evidence for the origin of water in the Boulder batholith and the Butte ore deposits: Econ. Geol., v. 69, p. 926–946.

Smith, D. C., 1984, Coesite in clinopyroxene in the Caledonides and its implications for geodynamics: Nature, v. 310, p. 641–644.

Sobolev, N. V., and Shatsky, V. S., 1990, Diamond inclusions in garnets from metamorphic rocks: Nature, v. 343, p. 742–746.

Svensen, H., Jamtveit, B., Yardley, B., Engvik, A. K., Austrheim, H., and Broman, C., 1999, Lead and bromine enrichment in eclogite-facies fluids: Extreme fractionation during lower-crustal hydration: Geology, v. 27, p. 467–470.

Tatsumi, Y., 1989, Migration of fluid phases and genesis of basalt magmas in subduction zones: Jour. Geophys. Res., v. 94, p. 4697–4707.

Thomas, S., 1991, Zur Kluftbildung in Gesteinen der Eklogitzone (Hohe Tauern, Osterreich): Unpubl. Ph.D. thesis, Technisches Universität Berlin.

Ulmer, P., and Trommsdorff, V., 1995, Serpentinite stability to mantle depths and subduction-related magmatism: Science, v. 268, p. 858–861.

Vallis, F., 1997, Stuglio degli isotopi stabili (H-O) nelle rocce ultrafemiche di ambiente eclogitico dell'unità Erro-Tobbio (Alpi Occidentali): Implicazioni sul rilascio di fluidi durante la subduzione: Unpubl. Ph.D. thesis, Università di Genova, 204 p.

Vallis, F., Frueh-Green, G.L., Scambelluri, M., and Piccardo, G.B., 1997, Eclogitization of the Erro-Tobbio serpentinized peridotite: A stable isotope study [abs.], *in* Fifth Int. Eclogite Conf., v. 9: Ascon, Switzerland, Terra Nova Abs. Supplement, p. 39.

Vallis, F., and Scambelluri, M., 1996, Redistribution of high-pressure fluids during retrograde metamorphism of eclogite facies rocks (Voltri Massif, Italian Western Alps): Lithos, v. 39, p. 81–92.

Valley, J. W., 1986, Stable isotope geochemistry of metamorphic rocks, in Valley, J. W., Taylor, H. P., Jr., and O'Neil, J. R., eds., Stable isotopes in high temperature geological processes: Mineral. Soc. Amer., Rev. Mineral., v. 16, p. 445–489.

Valley, J. W., and O'Neil, J. R., 1984, Fluid heterogeneity during granulite facies metamorphism in the Adirondacks: Stable isotope evidence: Contrib. Mineral. Petrol., v. 85, p. 158–173.

van Roermund, H. L. M., and Drury, M. R., 1998, Ultra-high pressure (P>6 GPa) garnet peridotites in western Norway: Exhumation of mantle rocks from >185 km depth: Terra Nova, v. 10, p. 295–301.

Wang, X., Zhang, R., and Liou, J. G., 1995, UHPM terrane in East Central China, *in* Coleman, R. G., and Wang, X., eds., Ultrahigh pressure metamorphism: Cambridge, UK, Cambridge Univ. Press, p. 356–390.

Watson, E. B., and Wark, D. A., 1997, Diffusion of dissolved SiO_2 in H_2O at 1 GPa, with implications for mass transport in the crust and upper mantle: Contrib. Mineral. Petrol., v. 130, p. 66–80.

Xu, J., 1993, Basic characteristics and tectonic evolution of the Tanchung-Lujiang fault zone, *in* Xu, J., ed., The Tancheng-Lujiang wrench fault system: New York, John Wiley and Sons, p. 17–50.

Yui, T. F., Rumble, D., Chen, C. H., and Lo, C. H., 1997, Stable isotope characteristics of eclogites from the ultra-high-pressure metamorphic terrain, east-central China: Chem. Geol., v. 137, p. 135–147.

Yui, T. F., Rumble, D., and Lo, C. H., 1995, Unusually low $\delta^{18}O$ ultrahigh-pressure metamorphic rocks from the Sulu terrain, eastern China: Geochim. et Cosmochim. Acta, v. 59, p. 2859–2864.

Zhang, R. Y., Rumble, D., Liou, J. G., and Wang, Q. C., 1998, Low $\delta^{18}O$, ultrahigh-P garnet-bearing mafic and ultramafic rocks from Dabie Shan, China: Chem. Geol., v. 150, p. 161-170.

Zheng, Y. F., Fu, B., Gong, B., and Li, S., 1996, Extreme ^{18}O depletion in eclogite from the Su-Lu terrane in east China: Eur. Jour. Mineral., v. 8, p. 317–323.

Zheng, Y. F., Fu, B., Gong, B., and Wang, Z.-R., 1998, Carbon isotope anomaly in marbles associated with eclogites from the Dabie Mountains in China: Jour. Geol., v. 106, p. 97–104.

Zheng, Y. F., Fu, B., Li, Y., Xiao, Y., and Li, S., 1998, Oxygen and hydrogen isotope geochemistry of ultra-high-pressure eclogites from the Dabie Mountains and the Sulu terrane: Earth Planet. Sci. Lett., v. 155, p. 113–129.

Zheng, Y. F., Fu, B., Xiao, Y., Li, Y., and Gong, B., 1999, Hydrogen and oxygen isotope evidence for fluid-rock interactions in the stages of pre- and post-UHP metamorphism in the Dabie Mountains: Lithos, v. 46, p. 677–693.

Geochemical, Isotopic, and SHRIMP Age Data for Precambrian Basement Rocks, Permian Volcanic Rocks, and Sedimentary Host Rocks to the Ore-bearing Intrusions, Noril'sk-Talnakh District, Siberian Russia

GERALD K. CZAMANSKE,

750 West Greenwich Place, Palo Alto, California 94303

JOSEPH L. WOODEN,

U.S. Geological Survey, Stanford USGS Micro-Isotopic Analytical Center, Stanford University, Stanford, California 94305

RICHARD J. WALKER,

University of Maryland, Department of Geology, College Park, Maryland 20742

VALERI A. FEDORENKO,

Central Research Institute of Geological Prospecting for Base and Precious Metals, Varshavskoye Shosse 129B, Moscow 113545, Russia

OLEG N. SIMONOV,

Taymyrian Committee for Natural Resources, Post Box 1374, Noril'sk 663300, Russia

JAMES R. BUDAHN, AND DAVID F. SIEMS

U.S. Geological Survey, Denver Federal Center, Denver, Colorado 80225

Abstract

Petrographic, geochemical, and isotopic data have been obtained for 33 samples selected to provide constraints on contamination models for the volcanic and intrusive components of the Late Permian to Early Triassic, Siberian flood-volcanic province. Twenty-one of these samples were carried from great depth in an explosive diatreme of Triassic age, whereas 12 were collected from drill core from depths of tens to 2000 m. The studied diatreme xenoliths are: (1) fragments of the crystalline basement; and (2) fragments of a basaltic-to-rhyolitic volcanic suite.

Prompted by an unexpected, Late Paleozoic, Rb-Sr isochron age for this compositionally diverse volcanic suite, a SHRIMP U-Pb zircon age of ~270 Ma was obtained for a rhyodacite xenolith. Previously, a SHRIMP zircon U-Pb age of ~910 Ma had been determined for a leucogranite xenolith from the crystalline basement; this sample also contains substantial amounts of inherited, Early Proterozoic and Archean zircon.

The presence of this volcanic suite, only ~20 m.y. older than the 251 Ma, flood-volcanic sequence, is an extremely provocative result, inasmuch as hundreds of exploration drill holes in the Noril'sk area, and throughout the Siberian platform, have encountered only Tungusskaya Series coal-bearing sedimentary rocks in this stratigraphic/time interval. These data support arguments that subduction/underthrusting from the West Siberian Lowland under the northwest margin of the Siberian craton took place in Late Permian time.

The isotopic data obtained for the xenolith suite indicate that the upper part of the crystalline basement under the northwest margin of the Siberian craton is composed of Late Proterozoic (Riphean) rocks—alkaline granites, trondhjemites, crystalline schists, gneisses, and amphibolites—with much in common with rocks of the Central zone of the Taymyr folded area, which has been interpreted as an accretionary block formed and joined to Siberia in Late Riphean to Vendian time.

Measured isotopic characteristics for the Precambrian crystalline basement, and the Paleozoic sedimentary rocks that host the ore-bearing intrusions in the Noril'sk region, provide parameters for quantitative modeling of crustal contamination during evolution of the Siberian flood-volcanic rocks and related intrusions, both while en route to the surface and at the site of intrusion emplacement.

Introduction

THE NORIL'SK-TALNAKH mining district of Siberia is one of the outstanding mineral districts in the world. Cu-Ni resources are comparable to those of the deposits at Sudbury, Ontario, Canada, and the platinum-group-element (PGE) resource is second only to that of the Bushveld Complex, South Africa. The deposits are highly unusual in their association with a continental-flood-volcanic province, in the vast amount of massive, magmatic sulfide found in association with hypabyssal mafic intrusions, and in the exceptionally high concentrations of PGE, Cu, and Ni in the ores.

The world-class significance of both the flood-volcanic province and the ore deposits, and the possibility that construction of a credible ore-deposit model could lead to discovery of similar deposits elsewhere in the world, have inspired intense study of the ores, mafic intrusions, and flood-volcanic rocks, for which extensive data sets and discussions have been published (e.g., Lightfoot et al., 1990, 1993, 1994; Sharma et al., 1991, 1992; Naldrett et al., 1992, 1994, 1995; Wooden et al., 1992, 1993; Arndt et al., 1993, 1995, 1998; Czamanske et al., 1994, 1995, 1998; Fedorenko, 1994; Walker et al., 1994; Basu et al., 1995; Hawkesworth et al., 1995; Horan et al., 1995; Fedorenko et al., 1996; Kamo et al., 1996, 2000; Fedorenko and Czamanske, 1997; Lassiter and DePaolo, 1997; Lightfoot and Hawkesworth, 1997; Sharma, 1997).

Many of these studies have called on crustal contamination of mantle-derived melts to generate the geochemical and isotopic properties of the intrusions and ores. Thus, to constrain models for the overall evolution of the mafic magmatic system that produced the flood basalts, ore-bearing intrusions, and ores of the Noril'sk area, it is essential to consider the geochemical and isotopic compositions of those rocks that represent the most likely sources of contamination for the mantle-derived, mafic melts—(1) the crystalline basement, (2) strata traversed during magma ascent, and (3) those sedimentary host rocks into which the ore-bearing intrusions were emplaced. At present there are essentially no relevant data available for these rocks, and this work was undertaken as a preliminary attempt to fill that void with respect to their Nd, Sr, and Pb isotopic characteristics.

Because the crystalline basement is not exposed or reached by drilling in the Noril'sk area and its surroundings, Maslov's diatreme, a Triassic explosive structure located only 25 km south of the Noril'sk-1 ore-bearing intrusion, represents a unique opportunity to sample the basement. Triassic tuffs are abundant, and related diatremes numerous in the Siberian flood-volcanic province, but fragments carried from the platform basement are extremely unusual and sparse within them (see references in Maslov and Nesterovskiy, 1961). Only Maslov's diatreme is known to contain supposed basement rock fragments in abundance, and of great diversity (Maslov and Nesterovskiy, 1961; Nemenyonok, 1978).

In addition to constraining the characteristics of possible crustal contaminants, geochemical and isotopic study of rock units subjacent to the Noril'sk area provides information of significance in constraining the construction of the platform basement and the paleotectonic evolution of north-central Siberia. Zonenshain et al. (1990) and Şengör et al. (1993) suggest that the Noril'sk area lies above the convergent, western margin of the Siberian platform. The present study gives support to this proposal and provides a time frame for a collisional event.

Geologic Setting

The Noril'sk-Talnakh mining district is situated near the extreme northwestern corner of both the Siberian craton and the most extensive, relatively undisturbed exposures of the flood-volcanic sequence. It lies nearly 700 km west of Precambrian crystalline rock exposures in the Anabar shield and 500 km south of Precambrian exposures in the central Taymyr Peninsula (Fig. 1). The intervening areas are covered by sedimentary and flood-volcanic rocks, as are the areas for well over 1000 km to the south and west of Noril'sk. To readily allow the reader additional perspective on our discussion of the diverse rock assemblage found in Maslov's diatreme, brief outlines of the stratigraphy of the Siberian platform cover, the geology of the Taymyr Peninsula, and the geochemistry of the Siberian flood-volcanic sequence are presented in Appendix 1.

The ore-bearing ("Noril'sk-type") intrusions of the region are elongate mafic-ultramafic intrusions 15–18 km long, 1–3 km wide, and less than 350 m thick; descriptions of the morphology and petrology of these intrusions were presented by Zen'ko and Czamanske (1994) and Czamanske et al. (1995). These unusual ore-bearing intrusions constitute one of 15 recognized intrusion types that are related to

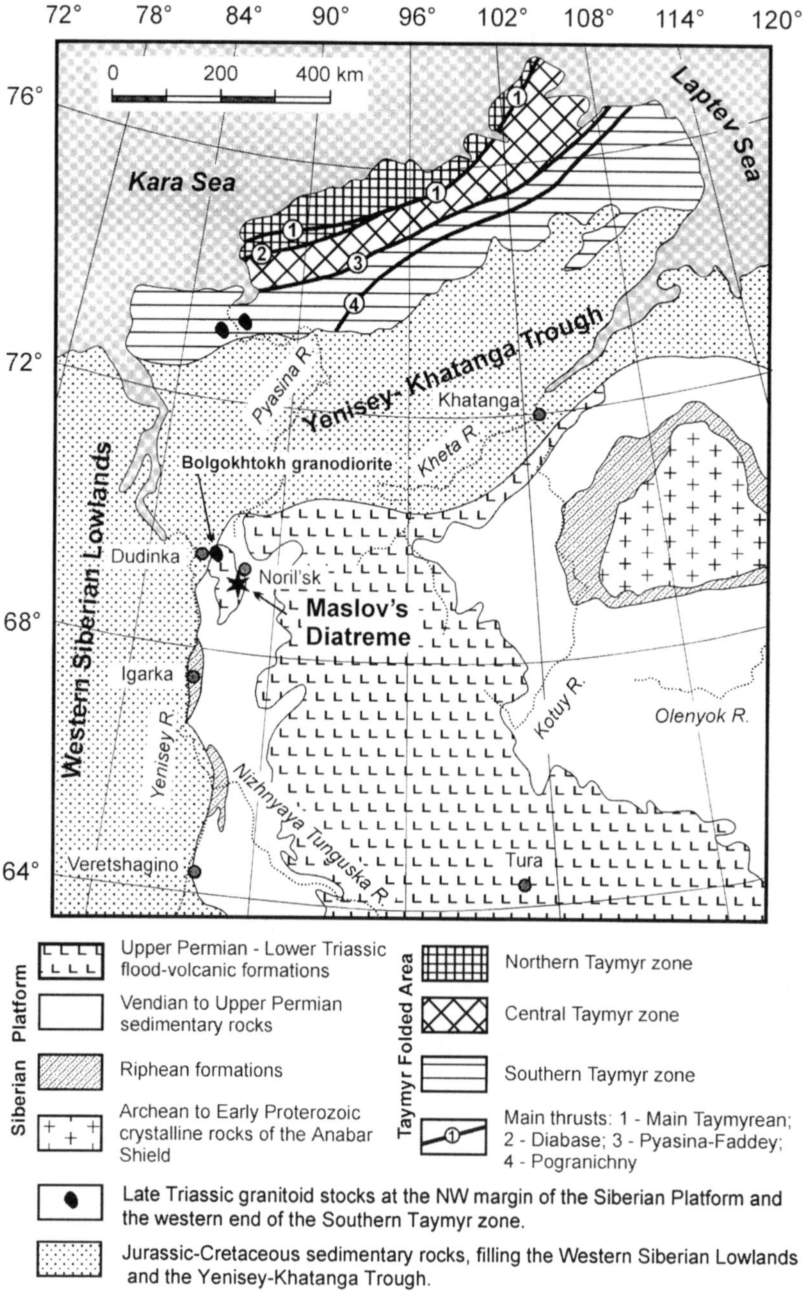

FIG. 1. Index map for north-central Siberia showing the relation of Maslov's diatreme and the Bolgokhtokh granodiorite to present exposures of the Permian–Triassic flood-volcanic province and major structural elements, including the West Siberian Lowland, the Yenisey-Khantanga trough, the Taymyr Peninsula, and the Anabar shield.

the ~251 Ma Siberian flood-volcanic province, apparently the world's largest flood-volcanic event. According to Masaitis (1983), ~4 million km^3 of lava once covered an area approaching 7 million km^2. The flood-volcanic rocks of the area proximal to the mining district are the best studied, as a result of intense exploration drilling. The volcanic sequence of this area is ~3500 m thick and includes the most complete sequence of the early volcanic rocks, comprising compositions that are among the most diverse to be found. In contrast to most other continental-flood-volcanic provinces, intermittent explosive volcanism contributed ~10 vol% to the volcanic sequence in the Noril'sk area. The lavas, tuffs, and intrusions (in approximate proportions of 84:10:6 in the Noril'sk region) were deposited on, and emplaced within, a 4000 to 5000 m thick Cambrian through Upper Permian sedimentary sequence. The ore-bearing intrusions are found in Lower Devonian to Upper Permian strata within the upper 1000 meters of this sequence and in the lower few hundred meters of the flood-volcanic sequence (see Zen'ko and Czamanske, 1994, their Fig. 1).

The age of Siberian flood-volcanic magmatism is now well constrained. Kamo et al. (1996) obtained an age of 251.2 ± 0.3 (2σ) Ma for the Noril'sk-1 intrusion by application of the conventional U-Pb dating technique to zircons and baddeleyite. This age postdates the initiation of flood-volcanic eruption because the Noril'sk-1 intrusion cuts the lower suites of the flood-volcanic sequence. On the basis of a variety of considerations, Fedorenko (1994) proposed that emplacement of the ore-bearing intrusions was contemporaneous with eruption of the 100 m thick tuff horizon at the boundary of the Lower and Upper subsuites of the Morongovsky basalt suite, after accumulation of about one-third of the flood-volcanic sequence in the Noril'sk region. More recently, Kamo et al. (2000) reported ages of 252.1 ± 0.4 (2σ) Ma and 251.1 ± 0.5 (2σ) Ma for samples from the Maymecha-Kotuy area of the province, which are believed to encompass virtually the entire duration of Siberian flood volcanism. These ages place Siberian flood volcanism precisely at the Permian–Triassic boundary, fixed at 251.4 ± 0.3 (2σ by Bowring et al. [1998]), supporting the possiblity that this volcanism contributed to the unprecedented, end-Permian mass extinction.

Opportunely, the Precambrian basement in the Noril'sk area was sampled by Maslov's diatreme (Fig. 1), situated less than 25 km due south of the town of Noril'sk and the Noril'sk-1, ore-bearing intrusion, just east of the major, Noril'sk-Kharaelakh fault. Volcanic stratigraphy suggests that this diatreme formed almost contemporaneously with emplacement of the ore-bearing intrusions.

The outcrop of Maslov's diatreme is located on a gentle slope in a part of the Noril'sk Plateau; the surrounding several hundred square meters are covered by debris of silty to sandy tuff. Exposures and trenches made during investigations by Nemenyonok (1978) indicate the presence of a round funnel (~5 m in diameter), which contains about 50 vol% of exotic rock fragments (subrounded, from gravel to 20–30 cm size). Extending to the southwest is a marginal phase of ~6 × 9 m, which contains 10 to 50 vol% fragments. A few fragments are dispersed 15 to 20 m from the funnel, and rare fragments are found a few tens to hundreds of meters from the funnel. The fragments appear fresh, with the majority in the size range from 5 to 10 cm. Rarely, the fragments are partly rimmed by a few millimeters of devitrified glass, but none show incipient melting. Detailed field inventory (supported by the petrographic and geochemical study involved in this report) indicates an approximate distribution of 49% volcanic rocks (basaltic, andesitic, and felsic), 18% leucogranites (roughly equally distributed between alkalic and trondhjemitic varieties), 16% red quartzite-like sandstones (similar to rocks of the Riphean Gubinsky Series of the Igarka area; see Appendix 1), 4% crystalline schists and gneisses, 4% amphibolites, and 9% altered rocks of indeterminate nature. Muscovite-bearing rocks are abundant, whereas biotite-bearing granitoids are absent and gneisses rarely contain biotite. There appear to be no fragments representing the uppermost 6000 m of Vendian (~650–550 Ma) and Paleozoic sedimentary rocks, largely carbonates and evaporites (for the regional stratigraphic column, see Fig. 15). This is remarkable because carbonate rocks are resistant to weathering in the Noril'sk area and sometimes are found in Triassic tuff breccias (I. A. Natorkhin, unpubl. manuscript, 1970).

The Bolgokhtokh granodiorite intrudes Upper Silurian to Upper Devonian sedimentary rocks about 50 km west of Noril'sk (Fig. 1; Komarova, 1968). It was discovered in 1965 during oil exploration, and is known only from drill-core data. A recent drill hole shows little variation in chemical composition over 2 km depth. It is compositionally quite distinct from other known intrusive bodies in the Noril'sk region, but seems analogous to several small granitoid intrusions of the western Taymyr area described by Ravich and Chayka (1956). The age of the Bolgokhtokh granodiorite, which metamorphoses Triassic volcanic rocks, has been established as 229.2 ± 0.3 (2σ) Ma (Kamo et al., 2000). Although younger than the flood-volcanic sequence, the source characteristics of the granodiorite provide useful information regarding the isotopic characteristics of the underlying continental crust.

Studied Sample Collection

We present extensive chemical and isotopic data for nine samples from the Middle–Upper Paleozoic sedimentary sequence that hosts the ore-bearing intrusions, and 21 representative rocks from the diatreme at our 19-F locality—12 represent the crystalline basement and 9 a volcanic suite. In addition, we analyzed two samples from the Late Triassic Bolgokhtokh granodiorite for comparison with the basement granitoids, and a sample of Late Precambrian leucobasalt from the Igarka area (~200 km south of Noril'sk; Fig. 1) for comparison with the volcanic rocks from the diatreme. Geochemical data for these samples are presented in Table 1, isotopic data in Table 2, and brief comments on their petrography in Appendix 2.

Rock types

Samples of the crystalline basement include amphibolitic greenstones (19F-3, 4, and 21); foliated, muscovite-bearing, metapelites (19F-6 and 7); and a variety of gneisses and granitoids. Samples 19F-11 and 12 are undeformed trondhjemites with K_2O contents of 0.61 and 0.48 wt%, respectively, whereas leucogranites 19F-8, 10, and 69 have higher K_2O contents. The trondhjemites and leucogranites are quite comparable petrographically. A muscovite gneiss (19F-9) and a pink augen gneiss (19F-13) complete the collection of crystalline basement rocks. Samples of volcanic rocks are unmetamorphosed, but may be either relatively fresh or more altered. The fresher suite includes two basalts (19F-51 and 62), a leucocratic basalt (19F-60), an andesite (19F-52), a dacite (19F-77), a rhyodacite (19F-57), and a rhyolite/trachyrhyolite (19F-76). The three more siliceous rocks each contain plagioclase phenocrysts. More altered samples include basalt (19F-1) and rhyolite (19F-22).

Samples of the metasedimentary host rocks to the intrusions and the Bolgokhtokh granodiorite were taken from drill cores; the last digits of each sample number represent the sampled depth in meters. Metasedimentary rock samples taken immediately adjacent to Noril'sk-type, ore-bearing intrusions include KZ-1799/1359.1 and 1366.1 after Carboniferous–Permian, Tungusskaya Series siltstone from 4.8 and 11.8 m, respectively, beneath the northeastern branch of the Talnakh intrusion, and KZ-1879/1659.4 and 1661 after Razvedochninsky (Lower–Middle Devonian) argillite from 35.3 and 33.7 m, respectively, above the roof of the Kharaelakh intrusion. Tungusskaya Series metasediments are also represented by sample SG-28/1993.9, taken 12.7 m above the base of the Lower Talnakh intrusion from a metasedimentary lens that occupies the interval 1987.2–1995.1 m within the body of the intrusion, and sample SG-28/2018.9, taken from 2.3 m below the base of the intrusion. Sample SG-28/2030.1, taken from 13.5 m below the base of the intrusion, represents a Middle Devonian, Upper Manturovsky Subsuite carbonate. (The Lower Talnakh intrusion has a close genetic connection with the ore-bearing intrusions; it lies just below the Talnakh intrusion, and apparently was emplaced shortly before it. For clarification regarding these intrusions and their stratigraphic placement, consult Zen'ko and Czamanske, 1994 and Czamanske et al., 1995.) Lower–Middle Devonian, Razvedochninsky argillite, well removed from the intrusions, is represented by samples K-1010-612.4 and 654.5. Samples BG-26-80.5 and 199 are of the Bolgokhtokh granodiorite. Sample IP-150/930 is a leucobasalt from the Late Precambrian, Rzhevsky Suite of the Igarka area.

Analytical methods

Details of sample preparation and all analytical procedures other than those for Nd-isotopic compositions are given by Wooden et al. (1993). Procedures for Nd-isotopic compositions are given by Walker et al. (1994). Uranium-Pb zircon analyses were conducted on SHRIMP I at Australian National University (ANU) in 1995 and on the Stanford-USGS SHRIMP-RG in 1999 and 2000, using standard operating procedures (Ireland, 1992). Uranium and Th concentrations were calculated relative to SHRIMP zircon standard SL-13, and U and Pb atomic ratios and ages were calculated relative to SL-13 for the ANU data and AS57 for the Stanford-USGS data.

Results and Discussion

Geochemistry

Host rocks of the ore-bearing intrusions. The overall geochemistry of the argillite and hornfels samples is consistent with their detrital sedimentary

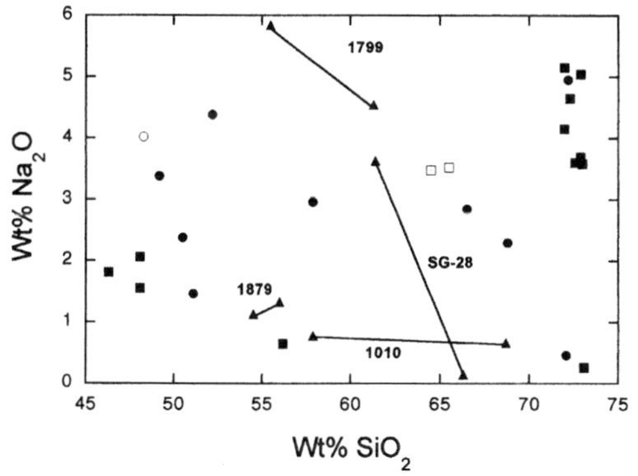

FIG. 2. Plot of SiO_2 against Na_2O for all samples. In this and most other figures, solid triangles represent argillite and hornfels, dots represent diatreme volcanic rocks, solid squares represent crystalline basement rocks, a circle represents Igarka leucobasalt IP-150/930, and open squares represent the Bolgokhtokh granodiorite. Lines in this and following figures typically connect argillite and hornfels samples from the same drill hole.

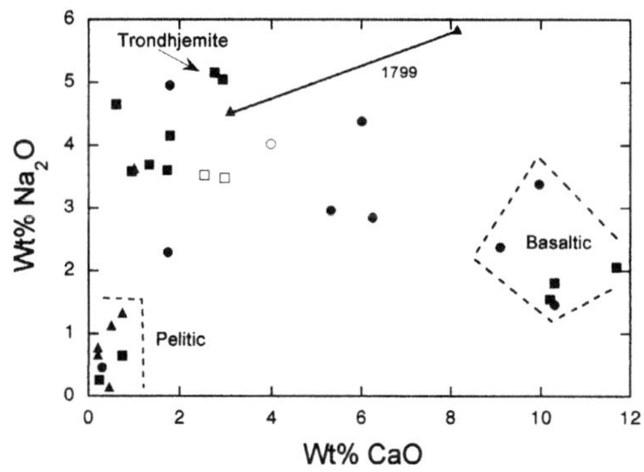

FIG. 3. Plot of CaO against Na_2O for all samples. Dashed polygons in this and following figures emphasize lithologic groupings. Trondhjemite refers to two crystalline basement samples with this composition.

origin. The unmetamorphosed Devonian argillite samples from drill hole K-1010 are relatively silica rich (58 and 69 wt% SiO_2), and have the characteristic very low CaO and Na_2O contents of clay-rich rocks (Figs. 2 and 3). The high total FeO contents (FeOT; 10–12 wt%) of the K-1010 samples are also distinctive. The KZ-1879 hornfels samples are clay rich, as denoted by the relatively low SiO_2 (54–56 wt%, Fig. 2), very low CaO and Na_2O (under 1.5 wt%, Fig. 3), and very high Al_2O_3 contents (22 wt%, Table 1). The geochemistry of the Tungusskaya Series hornfels and metasiltsones is different. The geochemistry of the KZ-1799 samples is indicative of a mafic graywacke, as suggested by the low SiO_2 (55–61 wt%, Fig. 2), and higher CaO and Na_2O contents (3–8, and 4.5–5.8 wt%, respectively, Figs. 3, 4, and 5), whereas the geochemistry of the SG-28 samples suggests that lithologically they lie between the wacke and argillitic end-members (Figs. 2 and 3). SG-28/2030.1 is the only carbonate sample analyzed (Table 1). Figure 7A shows the REE patterns for these rocks.

Volcanic rocks from the diatreme. The geochemistry of the volcanic samples reflects the lithologic range of these samples from basalt to rhyolite (Figs. 2 and 3). The basalts 19F-1, 51, and 62 represent modestly evolved magmas, with MgO = 5–7 wt%, moderately low TiO_2 (1.2–1.6 wt%), a range in total FeO (7–14 wt%, Fig. 6), and low Ni and Cr contents (<120 and <220 ppm, respectively). The REE patterns of these samples show slight LREE enrichment and minor negative Eu anomalies (Fig. 7B), which are interpreted to mean that these samples did not experience extensive plagioclase fractionation. The andesite 19F-52 is MgO and FeO rich (5.6 and 10 wt%, respectively), and its REE pattern is essentially the same as those for the basalts in abundance and shape. The silica content of 19F-60 (52.2 wt%, Fig. 2) would seem to place it with the basaltic samples as a leucobasalt, but its low MgO (2.5 wt%) and high Na_2O, K_2O, and TiO_2 contents (4.4, 3.6, and 2.6 wt%, respectively, Figs. 3 and 6) indicate a more potassic and/or fractionated petrogenesis. The REE content of 19F-60 is significantly higher than those of the basalts and the andesite, but the REE pattern shows only modest LREE enrichment. 19F-77 and 22 are relatively low in K_2O with respect to their SiO_2 and Na_2O contents (Fig. 4), and have low Rb (20, 33 ppm) and moderate Sr (405, 305 ppm) contents. These Sr contents distinguish these two samples from all other samples of this suite, which have lower Sr contents (Fig. 5). Both samples have low total REE contents, with 19F-77 having only modest LREE enrichment and 19F-22 showing stronger LREE enrichment, as a result of its very low HREE content (Fig. 7B). These characteristics suggest that the rocks represent sodic-suite magmatism, in contrast to the potassic nature of 19F-57, which has a strong, negative Eu anomaly, but only modest LREE enrichment. Sample 19F-76, with its unusually high K_2O (10 wt%) and low Na_2O contents (0.46 wt%), must have experienced potassic alteration. Sample 19F-76 is also distinctive for its large, negative Eu anomaly, which suggests

TABLE 1. Major- and Trace-Element Data for the Studied Sample Collection[1]

	Greenstone			Meta-pelite		Gneiss		Trondhjemite		Leucogranite			Granodiorite	
	19F-3	19F-4	19F-21	19F-7	19F-6	19F-13	19F-9	19F-11	19F-12	19F-8	19F-10	19F-69	BG-26 80.5	BG-26 199
							wt%							
SiO_2	48.1	46.3	48.1	56.2	73.1	72.3	72.9	72.0	72.9	72.0	72.6	73.0	64.5	65.5
Al_2O_3	14.0	14.3	12.7	19.5	12.4	13.9	14.1	15.6	15.0	14.9	15.3	15.0	14.3	14.6
Fe_2O_3	3.41	2.52	2.22	2.76	1.25	0.86	0.65	0.42	0.35	0.62	0.40	0.87[2]	2.08	1.81
FeO	11.85	12.02	8.34	8.57	4.17	1.33	0.96	1.02	0.84	1.18	0.46	n.d.	1.68	1.39
MgO	4.45	7.36	8.40	2.12	1.22	0.70	0.43	0.60	0.44	0.61	0.31	0.22	2.10	1.83
CaO	10.2	10.3	11.7	0.74	0.24	0.61	1.33	2.77	2.95	1.79	1.73	0.95	2.99	2.55
Na_2O	1.55	1.81	2.06	0.65	0.26	4.65	3.69	5.15	5.04	4.15	3.60	3.58	3.48	3.52
K_2O	0.19	0.73	0.39	3.11	2.57	3.16	4.03	0.67	0.48	1.51	3.91	4.38	4.91	4.88
TiO_2	2.65	1.88	1.11	1.14	0.85	0.29	0.19	0.17	0.12	0.28	0.10	0.09	0.54	0.48
P_2O_5	0.30	0.24	0.12	0.26	0.06	0.09	0.10	0.05	<0.05	<0.05	0.06	0.11	0.41	0.34
MnO	0.23	0.23	0.19	0.10	< 0.02	0.03	0.03	0.02	< 0.02	< 0.02	< 0.02	0.02	0.05	0.03
H_2O^+	2.87	2.48	2.72	4.39	2.52	1.02	0.71	0.96	0.84	1.35	0.77	0.86[2]	0.55	0.63
H_2O^-	0.1	0.14	0.11	0.33	0.29	0.17	0.19	0.14	0.15	0.12	0.11	n.d.	0.37	0.49
CO_2	0.26	0.03	1.93	0.48	0.02	0.02	0.03	0.01	0.17	1.61	0.05	n.d.	0.33	0.39
S	0.02	<0.01	<0.01	n.d.	n.d.	n.d.	n.d.	n.d.	n.d.	n.d.	n.d.	n.d.	n.d.	n.d.
							ppm							
Ba	31	186	146	550	2700	500	730	132	96	560	1000	570	4100	4000
Co	41	51	39	28	15.2	3.2	2.0	2.2	1.2	4.8	0.6	0.6	9	7
Cr	36	225	365	152	51	11	4.8	6.5	2	19	1	1	<20	25
Cs	0.14	1.1	0.32	2.9	0.57	1.9	4.0	0.85	0.49	0.96	1.9	2.3	3.1	3.2
Cu	320	17	178	40	21	11	12	<10	20	23	12	10	148	80
Hf	5.24	3.39	1.91	7.35	7.92	4.43	3.93	2.93	2.59	3.91	2.37	1.43	7.68	7.08
Nb	17	17	8	24	13	13	12	<5	<5	7	9	14	31	32
Ni	66	132	120	66	29	9.5	2.9	4	4.5	<10	2.4	<0.3	15	26
Pb	3.40	1.81	1.17	4.35	3.05	26.9	42.5	3.67	2.65	31.3	22.2	??//??	34.94	26.58
Rb	5	27	<5	114	51	172	172	27	22	47	128	169	150	156
Sc	39	47	50	21	14	5	3	3	2	3	2	3	8	7
Sr	170	220	194	87	72	128	360	305	250	245	430	250	1700	1600
Ta	0.859	0.702	0.308	1.44	0.716	1.57	1.81	0.085	0.051	0.426	0.504	0.641	3.05	2.83
Th	1.78	0.955	0.586	18.9	15.4	18.0	9.02	0.807	0.396	21.7	3.45	3.48	21.9	21.5
U	0.626	0.302	0.227	3.15	1.59	2.03	1.66	0.238	0.145	1.25	0.610	0.911	7.06	5.34
Y	51	37	16	35	18	20	16	<5	7	8	13	28	17	16
Zn	116	112	65	86	42	63	51	29	34	53	51	38	48	48
Zr	210	154	76	270	300	160	120	110	94	152	88	60	300	300
La	14.7	15.3	8.41	55.4	51.7	27.6	23.8	3.97	2.80	63.5	14.7	12.5	107	103
Ce	35.7	36.4	18.8	128	112	63.5	47.4	8.40	6.24	130	27.2	23.7	203	190
Nd	23.9	22.0	12.0	49.1	39.5	23.2	19.3	4.04	3.30	46.6	9.34	7.23	75.7	70.8
Sm	6.87	5.52	3.12	9.50	7.01	4.47	3.30	0.893	0.805	7.01	1.88	1.83	11.7	10.9
Eu	2.19	1.73	1.12	1.78	2.50	0.591	0.57	0.313	0.278	3.15	0.441	0.383	2.48	2.26
Gd	8.17	5.68	3.35	7.98	6.19	3.60	2.14	0.900	0.818	4.12	1.65	2.16	7.62	6.64
Tb	1.35	0.895	0.569	1.18	0.701	0.497	0.247	0.120	0.118	0.368	0.226	0.381	0.781	0.711
Ho	1.93	1.28	0.807	1.46	0.804	0.599	–	0.164	–	0.289	–	–	0.854	0.749
Tm	0.768	0.517	0.333	0.580	0.306	0.224	–	0.069	0.069	0.051	0.106	0.276	–	–
Yb	5.01	3.28	2.14	3.66	1.87	1.40	0.690	0.412	0.426	0.273	0.599	1.81	1.55	1.39
Lu	0.739	0.458	0.319	0.541	0.277	0.196	0.100	0.060	0.064	0.033	0.081	0.245	0.220	0.200

(Continues)

TABLE 1. *(continued)*

	Volcanic rocks										Hornfels			
	19F-1	19F-51	19F-62	19F-60	19F-52	19F-77	19F-57	19F-76	19F-22	IP 150-930	KZ-1799 1359.1	KZ-1799 1366.1	KZ-1879 1659.4	KZ-1879 1661
								wt%						
SiO_2	49.2	50.5	51.1	52.2	57.9	66.5	68.8	72.1	72.2	48.3	55.5	61.3	54.5	56.0
Al_2O_3	13.1	13.4	14.5	13.5	14.3	14.8	13.3	12.6	14.8	17.2	12.7	14.8	22.2	22.7
Fe_2O_3	3.67	6.37	8.45	11.15	5.03	2.33	3.11	3.48	0.44	11.10	2.34	1.83	2.17	1.34
FeO	4.18	7.90	5.63	2.50	5.18	2.90	2.99	0.28	0.96	3.03	4.95	1.96	6.85	5.93
MgO	5.40	7.25	6.47	2.86	5.64	2.69	1.68	0.25	0.52	3.77	2.85	1.49	2.25	2.40
CaO	9.96	9.10	10.3	6.01	5.32	6.25	1.74	0.30	1.78	3.99	8.15	3.11	0.50	0.74
Na_2O	3.38	2.37	1.46	4.38	2.96	2.84	2.29	0.46	4.95	4.02	5.84	4.54	1.13	1.33
K_2O	0.46	0.98	0.53	3.58	2.24	0.94	4.92	10.04	1.40	2.38	1.63	4.86	4.38	2.77
TiO_2	1.17	1.63	1.25	2.59	1.20	0.56	0.87	0.44	0.16	2.16	0.66	0.69	1.07	1.28
P_2O_5	0.17	0.29	0.15	0.98	0.17	0.08	0.26	0.03	0.06	0.48	0.24	0.08	0.24	0.43
MnO	0.10	0.24	0.18	0.19	0.15	0.11	0.09	0.03	< 0.02	0.22	0.15	< 0.02	< 0.02	< 0.02
H_2O^+	4.06	3.72	5.55	1.22	3.05	2.89	2.24	0.50	1.13	3.14	0.95	1.06	4.45	4.82
H_2O^-	0.39	1.58	1.30	0.59	1.15	1.66	0.61	0.29	0.18	0.38	0.30	0.17	0.41	0.22
CO_2	5.98	0.06	0.02	0.07	0.19	0.04	0.05	0.09	0.81	2.74	0.43	0.27	0.04	0.04
S	<0.01	0.01	0.01	0.01	0.01	0.02	0.01	0.01	n.d.	0.01	n.d.	n.d.	n.d.	n.d.
								ppm						
Ba	<30	390	114	810	425	520	970	1500	395	265	160	230	148	126
Co	29	43	49	20	34	18	7	1	2.1	28	2	8	23	14
Cr	136	24	220	< 20	144	67	< 20	< 20	2	225	102	91	114	126
Cs	0.32	0.16	0.47	0.06	0.26	0.15	3.2	1.4	0.56	3.5	0.31	0.15	2.2	0.86
Cu	78	148	124	22	63	66	10	14	<10	<10	25	76	17	102
Hf	2.66	3.58	2.19	9.20	4.17	2.32	10.3	22.3	2.96	4.35	5.56	5.95	5.38	5.85
Nb	12	12	<10	33	13	5	24	84	<5	–	17	19	15	17
Ni	70	52	122	16	46	44	< 10	< 10	<3.2	50	20	27	78	33
Pb	6.95	3.75	1.79	9.38	7.18	11.5	20.4	33.6	4.65	–	0.28	0.70	0.22	3.37
Rb	14	26	14	73	54	20	146	310	33	84	24	54	116	92
Sc	35	39	38	21	31	15	11	3	2	26	7	7	25	28
Sr	128	182	126	100	140	405	122	61	305	80	196	164	85	110
Ta	0.519	0.523	0.253	1.78	0.743	0.366	1.60	5.66	0.103	4.40	0.907	1.02	0.828	1.09
Th	1.42	2.72	1.10	5.61	4.68	3.18	15.40	23.70	2.66	5.13	11.3	10.7	7.30	9.69
U	0.522	1.32	0.377	1.46	1.45	1.45	3.07	4.92	0.603	4.78	1.70	2.14	3.07	4.09
Y	29	34	27	54	32	22	43	64	<5	28	40	19	38	53
Zn	34	110	78	152	83	54	112	76	39	270	37	16	18	25
Zr	110	156	97	415	176	106	425	990	114	196	225	255	220	240
La	12.0	16.0	8.33	51.1	23.1	14.1	64.5	111	9.53	49.3	42.6	5.83	34.8	43.6
Ce	24.6	35.3	18.0	121	47.5	27.1	144	234	19.0	101	128	17.7	58.0	95.7
Nd	14.4	21.7	12.4	61.1	23.4	11.5	63.2	87.9	8.77	42.8	67.2	9.64	24.1	43.0
Sm	3.48	5.34	3.43	12.9	5.24	2.44	12.3	16.5	1.74	8.13	12.7	2.10	4.55	9.6
Eu	1.12	1.37	0.906	3.24	1.23	0.691	1.66	0.961	0.503	2.10	1.60	0.780	1.35	2.17
Gd	3.91	5.62	3.56	13.0	5.15	2.50	10.4	13.1	1.25	6.52	9.89	2.26	4.85	9.90
Tb	0.641	0.937	0.567	1.840	0.810	0.372	1.40	2.00	0.180	0.905	1.27	0.311	0.841	1.460
Ho	0.906	1.56	0.872	2.36	1.13	0.529	1.49	2.55	–	–	1.33	0.443	1.25	1.93
Tm	0.392	0.613	0.393	0.819	0.458	0.207	0.643	1.16	0.055	–	0.464	-	0.559	0.774
Yb	2.59	3.79	2.51	4.75	2.91	1.24	3.86	7.14	0.330	2.06	2.86	1.46	3.65	4.59
Lu	0.385	0.555	0.363	0.669	0.430	0.174	0.543	0.965	0.043	0.284	0.430	0.236	0.545	0.679

(Continues)

TABLE 1. (*continued*)

	Hornfels		Carbonate	Argillite	
	SG-28	SG-28	SG-28	K-1010	K-1010
	1993.9	2018.9	2030.1	612.4	654.5
			wt%		
SiO_2	61.4	66.3	2.59	68.7	57.9
Al_2O_3	14.4	14.8	0.50	13.5	18.3
Fe_2O_3	1.08	0.73	1.16	7.19	8.66
FeO	6.77	4.98	0.88	3.86	3.74
MgO	4.21	3.33	19.2	1.32	2.42
CaO	1.00	0.46	30.4	0.21	0.21
Na_2O	3.63	< 0.15	<0.15	0.66	0.78
K_2O	1.47	3.24	0.16	1.48	2.92
TiO_2	0.82	0.84	<0.02	1.71	1.35
P_2O_5	0.14	0.15	<0.05	0.09	0.08
MnO	0.05	< 0.02	0.21	0.10	0.04
H_2O^+	3.47	4.04	0.50	4.23	5.40
H_2O^-	0.39	0.53	0.13	0.44	0.85
CO_2	0.55	0.21	41.3	0.63	0.67
S	n.d.	n.d.	n.d.	n.d.	n.d.
			ppm		
Ba	102	215	14	61	92
Co	42	9	1.1	13.5	34.4
Cr	118	124	4.7	96	112
Cs	0.98	1.6	0.2	2.8	5.1
Cu	19	19	20	<10	160
Hf	4.58	5.11	0.17	6.5	5.49
Nb	12	11	5	23	16
Ni	47	36	35	45	60
Pb	6.78	1.23	7.29	5.72	14.4
Rb	35	77	6	48	100
Sc	15	15	1	17	22
Sr	66	34	116	114	104
Ta	0.720	0.801	0.034	1.53	1.12
Th	6.48	6.96	0.242	9.99	9.11
U	3.33	3.73	0.642	4.20	4.48
Y	25	24	5	34	27
Zn	82	61	26	42	74
Zr	188	200	9	280	230
La	37.0	31.6	5.34	38.1	25.8
Ce	80.4	68.3	10.4	75.9	46.8
Nd	34.2	30.5	3.9	33.1	23.4
Sm	6.40	6.08	0.779	6.77	5.27
Eu	0.891	0.944	0.17	1.32	1.01
Gd	5.07	4.45	0.642	5.38	4.56
Tb	0.736	0.623	0.086	0.753	0.660
Ho	1.01	0.878	n.d.	1.21	1.05
Tm	0.423	0.369	0.038	0.632	0.463
Yb	2.58	2.33	0.228	4.17	3.03
Lu	0.385	0.362	0.035	0.659	0.454

[1]All major-element data are listed as measured; total iron is given as FeOT.

[2]For 19F-69, Fe_2O_3 = total iron and H_2O^+ = LOI.

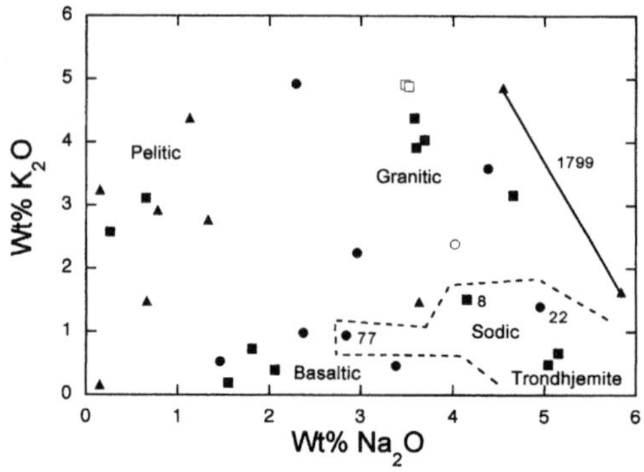

FIG. 4. Plot of Na_2O against K_2O for all samples. General compositional fields include samples from all groups. Within the sodic field, samples 19F-8, 22, and 77 are labeled.

extensive fractionation involving feldspar and modest LREE enrichment. The REE patterns of 19F-57 and 76 are in sharp contrast to the those of sodic-suite samples 19F-22 and 77 (Fig. 7B). Because the REE are more resistant to alteration processes that the alkaline elements, these differences imply, along with the other contrasting geochemical features mentioned above, that these two pairs of samples are not genetically linked.

Late Precambrian leucobasalt from the Igarka region. The geochemistry of Igarka sample IP-150/930 is most similar to that of 19F-60, in that both are iron enriched, with moderately high TiO_2 contents (Fig. 6), and have high Na_2O and low Sr concentrations (Fig. 5). Neither sample fits well with the more typical basaltic or felsic samples in the volcanic xenolith group, and their compositions appear to be the result of extensive, tholeiitic fractionation (Fig. 6). The REE patterns of these samples are generally similar, showing much higher total REE contents than the basaltic samples and more fractionated REE patterns (Fig. 7B).

Crystalline basement rocks from the diatreme. Like the volcanic rocks, the variable lithologic nature of these samples dictates a wide range in major- and trace-element geochemistry (Figs. 2 through 6). The three basaltic amphibolite samples (19F-3, 4, and 21) contain from 4.4–8.4 wt% MgO, are generally iron enriched, and have normal to moderately high TiO_2 contents that follow the level of iron enrichment (Fig. 6). REE patterns show only slight LREE enrichment and no Eu anomalies (Fig. 7D). Metapelite samples 19F-6 and 7 are both strongly depleted in Na_2O (0.26 and 0.65 wt%) and CaO (1.22 and 2.12 wt%; Fig. 3), as expected for

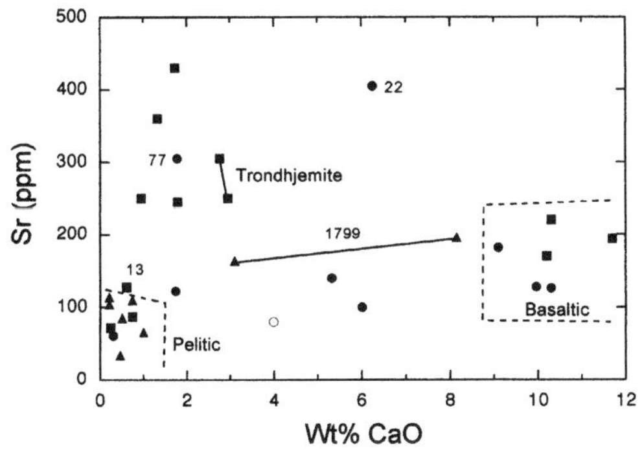

FIG. 5. Plot of CaO against Sr for all samples.

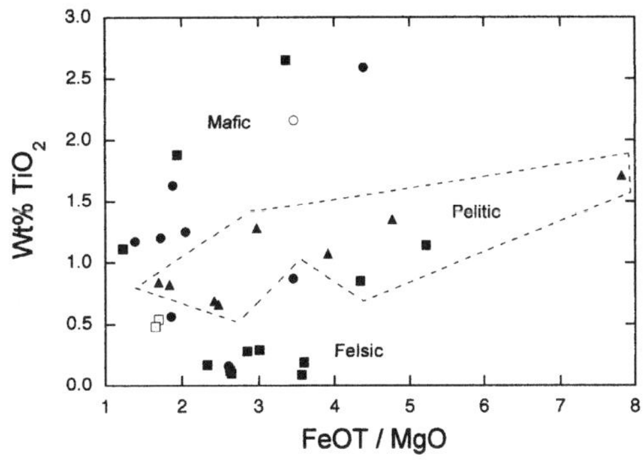

FIG. 6. Plot of Total FeO/MgO against TiO_2 for all samples. All samples are divided into three broad compositional types (mafic, felsic, pelitic).

samples originally dominated by clay minerals (e.g., Pettijohn, 1975). Their REE patterns are typical of pelitic rocks in showing modest LREE enrichment, but only 19F-7 shows the more typical, negative Eu anomaly (Fig. 7D), which generally has been attributed to weathering of evolved, granitic upper crust (e.g., Taylor, 1985). Gneissic samples 19F-13 and 9 have granitic compositions, with normal Na_2O and K_2O contents (Fig. 4). Both samples are modestly LREE enriched, with 19F-13 having a significant negative Eu anomaly and 19F-9 a modest negative Eu anomaly (Fig. 7C). Leucogranite samples 19F-10 and 69 have SiO_2 (72–73 wt%, Fig. 2), Na_2O, and K_2O contents similar to those in the gneissic samples (Fig. 4). The REE pattern of 19F-10 shows modest LREE enrichment and a slight, negative Eu anomaly (Fig. 7C). Leucogranite 19F-8 has a low K_2O content (1.51 wt%), indicative of a sodic-suite rock, yet differs from the trondhjemitic samples in having much higher REE contents, strong LREE enrichment, and a modest positive Eu anomaly. Samples 19F-11 and 12 are true trondhjemites (e.g., Barker, 1979) with 72–73 wt% SiO_2 (Fig. 2), 5 wt% Na_2O, very low K_2O (0.48 and 0.67 wt%, Fig. 4) and Rb (22 and 27 ppm), and moderate Sr (250–305 ppm) contents (Fig. 5). The trondhjemites have the lowest REE contents of all the samples studied (La ~10X chondrite), and show no Eu anomaly and modest LREE enrichment (Fig. 7C).

Late Triassic granodiorite from the Bolgokhtokh stock. The Bolgokhtokh granodiorite is distinguished from the basement-suite granitic rocks by the fact that it is a true granodiorite, with SiO_2 ~65 wt% (Fig. 2), and has high Sr and Ba concentrations of about 1650 and 4050 ppm, respectively. The REE patterns of these samples are strongly fractionated, with LREE enrichment (La ~300× chondrite), a modest, negative Eu anomaly, and modest HREE depletion (Fig. 7D).

Summary. In summary, the geochemistry of the volcanic and crystalline basement sample groups from the diatreme is in keeping with the varied lithologic nature of the samples in each group. Each group contains basaltic samples that are more properly classified as tholeiitic, rather than calc-alkaline or alkaline. There are no obvious, strongly alkaline rocks in the entire xenolith suite. The more silicic samples in each group can be divided generally into sodic versus normal granitic types. Two of the sodic samples in the crystalline basement group are classic trondhjemites, most commonly associated with Precambrian terranes (e.g., Barker, 1979). The basement group contains metapelitic samples, whereas no unmetamorphosed, sedimentary or volcaniclastic rocks were identified in the xenolith suite that would be placed in the volcanic group on the basis of common petrographic characteristics. The Paleozoic sedimentary rocks onto and into which the Permian–Triassic volcanic rocks and intrusions were emplaced are not present in the xenolith suite, but are represented by the argillites and hornfelsed sediments collected from drill cores. The geochemistry of these samples reflects a range from clay-dominated, to average graywacke, to mafic graywacke, to carbonate lithologies. Sufficient differences exist in the overall geochemistry of the volcanic suite versus that of the crystalline basement suite to indicate that a significant difference in age and/or petrogenesis must be considered.

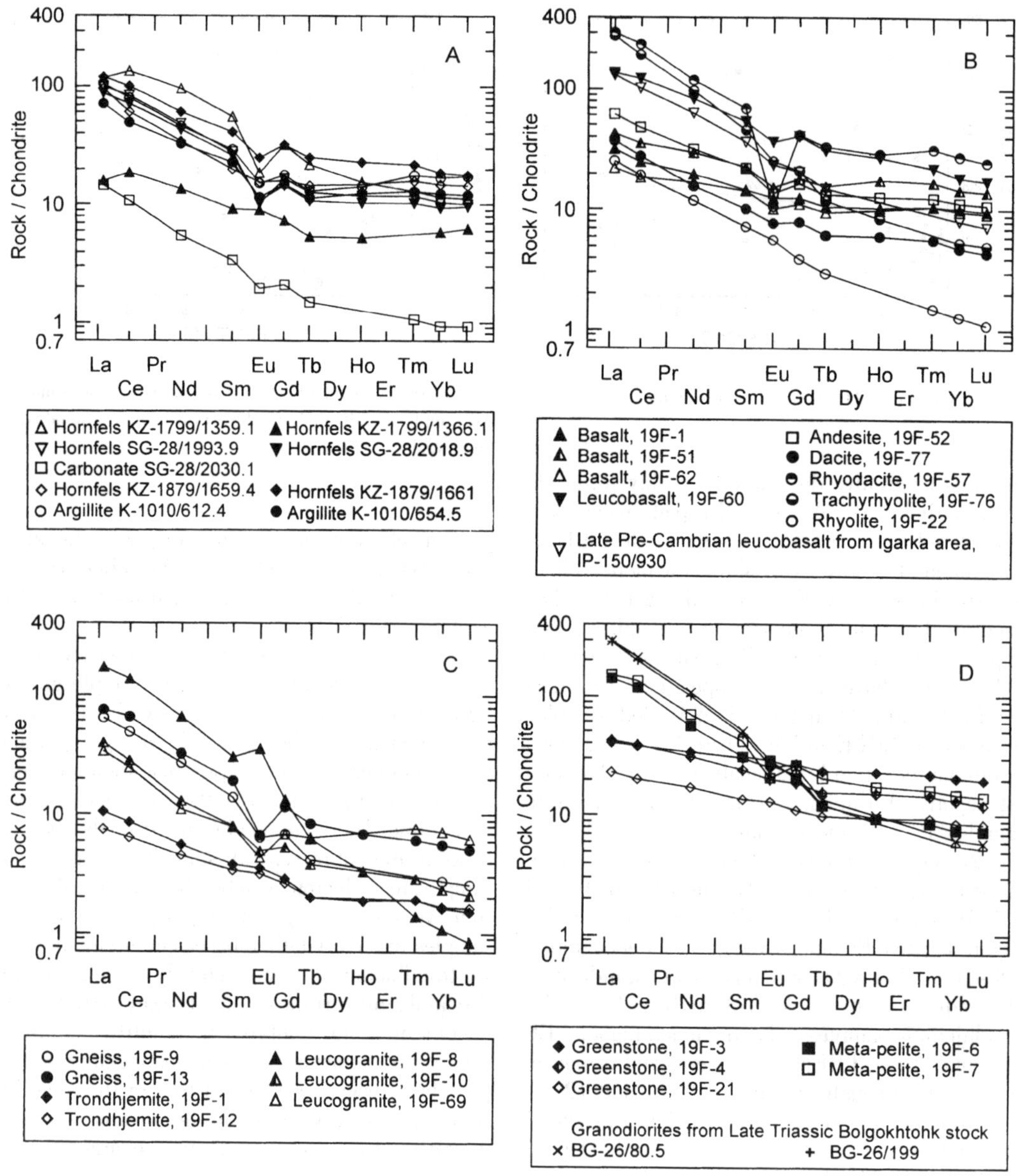

FIG. 7. Rare-earth-element patterns for groups of samples. A. Argillites and hornfels. B. Volcanic rocks. C. Crystalline basement rocks. D. Metapelites, greenstones, and granodiorites.

Rb-Sr isotopic data and chronologic groups

The studied suite of samples is most easily divided into general age groups on the basis of the whole-rock, Rb-Sr isotopic data (Table 2; Figs. 8A and 8B). Rb-Sr isotopic data are available for all samples, and many samples have Rb/Sr ratios high enough that model ages can be calculated by assuming an initial Sr isotopic ratio.

Data for the argillite and hornfels samples indicate Paleozoic ages generally consistent with their known depositional ages. Two-point Rb-Sr isochrons constructed on the basis of the two samples obtained

TABLE 2. Pb, Sr, and Nd Isotopic Data for the Studied Sample Collection, with Calculated Model Ages

	Greenstone			Meta-pelite		Gneiss		Trondhjemite		Leucogranite			Granodiorite		Volcanic rocks		
	19F-3	19F-4	19F-21	19F-7	19F-6	19F-13	19F-9	19F-11	19F-12	19F-8	19F-10	19F-69	BG-26 80.5	BG-26 199	19F-1	19F-51	19F-62
ppm Pb	3.40	1.81	1.17	4.35	3.05	26.88	42.47	3.67	2.65	31.33	22.18	25.57	34.94	26.58	6.95	3.75	1.79
ppm U	0.626	0.302	0.227	3.15	1.59	2.03	1.66	0.238	0.145	1.25	0.61	0.911	7.06	5.34	0.522	1.32	0.377
ppm Th	1.78	0.955	0.586	18.9	15.4	18	9.02	0.807	0.396	21.7	3.45	3.48	21.9	21.5	1.42	2.72	1.1
$^{232}Th/^{204}Pb$	34.59	34.68	31.50	338.14	419.52	45.30	13.77	14.06	9.51	45.62	10.02	8.80	39.31	50.76	13.25	47.61	40.35
$^{238}U/^{204}Pb$	11.86	10.69	11.89	54.94	42.22	4.98	2.47	4.04	3.40	2.56	1.73	2.24	12.35	12.29	4.75	22.52	13.48
$(^{206}Pb/^{204}Pb)_M$	18.879	18.482	17.154	23.197	23.179	20.095	18.433	17.703	17.711	18.012	17.901	18.161	16.708	16.728	18.085	18.747	18.642
$(^{206}Pb/^{204}Pb)_I$	18.410	18.059	16.683	21.025	21.509	19.898	18.336	17.543	17.577	17.910	17.833	18.072	16.279	16.301	17.897	17.857	18.109
$(^{207}Pb/^{204}Pb)_M$	15.642	15.606	15.463	15.971	15.993	15.738	15.621	15.477	15.448	15.648	15.561	15.604	15.324	15.324	15.542	15.533	15.532
$(^{207}Pb/^{204}Pb)_I$	15.618	15.585	15.439	15.859	15.907	15.728	15.616	15.469	15.441	15.643	15.558	15.600	15.303	15.303	15.533	15.487	15.505
$(^{208}Pb/^{204}Pb)_M$	38.457	38.489	36.790	46.558	52.821	38.901	37.518	37.394	37.110	39.072	37.648	37.609	37.171	37.200	37.966	38.201	38.337
$(^{208}Pb/^{204}Pb)_I$	38.027	38.057	36.298	42.749	47.600	38.338	37.347	37.219	36.991	38.504	37.523	37.499	36.741	36.664	37.801	37.609	37.835
ppm Rb	5	27	3	114	51	172	172	27	22	47	128	173	150	156	14	26	14
ppm Sr	170	220	194	87	72	128	360	305	250	245	430	250	1700	1600	128	182	126
$^{87}Rb/^{86}Sr$	0.085	0.355	0.045	3.792	2.050	3.888	1.382	0.256	0.255	0.555	0.861	2.002	0.255	0.282	0.316	0.413	0.322
$(^{87}Sr/^{86}Sr)M$	0.70661	0.71246	0.70612	0.76949	0.75174	0.77387	0.73357	0.70790	0.70658	0.73504	0.72270	0.73699	0.70719	0.70718	0.72821	0.70722	0.70688
$(^{87}Sr/^{86}Sr)I$	0.7063	0.7112	0.7060	0.7560	0.7445	0.7600	0.7287	0.7070	0.7057	0.7331	0.7196	0.7299	0.7064	0.7063	0.7271	0.7057	0.7057
Sm	6.867	–	3.12	9.498	7.011	4.468	3.302	0.893	–	7.012	–	2.24	–	10.9	–	5.339	3.431
Nd	23.90	–	11.98	49.12	39.45	23.21	19.31	4.04	–	46.64	–	9.23	–	70.80	–	21.66	12.41
$^{147}Sm/^{144}Nd$	0.1738	–	0.1575	0.1169	0.1075	0.1164	0.1034	0.1337	–	0.0909	–	0.1400	–	0.0931	–	0.1491	0.1672
$^{143}Nd/^{144}Nd$	0.512656	–	0.512277	0.511521	0.511857	0.511662	0.511712	0.511956	–	0.511538	–	0.511796	–	0.511738	–	0.512511	0.51264
εNd (251 Ma)	1.12	–	-5.75	-19.2	-12.3	-16.4	-15.0	-11.3	–	-18.0		-14.6	–	–14.2	–	-0.91	1.02
Nd TDM model age	1.46		2.07	2.4	1.7	2.16	1.84	2.07		1.87		2.6		1.65		1.25	1.31
Sr model age	Low Rb/Sr	1.66	Low Rb/Sr	1.21	1.62	1.25	1.49	1.06	0.71	3.83	1.51	1.15	0.7064	0.7063	5.19	0.55	0.63

(Continues)

Table 2. *(continued)*

	Volcanic rocks							Hornfels							Carbonate	Argillite	
	19F-60	19F-52	19F-77	19F-57	19F-76	19F-22	IP150/930	KZ-1799	KZ-1799	KZ-1879	KZ-1879	SG-28	SG-28	SG-28	SG-28	K-1010	K-1010
								1359.1	1366.1	1659.4	1661	1993.9	2018.9		2030.1	612.4	654.5
ppm Pb	9.38	7.18	11.46	20.39	33.62	4.65	23.03	0.28	0.70	0.22	3.37	6.78	1.23		7.29	5.72	14.35
ppm U	1.46	1.45	1.45	3.07	4.92	0.603	4.78	1.7	2.14	3.07	4.09	3.33	3.73		0.642	4.2	4.48
ppm Th	5.61	4.68	3.18	15.4	23.7	2.66	5.13	11.3	10.7	7.3	9.69	6.48	6.96		0.242	9.99	9.11
$^{232}Th/^{204}Pb$	39.00	42.67	18.04	49.77	45.59	37.00	14.71	3557.35	1322.18	3232.21	198.15	62.82	408.14		2.16	119.23	42.22
$^{238}U/^{204}Pb$	9.89	12.89	8.02	9.67	9.23	8.18	13.36	521.68	257.77	1325.03	81.53	31.47	213.21		5.60	48.86	20.24
$(^{206}Pb/^{204}Pb)_M$	18.297	18.491	18.211	18.519	17.904	17.976	19.249	26.096	27.320	37.971	20.712	18.811	22.618		18.545	20.333	19.213
$(^{206}Pb/^{204}Pb)_I$	17.906	17.981	17.894	18.136	17.539	17.653	18.721	5.467	17.127	−14.424	17.489	17.567	14.186		18.323	18.401	18.413
$(^{207}Pb/^{204}Pb)_M$	15.515	15.541	15.627	15.564	15.449	15.476	15.646	15.973	16.074	16.563	15.668	15.555	15.783		15.554	15.654	15.621
$(^{207}Pb/^{204}Pb)_I$	15.495	15.515	15.611	15.545	15.430	15.459	15.619	14.916	15.552	13.881	15.502	15.491	15.351		15.543	15.555	15.580
$(^{208}Pb/^{204}Pb)_M$	38.194	38.253	37.962	38.682	38.047	37.957	38.033	55.596	52.444	53.578	39.776	38.220	41.362		37.916	39.449	38.629
$(^{208}Pb/^{204}Pb)_I$	37.709	37.722	37.738	38.063	37.479	37.497	37.850	11.322	35.988	13.351	37.310	37.439	36.283		37.889	37.965	38.104
ppm Rb	73	54	20	146	310	33	84	24	54	116	92	35	77		6	48	100
ppm Sr	100	140	405	122	61	305	80	196	164	85	110	66	34		116	114	104
$^{87}Rb/^{86}Sr$	2.112	1.116	0.143	3.463	14.705	0.313	3.038	0.354	0.953	3.949	2.420	1.534	6.553		0.150	1.218	2.782
$(^{87}Sr/^{86}Sr)M$	0.71333	0.71140	0.71263	0.72220	0.76546	0.70832	0.73899	0.71009	0.71287	0.72745	0.72212	0.71462	0.73822		0.70872	0.71521	0.72258
$(^{87}Sr/^{86}Sr)I$	0.7058	0.7074	0.7121	0.7099	0.7132	0.7072	0.7282	0.7088	0.7095	0.7134	0.7135	0.7092	0.7149		0.7082	0.7109	0.7127
Sm	–	5.238	2.435	12.25	16.54	1.743	8.613	–	2.097	–	9.567	6.397	–		–	–	–
Nd	–	23.39	11.53	63.19	87.89	8.77	46.11	–	9.64	–	43.00	34.16	–		–	–	–
$^{147}Sm/^{144}Nd$	–	0.1354	0.1278	0.1172	0.1138	0.1202	0.1129	–	0.1316	–	0.1346	0.1132	–		–	–	–
$^{143}Nd/^{144}Nd$	–	0.512201	0.512094	0.51209	0.512178	0.512389	0.512147	–	0.51219	–	0.512555	0.512358	–		–	–	–
εNd (251 Ma)	–	−6.53	−8.37	−8.11	−6.28	−2.37	−6.86	–	−6.62		0.41	−2.75	–		–	–	–
Nd TDM model age		1.63	1.68	1.5	1.32	1.07	1.35		1.58		0.94	1.04					
Sr model age	0.31	0.46	4.13	0.37	0.29	0.96	0.81	0.33(I)	0.33(I)	0.24(I)	0.24(I)	0.33(I)	0.33(I)		NA	0.33(I)	0.33(I)

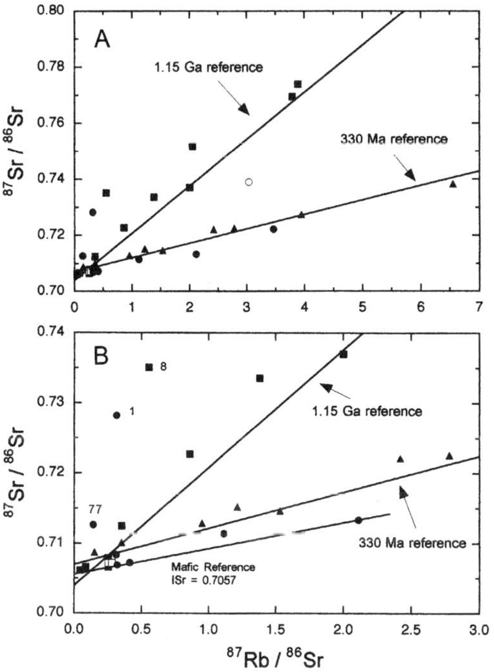

FIG. 8. A and B. Rb-Sr isochron diagrams for all samples at differing scales. Crystalline basement rocks generally have Proterozoic model ages (note 1.15 Ga reference isochron) that distinguish them from the volcanic and sedimentary rocks, which have Paleozoic model ages. As more clearly seen in 8B, the sedimentary rocks scatter about a 330 Ma reference isochron. Most of the volcanic rocks lie below the 330 Ma reference isochron, with three mafic samples suggesting a limiting initial Sr isotopic ratio of 0.7057.

from each of four drill holes yield the following model ages and initial $^{87}Sr/^{86}Sr$ ratios (ISr): K-1010, 331 Ma, ISr = 0.710; KZ-1799, 326 Ma, ISr = 0.7085; SG-28, 330 Ma, ISr = 0.7085; KZ-1879, 245 Ma, ISr = 0.715. The four samples of the Tungusskaya Series obtained from drill holes KZ-1799 and SG-28 have an average Rb-Sr age of 327 Ma. Sample SG-28-2030.1, a dolomite, indicates an initial ISr of about 0.7080 at the time of deposition. Only the KZ-1879 samples appear to have had their Rb-Sr isotopic systems reset by emplacement of the ore-bearing intrusions at ~251 Ma (Kamo et al., 1996). We suspect that this is because the KZ-1879 samples were collected above an ore-bearing intrusion, and these intrusions are overlain by abnormally thick alteration halos (e.g., Czamanske et al., 1995).

Except for altered basalt 19F-1, the volcanic rock samples plot along with, or below, the argillite and hornfels samples on the whole-rock isochron diagram (Fig. 8). A general conclusion is that these samples are roughly of Late Paleozoic and/or Early Mesozoic age. For example, the three basaltic samples, 19F-51, 60, and 62 have ISr = 0.7057 ± 0.0001 for an age of 251 Ma. The more felsic samples have model ISr values at 251 Ma, which generally increase with silica content; e.g., 19F-52, an andesite, 0.7074; 19F-57, a rhyodacite, 0.7099; and K-rich rhyolite 19F-76, 0.7132. Sample 19F-77, an altered dacite, seems to have an unusually low Rb/Sr ratio, but its model ISr value is 0.7121, in keeping with the trend of higher ISr values with increasing silica content; 19F-22, a sodic-suite sample, is an exception to this trend, having a model ISr value of 0.7072 at 251 Ma. The trend of increasing ISr with silica content may imply that the petrogenesis of these rocks involved increasing interaction with an older crust to produce more silicic rocks.

The Rb-Sr isotopic data for the Igarka leucobasalt lie between the data arrays for the volcanic and crystalline basement samples on an Rb-Sr isochron diagram (Fig. 8A). With a single sample, it is only possible to calculate a range of model ages. The Rb-Sr model age of the leucobasalt ranges from 806 to 623 Ma for ISr values ranging from 0.704 to 0.712, respectively. Our prejudice is that the ISr lies toward the lower end of this range (0.704–0.708), with a corresponding age range of 760 ± 50 Ma. M. P. Savushkin (unpubl. manuscript, 1994) reported Rb-Sr ages of 830–925 Ma for volcanic rocks from the Rzhevsky Suite and 842–935 Ma for the Ludovsky Series (see Appendix 1). However, our results indicate that the probable age of sample IP-150-930 from the Rzhevsky Suite is somewhat younger.

Crystalline basement samples 19F-3 through 21 and 19F-69 are either granitoids or have metamorphic textures. The Rb-Sr data for samples 19F-6, 7, 9, 10, 13, and 69 indicate Middle to Late Proterozoic ages. Silicic, low-K sample 19F-8 appears to have a disturbed Rb-Sr isotopic system that yields a model age > 3 Ga. The very low Rb/Sr ratios of basaltic amphibolite samples 19F-3, 4, and 21, and trondhjemitic samples 19F-11 and 12, do not yield meaningful Rb-Sr model ages. However, these samples are tentatively considered to be of Precambrian age because they are texturally similar to other samples with Precambrian model ages.

The two Bolgokhtokh granodiorite samples have identical Rb-Sr isotopic characteristics within analytical error. This body has an age of 229.2 ± 0.3 Ma (Kamo et al., 2000) and a calculated ISr of 0.7063 (Table 2). This ISr value puts this body within the lower part of the range calculated for the volcanic xenolith samples. Fedorenko (1994) proposed that the Bolgokhtokh granodiorite represents partial melting of the lower crust under amphibolite conditions. The ISr of 0.7063 is consistent with such an interpretation, because it is not unusual for the lower crust to have low Rb/Sr ratios.

U-Pb zircon data

The credibility of the Rb-Sr model ages is substantially enhanced by the U-Pb zircon systematics of three samples examined with the SHRIMP-RG ion microprobe in the cooperative Stanford-USGS facility. A preliminary U-Pb zircon study of 19F-69 was done with the SHRIMP I microprobe at Australian National University in 1995. For latest Precambrian and Phanerozoic samples, the $^{206}Pb/^{238}U$ ratio, corrected for common Pb using the $^{207}Pb/^{206}Pb$ ratio, yields the most reliable age, because too few ^{207}Pb ions are produced for good counting statistics on an ion microprobe, and $^{207}Pb/^{235}U$ and $^{207}Pb/^{206}Pb$ ratios have large uncertainties as a result. $^{206}Pb/^{238}U$ ages produced in this way are equivalent to treatment of conventional U-Pb data on a Terra-Wasserburg concordia diagram. For older samples, the most reliable age using ion microprobe data comes from the ^{204}Pb corrected $^{207}Pb/^{206}Pb$ ratio, because it is not subject to the uncertainties of determining the Pb/U ratio by comparison to "standard" zircons. Pb-U ages corrected for ^{204}Pb are also reported for these older samples, and have analytical errors, including standard uncertainty, of 1 to 2%.

Rhyodacite sample 19F-57, for which Rb-Sr data indicate a Late Paleozoic age, yielded about two dozen zircons from a fist-size specimen. Of the 21 grains examined with the ion microprobe (Table 3, Fig. 9A), 18 yielded ^{207}Pb corrected $^{206}Pb/^{238}U$ ages from 258–374 Ma (Fig. 9A). Three grains had Proterozoic, $^{207}Pb/^{206}Pb$ ages of 1000, 1660, and 1998 Ma (Table 3). These data may be interpreted to indicate that the crystallization age of this sample was 270 ± 3 Ma, as defined by the data from 14 grains with an age range of 258–279 Ma. Data for four other grains give ages above this age range (288, 293, 303, and 358-374 Ma) and are interpreted to indicate inheritance from slightly older Paleozoic zircons. Grain #1 (core and rim spots have ages of 374 ± 7 and 358 ± 19 Ma) may be representative of these older zircons that experienced little or no younger overgrowth. Inherited zircon grains also include Proterozoic material, consistent with Rb-Sr model ages of the crystalline basement samples and U-Pb zircon data from 19F-69, as presented below. Fist-size, altered dacite sample 19F-77 yielded only two zircon grains, with U-Pb ages of 355 ±7 and 998 ±16 Ma (both ^{207}Pb corrected $^{206}Pb/^{238}U$ ages). It is possible that these two grains result from cross contamination from another sample during the sample processing. However, the two ages are consistent with the larger data sets from samples 19F-57 and 69, and may be taken as weak confirmation of the results obtained from those samples.

Sample 19F-69, a leucogranite with a Rb-Sr model age of 1.15 Ga, yielded about three dozen zircon grains. Whereas U-Pb ages from these grains range broadly from 619 to 3012 Ma (Table 4, Fig. 9B), some obvious conclusions can be drawn. Sixteen spots (including both cores and rims) yielded ^{204}Pb corrected $^{206}Pb/^{238}U$ ages from 873 to 944 Ma. The average of these ages (~910 Ma) most likely represents the crystallization age of this sample (Fig. 9C). The four ages from 619–838 Ma probably represent Pb loss from the ~910 Ma grains in response to younger events experienced by this sample, including the magmatic event that resulted in it becoming a xenolith. Over half of the analyses yielded ages older than the crystallization age. These ages represent zircons inherited from the source of the ~910 Ma granite and, in at least three cases, these older zircon cores are mantled by ~910 Ma zircon (grains 26, 27, 28; e.g., Fig. 10). Given the complex history of this sample, it is difficult to judge whether or not these older ages represent actual geologic events (crystallization, metamorphism, etc.) or minimum ages associated with Pb loss. Four spots define a tight age range of 2089–2114 Ma ($^{207}Pb/^{206}Pb$). Seven spots yielded ages ranging from 1865 to 1999 Ma ($^{207}Pb/^{206}Pb$). This group of 11 ages constitutes very strong evidence than 19F-69 was melted from a source with a major Early Proterozoic component. This conclusion is even more evident if discordia lines are drawn from 910 Ma through the older data points, inasmuch as several discordant points appear to lie along or near a field with upper intercepts between 1.9 and 2.1 Ga (Fig. 9C). Two of the six ages between 1217 and 1764 Ma are concordant within analytical error (1217 and 1432 Ma), and could indicate younger source materials.

TABLE 3. SHRIMP U-Th-Pb Geochronologic Data for Felsic Volcanic Samples 19F-57 and 19F-77

Spot	Concentrations (ppm) U	Th	Isotopic ratios (^{207}Pb corrected) $^{208}Pb/^{232}Th$	±	$^{206}Pb/^{238}U$	±	$^{238}U/^{206}Pb$	±	Calculated ages (Ma) $^{208}Pb/^{232}Th$	±	$^{206}Pb/^{238}U$	±
19F57-1.1	461	91	0.02043	0.00197	0.05965	0.00122	16.763	0.342	409	39	374	7
19F57-1.2	245	50	0.01785	0.00195	0.05707	0.00313	17.522	0.961	358	39	358	19
19F57-2.1	634	293	0.01298	0.00043	0.04289	0.00055	23.313	0.301	261	8	271	3
19F57-3.1	102	54	0.01398	0.00103	0.04194	0.00111	23.844	0.63	281	21	265	7
19F57-4.1	1496	409	0.01319	0.00048	0.04233	0.00059	23.624	0.328	265	10	267	4
19F57-5.1	944	133	–	–	0.16449	0.00324	6.08	0.12	–	–	982	18
19F57-6.1	149	109	0.01359	0.0005	0.04407	0.00076	22.691	0.39	273	10	278	5
19F57-7.1	156	104	0.01352	0.00059	0.04385	0.00078	22.807	0.408	271	12	277	5
19F57-8.1	1229	442	0.01274	0.00033	0.04112	0.00053	24.317	0.316	256	7	260	3
19F57-9.1	1330	675	0.01318	0.00035	0.04239	0.00048	23.593	0.269	265	7	268	3
19F57-10.1	348	173	0.01368	0.00083	0.04244	0.00134	23.561	0.747	275	16	268	8
19F57-11.1	388	185	0.06648	0.0028	0.28632	0.00629	3.493	0.077	1301	53	1623	32
19F57-12.1	266	49	0.05369	0.00234	0.17674	0.00265	5.658	0.085	1057	45	1049	15
19F57-13.1	1010	363	0.01399	0.00037	0.04426	0.00065	22.594	0.333	281	7	279	4
19F57-14.1	355	236	0.01364	0.00042	0.04277	0.00075	23.383	0.408	274	8	270	5
19F57-15.1	137	86	0.01322	0.00062	0.04315	0.00086	23.173	0.46	265	12	272	5
19F57-16.1	882	289	0.01566	0.00047	0.04811	0.00067	20.785	0.289	314	9	303	4
19F57-17.1	1336	772	0.01837	0.00034	0.04368	0.00058	22.895	0.303	368	7	276	4
19F57-18.1	275	201	0.01342	0.00046	0.04309	0.00068	23.209	0.367	269	9	272	4
19F57-19.1	361	167	0.0145	0.00048	0.04568	0.00075	21.89	0.361	293	10	288	5
19F57-20.1	629	288	0.01496	0.00045	0.0465	0.00074	21.506	0.342	300	9	293	5
19F57-21.1	193	131	0.01248	0.00099	0.0408	0.00119	24.511	0.718	251	20	258	7
19F77-1.1	275	32	0.01706	0.00215	0.05658	0.00108	17.673	0.338	342	43	355	7
19F77-2.1	151	70	0.05535	0.00199	0.16748	0.00293	5.971	0.104	1089	38	998	16

(Continues)

TABLE 3. (*continued*)

Spot	Isotopic ratios (^{204}Pb corrected)						Calculated ages (Ma)					
	$^{206}Pb/^{238}U$	±	$^{207}Pb/^{235}U$	±	$^{207}Pb/^{206}Pb$	±	$^{206}Pb/^{238}U$	±	$^{207}Pb/^{235}U$	±	$^{207}Pb/^{206}Pb$	±
19F57-1.1	0.05951	0.00121	0.4276	0.01835	–	–	373	7	361	13	–	–
19F57-1.2	0.05713	0.00313	0.42945	0.03369	–	–	358	19	363	24	–	–
19F57-2.1	0.04287	0.00055	0.30302	0.01178	–	–	271	3	269	9	–	–
19F57-3.1	0.04207	0.0011	0.313	0.01845	–	–	266	7	277	14	–	–
19F57-4.1	0.04227	0.00059	0.29422	0.01336	–	–	267	4	262	11	–	–
19F57-5.1	0.17037	0.00331	2.3953	0.08383	0.10197	0.00273	1014	18	1241	25	1660	50
19F57-6.1	0.04404	0.00075	0.31171	0.01232	–	–	278	5	275	10	–	–
19F57-7.1	0.04367	0.00079	0.29217	0.0255	–	–	276	5	260	20	–	–
19F57-8.1	0.0412	0.00054	0.30049	0.00859	–	–	260	3	267	7	–	–
19F57-9.1	0.04243	0.00049	0.30717	0.01261	–	–	268	3	272	10	–	–
19F57-10.1	0.04235	0.00134	0.29121	0.0172	–	–	267	8	260	14	–	–
19F57-11.1	0.29338	0.00643	4.9685	0.12788	0.12283	0.00136	1658	32	1814	22	1998	20
19F57-12.1	0.17638	0.00264	1.7635	0.03758	0.07252	0.00097	1047	14	1032	14	1000	27
19F57-13.1	0.04423	0.00065	0.31336	0.0071	–	–	279	4	277	6	–	–
19F57-14.1	0.04297	0.00076	0.32855	0.01711	–	–	271	5	288	13	–	–
19F57-15.1	0.04283	0.00092	0.27011	0.04461	–	–	270	6	243	36	–	–
19F57-16.1	0.04793	0.00067	0.32567	0.01085	–	–	302	4	286	8	–	–
19F57-17.1	0.04363	0.00058	0.30595	0.00801	–	–	275	4	271	6	–	–
19F57-18.1	0.04308	0.00098	0.30765	0.08306	–	–	272	6	272	67	–	–
19F57-19.1	0.04561	0.00075	0.31944	0.01196	–	–	288	5	281	9	–	–
19F57-20.1	0.04635	0.00074	0.31728	0.00899	–	–	292	5	280	7	–	–
19F57-21.1	0.04077	0.00118	0.28604	0.02548	–	–	258	7	255	20	–	–
19F77-1.1	0.05653	0.0011	0.41186	0.02773	–	–	354	7	350	20	–	–
19F77-2.1	0.16709	0.00291	1.6226	0.04495	0.07043	0.00137	996	16	979	18	941	40

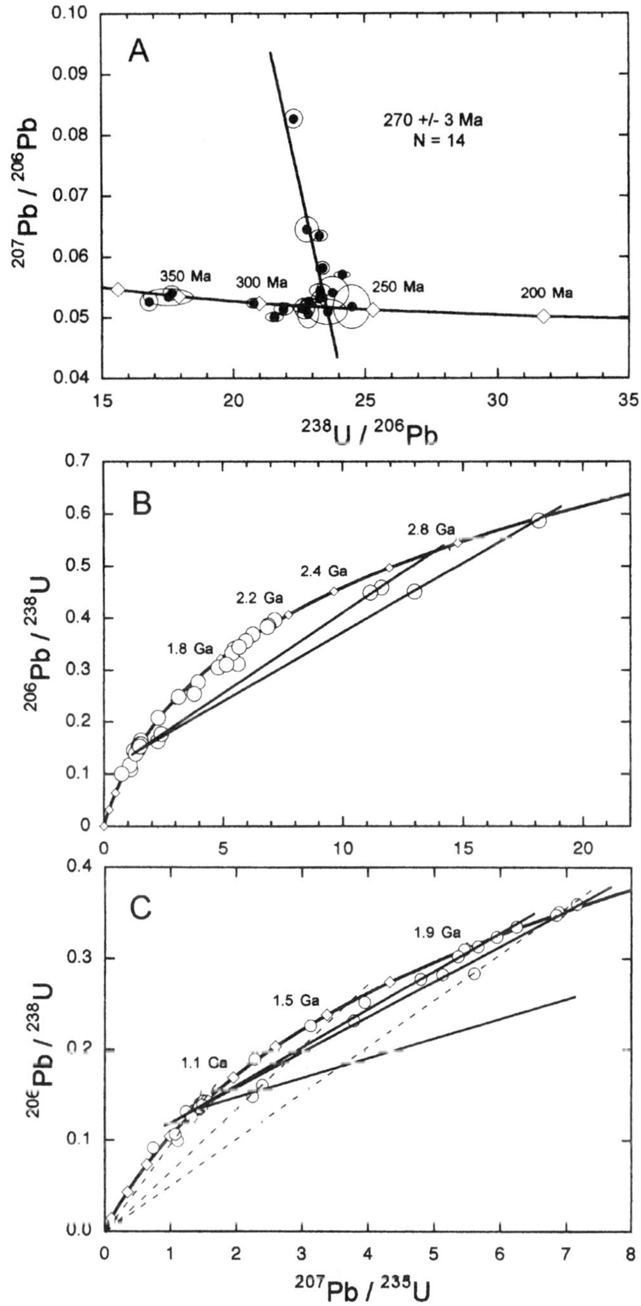

FIG. 9. U-Pb concordia diagrams for samples 19F-57 and 77 (A, Terra-Wasserburg) and 19F-69 (B and C, Wetherill). A. 19F-57 has an age of ~270 Ma, but shows minor Paleozoic and Proterozoic inheritance (see text). The steeply trending line represents a chord to a common Pb component. B and C. 19F-69 has a crystallization age of ~910 Ma, but contains abundant, inherited zircon. Inheritance ages are concentrated in the range from 1.8 to 2.1 Ga, but extend to 3.0 Ga. Solid lines on B and C represent mixing lines between 910 Ma and measured isotopic data and provide a maximum upper intercept. Dotted lines shown on C represent recent Pb loss and provide minimum, upper intercept ages.

TABLE 4. SHRIMP U-Th-Pb Geochronologic Data for Leucogranite Sample 19F-69

	Concentrations, ppm		Isotopic ratios, ^{204}Pb corrected								Calculated ages, Ma							
Spot[1]	U	Th	$^{208}Pb/^{232}Th$	± 8/32	$^{206}Pb/^{238}U$	±	$^{207}Pb/^{235}U$	±	$^{207}Pb/^{206}Pb$	±	$^{208}Pb/^{232}U$	±	$^{206}Pb/^{238}U$	±	$^{207}Pb/^{235}U$	±	$^{207}Pb/^{206}Pb$	±
1T	221	152	0.04361	0.00109	0.14786	0.00203	1.44187	0.03655	0.07072	0.00139	863	21	889	11	907	15	949	41
2C	400	18	0.04829	0.00144	0.15473	0.00185	1.44902	0.02156	0.06792	0.00051	953	28	927	10	909	9	866	16
3T	100	35	0.04785	0.00142	0.15591	0.00257	1.47985	0.04894	0.06884	0.00184	945	27	934	14	922	20	894	56
4T	172	85	0.05584	0.0061	0.34101	0.00751	5.45544	0.24298	0.11603	0.00418	1098	117	1892	36	1894	39	1896	66
5C	150	63	0.15729	0.00275	0.58705	0.00795	18.15317	0.28488	0.22427	0.00144	2953	48	2977	32	2998	15	3012	10
6C	588	164	0.02004	0.0025	0.15275	0.00164	1.47504	0.04707	0.07003	0.00201	401	50	916	9	920	20	929	60
7C	571	179	0.07348	0.00482	0.277	0.00308	3.95313	0.10686	0.1035	0.00241	1433	91	1576	16	1625	22	1688	44
8C	665	162	0.09843	0.00505	0.20784	0.00268	2.28092	0.07878	0.0796	0.00243	1898	93	1217	14	1206	25	1187	61
9T	642	81	0.0135	0.0067	0.14599	0.00173	1.36056	0.05593	0.06759	0.00256	271	134	878	10	872	24	856	81
10T	3989	1843	0.05299	0.00476	0.16465	0.00246	1.54837	0.12947	0.0682	0.00549	1044	92	983	14	950	53	875	176
11T	2292	130	0.03562	0.01919	0.10915	0.01733	1.10553	0.20076	0.07346	0.00518	707	378	668	102	756	102	1027	150
11T	2547	192	0.0292	0.00637	0.11725	0.00192	1.07378	0.03523	0.06642	0.00176	582	126	715	11	741	17	820	56
12C	228	170	0.0499	0.00237	0.36839	0.00505	6.24316	0.15165	0.12291	0.00226	984	46	2022	24	2011	21	1999	33
13T	1555	1928	0.01482	0.00179	0.10074	0.00169	0.7314	0.12783	0.05266	0.00905	297	36	619	10	557	78	314	350
14C	42	71	0.09817	0.00348	0.312	0.0062	5.60618	0.29054	0.13032	0.00592	1893	64	1751	31	1917	46	2102	82
15C	875	612	0.05163	0.00061	0.16319	0.00162	2.25093	0.02766	0.10004	0.00061	1017	12	974	9	1197	9	1625	11
16T	1000	97	0.00174	0.02441	0.14496	0.00224	1.23638	0.14645	0.06186	0.00715	35	487	873	13	817	69	669	269
17C	1604	599	0.08635	0.00257	0.24867	0.00253	3.13071	0.06576	0.09131	0.00157	1674	48	1432	13	1440	16	1453	33
18T	327	79	0.11001	0.00701	0.33275	0.00432	5.36361	0.13466	0.11691	0.00233	2110	128	1852	21	1879	22	1909	36
18C	425	100	n.a.[2]	n.a.	0.3052	–	4.799	–	0.114	–	n.a.	n.a.	1717	30	1784	35	1865	62
19T	284	171	0.04565	0.00089	0.15142	0.00221	1.43839	0.02753	0.0689	0.00074	902	17	909	12	905	12	896	22
19C	443	160	n.a.	n.a.	0.1495	–	1.431	–	0.0694	–	n.a.	n.a.	898	21	902	21	911	44
20T	588	128	0.08026	0.00892	0.31101	0.0037	5.12918	0.1422	0.11961	0.00282	1560	168	1746	18	1841	24	1950	43
20C	195	96	n.a.	n.a.	0.356	–	5.947	–	0.1212	–	n.a.	n.a.	1963	38	1968	25	1973	28
21T	316	178	0.04727	0.00149	0.15069	0.00197	1.46043	0.04764	0.07029	0.00199	933	29	905	11	914	20	937	59
22C	70	51	0.11301	0.00334	0.39588	0.00879	7.1604	0.17922	0.13118	0.00119	2164	61	2150	41	2132	23	2114	16
22C	71	52	n.a.	n.a.	0.386	–	6.884	–	0.1293	–	n.a.	n.a.	2104	67	2097	40	2089	33
23C	188	80	0.04489	0.00082	0.1492	0.00176	1.41847	0.03729	0.06895	0.00152	888	16	897	10	897	16	897	46
23C	209	76	n.a.	n.a.	0.1501	–	1.452	–	0.0702	–	n.a.	n.a.	901	22	910	25	933	63
24T	482	179	0.04638	0.00373	0.25459	0.00288	3.78672	0.09681	0.10787	0.00233	916	72	1462	15	1590	21	1764	40
25C	622	403	0.03407	0.0012	0.15599	0.00189	1.51828	0.04841	0.07059	0.00198	677	24	934	11	938	20	946	58
26T	1680	114	0.05276	0.00339	0.15766	0.00153	1.50899	0.02216	0.06941	0.00069	1039	65	944	9	934	9	911	20
26C	86	52	n.a.	n.a.	0.4582	–	11.619	–	0.1839	–	n.a.	n.a.	2432	76	2574	61	2688	81
27T	462	195	0.05572	0.00146	0.17675	0.00202	2.39523	0.0459	0.09829	0.00138	1096	28	1049	11	1241	14	1592	26
27C	218	77	n.a.	n.a.	0.4506	–	12.978	–	0.2089	–	n.a.	n.a.	2398	44	2678	22	2897	10
28T	720	93	0.047	0.0026	0.15747	0.00195	1.54999	0.03062	0.07139	0.00099	928	50	943	11	950	12	969	29
28C	140	110	n.a.	n.a.	0.3828	–	6.846	–	0.1297	–	n.a.	n.a.	2090	41	2092	23	2094	15
29C	772	632	n.a.	0.00171	0.34463	0.00347	5.66137	0.08634	0.11914	0.00122	2012	31	1909	17	1926	13	1943	18
30C	215	83	n.a.	n.a.	0.1494	–	1.437	–	0.0698	–	n.a.	n.a.	897	37	904	42	921	101
31C	286	153	n.a.	n.a.	0.449	–	11.145	–	0.18	–	n.a.	n.a.	2391	60	2535	30	2653	11
32C	321	135	n.a.	n.a.	0.1388	–	1.32	–	0.069	–	n.a.	n.a.	838	17	855	31	898	98
33C	1384	109	n.a.	n.a.	0.1529	–	1.489	–	0.0706	–	n.a.	n.a.	917	15	926	11	946	11

[1]Numeral indicates number assigned to grain in mount; T = tip; C = core of grain.

[2]Analysis obtained at ANU, where insufficient data for ^{208}Pb preclude meaningful calculation.

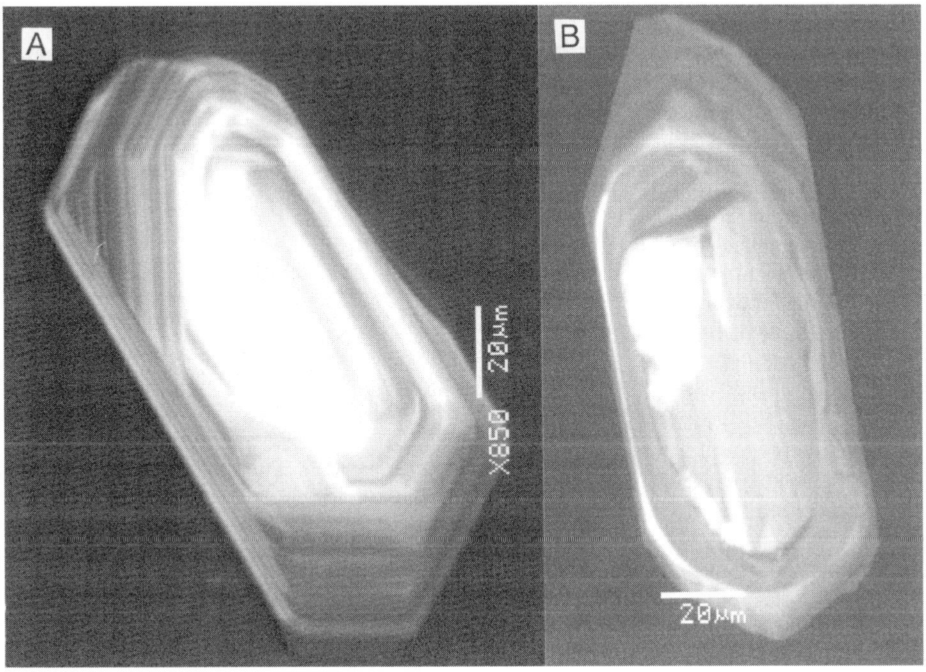

FIG. 10. Cathode luminescent images of zircons from xenolith samples 19F-57 and 69. A. Zoning patterns for rhyodacite 19F-57 indicate simple crystallization from a magma. B. Images for leucogranite 19F-69 show growth of younger zircon over an older core, as verified by U-Pb isotopic data for several grains.

Analyses of the remaining four spots that give ages between 1217 and 1764 Ma are discordant and difficult to interpret as anything but minimum ages. Finally, the four ages from 2653 to 3012 Ma provide minimum ages for Archean rocks in the source of 19F-69. In summary, it is clear that 19F-69 represents a granite crystallized at ~910 Ma that was melted from a complex source region dominated by Early Proterozoic (1900–2100 Ma) rocks, but also containing Archean, and possibly Middle and Late Proterozoic, rocks. The zircon U-Pb data verify the Proterozoic Rb-Sr model age of sample 19F-69 and, indirectly, the Proterozoic model ages of the other crystalline basement samples.

Sm Nd isotopic data

Sm-Nd isotopic data were obtained for representative samples from each of the age, lithologic, and chemical groups defined in this study (Table 2; Fig. 11; see also Fig. 14B). As might be expected for such a heterogeneous family of samples, depleted-mantle model ages (T_{DM}) are extremely variable; they range from 0.94 to 2.60 Ga and show no consistent value within groups (Table 2). Nd model ages are older than Sr model ages for the vast majority of the samples; exceptions are generally related to samples with such low Rb concentrations that Rb-Sr model ages cannot be meaningfully calculated, or Rb loss is suspected. In general, Nd isotopic data for xenolith samples such as these are expected to reflect both the formation age of the samples and the overall, average age of the source materials from which each sample was derived. Sr model ages are more likely to reflect the time since a magma was formed, or closure of an open system in which Rb and Sr were mobile, e.g., a metamorphic or fluid-alteration event.

Some general comments can be made regarding the relation of the Sm-Nd data to the sample age groups. The crystalline basement samples have the oldest T_{DM} model ages, ranging from 2.60 to 1.46 Ga. Only the youngest of these model ages overlaps the model ages for the volcanic and argillite/hornfels samples, which range from 1.68 to 0.94 Ga. This pattern is consistent with the age difference between these groups, i.e., Proterozoic versus Paleozoic. Graphically, this pattern manifests itself in a plot of $^{143}Nd/^{144}Nd$ against $^{147}Sm/^{144}Nd$ (Fig.11), with the crystalline basement plotting at $^{143}Nd/^{144}Nd$ values lower than those for samples from the other two groups. Exceptions to this pattern are basaltic amphibolite samples 19F-3 and 21, which lie to the left of the field defined by the younger samples. It should be noted that the Nd isotopic data strengthen assignment of the trondhjemitic and basaltic amphibolite samples to the older crystalline basement group, an assignment originally made only on the basis of textural features, because the Sr model ages are not meaningful (see above).

The Sm-Nd isotopic datum for the Igarka leucobasalt lies within the field for the volcanic and sedimentary rock samples, whereas that for the Bolgokhtokh granodiorite is more closely associated with the crystalline basement samples (Fig. 11). In the case of the granodiorite, these data are consistent with its generation from an older Precambrian basement.

The contrast of the Nd model age of 2.60 Ga for 19F-69 with its crystallization age of ~910 Ma is consistent with the large population of inherited Early Proterozoic and Archean zircons found in this sample. However, given the preponderance of Early Proterozoic versus Archean inherited grains, it is surprising that the Nd model age, which should reflect an average source age, is as old as Late Archean. The Nd model ages (2.16–1.46 Ga; excluding the 2.40 Ga age of metapelite sample 19F-7) of other samples in the crystalline basement group are more in line with the inherited zircon pattern of 19F-69. The fact that the Nd model ages are older than the probable formation ages of most samples, and average about 2.1 Ga, strongly indicates that the crust from which these xenoliths was derived has a major Early Proterozoic and Archean component which was reworked in Late Proterozoic time.

The dominantly Middle Proterozoic, Nd model ages of the Paleozoic volcanic xenolith suite also suggests that Precambrian crust played a major role in the evolution of this volcanic suite. However, the distinctively younger Nd model ages for this suite of samples, compared to the crystalline basement suite, indicates that the volcanic rocks were not simply melts of the basement suite. Either the average age of the crust had become significantly younger by Late Paleozoic time, or the volcanic rocks were the product of mixing between the older basement crust and Paleozoic magmas with εNd values of 0 and higher. It is not possible to choose between these two models. However, the later case is true for some of

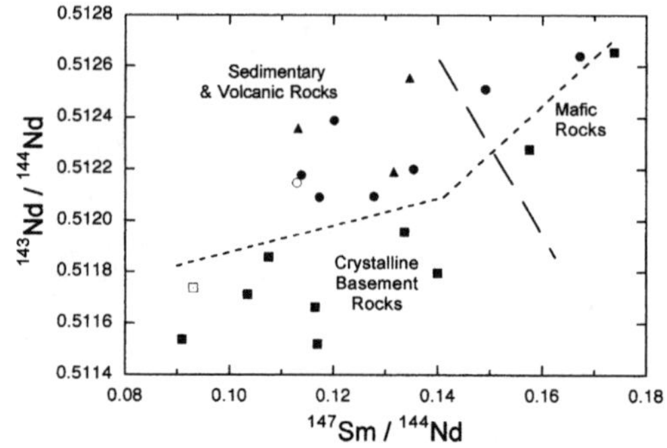

FIG. 11. Sm-Nd isochron diagram for all samples. Crystalline basement rocks are distinguished from the volcanic and metasedimentary rocks by having lower $^{143}Nd/^{144}Nd$ with respect to $^{147}Sm/^{144}Nd$. This pattern is a reflection of the older ages of the basement rocks and indicates that the younger rocks are not simple melts or mixtures of the basement rocks. Mafic samples from both the volcanic and crystalline basement groups are distinguished by their higher $^{147}Sm/^{144}Nd$ ratios, a reflection of their less fractionated REE patterns.

the volcanic rocks associated with the 251 Ma Siberian flood-volcanic province, and is commonly observed in volcanic suites of many ages and environments. The Nd isotopic data for the argillites and hornfelsed sedimentary rocks give Nd model ages of 0.94 to 1.58 Ga, quite similar to those for the volcanic suite from the diatreme. These Paleozoic sedimentary rocks probably sampled regionally exposed crust with a Middle to Late Proterozoic average age.

Common Pb isotopic systematics

Whole-rock, Pb isotopic data for the studied collection provide additional distinctions for the three groups of samples (Table 2, Figs. 12 and 13). The volcanic rock samples from the diatreme (with the exception of 19F-77) have the most coherent pattern, and lie along a linear trend on a uranogenic-Pb plot, with a slope that corresponds to an age of about 2.2 Ga (Fig. 12). This linear trend lies below the average-crust model of Stacey and Kramers (1976), indicating a source with a long-term, U/Pb ratio significantly less than that of average crust. On a thorogenic-Pb diagram (Fig. 13), these samples cluster along the average-crust-model line, indicating a long-term, Th/U ratio of about 4.

The Igarka leucobasalt has uranogenic-Pb isotopic systematics that fit with the volcanic xenolith samples, but its thorogenic data show it to lie below the average-crust model. The uranogenic-Pb data, which lie just off the end of the average-crust-model

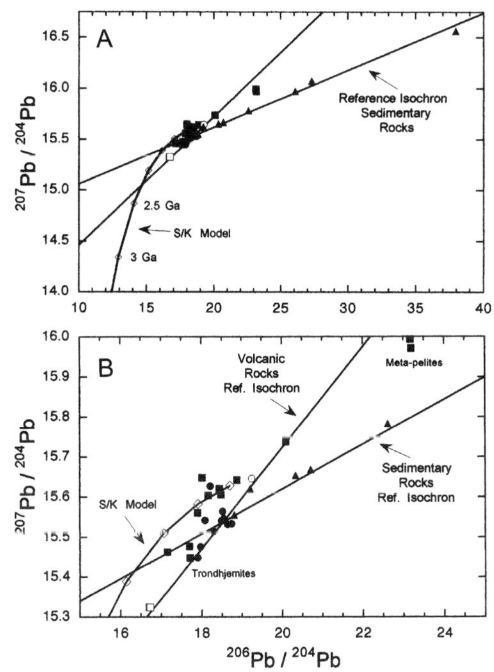

FIG. 12. A and B. Plots of $^{206}Pb/^{204}Pb$ against $^{207}Pb/^{204}Pb$ for all samples. The metasedimentary rocks scatter about a Paleozoic, Pb-Pb reference isochron. Many of the volcanic rocks and some of the crystalline basement rocks scatter about a limiting Pb-Pb reference isochron of about 2 Ga. On this and following plots of Pb isotopic data, the Stacey and Kramers (1976) average-crust evolution model is shown.

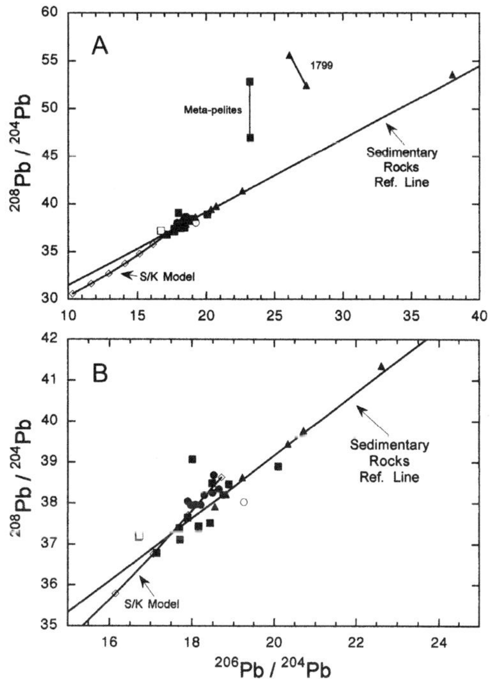

FIG. 13. A and B. Plots of $^{206}Pb/^{204}Pb$ against $^{208}Pb/^{204}Pb$ for all samples. Most of the Paleozoic sedimentary rocks lie along a trend indicating a Th/U ratio less than the crustal average. The samples from drill hole KZ-1799 are an exception, and require Th/U ratios much higher than the crustal average, as do the two metapelitic samples in the crystalline basement group.

curve, require some contribution of Early Proterozoic and/or Archean crust in the source of the parental magma or as a crustal contaminant. The Bolgokhtokh granodiorite has the lowest $^{206}Pb/^{204}Pb$ and $^{207}Pb/^{204}Pb$ ratios and among the lowest $^{208}Pb/^{204}Pb$ ratios measured for any of the samples analyzed for this paper. On the uranogenic-Pb diagram, it lies well below the average-crust-model curve, but possibly along the array defined by the volcanic samples. On the thorogenic Pb diagram, the granodiorite data lie above the average-crust-model curve. These Pb data require a source for the granodioritic parental magma that had long-term bulk U/Pb and Th/Pb ratios significantly less than average crust. These geochemical features are typically associated with a Proterozoic/Archean lower crust, and are consistent with the relatively low ISr value calculated for this intrusive body.

The argillite and hornfels samples display a wide range of Pb isotopic values ($^{206}Pb/^{204}Pb$, 19–38; $^{208}Pb/^{204}Pb$, 38–41). On the uranogenic-Pb isotopic diagram, the slope of the array defined by these samples is consistent with their Paleozoic depositional ages. Both of these features are consistent with the clay-rich character of these samples—i.e., low Pb concentrations and consequently high U/Pb and Th/Pb ratios at the time of deposition. Radiogenic growth since the time of deposition, therefore, dominates the present-day Pb isotopic systematics of these samples. The samples from drill hole KZ-1799 have high $^{208}Pb/^{204}Pb$ ratios that lie off the trend of the other samples on the thorogenic diagram (Fig. 13A). This indicates a higher Th/U ratio for these samples since the time of deposition, and is consistent with the graywacke versus argillitic compositional distinction of these samples, as discussed above. Note that the Pb isotopic arrays defined by this group of samples terminate in the field defined by the volcanic samples, thus allowing a general connection between the crustal sources of the argillite/hornfels and volcanic rock groups.

The Pb isotopic systematics of the crystalline basement samples are the most scattered of the three groups. No single, linear arrays are defined on either the uranogenic-Pb or thorogenic-Pb isotopic plots. Pb isotopic ratios show considerable ranges ($^{206}Pb/^{204}Pb$, 17–23; $^{208}Pb/^{204}Pb$, 37–53). These characteristics can be related to the extreme range in composition of these samples from basaltic, to granitic, to pelitic; to their Proterozoic ages; and to the probability that Archean crust is involved in the history of at least some of the samples. The uranogenic systematics of most of the basement samples (the trondhjemitic samples are an exception) are more radiogenic than those of the other two groups, and scatter close to the average-crust-model curve. The Pb isotopic data indicate that the crystalline basement samples as a group could not have served as the sole source of either the volcanic or argillite/hornfels sample groups.

Implications

Tectonic implications

Maslov and Nesterovskiy (1961), and Nemenyonok (1978) considered the crystalline rock fragments of the diatreme to represent the platform basement and the volcanic rocks to be representative of the Permian–Triassic flood volcanism. However, it was noted during our field study and sample collection that the volcanic rocks had no direct analogues in the flood-volcanic sequence. Petrographic and geochemical investigation showed that about half of the volcanic fragments are of dacitic to rhyolitic composition, lithologies not found in the flood-volcanic sequence near Noril'sk. Moreover, all of the basaltic samples display a small negative Eu anomaly (Fig. 7B), a characteristic totally absent from the Permian–Triassic lavas (Fig. 14A; see Lightfoot et al., 1990; Wooden et al., 1993; Fedorenko et al., 1996; Fedorenko and Czamanske, 1997) or in leucobasalt sample IP-150/930 of the Riphian, Rzhevsky Suite from the Igarka region.

Prior to completion of the isotopic and age components of this study, we had thought these unmetamorphosed volcanic rocks to be of Late Riphean–Vendian age, by analogy with petrographically and geochemically similar, rift-related lavas of the Central Taymyr Zone (Fig. 1 and Appendix 1; Vernikovskiy, 1996). Upon our request, V. F. Proskurnin of VNIIOkeangeologiya, St Petersburg, graciously made a petrographic comparison of the volcanic suite from Maslov's diatreme with the volcanic formations of Taymyr Peninsula (see Appendix 1). He determined that although our samples were fresher, some of them closely resemble the Precambrian

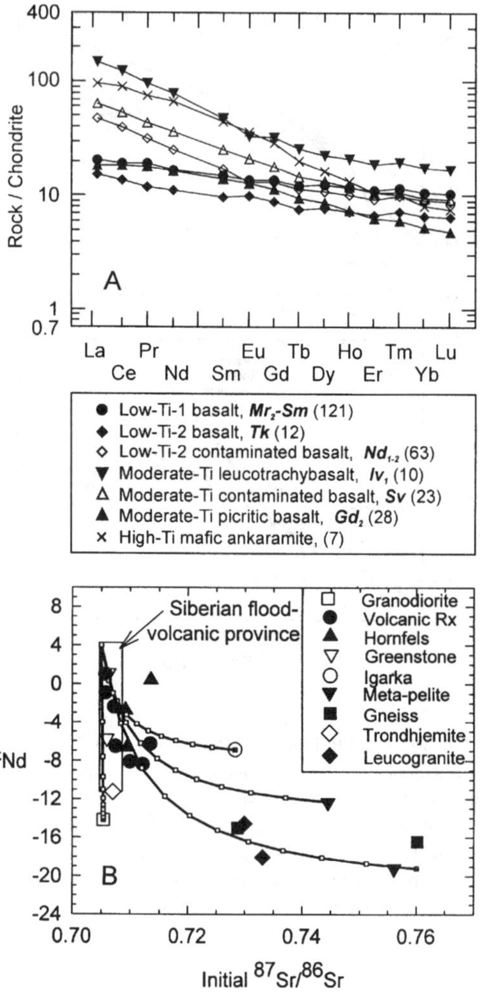

FIG. 14. A. REE plot for the Siberian, Permian–Triassic, flood-volcanic rocks. Value in parentheses is the number of samples represented by the plotted data. B. Plot of εNd against initial $^{87}Sr/^{86}Sr$ for the studied sample collection and rocks of the Siberian flood-volcanic province. A variety of possible mixing/contamination lines are shown. Isotopic values are calculated for 251 Ma, the age of the flood-volcanic event. Mixing parameters are discussed in the text. The legend entry Igarka is for leucobasalt IP-150/930.

lavas of the Taymyr. Moreover, the Central Taymyr accretionary block contains examples of most of the Precambrian crystalline basement rock types "sampled" by Maslov's diatreme—alkali and trondhjemitic granitoids, schists, gneisses, and amphibolites (see Appendix 1).

Thus, the most unexpected and controversial aspect of this investigation is the determination that the volcanic rock fragments from the diatreme represent a compositionally diverse suite of Upper Permian lavas. In fact, the SHRIMP zircon age study that yielded the age of 270 ± 3 Ma for sample 19F-57 was initiated largely because the Rb-Sr isochron age for this volcanic suite seemed quite improbable. The Carboniferous–Permian, terrigenous, coal-bearing Tungusskaya Series is well exposed and has been thoroughly sampled by hundreds of drill holes in the Noril'sk area and surroundings; only sedimentary rocks occur in the Middle Carboniferous to Late Permian section.

However, volcanism (including felsic volcanism) did take place in the southeastern part of the West Siberian Lowland (Fig. 1) during the Late Paleozoic, and evidence of Late Paleozoic volcanic activity just west of the Noril'sk area is present in the sedimentary sequence of this region. O. Simonov (unpubl.) has described tuffs containing fragments of andesite and basaltic andesite as large as 2–3 cm in that part of the Tungusskaya Series belonging to the lower half of the Upper Permian. Well-rounded pebbles of felsic lava (1–2 cm, a few 10 cm across) are known in the Upper Permian suites of the Tungusskaya Series, including its uppermost part. Moreover, the graywacke-like character of parts of the Tungusskaya Series (see Appendix 1 and earlier discussion of sample geochemistry) supports the proximity of volcanic formations which could be the source of the volcanic-suite xenoliths from the diatreme.

Thus, the results of this investigation place constraints on speculation regarding the tectonic development of the western edge of the Siberian platform. Insight into this problem is briefly considered here, but will be the subject of a future report by Fedorenko, Simonov, Proskurnin, Savushkin, and others. Sengör et al. (1993) documented accretionary activity and crustal growth between Siberia and Baltica from the Early Cambrian through the Permian. Most of what they termed the "Altaid collage" evolved west and south of the Angara craton, and involved northward movement against the craton. However, the final (Late Permian) event in this evolution was a reversal of the shear sense between Baltica and Siberia, such that Baltica swung south and east with respect to Siberia. They noted that this reversal stretched the Nadym basin (one of the deepest parts of the West Siberian basin), led to collision in the Taymyr, and may have been responsible for the onset of the Siberian flood-volcanic eruptions. We now suggest that our study provides evidence for subduction/underthrusting under the western edge of the Siberian platform, of a pre–251 Ma accretionary block of West Siberian rocks containing the ~270 Ma volcanic suite. Tectonic activity of this nature appears to afford the only possibility for introducing the volcanic suite into the stratigraphic section sampled by the diatreme, and could well have been associated with initiation of the flood-volcanic activity.

Implications for the crustal section at the northwestern margin of the Siberian craton and potential crustal contaminants of the Siberian flood-volcanic province

The stratigraphy of Figure 15 represents the platform cover in the Noril'sk and Igarka regions down through the quartzite-like sandstones of the Gubinsky Series. Our study shows that somewhere beneath the Gubinsky Series lies an underthrust block containing parts of a Late Permian volcanic suite. We have also shown that the crustal section also contains a diverse assemblage of Late Proterozoic (Riphean) crystalline rocks—alkali granites, trondhjemites, crystalline schists, gneisses, and amphibolites—that have much in common with lithologies in the Central zone of the Taymyr folded area (see Appendix 1). Inherited zircons in 19F-69 and Nd model ages for several other samples show that these Riphean rocks contain Early Proterozoic and Late Archean components. Most likely, these rocks had a source that resembled the Precambrian crust of the Anabar shield, which is composed of the 2.7 Ga Anabar Granulitic Complex and the 1.96 Ga Lamuyka Complex (Rosen, 1989). Although these ages are reflected in the inherited component of 19F-69, no representatives of these older rocks were identified among our samples from the diatreme; thus we have no direct evidence concerning the actual presence or nature of Archean crust in the Noril'sk area. However, the data strongly indicate that Late to Middle Proteozoic crust, ± Early Proterozoic and Archean crust, was present in the region, because the Nd isotopic data show that such a crust contributed to the formation of the Paleozoic sedimentary and volcanic rocks in the Noril'sk

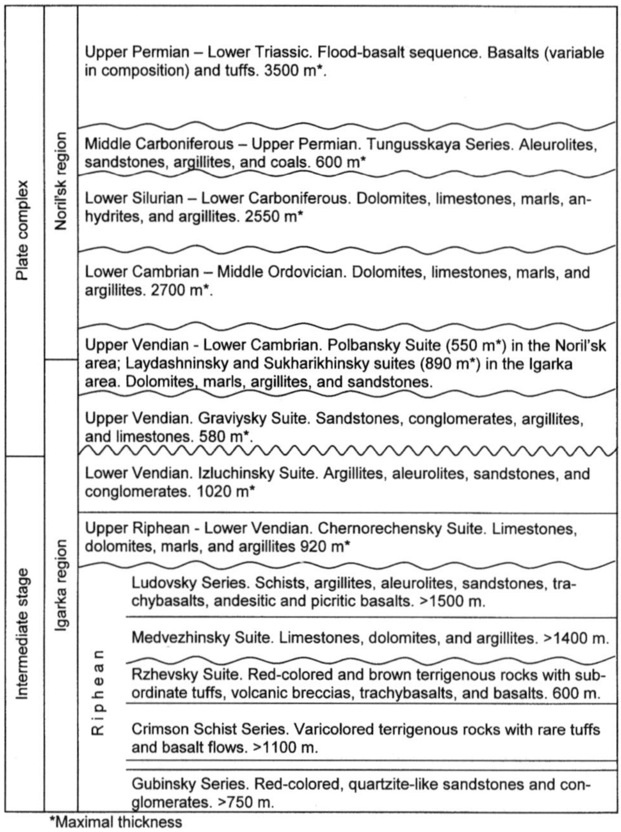

FIG. 15. Stratigraphic column for the observed platform cover beneath the Noril'sk region. Straight lines indicate a depositional contact; long wavelength lines indicate minor unconformities; short wavelength lines indicate a major unconformity above the lower assemblage of the platform cover; double straight lines indicate that the nature of the contact is not known.

crustal section. The uncertainty about the presence of the Archean crust below Noril'sk exists because the inherited zircon population might represent either: (a) a deeper crustal section not included in the underthrust crustal block, or (b) Middle to Late Proterozoic sedimentary rocks containing detrital Early Proterozoic and Archean zircons.

On a small scale, the variability of the Rb/Sr, Sm/Nd, U/Pb, and Th/Pb ratios in this crustal section produced a very isotopically heterogeneous crust over a period of two billion years or more. Given this evident complexity, the isotopic character of small portions of this crust would be difficult to constrain for geochemical modeling. However, large-scale interactions with this crust would be expected for: (1) a major magmatic event such as the Siberian flood volcanism, or (2) a large partial melt of the crust such as the one that produced the Bolgokhtokh granodiorite, and would more likely reflect the average isotopic characteristics associated with an Early to Middle Proterozoic craton. In this respect, the Bolgokhtokh granodiorite provides unique information about the crust below the Noril'sk region. Its initial isotopic ratios for Sr and Pb, combined with a Nd model age of 1.6 Ga, document a provenance that is dominated by Proterozoic crust with long-term, low Rb/Sr, U/Pb, and Th/Pb ratios. Only the trondhjemitic xenoliths of the crystalline basement group have isotopic characteristics approaching those indicated for the crustal source of the Bolgokhtokh granodiorite (Fig. 14B). On a plot of εNd against initial $^{87}Sr/^{86}Sr$, basaltic amphibolites of the crystalline basement define a nearly vertical trend that extends downward to the Bolgokhtokh granodiorite (Fig. 14B). As with the source of the granodiorite, this mafic crust retained generally low

Rb/Sr ratios to provide for these long-term, low $^{87}Sr/^{86}Sr$ ratios. Unlike the source of the granodiorite, however, they evolved with a significantly higher Sm/Nd ratio, leading to εNd values closer to zero, than the εNd value of approximately –14 for the Bolgokhtokh granodiorite. Rosen (1989, Table 1) documented low Rb/Sr ratios for the 2.7 Ga Anabar Granulitic Complex, which are consistent with the type of source characteristics required for the Bolgokhtokh granodiorite.

The majority of the crystalline basement samples have $^{87}Sr/^{86}Sr$ ratios greater than 0.72 at 251 Ma, combined with initial εNd values of –12 or less. With regard to the Sr-Nd isotopic systematics of the Siberian flood-volcanic rocks and related intrusions, these samples provide a broad spectrum of compositions that would have been capable of mixing with the high εNd-low initial $^{87}Sr/^{86}Sr$ end-member of the basalt series to generate much of the range of isotopic compositions that has been observed in the flood-volcanic sequence. Curves are calculated (Fig. 14B) for mixing between the high εNd–low initial $^{87}Sr/^{86}Sr$ end-member of the flood-volcanic series εNd = +4, Nd = 10 ppm, $^{87}Sr/^{86}Sr$ = 0.705, Sr = 150 ppm) and: (1) the Bolgokhtokh granodiorite, approximating the characteristics of its presumed crustal source (εNd = –14.2, Nd = 70.8 ppm, $^{87}Sr/^{86}Sr$ = 0.70534, Sr = 1600 ppm); (2) a composition similar to the metapelite with the most radiogenic Sr isotopic composition (19F-7; εNd = –19.2, Nd = 49.1 ppm, $^{87}Sr/^{86}Sr$ = 0.760, Sr = 87 ppm); (3) the metapelite with the highest εNd value (19F-6; εNd = –12.3, Nd = 39.4 ppm, $^{87}Sr/^{86}Sr$ = 0.7445, Sr = 72 ppm); and (4) the Igarka leucobasalt (εNd = –6.9, Nd = 46.1 ppm, $^{87}Sr/^{86}Sr$ = 0.72823, Sr = 80 ppm).

Our results provide permissive evidence that at least some of the isotopic variation within the flood-volcanic suite can be explained by crustal assimilation. In brief, several of the possible contaminants have the relatively low εNd, high Nd, and low Sr concentrations necessary to generate the range of Nd-Sr isotopic compositions observed in this Permian–Triassic volcanic sequence, if contamination was a major factor in the generation of the sequence. For example, even moderate additions (1 to 10 vol%) of a crustal component, such as metapelite 19F-7, to a mantle-derived melt with the characteristics of the least evolved samples of the Siberian flood-volcanic province could produce much of the observed Nd and Sr isotopic variation (Fig. 14B). Mixing with material similar to that of the presumed protolith of the Bolgokhtokh granodiorite could produce the volcanic rocks with low εNd and low initial $^{87}Sr/^{86}Sr$. Thus, contamination models provide an alternate mechanism to mantle heterogeneities for explaining the range of isotopic compositions observed in the Siberian flood-volcanic sequence.

Acknowledgments

V. F. Proskurnin of VNIIOkeangeologiya, St. Petersburg graciously made a petrographic comparison of the volcanic suite from Maslov's diatreme with volcanic formations of Taymyr Peninsula. A. B. Ford of the U.S. Geological Survey assisted in the petrographic description of the samples. We are most grateful to M. P. Savushkin of the Noril'sk Mining Company, Noril'sk, who provided us with the most recent unpublished data on the stratigraphy of the Igarka region.

REFERENCES

Arndt, N., Chauvel, C., Czamanske, G., and Fedorenko, V., 1998, Two mantle sources, two plumbing systems: Tholeiitic and alkaline magmatism of the Maymecha River basin, Siberian flood volcanic province: Contrib. Mineral. Petrol., v. 133, p. 297–313.

Arndt, N. T., Czamanske, G. K., Wooden, J. L., and Fedorenko, V. A., 1993, Mantle and crustal contributions to continental flood volcanism: Tectonophysics, v. 223, p. 39–52.

Arndt, N., Lehnert, K., and Vasiliev, Y., 1995, Meimechites: Highly magnesian lithosphere-contaminated alkaline magmas from deep subcontinental mantle: Lithos, v. 34, p. 41–59.

Barker, Fred, ed., 1979, Trondhjemites, dacites, and related rocks: Amsterdam, Elsevier Scientific, Developments in petrology series, no. 6, 659 p.

Basu, A. R., Poreda, R. J., Renne, P. R., Teichmann, F., Vasiliev, Yu. R., Sobolev, N. V., and Turrin, B. D., 1995, High-^{3}He plume origin and temporal-spatial evolution of the Siberian flood basalts: Science, v. 269, p. 822–825.

Bochkarev, V. S., and Krinochkin, V. G., 1988, Precambrian and Paleozoic formations of Western Siberia: Trudy IGIG SO AN SSSR, v. 728, p. 80–103 (in Russian).

Bowring, S. A., Erwin, D. H., Jin, Y. G., Martin, M. W., Davidek, K., and Wang, W., 1998, U/Pb zircon geochronology and tempo of the end-Permian mass extinction: Science, v. 280, p. 1039–1045.

Czamanske, G. K., Gurevich, A. B., Fedorenko, V., and Simonov, O., 1998, Demise of the Siberian plume: Paleogeographic and paleotectonic reconstruction from the prevolcanic and volcanic records, North-Central Siberia: INT. GEOL. REV., v. 40, p. 95–115.

Czamanske, G. K., Wooden, J. L., Zientek, M. L., Fedorenko, V. A., Zen'ko, T. E., King, Bi-S. W., Knight, R. J., and Siems, D. F., 1994, Geochemical Pb and Sr-isotopic constraints on the petrogenesis of the Noril'sk-Talnakh ore-forming system, *in* Lightfoot, P. C., and Naldrett, A. J., eds., Proc. of the Sudbury-Noril'sk Symposium: Ontario Geol. Surv. Spec. Vol. 5, p. 313–342.

Czamanske, G. K., Zen'ko, T. E., Fedorenko, V. A., Calk, L. C., Budahn, J. R., King, Bi-S. W., and Siems, D. F., 1995, Petrographic and geochemical characterization of ore-bearing intrusions of the Noril'sk type, Siberia, with discussion of their origin, *in* Ishihara, S., and Czamanske, G. K., eds., Proc. of the Sapporo International Conference on "Mineral Resources of the NW Pacific Rim," 1994: Society of Resource Geology, Tokyo, Japan, Resource Geology Special Issue, no. 18, p. 1–48.

Distler, V. V., and Kunilov, V. E., eds., 1994, Geology and ore deposits of the Noril'sk region, *in* VII International Platinum Symposium Guidebook, Moscow, 67 p.

Fedorenko, V. A., 1994, Evolution of magmatism as reflected in the volcanic sequence of the Noril'sk region, *in* Lightfoot, P. C., and Naldrett, A. J., eds., Proceedings of the Sudbury Noril'sk Symposium: Ontario Geol. Surv. Spec. Vol. 5, p. 171–183.

Fedorenko, V. A., Lightfoot, P. C., Naldrett, A. J., Czamanske, G. K., Hawkesworth, C. J., Wooden, J. L., and Ebel, D. S., 1996, Petrogenesis of the Siberian flood-basalt sequence at Noril'sk: INT. GEOL. REV., v. 38, p. 99–135.

Fedorenko, V. A., and Czamanske, G. K., 1997, Results of new field and geochemical studies of the volcanic and intrusive rocks of the Maymecha-Kotuy area, Siberian flood-basalt province, Russia: INT. GEOL. REV., v. 39, p. 479–531.

Fedorenko, V., Czamanske, G., Zen'ko, T., Budahn, J., and Siems, D., 2000, Field and geochemical studies of the melilite-bearing Arydzhangsky Suite, and an overall perspective on the Siberian alkaline-ultramafic flood-volcanic rocks: INT. GEOL. REV., v. 42, p. 769–804.

Hawkesworth, C. J., Lightfoot, P. C., Fedorenko, V. A., Blake, S., Naldrett, A. J., Doherty, W., and Gorbachev, N. S., 1995, Magma differentiation and mineralization in the Siberian continental flood basalts: Lithos, v. 34, p. 61–88.

Horan, M. F., Walker, R. J., Fedorenko, V. A., and Czamanske, G. K. , 1995, Os and Nd isotopic constraints on the temporal and spatial evolution of flood basalt sources, Siberia: Geochim. et Cosmochim. Acta, v. 59, p. 5159–5168.

Ireland, T. R., 1992, Crustal evolution of New Zealand: Evidence from age distributions of detrital zircons in western province paragneiss and Torlesse graywacke: Geochim. et Cosmochim. Acta, v. 56, p. 911–920.

Kamo, S. L., Czamanske, G. K., Amelin, Y., Fedorenko, V. A., and Trofimov, V. R., 2000, U-Pb zircon and baddeleyite, and U-Th-Pb perovskite ages for Siberian flood volcanism, Maymecha-Kotuy area, Siberia [abs.], *in* Goldschmidt 2000, Oxford, U.K., "An International Conference for Geochemistry," European Association for Geochemistry and the Geochemical Society, Journal of Conference Abstracts, vol. 5, no. 2, p. 569.

Kamo, S. L., Czamanske, G. K., and Krogh, T. E., 1996, A minimum U-Pb age for Siberian flood-basalt volcanism: Geochim. et Cosmochim. Acta, v. 60, p. 3505–3511.

Komarova, M. Z., 1968, About a granitoid intrusion in the Noril'sk area: Geologiya i geofizika, no. 5, p. 104–107 (in Russian).

Lassiter, J. C., and DePaolo, D. J., 1997, Plume/lithosphere interaction in the generation of continental and oceanic flood basalts: chemical and isotopic constraints, *in* Mahoney, J. J., and Coffin, M. F., eds., Large igneous provinces: Continental, oceanic, and planetary flood volcanism: Washington, DC, Amer. Geophys. Union, Geophys. Monog. no. 100, p. 273–295.

Lightfoot, P. C., and Hawkesworth, C. J., 1997, Flood basalts and magmatic Ni, Cu, and PGE sulfide mineralization: Comparative geochemistry of the Noril'sk (Siberian traps) and West Greenland sequences, *in* Mahoney, J. J., and Coffin, M. F., eds., Large igneous provinces: Continental, oceanic, and planetary flood volcanism: Washington, DC, Amer. Geophys. Union, Geophys. Monog. no. 100, p. 357–380.

Lightfoot, P. C., Hawkesworth, C. J., Hergt, J., Naldrett, A. J., Gorbachev, N. S., Fedorenko, V. A., and Doherty, W., 1993, Remobilization of the continental lithosphere by a mantle plume: Major-, trace-element, and Sr-, Nd- and Pb-isotopic evidence from picritic and tholeiitic lavas of the Noril'sk district, Siberia: Contrib. Mineral. Petrol., v. 114, p. 171–188.

______, 1994, Chemostratigraphy of Siberian trap lavas, Noril'sk district, Russia: Implications for the evolution of flood basalt magmas, *in* Lightfoot, P. C., and Naldrett, A. J., eds., Proc. of the Sudbury-Noril'sk Symposium: Ontario Geol. Surv. Spec. Vol. 5, p. 283–312.

Lightfoot, P. C., Naldrett, A. J., Gorbachev, N. S., Doherty, W., and Fedorenko, V. A., 1990, Geochemistry of the Siberian trap of the Noril'sk area, USSR, with implications for the relative contributions of crust and mantle to flood basalt magmatism: Contrib. Mineral. Petrol., v. 104, p. 631–644.

Masaitis, V. L., 1983, Permian and Triassic volcanism of Siberia: Zapiski VMO, part CXII, issue. 4, p. 412–425 (in Russian).

Masaitis, V. L., Maschak, M. S., Raykhlin, A. I., Selivanovskaya, T. V., and Shafranovskiy, G. I., 1998, Diamond-bearing impactites of the Popigay astrobleme: St. Petersburg, VSEGEI Press, 178 p. (in Russian).

Maslov, G. D., and Nesterovskiy, V. S., 1961, Explosive fragments of the crystalline basement rocks in the Triassic tuffs: Geologiya i geofizika, no. 12, p. 128–130 (in Russian).

Naldrett, A. J., Asif, M., Gorbachev, N. S., Kunilov, V. E., Stekhin, A. I., Fedorenko, V. A., and Lightfoot, P. C., 1994, The composition of the Ni-Cu ores of the Oktyabr'sky deposit, Noril'sk region, *in* Lightfoot, P. C., and Naldrett, A. J., eds., Proc. of the Sudbury-Noril'sk Symposium: Ontario Geol. Surv. Spec. Vol. 5, p. 357–371.

Naldrett, A. J., Fedorenko, V. A., Lightfoot, P. C., Kunilov, V. E., Gorbachev, N. S., Doherty, W., and Johan, Z., 1995, Ni-Cu-PGE deposits of the Noril'sk region, Siberia: Their formation in conduits for flood basalt volcanism: Trans. Inst. Min. Metall. London, v. 104, p. B18–B36.

Naldrett, A. J., Lightfoot, P. C., Fedorenko, V., Doherty, W., and Gorbachev, N. S., 1992, Geology and geochemistry of intrusions and flood basalts of the Noril'sk region, USSR, with implications for the origin of the Ni-Cu ores: Econ. Geol., v. 87, p. 975–1004.

Nemenyonok, T. I., 1978, Explosive magmatism, *in* Zolotukhin, V. V., and Vilenskiy, A. I., eds., Petrology and prospects for mineralization of the traps from the northern Siberian platform: Novosibirsk, Nauka Press, p. 8–43 (in Russian).

Pettijohn, F. J., 1975, Sedimentary rocks: New York, Harper and Row, 718 p.

Ravich, M. G., and Chayka, L. A., 1956, Small intrusions of the Byrranga Range (Taymyr Peninsula): Trudy NIIGA, v. 88, 105 p. (in Russian).

Rosen, O. M., 1989, Two geochemically different types of Precambrian crust in the Anabar Shield, North Siberia: Precamb. Res., v. 45, p. 129–142.

Sengör, A. M. C., Natal'in, B. A., and Burtman, V. S.,1993, Evolution of the Altaid tectonic collage and Paleozoic crustal growth in Eurasia: Nature, v. 364, p. 299–307.

Sharma, M., 1997, Siberian traps, *in* Mahoney, J. J., and Coffin, M. F., eds., Large igneous provinces: Continental, oceanic, and planetary flood volcanism: Washington, DC, Amer. Geophys. Union, Geophys. Monog. no. 100, p. 273–295.

Sharma, M., Basu, A. R., and Nesterenko, G. V., 1991, Nd-Sr isotopes, petrochemistry, and origin of the Siberian flood basalts, USSR: Geochim. et Cosmochim. Acta, v. 53, p. 1183–1192.

Sharma, M., Basu, A. R., and Nesterenko, G. V., 1992, Temporal Sr-, Nd-, and Pb-isotopic variations in the Siberian flood basalts: Implications for the plume-source characteristics: Earth Planet. Sci. Lett., v. 113, p. 365–381.

Simonov, O. N., Lul'ko, V. A., Amosov, Yu. N., and Salov, V. M., 1994, Geological structure of the Noril'sk region, *in* Lightfoot, P. C., and Naldrett, A. J., eds., Proc. of the Sudbury-Noril'sk Symposium: Ontario Geol. Surv. Spec. Vol. 5, p. 161–170.

Stacey, J. S., and Kramers, J. D., 1976, Approximation of terrestrial lead isotopic evolution by a two-stage model: Earth Planet. Sci. Lett., v. 113, p. 207–221.

Taylor, S. R., 1985, The continental crust, its composition and evolution: An examination of the geochemical record preserved in sedimentary rocks: Oxford, Blackwell Scientific, 312 p.

Vernikovskiy, V. A., 1996, Geodynamic evolution of the Taymyr folded area: Novosibirsk, Siberian Branch, RAS SPC UIGGM, 203 p. (in Russian).

Walker, R. J., Morgan J. W., Horan, M. F., Czamanske, G. K., Krogstad, E. J., Fedorenko, V., and Kunilov, V. E., 1994, Re-Os isotopic evidence for an enriched-mantle source for the Noril'sk-type ore-bearing intrusions, Siberia: Geochim. et Cosmochim. Acta, v. 58, p. 4179–4197.

Wooden, J. L., Czamanske, G. K., Bouse, R. M., Knight, R. J., and Budahn, J. R., 1992, Pb isotopic data indicate a complex, mantle origin for the Noril'sk-Talnakh ores: Econ. Geol., v. 87, p.1153–1165.

Wooden, J. L., Czamanske, G. K., Fedorenko, V. A., Arndt, N. T., Chauvel, C., Bouse, R. M., King, Bi-S. W., Knight, R. J., and Siems, D. F., 1993, Isotopic and trace-element characterization of the Siberian continental flood basalts of the Noril'sk area: Geochim. et Cosmochim. Acta., v. 57, p. 3677–3704.

Zen'ko, T. E., and Czamanske, G. K., 1994, Tectonic controls on ore-bearing intrusions of the Talnakh ore junction: Position, morphology, and ore distribution: INT. GEOL. REV., v. 36, p. 1033–1057.

Zhuravlev, E. G., 1986, The trap formation of the Western Siberian plate: Izvestiya AN SSSR, ser. geol., no. 7, p. 26–32 (in Russian).

Zonenshain, L. P., Kuzmin, M. I., and Natapov, L. M., 1990, Geology of the USSR: A plate tectonic synthesis: Washington, DC, Amer. Geophys. Union, Geodynamics Series, vol. 21, 242 p. [English translation of Lithosphere plate tectonics in the USSR Territory, v. 2, Moscow, Nedra Press, 334 p.].

Appendix 1

The northern Siberian platform

The crystalline basement is exposed in the Anabar shield, nearly 700 km east of the Noril'sk area (Fig. 1). According to Rosen (1989), the Anabar shield is composed mainly of gneisses and granitoids of the 2.7 Ga Anabar Granulitic Complex and the 1.96 Ga Lamuyka Complex. Masaitis et al. (1998) reported comparable ages for rocks from the northern part of the Anabar shield, where metamorphosed rocks 2.5–2.98 Ga in age are cut by younger 1.8–2.2 Ga granitoids.

The Siberian platform cover is broadly exposed throughout the platform and along its northwestern margin it is especially well studied on the basis of: (1) drilling in the Noril'sk region that can be projected to depths of 5900 m, and (2) drilling in the Igarka region (200 km south of Noril'sk) that can be projected to ~8000 m depth. This cover has been subdivided into two structural assemblages—a rather strongly deformed, but weakly metamorphosed, lower assemblage, overlain by a typical "plate complex" (Distler and Kunilov, 1994; Simonov et al., 1994). The lower assemblage, of Riphean–Vendian age, is known to exceed 7000 m in thickness (Fig. 15). The lowermost Gubinsky Series (base undetermined) is composed of red, quartzite-like sandstones and conglomerates. Sedimentary terrigenous and carbonate rocks with subordinate tuffs, volcanic breccias, trachybasalts, and basalts lie above this sequence. These rocks are interpreted to have formed in a foredeep, situated between the Siberian craton and a folded area to the west (M.P. Savushkin, unpubl. manuscript, 1999). Savushkin (unpubl. manuscript, 1994) reported Rb-Sr ages of 830–925 Ma for volcanic rocks from the Rzhevsky Suite, and 842–935 Ma for the Ludovsky Series (see Fig. 15).

Accumulation of the lower assemblage was terminated by orogeny, with formation of the Baikalian mountain system west of the Yenisey River. The foredeep migrated to the east, where molasse of the Upper Vendian Graviysky Suite accumulated. This molasse, the first component of the "plate complex," is overlain by the silty-pelitic Laydashninsky Suite, then by the Sukharikhinsky Suite, which is mainly carbonate in the Igarka area. The Upper Vendian–Lower Cambrian, Polbansky Suite of the Noril'sk area is the analog of the Sukharinsky Suite (Fig. 15). In the Noril'sk area, this "plate" sequence continues with a thick (~5000 m) Cambrian to Lower Carboniferous succession, composed of carbonates with subordinate evaporitic and terrigenous (mainly argillites) rocks.

Following uplift and mild folding in the Early–Middle Carboniferous, the terrigenous, coal-bearing Tungusskaya Series (as much as 600 m thick) was deposited in Middle Carboniferous to Late Permian time. The Upper Permian–Lower Triassic flood-volcanic sequence overlies the Tungusskaya Series with an erosional break of as much as several hundred meters. The thickness of this volcanic sequence reaches ~3500 m in the Noril'sk area. Fedorenko and Czamanske (1997) correlated the volcanic piles of the Noril'sk and Maymecha-Kotuy areas and concluded that the combined thickness of the north Siberian flood-volcanic sequence is ~6500 m in northern Siberia. This volcanism was accompanied by significant intrusive activity. A brief overview of the geochemistry of the flood-volcanic and associated intrusive rocks is given below.

Uplift and moderately intense platform-style tectonogenesis occurred in the Middle to Late Triassic, as flood volcanism ceased, leading to the recent structural plan of the Siberian platform (Fig. 1). Isometric and mildly elongate folds of several thousand meters amplitude, and numerous faults of 1200–1300 m slip developed. Small granitoid stocks were emplaced in the western Taymyr (Ravich and Chayka, 1956), and the Bolgokhtokh granodiorite stock appears to be comparable.

Jurassic–Cretaceous terrigenous sedimentary rocks are found at the periphery of the Siberian platform. They reach thicknesses of ~5000 m the West Siberian Lowland and the Yenisey-Khatanga trough (Fig. 1).

The Taymyr folded area

The Taymyr folded area is subdivided into the Southern, Central, and Northern zones, separated by large thrust faults (Zonenshain et al., 1990; Vernikovskiy, 1996). The Southern Taymyr Zone is a deep depression filled with Ordovician–Lower Carboniferous carbonate rocks (to 6000 m), Middle Carboniferous–Permian terrigenous rocks (to 7000 m), and Upper Permian–Lower Triassic sedimentary and volcanic rocks (to 4000–6000 m). The volcanic rocks have much in common with the flood basalts of the Siberian platform. This sequence is folded, with the degree of folding increasing from south to north (Vernikovskiy, 1996).

The Central Taymyr Zone was considered by Vernikovskiy (1996) to be an accretionary block consisting of Pre-Riphean(?) and Riphean sedimentary, volcanic, and intrusive formations that have undergone metamorphism of different facies and types. These accretionary formations are cut by a volcanic belt (supposedly of rift genesis) of Late Riphean to Vendian(?) age and covered by Vendian to Lower Carboniferous terrigenous rocks. Among the accretionary formations are two terrains of highly metamorphosed and granitoid rocks, as well as several terrains and plates composed of island-arc complexes, ophiolites, olistostromes, collisional molasses, and carbonate units. The Rb-Sr age of metamorphism of the gneisses in the highly metamorphosed terrains is 763 ± 12 Ma. Both normal and subalkaline granites have zircon U-Pb ages of 846 ± 11 Ma and Rb-Sr ages of 840 ±170 Ma. Plate-shaped bodies (observed thicknesses to 1 km) of plagiogranite of the tonalite-trondhjemite series occur in one ophiolite terrain; they have a zircon U-Pb age of 740 ± 38 Ma and a Rb-Sr age of 727 ± 83 Ma (Vernikovskiy, 1996). The rift(?) formations lie in a discrete belt of volcanic rocks, conformable to the strike of the eastern part of the Central Taymyr Zone. This belt is composed of trachybasaltic and basaltic tuffs (K_2O contents of 2–3, locally to 7 wt%; TiO_2 contents to 2–4 wt%), lavas with pillow jointing, dike-like bodies of gabbro-diabase, and sparse felsic tuffs. Felsic lavas, including alumina-supersaturated rhyodacites and dacites, are present in the extreme northeastern part of the belt. The total thickness of these rift(?) formations is no more than 200–220 m, and no age determinations are available for these rocks. Vernikovskiy (1996) believed that they represent a distinct, bimodal volcanic complex formed during rifting at an active continental margin in Late Riphean–Vendian time.

The Northern Taymyr Zone is composed mainly of flyschoidal rocks, considered to represent sedimentation along a continental slope. The rocks are metamorphosed from low greenschist to amphibolite facies; migmatization and granitization characteristically accompany the amphibolite-facies metamorphism. Small (to a few tens of km^2), elongate bodies of autochthonous gneissic granites formed, and autochthonous, parautochthonous, and allochthonous granitoid intrusions were emplaced. Vernikovskiy (1996) reported a monazite U-Pb age of 306 ± 2 Ma for an autochthonous granite, and a zircon U-Pb age of 264 ± 8 Ma for an allochthonous granodiorite. Sm-Nd isotopic data for the same sample of autochthonous granite show that the crustal source of the granite had a Riphean age (T_{DM} = 1170 Ma).

Vernikovskiy (1996) proposed two main stages of crustal formation in the Taymyr folded area. The first stage was Late Riphean–Vendian, when the Central Taymyr accretionary block was formed and joined to Siberia. The second stage related to collision between the Kara and Siberian paleocontinents in Late Paleozoic time. The absence of Early to Middle Paleozoic ophiolites and subducted complexes is remarkable; these should have developed if oceanic crust existed between the Siberian and Kara paleocontinents. A possible explanation is an oblique collision or a major shift along the edge of Siberia.

The West Siberian Lowland

Limited information concerning the pre-Jurassic rocks of the West Siberian Lowland is available from oil-exploration drilling. Few boreholes have reached these rocks, and those that do typically sample only several tens or a few hundred meters of section (Bochkarev and Krinochkin, 1988). Supposed Riphean crystalline schists, and Vendian to Upper Paleozoic terrigenous and carbonate rocks with rare volcanic rocks, are known in the eastern part of the West Siberian province (Bochkarev and Krinochkin, 1988, their Fig. 2, column X). However, this information relates mainly to areas located hundreds to 1000 km south of the Noril'sk region. Upper Paleozoic volcanic formations are broadly developed in the southeastern part of the province (~1000 km SSW of Noril'sk); they are composed mainly of rhyolitic and dacitic-rhyolitic rocks, with K-Ar ages of 287 to 318–331 Ma (Bochkarev and Krinochkin, 1988). Zhuravlev (1986) described basalts as much as 1150 m thick that probably correspond to the Siberian flood basalts. These basalts fill a series of grabens below Jurassic–Cretaceous sedimentary rocks over an area of $\sim 0.75 \times 10^6$ km^2 (Zhuravlev, 1986).

The Siberian Permian–Triassic flood-volcanic province

The geochemistry of the flood-volcanic and related intrusive rocks, including their Sr, Pb, and Nd isotopic characteristics, has been studied in detail for the Noril'sk region (e.g., Lightfoot et al., 1990; Wooden et al., 1993; Czamanske et al., 1995; Hawkesworth et al., 1995; Naldrett et al., 1995) and for the Maymecha-Kotuy region, located northwest of the Anabar shield (Fedorenko and Czamanske, 1997; Fedorenko et al., 2000). Fedorenko et al. (1996, 2000) and Fedorenko and Czamanske (1997) subdivided the Siberian Permian–Triassic flood-volcanic rocks into four geochemical rock/magma types – low-Ti-1, low-Ti-2, moderate-Ti, and high-Ti (average compositions for typical rocks of these types are presented by Fedorenko et al., 2000, their Table 1). The low-Ti and moderate-Ti rock types are associated with the tholeiitic rock series and the high-Ti type with the alkaline-ultramafic series. All representatives of the types are easily recognized in plots of Th/U against Sm/Yb with no dependence on degree of fractionation and little dependence on crustal contamination (Fedorenko et al., 2000, their Fig. 3). Uncontaminated varieties of the rocks have unique isotopic characteristics (Fedorenko et al., 2000, their Fig. 4), showing that the parental magmas of the rock types were not related to one another through fractionation, continuous melting, or magma mixing; i.e., each of the types requires a mantle source of distinct composition.

The low-Ti-1 type is the most voluminous: it composes the entire upper part of the volcanic pile in the Noril'sk region (from the Upper Morongovsky Subsuite to the Samoedsky Suite) with thicknesses to 2200 m. It is represented everywhere by monotonous basalts with SiO_2 = 49–50.5 wt% and MgO = 6–8 wt%. These basalts have TiO_2 contents of 1–2 wt% and low absolute abundances of incompatible elements; Sm/Yb = 1.2–1.5 and Th/U = 1.7–3. Relatively high Pb isotopic ratios, low Sr isotopic ratios, and positive εNd values are typical.

The-low-Ti-2 type (TiO_2 = 0.8–1.3 wt%) is represented by basalts and melanocratic (to picritic) basalts of the Tuklonsky and Nadezhdinsky suites, with thicknesses to 500–600 m. The Tuklonsky basalts and melabasalts have SiO_2 contents of 47.5–50.5 wt% and MgO = 8.5–16.5 wt%; they are similar to the low-Ti-1 type, with regard to low abundances of incompatible elements and Sm/Yb = 1.2–1.6, but are distinguished by Th/U = 3.5–4.3. The isotopic composition of the-low-Ti-2 type is quite distinctive, with low Pb isotopic ratios and negative εNd values. The Nadezhdinsky basalts, especially of the lower and middle subsuites (Nd_{1-2}), are distinguished by elevated SiO_2 (50.6–54.3 wt%) and K_2O contents, as well as REE enrichment. Most recent investigators (e.g., Lightfoot et al., 1990, 1993, 1994; Naldrett et al., 1992, 1995; Wooden et al., 1993; Arndt et al., 1993; Hawkesworth et al., 1995; and Fedorenko et al., 1996) are in agreement that the Nd_{1-2} rock compositions reflect strong crustal contamination.

The moderate-Ti type, which composes the three lower volcanic suites in the Noril'sk region, with thicknesses to 500 m, is variable in composition, being represented by leucocratic basalts and trachybasalts, basalts of typical MgO content, and picritic basalts. It is characterized by moderate to high incompatible-element concentrations, except in a few lavas where plagioclase phenocrysts are abundant. TiO_2 contents range from 1.1 to 1.8 wt% in picritic basalts, through 1.4 to 2.6 wt% in basalts and leucocratic basalts, to 3–4 wt% in leucocratic trachybasalts. Sm/Yb ratios are 2.2–3.0, and Th/U ratios are variable. Among presumably uncontaminated rocks of this type, Pb and Sr isotopic characteristics are relatively radiogenic and εNd values range from –0.20 in leucocratic trachybasalts to +4.63 in picritic basalts. Contaminated rocks have lower εNd values at relatively constant $^{87}Sr/^{86}Sr$ ratios; i.e., the contaminant had a different composition than that associated with contamination of the low-Ti-2 lavas.

High-Ti type rocks are abundant in the Maymecha-Kotuy region, and range from felsic to ultramafic in composition, with the predominance of mafic varieties. In the Noril'sk region, they are sparsely represented only by the mafic to melanocratic, Kaltaminsky ankaramites, found near or at the top of the Morongovsky Suite; their thickness is only several tens of meters. These rocks have high incompatible-element concentrations, with typical TiO_2 contents in mafic rocks of 3–5 wt%. REE profiles are steep, with La/Yb ranging from 15 to 170; Sm/Yb is always >4.5 (providing there is no admixture of low-Ti magma); and Th/U is typically is 3–4 or higher. Pb and Sr isotopic characteristics are relatively unradiogenic, and εNd values are positive.

Appendix 2. Brief Petrographic Description of the Studied Samples

Note that in contrast to Table 1, all values for SiO_2 content presented in this appendix have been recalculated to 100 wt% anhydrous oxides. Where so indicated, V. F. Proskurnin, VNIIOkeangeologiya, St. Petersburg, has kindly examined the sample in thin section and named it according to its petrographic and geochemical characteristics on the basis of petrographically and geochemically comparable Riphean volcanic rocks which he has studied in the Taymyr.

19F-3 (49.6 wt% SiO_2): Epidote-bearing amphibolite after an iron-rich, mafic intrusive rock.

19F-4 (47.4 wt% SiO_2): Amphibolite after hornblendite(?). Coherent, relatively fresh, patchy-zoned amphibole grains 2–4 mm across show no preferred orientation and are typically separated by crush zones <1 mm thick composed of fine-grained amphibole and plagioclase. Sparse quartz stringers <0.1 mm thick cut the rock.

19F-21 (50.5 wt% SiO_2): Heavily altered (chloritized) amphibolite after a mafic intrusive rock.

19F-7 (59.1 wt% SiO_2): Garnetiferous muscovite schist (metapelitic?). Rounded to euhedral garnet grains (typically 2–3 mm across) are highly fractured; they are commonly "wrapped" in a chlorite selvage and contain chlorite veining. Interstitial material includes fine-grained, undeformed, as well as coarse-grained, deformed muscovite; mosaics of fine-grained quartz; and chlorite. Coarser quartz has crystallized in strain shadows adjacent to garnet grains. Fine, irregular oxide grains are plentiful.

19F-6 (76.1 wt% SiO_2): Quartz, muscovite schist. Muscovite and quartz are strongly deformed; quartz generally shows strongly undulatory extinction. Quartz and K-feldspar grains display irregular contacts. Irregular stringers of blue-birefringent chlorite give a "streaky" appearance to thin sections and may host abundant rutile(?) needles. These stringers (and muscovite) isolate tapering lenses of felsic material 2–4 mm thick.

19F-13 (73.8 wt% SiO_2): Microcline orthogneiss. Highly irregular aggregates of quartz (typically 1.5–2 mm × 4–8 mm) are enveloped by recrystallized, pink feldspar; the largest microcline remnant measures 2 × 3 mm. More typical is a matrix of grains only 0.2–0.3 mm across in a polyhedral texture interpreted as annealed, ductile shear-banding. Webs of dark, discontinuous, chloritic alteration products <0.5 mm wide impart a dark coloration. Myrmekite is common. One small pyrite grain and three grains of limonite after euhedral pyrite were noted.

19F-9 (74.1 wt% SiO_2): Granitic gneiss in which generally undeformed quartz grains (as large as 1.5 × 4 mm) compose elongate, irregular patches 1–4 mm thick and form bands that define the gneissic texture. Interfingering material consists of finer-grained quartz + K-feldspar (microcline?), which may represent ductile shear bands. Muscovite is not abundant and may locally be foliated.

19F-11 (73.1 wt% SiO_2): Leucogranite. Highly irregular, interconnecting aggregates of quartz grains (as much as 2 mm across) may reach 3 cm. Quartz is interstitial to and crystallized after subhedral to euhedral plagioclase grains (1 mm to 2 × 3 mm). Plagioclase is almost completely altered, with only narrow (>0.1 mm) unaltered rims remaining on most grains; only chlorite can be found after primary biotite(?). Minor sericite is present.

19F-12 (74.3 wt% SiO_2): Leucogranite. Highly irregular aggregates of dusty quartz grains may reach 6 × 8 mm. Subhedral to euhedral plagioclase grains are completely altered with minor development of epidote. Primary biotite(?) is represented by chlorite with deep blue interference colors. Tiny flakes of sericite are common among plagioclase alteration products, and somewhat larger grains are present in interstices with chlorite.

19F-8 (74.2 wt% SiO_2): Weakly foliated, muscovite-bearing granite. Subrounded, granular-textured aggregates of moderately altered, undeformed, white plagioclase (typically 2–4 mm across) and quartz (<2 mm) are characteristically separated by thin, dark partings of muscovite and alteration products. Incipient foliation shown locally by preferred alignment of secondary muscovite and alteration products. Relatively fine-grained brownish areas several cm across are composed of irregular, fractured quartz aggregates in a mass of chlorite and lesser muscovite. The chlorite may be clouded with rutile needles. Flakes of graphite (<0.2 mm long) are most abundant in the larger alteration patches. Noted one subhedral zircon 0.14 x 0.33 mm.

19F-10 (73.7 wt% SiO_2): Muscovite, microcline leucogranite containing well-defined, randomly oriented, fresh tablets of muscovite as thick as 2 mm and as long as 6 mm. Quartz and microcline are granular textured, generally aggregated, and have grain sizes measured in millimeters. Quartz typically has undulatory extinction. Minor myrmekite and secondary muscovite are present.

19F-69 (75.0 wt% SiO_2): Muscovite-bearing, microcline leucogranite. Anhedral quartz grains are commonly several millimeters across, whereas blocky, subhedral to anhedral microcline grains are typically smaller. Interstitial muscovite occurs as aggregates of randomly oriented grains, no larger than 0.5 × 2 mm. Some small muscovite grains have been trapped in quartz and microcline, but minor secondary growth of muscovite is present in altered feldspar. Minor myrmekite is present. Trace zircons were extracted for U-Pb analysis by SHRIMP.

BG-26/88.5 and 199 (avg. 67 wt% SiO_2): Granular-textured granodiorite containing porphyritic, perthitic K-feldspar (to 6 × 8 mm) and plagioclase (to 3 × 6 mm). Quartz grains may reach 2 mm in diameter, but are more commonly <<1 mm across. Amphibole grains (some are euhedral, to 1.5 × 5 mm) typically contain small oxide and apatite grains and are typically extensively replaced by biotite; biotite also occurs as clear, isolated grains as large as 0.5 × 1.5 mm. Euhedral titanite (to 1.5 mm) and subhedral (<1 mm) to rounded (to >1 mm) apatite grains are conspicuous. Sparse sulfide grains (<0.2 mm) are present.

19F-1 (54.2 wt% SiO_2): Altered basalt, which may originally have displayed subophitic texture. Abundant carbonate is now present.

19F-51 (50.5 wt% SiO_2): Basalt, with a coarser-grained variant of the intersertal-subophitic texture of samples 19F-52 and 62. Roughly 20% of the rock is composed of subrounded clots of relatively fresh, oikocrystic pyroxene 2–3 mm across, separated by areas of altered glass hosting plagioclase laths. Commonly, only three or four grains of pyroxene are present in each clot. There is no apparent distinction between the elongate plagioclase laths that extend from, or cross, oikocrysts and those hosted in altered glass (commonly 0.1–0.15 × 1.5–2 mm). (Proskurnin—gabbrodolerite.)

19F-62 (51.1 wt% SiO_2): Basalt, displaying intersertal texture of subophitic clinopyroxene (<0.5 mm across and plagioclase (commonly 0.05 x 1 mm) in altered dark glass. Vesicles (<1 mm) filled with "gunk" and well-formed quartz crystals complete diktytaxitic texture. (Proskurnin—gabbrodolerite.)

19F-60 (52.2 wt% SiO_2 at 2.9 wt% MgO and 2.6 wt% TiO_2): Leucocratic basalt, with an ultrafine-grained, dark groundmass suggestive of the quenched, glassy margin of a lava flow, and containing sparse, elongate, plagioclase phenocrysts (commonly 0.1 × 0.8 mm; rarely 0.2 × 2 mm). (Proskurnin—meta-trachybasalt.)

19F-52 (57.9 wt% SiO_2): Andesite, with a texture comparable to that of sample 19F-62, but with sparse, conspicuous round quartz xenocrysts (as much as 3 mm across) and some less elongate, twinned plagioclase laths. (Proskurnin—dolerite(?) with quartz phenocrysts.)

19F-57 (68.8 wt% SiO_2): Rhyodacite. Fine-grained groundmass with regular-spaced, blotchy, brownish green alteration patches (commonly 0.3–0.5 mm across) contains heavily resorbed and locally recrystallized plagioclase xenocrysts (as large as 2–3 mm). Former mafic phases totally altered; Fe-Ti oxides relatively notable. Trace zircons were extracted for U-Pb analysis by SHRIMP. (Proskurnin—albitophyre of dacite.)

19F-76 (72.1 wt% SiO_2): Trachyrhyolite. Ultrafine-grained groundmass (after aphanitic rhyolite glass?) contains sparse grains of altered potassium feldspar (as much as 0.7 × 2 mm) and rare small (<0.5 mm) vesicle fillings(?) of multigranular quartz. (Proskurnin—subalkaline trachyrhyodacite albitophyre, potassium-feldspathized.)

19F-77 (66.5 wt% SiO_2): Hyalopilitic textured dacite, with relatively fresh feldspar laths 0.02–0.06 × 0.1–0.4 mm set in altered brown glass. Heavily resorbed, fractured quartz xenocrysts (as large as 3 × 8 mm) and subhedral feldspar xenocrysts (as large as 2 × 3 mm) are now recrystallized to very fine grained mosaic, often with a thin, clear (albitic?) rim. A few small (~0.5 mm) vesicle fillings(?) of carbonate occur. Trace zircons were extracted for U-Pb analysis by SHRIMP. (Proskurnin—glomeroporphyritic andesite albitophyre. Albite in phenocrysts. Hyalopilitic plagioclase in altered groundmass.)

19F-22 (74.2 wt% SiO_2): Altered holocrystaline rhyolite containing ~7 modal% subhedral feldspar grains (to 0.4 × 0.8 mm); ~2 modal% quartz grains (to 0.4 mm); and a few modal% of well aligned, acicular (often 0.15 × 1.5 mm) amphibole grains (now altered to chlorite).

KZ-1799/1359.1 (58.4 wt% SiO_2): Fine-grained (typical grains <0.05 mm), laminated, hornfelsed argillite containing dark (carbonaceous?) material and sparse, fine-grained oxide minerals.

KZ-1799/1366.1 (64.8 wt% SiO_2): Fine-grained, strongly laminated, hornfelsed argillite containing irregular, metamorphic sericite and lamellar disseminations of fine-grained pyrite.

KZ-1879/1659.4 (57.2 wt% SiO_2): Relatively massive, very fine grained, meta-argillite containing ovoid bodies (to 3 × 8 mm) of somewhat coarser, recrystallized material. Diffuse, rounded-to-subhedral patches (typically 0.05–0.1 mm) occur throughout the very fine grained matrix and may represent incipient, porphyroblastic garnet crystallization or alteration products after such crystallization.

KZ-1879/1661 (59.0 wt% SiO_2): Very fine grained, meta-argillite showing extensive development of the diffuse, rounded to subhedral patches noted for sample KZ-1879-1659.4, but with only traces of recrystallization.

SG-28-1993.9 (64.7 wt% SiO_2): Fine-grained, hornfelsed, silty argillite containing poorly crystallized, irregular sericite aggregates and peppered with highly irregular aggregates (to 0.2 × 0.4 mm) of fine-grained pyrite.

SG-28-2018.9 (69.9 wt% SiO_2): Laminated, silty argillite containing conspicuous, angular quartz grains (to 0.2 mm) set in a turbid, clay-rich matrix. Foliation and recrystallization are absent.

Zeolite Parageneses in the North Atlantic Igneous Province: Implications for Geotectonics and Groundwater Quality of Basaltic Crust

PHILIP S. NEUHOFF, THRÁINN FRIDRIKSSON, AND DENNIS K. BIRD

Department of Geological and Environmental Science, Stanford University, Stanford, California 94305-2115

Abstract

Zeolites are among the most common products of chemical interaction between groundwaters and the Earth's crust during diagenesis and low-grade metamorphism. The unique crystal structures of zeolites result in large molar volumes, high cation-exchange capacities, and reversible dehydration. These properties influence both the stability and chemistry of zeolites in geologic systems, leading to complex parageneses and compositional relationships that provide sensitive indicators of physicochemical conditions in the crust. Observations of zeolite occurrence in Tertiary basaltic lavas in the North Atlantic region indicate that individual zeolite minerals are distributed in distinct, depth-controlled zones that parallel the paleosurface of the plateau basalts and transgress the lava stratigraphy. The zeolite zones are interpreted to have formed at the end of burial metamorphism of the lavas. Relative timing relations between various mineral parageneses and crustal-scale deformal features indicate that the minerals indicative of the zeolite zones formed within 1 million years after cessation of volcanism. Empirical correlation between the depth distribution of zeolite zones and the temperatures of formation of zeolites in geothermal systems provides estimates of regional thermal gradients and heat flow in flood-basalt provinces. Similarly, the orientations of zeolite zones can be used to distinguish synvolcanic and post-volcanic crustal deformation. Because zeolites that characterize the individual zones display different ion-exchange selectivities for various cations, reactions between groundwaters and zeolites in basaltic aquifers can result in depth-controlled zones where individual elements are concentrated in the crust. This is established for Sr, which is concentrated by at least an order of magnitude in heulandite, resulting in an overall Sr enrichment of lavas in the heulandite-stilbite zeolite zone.

Introduction

ZEOLITES ARE TECTOSILICATES characterized by open frameworks of Si and Al surrounding ~2–10 Å channels containing molecular water and charge-balancing cations of alkali and alkali-earth metals. Their distinctive crystal structures result in high cation-exchange capacities and molecular sieve capabilities—properties that have widespread industrial application in water softening, catalysis, and wastewater treatment (Mumpton, 1977a, 1988). The relatively large enthalpies of reaction for hydration and dehydration of zeolites is employed in heat pump technology, where the endothermic nature of zeolite dehydration is employed in energy generation and refrigeration (e.g., Kasai et al., 1994). Formation of zeolites with large molar volumes during diagenesis in petroleum reservoirs reduces porosity and permeability, thus decreasing production potential (e.g., Helmold and van de Kamp, 1984; Crossey et al., 1984). Ion exchange by zeolites is widely employed in water softening, wastewater treatment, and agriculture (e.g., Mumpton, 1977a, 1988), and *in situ* ion-exchange by zeolites in aquifers often exerts a strong control on the major- and minor-element chemistry of groundwaters, as demonstrated by studies of their potential to control radioactive waste migration (e.g., Smyth, 1982). Sorption of gases, water, and hydrocarbons by zeolites also is widely exploited for molecular sieves, heat pumps, and desiccants (e.g., Mumpton, 1977a, 1988). Although most zeolites used in industrial applications are synthetic, natural zeolites are mined worldwide, and increasing attention is focused on developing natural zeolite resources.

Geologically, zeolites are common products of diagenesis and the lowest grades of metamorphism. Zeolites form during reactions of aqueous fluids with marine sediments (e.g., Boles, 1977; Kastner, 1979), saline lake sediments (e.g., Hay, 1977; Surdam, 1977), volcanic tuffs (where most economic-

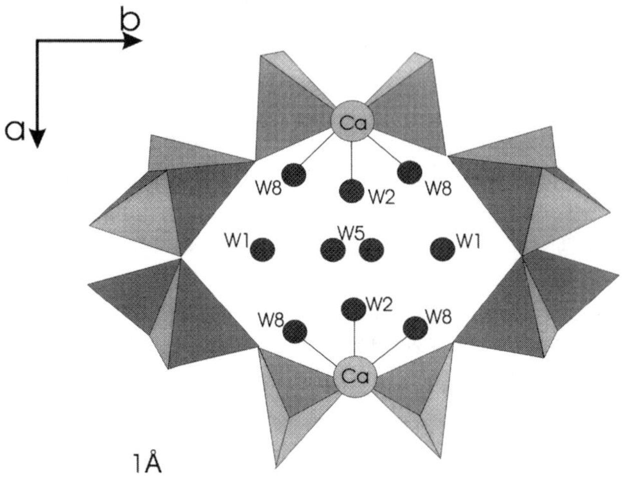

FIG. 1. Crystal structure of laumontite (Table 1), after Armbruster and Kohler (1992). Polyhedra represent tetrahedrally coordinated Si and Al (Al tetrahedra adjoin Ca sites). Four distinct water sites are present (labeled W1, W2, W5, and W8).

scale zeolite deposits are found; e.g., Hay and Sheppard, 1977; Broxton, et al., 1987), soils (e.g., Baldar and Whittig, 1968), volcanoclastic sandstones (e.g., Surdam and Boles, 1979; Merino, 1975a, 1975b), and basalts (e.g., Walker, 1951, 1959a, 1959b, 1960a, 1960b). Hydrothermal systems often produce zeolites as well (e.g., Bargar and Beeson, 1981; Bird et al., 1984; Kristmannsdóttir and Tómasson, 1978); hydrothermal alteration may overprint some diagenetic and metamorphic occurrences, producing complex parageneses (Gottardi, 1989; Neuhoff et al., 1999). Additionally, analcime[1] and pollucite possibly form as primary magmatic phases in silica-undersaturated environments and pegmatites (see review in Gottardi and Galli, 1985). The common formation of zeolites in geothermal systems, petroleum reservoirs, and groundwater aquifers, combined with their unique physical and chemical properties, has generated considerable research into the stability and chemistry of zeolites. Zeolites also are useful indicators of the physical and chemical conditions associated with petroleum (e.g., Iijima, 1988), geothermal resources (Kristmannsdóttir and Tómasson, 1978), and basalt-hosted ore deposits (e.g., Stoiber and Davidson, 1959) in the Earth's crust. This stems from the fact that zeolite stability is a sensitive function of pressure, temperature, and fluid composition. Continued application of zeolite petrogenesis for understanding thermal gradients in the Earth's crust (critical for petroleum and geothermal exploration) and the chemical evolution of hydrothermal and groundwater systems (for mineral exploration and environmental applications) requires a thorough understanding of zeolite stability.

In this paper, we review critical observations concerning the geologic occurrence and the physical chemistry of these phases, emphasizing our recent work on the occurrence, paragenesis, and thermochemistry of zeolites in lavas of the North Atlantic Tertiary Igneous Province. Basalt-hosted zeolite deposits, such as those in the North Atlantic region, are among the most common occurrences of these minerals. We conclude this communication with a discussion of the implications of zeolite parageneses in North Atlantic flood basalts for elucidating the petrotectonic development of large igneous provinces and the cycling of environmentally significant elements such as Sr in basaltic aquifer systems.

Zeolites, the Zeolite Facies, and Zeolite Zones

Since the discovery of zeolites by Cronstedt (1756),[2] naturally occurring zeolite minerals have received considerable attention for their unique

[1] Zeolite nomenclature in the present communication follows the suggestions of Coombs et al. (1997).

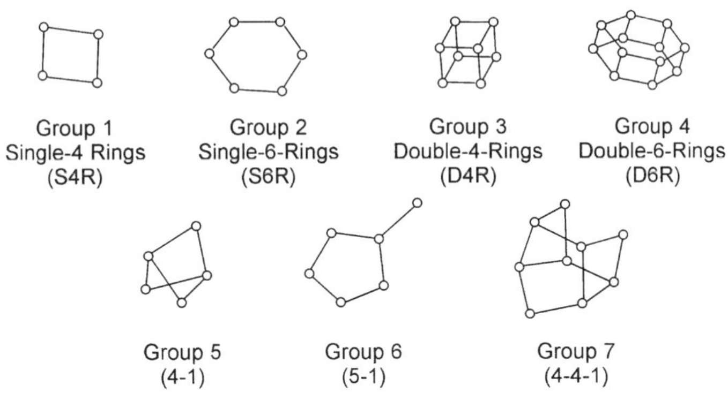

FIG. 2. Groups of fundamental tetrahedral arrangements in zeolites (after Breck, 1974). In each drawing, the circles represent tetrahedra (SiO_4^{4-} or AlO_4^{5-}) and the lines represent oxygens shared by the tetrahedra. Only groups 1, 4, 5, 6, and 7 include rock-forming zeolites.

physical and chemical properties. Zeolites are tectosilicates consisting of frameworks of tetrahedra containing cations coordinated by four oxygen atoms; the tetrahedra surround cages or channels containing non-framework (here called interchannel) cations and molecular water (Fig. 1; Coombs et al., 1997). Their compositions are represented by the structural formula

$$(A^{+z})_{y/z}(B^{+3})_y(Si)_xO_{2(x+y)} \cdot nH_2O, \quad (1)$$

where A represents interchannel cations (such as Na+, K^+, Ca^{+2}, Ba^{+2}, Sr^{+2}, Cs^+, and Mg^{+2}), B are tetrahedrally coordinated trivalent cations in the zeolite framework (Al^{+3} and Fe^{+3}), z is the charge on the interchannel cations, n is the number of moles of interchannel molecular water, and x and y are the stoichiometric coefficients for trivalent cations and Si^{+4} in tetrahedral sites, respectively. The quantities y/z and 2(x+y) represent the stoichiometries of the interchannel cations and framework oxygen, respectively, necessary for maintaining charge balance in the tectosilicate lattices of zeolites. The atomic frameworks of zeolites are grouped into seven structure types based on fundamental arrangements of tetrahedra making up their crystal lattices (e.g., Breck, 1974; Fig. 2). Over thirty[3] recognized zeolite species occur in nature, but only about a dozen are common rock-forming zeolites (Tschernich, 1992; Table 1) and those are members of only five structure-type groups (1, 4, 5, 6, and 7 in Table 1). Within these groupings, individual framework topologies are identified that may exist in several isomorphous forms (e.g., Meier, 1988; Table 1). The local environments (coordination, identities of nearest neighbors) of interchannel cations and H_2O molecules vary widely between and within structure types and topologies (e.g., Mortier, 1982).

Three types of solid solution in zeolites are consistent with the stoichiometry of formula (1). These solutions are not strictly coupled and can occur independently from the other substitutions as long as charge balance is maintained. The first of these is solid solution within the tetrahedral sites. Tetrahedral substitution of Fe^{+3} for Si^{+4} or Al^{+3} is limited and among the minerals in Table 1 is common only in clinoptilolite (Gottardi and Galli, 1985). The degree of Si and Al substitution observed in the zeolites is highly variable, ranging from virtually no variation in natrolite, mesolite, scolecite, yugawaralite, laumontite, and wairakite to variations of up to 50% in molar Si/Al ratios in analcime, phil-

[2]Most sources cite Cronstedt (1756) as the discoverer of zeolites and the word itself certainly originated with him. However, it is likely that they were known in prehistoric times and zeolite minerals are clearly mentioned in Horrebow (1752). See Betz (1981) and Mumpton (1977b) for historical reviews of zeolite science.

[3]The International Mineralogical Association (Coombs et al., 1997) recently broadened the definition of the term "zeolite mineral" to include several minerals formerly referred to as feldspathoids and phosphates, and further divided the zeolite minerals based on interchannel cation content. According to this new definition, 82 naturally occurring zeolite species exist.

TABLE 1. Crystal Chemistry and Stoichiometries of Common Rock-Forming Zeolites

Structure type[1]	Framework topology[2]	Mineral name	Si/Al ratio[3]	Dominant interchannel cations[3]	Ideal composition
Group 1	ANA	analcime	~1.45 – ~2.70	Na (Cs, Ca)	$NaAlSi_2O_6 \cdot H_2O$
(S4R)		wairakite	~1.85 – ~2.25	Ca (Na)	$CaAl_2Si_4O_{12} \cdot 2H_2O$
	LAU	laumontite	~1.80 – 2.35	Ca (K,Na)	$CaAl_2Si_4O_{12} \cdot 4.5H_2O$
	YUG	yugawaralite	~2.85 – ~3.15	Ca (Na, K, Sr)	$CaAl_2Si_6O_{16} \cdot 4H_2O$
	PHI	phillipsite	~1.27 – ~3.34	K, Ca, Na (Ba)	$K_2(Na, Ca_{0.5})_4Al_6Si_{10}O_{32} \cdot 12H_2O$
Group 4	CHA	chabazite	~1.40 – ~4.30	Ca, Na, K	$(Ca, Na_2)Al_2Si_4O_{12} \cdot 6H_2O$
(D6R)	ERI	erionite	~ 2.13 – ~3.76	K, Na, Ca (Mg)	$K_2(Na, Ca_{0.5})_8Al_{10}Si_{26}O_{72} \cdot 30H_2O$
Group 5	NAT	natrolite	1.50	Na	$Na_2Al_2Si_3O_{10} \cdot 2H_2O$
(4-1)		mesolite	1.50	Ca, Na	$Ca_{0.666}Na_{0.666}Al_2Si_3O_{10} \cdot 2.668H_2O$
		scolecite	1.50	Ca	$CaAl_2Si_3O_{10} \cdot 3H_2O$
		gonnardite	~1 – ~1.50	Na, Ca	$(Na_2,Ca)\ Al_2Si_3O_{10} \cdot 3H_2O$
	THO	thomsonite	~1 – ~1.27	Ca, Na	$Ca_2NaAl_5Si_5O_{20} \cdot 6H_2O$
Group 6	MOR	mordenite	~4 – ~6.15	Na, Ca, K	$(Ca_{0.5},Na)AlSi_5O_{12} \cdot 4H_2O$
(5-1)	EPI	epistilbite	~2.57 – ~3.34	Ca, Na	$(Ca, Na_2)Al_2Si_6O_{16} \cdot 4H_2O$
Group 7	HEU	heulandite	~2.45 – 4[4]	Ca, Na, K (Sr, Ba, Mg)	$(Ca, Na_2)Al_2Si_7O_{18} \cdot 6H_2O$
(4-4-1)	HEU	clinoptilolite	4 – ~5.25[4]	Ca, Na, K (Sr, Ba, Mg)	$(Ca_{1.5}, K_3,Na_3)Al_3Si_{15}O_{36} \cdot 12H_2O$
	STI	stilbite	~ 2.45 – ~3.45	Ca, Na (K)	$CaNa_2Al_5Si_{13}O_{36} \cdot 14H_2O$
	STI	stellerite	~3.00 – ~3.54	Ca	$CaAl_2Si_7O_{18} \cdot 7H_2O$

[1]After Breck, 1974.
[2]After Meier, 1988.
[3]After Coombs et al. (1997), except gonnardite, after Artioli and Galli (1999).
[4]Clinoptilolite and heulandite distinguished in terms of Si/Al ratio (Coombs et al., 1997); heulandite has Si/Al < 4.0; clinoptilolite has Si/Al > 4.0.

lipsite, chabazite, mordenite, heulandite, clinoptilolite, and stilbite.

Secondly, solid solution among interchannel cations is often quite extensive, as evidenced by the large ion-exchange capacities of some zeolites (e.g., Colella, 1996). Total interchannel ion charge is necessarily a function of Al^{3+} and Fe^{3+} content because the interchannel ions compensate for charge imbalances resulting from the presence of trivalent ions in tetrahedral sites. In zeolites exhibiting variable Si/Al ratios, samples with high Si/Al ratios commonly are more rich in monovalent interchannel cations than are more aluminous samples of the same species. However, strictly coupled divalent-Al/monovalent-Si substitutions have been reported only for thomsonite (Ca^{2+} and Al^{3+} for Na^+ and Si^{4+}; Ross et al., 1992). Ross et al. (1992) also proposed coupled substitution of Ca^{2+} and Al^{3+} for Na^+ and Si^{4+} in gonnardite; this has been discredited by Artioli and Galli (1999) based on new chemical analysis of gonnardite minerals. In some but not all zeolites, H_2O molecules and interchannel cations can occupy equivalent or competing sites within zeolite channels. Twice as many monovalent ions as divalent ions are necessary to compensate for charge imbalances caused by Al^{+3} in the framework, and the additional monovalent ions often occlude H_2O molecules present in isostructural zeolites with divalent interchannel cations (e.g., natrolite and scolecite, heulandite-Ca and heulandite-Na, clinoptilolite-Ca and clinoptilolite-Na).

The third type of solid solution in zeolites is a consequence of the loosely bound nature of molecular water in zeolites. The total water content of zeolites is a sensitive function of temperature, total pressure, and the partial pressure of water (e.g., van Reeuwijk, 1974). Hydration/dehydration of zeolites is reversible and generally rapid.

The importance of zeolites as low-temperature alteration phases in the Earth's crust was noted early on in the history of metamorphic petrology. However, Eskola (1939) rejected the concept of zeolites representing a separate metamorphic facies over concerns that zeolite assemblages do not represent true thermodynamic equilibrium at the conditions under which they form. The issue was revisited in the 1950s, with Rengarten (1950) proposing a "geochemical zeolite facies" in which zeolite assemblages represent alteration of sediments in contact with aqueous solutions of unusual composition (alkaline, highly oxidized colloidal solutions of silica and alumina). This idea never received widespread acceptance, and not until the benchmark papers of Fyfe et al. (1958) and Coombs et al. (1959) did the present concept of the zeolite facies as a necessary intermediate metamorphic grade between diagenesis and greenschist-facies metamorphism arise.[4] Although acknowledging the important role of fluid chemistry on zeolite stability, these authors presented a wealth of geologic evidence from regional metamorphic terrains in New Zealand that suggested that zeolites represent widespread and consistent indicators of specific temperature and pressure conditions, as noted in Figure 3. This concept of the zeolite facies has remained unchanged to the present day, although numerous authors, most notably Zen (1961), have pointed to the important role of mobile components (in the sense of Korzhinskii, 1959) such as H_2O and CO_2 in controlling zeolite stability.

Whether natural zeolite assemblages form in true thermodynamic equilibrium is still largely unanswered. Conclusions on both sides of the issue abound in the scientific literature. Nevertheless, the issues of zeolite metastability and the influence of pressure, temperature, and composition on zeolite stability remain, as do the important considerations of the influence of kinetic factors such as reaction rates and fluid flow. Several researchers have noted that the occurrence and mode of individual zeolite species appear to vary with geologic age. Coombs et al. (1959) noted that the diversity of zeolite species decreases with increasing geologic age. The general trend observed in diagenetically altered sediments is that the modal abundance of analcime increases with geologic age at the expense of more silica-rich and disorded zeolite species such as chabazite, phillipsite, clinoptilolite, and erionite (Hay, 1966). Analcime is the dominant zeolite in rocks of Mesozoic or older age. This trend appears to be coupled to an increase in non-zeolitic silicates (e.g., feldspars, searlesite) with increasing geologic age. In marine sediments, analcime and clinoptilolite become more abundant with age at the expense of phillipsite (Kastner, 1979). The paucity of zeolites in pre-Cenozoic sediments is often taken to suggest that many diagenetic zeolites are metastable (Morse and Casey, 1988). However, well-developed depth-

[4] At the time, no separate prehnite-pumpellyite facies was formally designated. Current nomenclature places prehnite-pumpellyite and prehnite-actinolite facies as intermediate between the zeolite facies and the greenschist facies (Coombs, 1960; Seki, 1961; Liou et al., 1985).

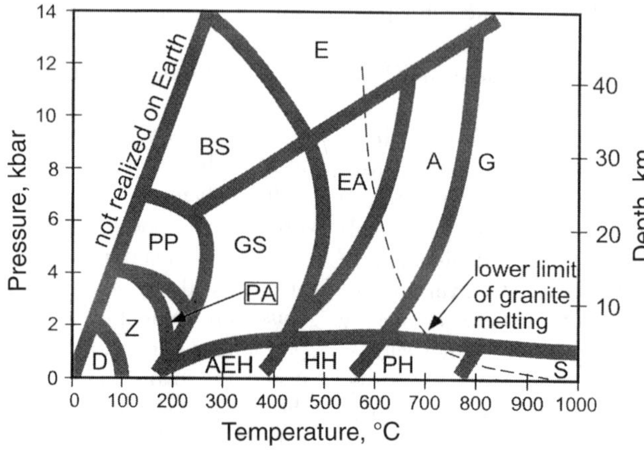

FIG. 3. Generalized petrogenetic grid (modified from Spear, 1993) depicting the temperature and pressure conditions for various metamorphic facies. Shown for reference are the pressure-temperature boundaries for conditions that are not realized on Earth and for melting of hydrous granitic melts. The depth scale corresponds to lithostatic loads of basaltic crust at the pressures depicted along the left axis. Abbreviations correspond to diagenesis (D), zeolite facies (Z), prehnite-pumpellyite facies (PP), prehnite-actinolite facies (PA), blueschist facies (BS), greenschist facies (GS), albite-epidote hornfels (AEH), hornblende hornfels (HH), pyroxene hornfels (PH), epidote amphibolite facies (EA), amphibolite facies (A), eclogite facies (E), granulite facies (G), and sanadinite facies (S).

controlled patterns of zeolite distribution observed in Tertiary basalts (Figs. 4A and 4B) and active basalt-hosted geothermal systems (Fig. 4D) appear to persist in rocks of Precambrian age (Fig. 4C; Schmidt, 1993).

The distribution of individual zeolite species within diagenetically altered or metamorphosed sediments and volcanic rocks is commonly characterized by isograds of first appearance of one or more zeolites bounding spatially restricted zones. Four common patterns of alteration, from different geologic environments (after Hay, 1977), are depicted in Figure 5. The distribution of zeolitic alteration of tuffs in saline lake environments (Fig. 5A) appears to be controlled by variations in salinity under isothermal/isobaric conditions. Similarly, zeolite formation in the vadose zone (Fig. 5B) is controlled by changes in fluid chemistry (likely an increase in pH with depth as a result of volcanic glass hydrolysis) during progressive interaction with the rock. On the other hand, zeolite distribution in hydrothermal systems (Fig. 5C) and burial diagenetic/metamorphic systems (Fig. 5D) is more readily correlated with temperature as determined by vitrinite reflectance (Iijima, 1988), coal rank (e.g., Kisch, 1987), and temperature measurements in active geothermal/metamorphic systems (Iijima, 1978, 1988; Kristmannsdóttir and Tómasson, 1978). The effect of pressure on zeolite isograds usually cannot be assessed independently from that of temperature; however, Iijima (1978, 1988) has demonstrated that the temperatures corresponding to the isograds depicted in Figure 4E are essentially independent of pressure. Minerals occurring in burial-related zonal sequences reflect bulk rock composition, as illustrated for basalts (Figs. 4A to 4D) and volcanoclastic sediments (Fig. 4E). The strong correlation between burial-related zeolite isograds and temperature is often interpreted as evidence that zeolite-zone assemblages represent a progression of thermally governed equilibrium assemblages (Walker, 1960b; Liou et al., 1991), where zone boundaries reflect progressive dehydration to less water rich assemblages with increasing temperature (Cornu, 1908; Miyashiro and Shido, 1970). This simple correlation with temperature is verified by phase equilibrium experiments for laumontite and wairakite (Liou, 1970, 1971b) that agree reasonably well with measured temperatures of zeolite isograds in active geothermal systems (Kristmannsdóttir and Tómasson, 1978). This concept rarely applies to other zeolites; for example, the experimentally determined stability fields for heulandite and stilbite (Liou, 1971a, Cho et al., 1987; Thompson, 1970) differ substantially from measured and inferred temperatures in geologic systems (Noh and Boles, 1993; Fig. 4F).

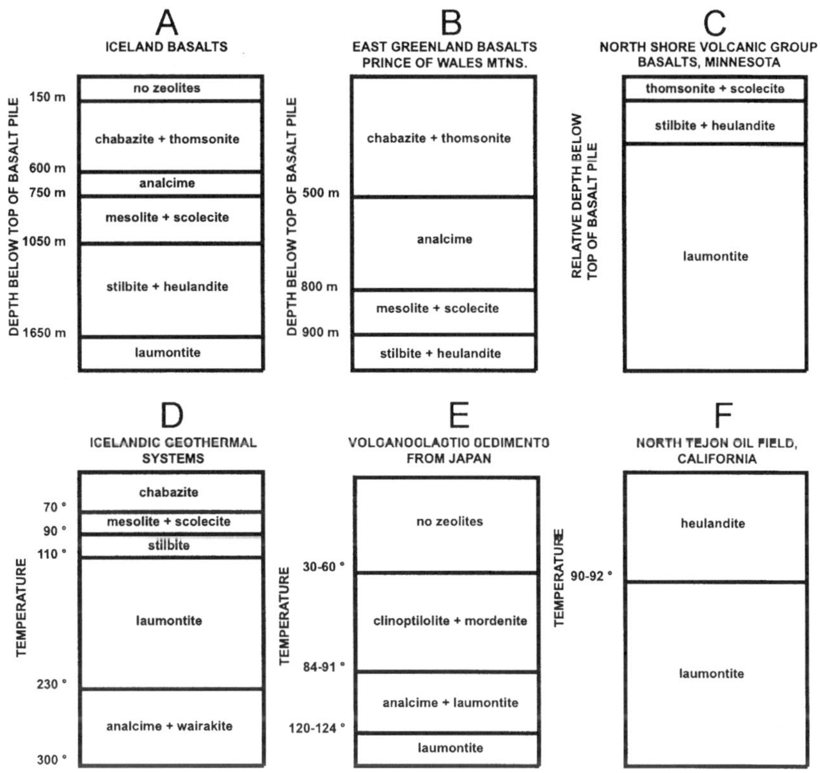

FIG. 4. Stratigraphic columns showing the distribution of zeolites as a function of depth and/or temperature in geothermal systems and regional metamorphic/diagenetic occurrences. Mineral names represent the predominant zeolites found in each zone; other minerals (including zeolites) are locally developed. Data taken from Walker (1960b; Iceland basalts), this study (Iceland and East Greenland basalts), Schmidt (1993; North Shore Volcanic Group basalts), Kristmannsdóttir and Tómasson (1978; Iceland geothermal systems), Iijima (1988; Japan sediments), and Noh and Boles (1993; California sediments).

These discrepancies can in some cases be explained by fluid chemistry as a variable controlling zeolite paragenesis (e.g., Donohoe et al., 1984; Bowers and Burns, 1990). Zeolite stability is a sensitive function of the activity of aqueous silica (Abercrombie et al., 1994), alkali, and alkali earth to hydrogen-ion activity ratios (Bowers and Burns, 1990), the activity ratio of CO_2/H_2O (Crossey et al., 1984), and the coupled effects of organic acid activities and fluid oxidation state (Surdam et al., 1984; Neuhoff and Bird, 1994). The correlation between zeolite and silica polymorph distribution in basalts (Fig. 6) and sediments (Iijima, 1988) is clear evidence that many zeolite-bearing assemblages are metastable. Thus, many zeolites may be intermediate reaction products favored by high saturation states during dissolution of extremely metastable phases such as volcanic glasses and/or kinetic inhibition of more stable phases (Dibble and Tiller, 1981; Senderov, 1983; Morse and Casey, 1988). This general phenomenon is known as the Ostwald (1897) Step Rule. The physical chemical basis for the Ostwald Step Rule remains a mystery, although most authors[5] favor the idea that the metastable intermediates precipitate more rapidly than thermodynamically stable phases (e.g., Morse and Casey, 1988). Consequently, precipitation- and dissolution-rate studies for zeolites are growing in number (e.g., Ragnarsdóttir, 1993; Savage et al., 1993; Murphy et

[5]Experimental synthesis of zeolites stimulated an interesting attempt to relate the Ostwald Step Rule to the phenomenon of minimum entropy production when systems are in a steady state (van Santen, 1984). van Santen's (1984) derivation has now been invalidated (Casey, 1988), although this concept deserves further attention.

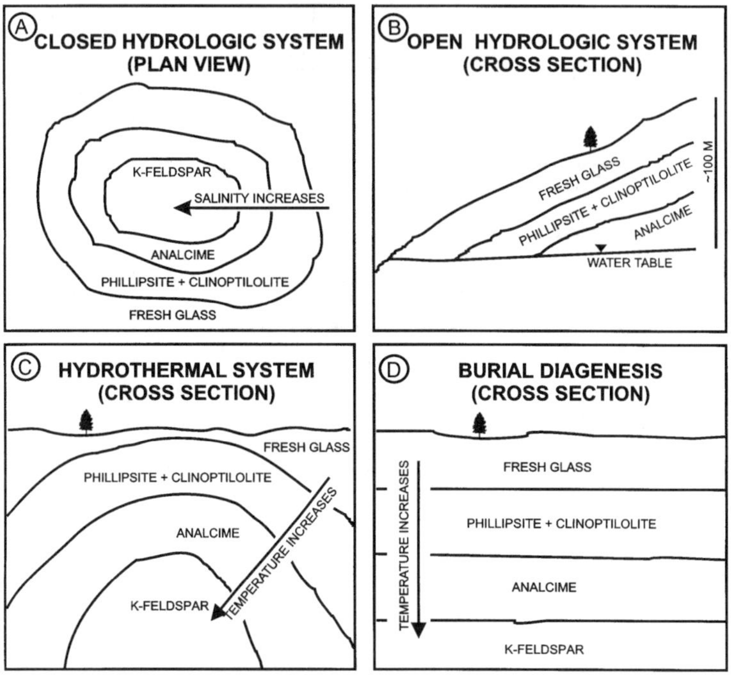

FIG. 5. Zonal patterns of zeolite distribution in various genetic environments (after Hay, 1977). Mineral names correspond to common assemblages observed in altered silicic sediments; variations in these patterns exist, particularly in other types of rocks. A. Plan (map) view of zeolite distribution in "closed" hydrologic systems, such as playa lakes. Mineral distribution in these systems is often taken to reflect variations in the salinity of alteration fluids. B. Cross-section view of zeolite distribution in "open" hydrologic systems. Mineral distribution is taken to reflect pH changes during progressive interaction with the host rock. C. Cross-section view of zeolite distribution in hydrothermal systems. Mineral distribution is taken to reflect temperature gradients during alteration. D. Cross-section view of zeolite distribution in sediments undergoing burial diagenesis. Mineral distribution is taken to reflect temperature gradients during alteration.

al., 1996). Fluid transport also has been shown to play an important role in zeolite formation. Modeling by Moore et al. (1986) indicated that rate-limiting reactions and fluid flow were responsible for the distribution of zeolites observed in Icelandic geothermal systems (see Fig. 4D). Similarly, hydrothermal flow-through experiments suggest that the rate of mass transport through reacting systems influences the formation of zeolites (e.g., Hawkins, 1981; Lane et al., 1989; Robert and Goffé, 1993).

In regional zeolite occurrences where age relations between geotectonic and mineral parageneses are well constrained, zeolite alteration proceeded rapidly after deposition of the host rocks. Radiometric dating (by K-Ar methods) of early celadonite and late K-feldspar in dacitic/rhyolitic tuffs of the John Day Formation, Oregon, gives approximately the same ages for these stages of alteration (22 to 23 Ma), corresponding to the cessation of tuff deposition (Hay, 1963). Age determinations for the base of the John Day Formation indicate that deposition started about 31 Ma, whereas the ages near the center of the formation are ~24 to 25 Ma. These relations indicate that alteration did not begin until much of the formation had been deposited, and then proceeded rapidly. Similarly, more recent tuffaceous sediments, such as in southern Italy (~10,000 years old; Capaldi et al., 1971) are frequently pervasively altered to zeolites and clays but show no signs of active alteration. Jakobsson and Moore (1986), studying hyaloclastites produced during the eruptions of Surtsey from 1963 to 1967, found that zeolitic alteration had already commenced less than 20 years after eruption. These relations are, in general, consistent with our observations of zeolite-zone parageneses in Iceland and Greenland, summarized below.

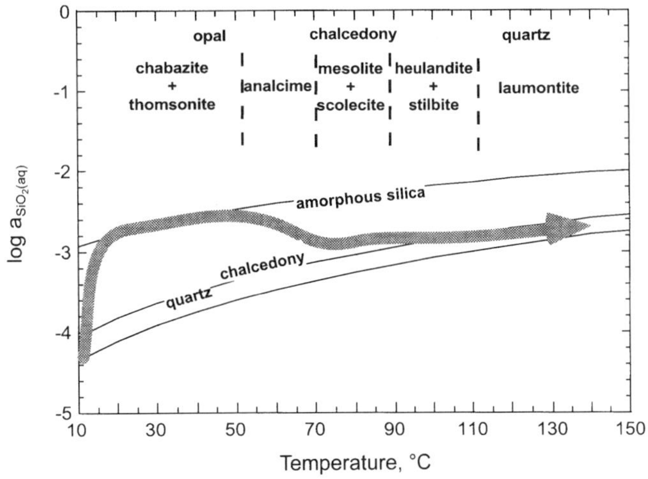

FIG. 6. Variation in the activity of aqueous silica as a function of temperature during regional zeolite-facies metamorphism of basaltic lavas in the North Atlantic region. Temperatures associated with various zeolite zones (see Figure 4) are from Kristmannsdóttir and Tómasson, 1978). Silica polymorph distribution is based on regional mapping in East Greenland (Neuhoff et al., 1997). Grey curve reflects silica activities as a function of temperature based on the occurrence of silica polymorphs. Shown for reference are activities of aqueous silica in equilibrium with opal, chalcedony, and quartz (black curves) calculated at Psat with SUPCRT92 (Johnson et al., 1992).

Physical Chemistry of Zeolite Paragenesis

The earliest experimental studies of zeolite formation were aimed at determining pressure and temperature conditions under which zeolites formed. Some of the more prominent studies involved "hydrothermal synthesis" in which zeolites were formed from mixtures of oxides, gels, glasses, or silicates (see review in Coombs et al., 1959; many other studies were subsequently conducted). While these studies were suggestive of the pressure and temperature (and sometimes bulk composition) conditions under which individual zeolites form, equilibrium frequently was not established and the conditions under which the experiments were conducted are too limited to be widely applicable. With the growing understanding that metamorphic facies (Fig. 3) are controlled by univariant phase equilibria among minerals and coexisting fluids, work was undertaken to establish the pressure and temperature conditions for univariant equilibria involving zeolites (see Frey et al., 1991). Some of this work has provided invaluable insights into the stability of certain zeolites; however, demonstration of reversibility for zeolite equilibria is often difficult because of sluggish reaction rates at the low temperatures at which they occur.

A major impetus for investigations of phase equilibria is the derivation of thermodynamic properties. The growth of equilibrium thermodynamics as a tool for understanding mineral stability[6] stimulated numerous calorimetric studies as well as several recent attempts to compile the results of phase equilibrium and calorimetric experiments into internally consistent thermodynamic databases for application to geochemical problems (e.g., Helgeson et al., 1978; Berman, 1988; Johnson et al., 1992; Holland and Powell, 1998). Thermodynamic evaluation of phase equilibrium experiments involving zeolites has found limited success (see Helgeson et al., 1978, Holland and Powell, 1998), reflecting factors such as partial dehydration of zeolites under hydrothermal conditions, small entropy and volume changes of reactions, and variations in ordering states and compositions of zeolites and feldspars in experiments. Nevertheless, thermodynamic data calculated from phase equilibrium and calorimetric

[6] An interesting sidenote to the study of mineral stability is the notion of energy and packing indices in silicates advocated by Gruner (1950), Fairburn (1943), and others. Kostov (1969) suggested that the zonal distribution of zeolites in hydrothermal systems can be explained by these variables; this research is largely ignored today.

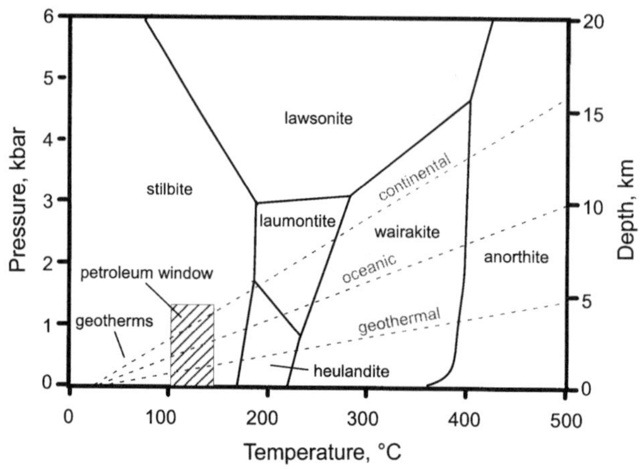

FIG. 7. Reactions governing the pressure-temperature stability of calcium zeolites (after Frey et al., 1991). Shown for reference is the temperature window for petroleum formation (Tissot and Welte, 1984) and three typical crustal geotherms.

experiments have been widely applied to calculation of univariant phase equilibria (e.g., Frey et al., 1991) and heterogeneous equilibria with aqueous electrolyte solutions (e.g., Bowers and Burns, 1990) involving zeolites. Below we review some of the major types of reactions influencing zeolite stability, focusing on the thermodynamic properties of reactions that control the distribution and composition of zeolites in geologic systems. Further discussion of the thermodynamic properties of specific zeolites and reactions can be found in Neuhoff (1999).

Univariant phase equilibria

A central tenet of the concept of the zeolite facies is that zeolites are thermodynamically stable with respect to other calcium and sodium aluminosilicates at low temperatures. Figure 7 depicts the temperature and pressure conditions of experimentally constrained univariant equilibria controlling the stability of zeolites under sub-greenschist, sub-blueschist conditions (based on calculations from Frey et al., 1991). In general, the maximum temperature and pressure limits of zeolite stability shown in Figure 7 are in agreement with geologic observations of zeolite occurrence in geothermal systems and metamorphic terrains (Kristmannsdóttir and Tómasson, 1978; Frey et al., 1991). Calcium zeolites characteristic of the zeolite facies (e.g., laumontite, heulandite, stilbite) are stable relative to lawsonite-bearing assemblages (typically considered indicative of the blueschist facies) below 3 kbar (Fig. 7). Above 400°C, anorthite is stable relative to wairakite, the calcium zeolite stable at highest temperatures. However, the phase relations shown in Figure 7 suggest that stilbite is stable at temperatures and pressures down to those prevailing at the Earth's surface, contradicting the geologic observations summarized in Figure 4.

Many of the reactions shown in Figure 7 are characterized by relatively small entropies and volumes (i.e., less than 4 cal/molK or 4 cc/mol, respectively; Neuhoff, 1999). This likely contributes to slow reaction rates in the laboratory (in addition to low temperatures). Unlike dehydration equilibria observed at higher metamorphic grades, entropies of zeolite facies reactions are considerably smaller because of the relatively small entropy of water at these conditions. In addition to making these reactions extremely difficult to study experimentally, the small magnitude of the entropies and volumes of many zeolite facies reactions makes the corresponding Clapeyron slopes extremely sensitive to small variations in these properties. For example, the configurational entropy of analcime ranges from 0.8 to 1.9 cal/molK (Phillips and Kirkpatrick, 1994), similar in magnitude to the entropy of the reaction

$$\text{analcime} + \text{quartz} \Leftrightarrow \text{albite} + H_2O. \qquad (2)$$

Changes in the molar volumes of zeolites as a result of compression or expansion of the zeolite lattice at elevated temperatures and pressures can dramatically affect the volume of these reactions given the small volumes of the reactions. These phenomena have only been explored in analcime, the molar volume of which at 25 °C is ~1.5% less at 5 kbar than at 1 bar (Yoder and Weir, 1960; Ott et al., 1986), as compared to ~0.5% for most common rock-forming silicates (see Helgeson et al., 1978). Additional data on the configurational entropies, thermal expansivities, and compressibilities of zeolites are required to accurately assess the stabilities of zeolites in geologic systems.

Difficulties also arise in attempts to use experimental observations of phase equilibria involving zeolites to assess thermobarometric conditions in zeolite-facies rocks. Quite often there is a significant disagreement between the temperature (at a given pressure) observed at the position of a given zeolite isograd reaction and the temperature determined from phase equilibrium experiments. For example, in the North Tejon oil field in California, heulandite and laumontite coexist at ~90°C, 1kbar (Fig. 4F; Noh and Boles, 1993), considerably lower than suggested by the equilibrium phase relations shown in Figure 7. This discrepancy could reflect formation of the zeolites under a different (higher) thermal regime than that measured downhole by Noh and Boles (1993). However, recent calculations (Neuhoff, 1999) have demonstrated that these disagreements between experimental and geologic observations of isograd temperature can be attributed largely to variations in order-disorder and composition between natural and synthetic assemblages.

As discussed above, considerable evidence suggests that true thermodynamic equilibrium is frequently not achieved during zeolite formation in natural and experimental systems. The lower temperature boundary where equilibrium prevails, although the necessary criterion for defining the lower boundary of the zeolite facies, has received relatively little attention. One characteristic that may be used is the presence of quartz. Most of the univariant equilibria that control zeolite stability (Fig. 7) contain quartz as part of the assemblage. Geologic observations suggest that quartz does not form at temperatures below ~100°C in many geologic systems (e.g., Gíslason and Arnórsson, 1990; Iijima, 1978, 1988). The first appearance of quartz in these systems often corresponds with geologic observations of analcime and albite coexistence, suggesting that this isograd reflects equilibrium with respect to reaction 2. It should be noted that phase rule analysis of alteration assemblages in low-grade rocks often indicates that significant degrees of freedom remain after metamorphism. Thus, equilibrium with respect to a given univariant reaction does not necessarily indicate true thermodynamic equilibrium with respect to all components in the system. In any event, assemblages bearing zeolites that form at temperatures below 100°C almost certainly do not reflect univariant equilibria. In some cases, isograds below 100°C can be attributed to local equilibrium with respect to silica activities buffered by the presence of metastable silica polymorphs (see below).

Solubility of zeolites and zeolite assemblages

Prediction of the consequences of chemical interaction between aqueous electrolyte solutions and volcanic rocks and sediments in systems such as petroleum reservoirs, geothermal systems, and nuclear waste repositories requires consideration of the solubility of zeolites. Recently, a number of studies have attempted to measure the solubility of individual zeolites over a range of temperatures and pressures (e.g., Murphy et al., 1996; Wilkin and Barnes, 1998). Alternatively, the solubilities of these phases can be calculated from the thermodynamic properties of reactions describing precipitation/dissolution between zeolites and aqueous solutions (using properties of the zeolites calculated from univariant phase equilibria and/or calorimetric experiments).

Figure 8 shows the solubility of stilbite (solid curve; calculated using data from Neuhoff, 1999) compared with calculated ion activity products (Q) of geothermal fluids for the reaction

$$\underset{\text{stilbite}}{CaAl_2Si_7O_{18}\cdot 7H_2O} \Leftrightarrow Ca^{+2} + 2\ AlO_2^- + 7\ SiO_{2(aq)} + 7\ H_2O\ . \qquad (3)$$

The relations depicted in Figure 8 are typical for zeolites in diagenetic and geothermal systems. It can be seen that the solubility of stilbite increases with increasing temperature up to ~300°C, consistent with the positive enthalpy of reaction (3; ΔH°_r) over this temperature range at Psat. As is the case for most zeolites, log Q evaluated from the compositions of low-temperature (near-surface) fluids is substantially larger than log K (represented by the solubility curve in Fig. 8). Gíslason and Arnórsson (1990) interpreted this supersaturation with respect to zeolites to be a consequence of the sluggish precipitation rates of aluminosilicates in low-temperature meteoric waters. The degree of supersaturation tends to decrease with increasing temperature and

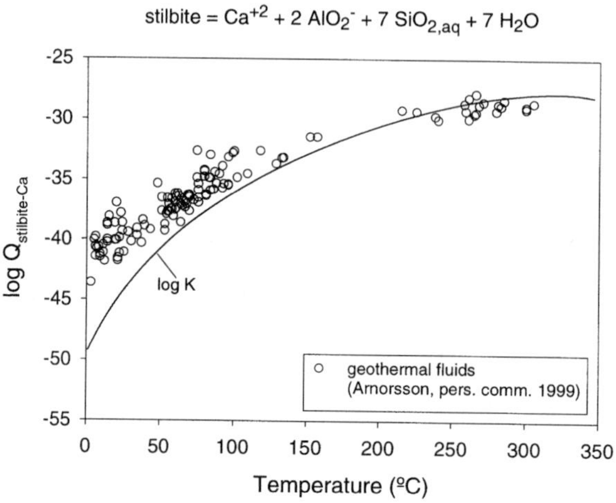

FIG. 8. Logarithms of ion activity products calculated from analyses of Icelandic geothermal fluids (symbols) (S. Arnórsson, pers. commun., 1999) and equilibrium constants (curves) as a function of temperature at Psat for the stilbite hydrolysis reaction (3).

in most cases fluids at temperatures > 200°C are undersaturated with respect to zeolites. Stilbite occurs at temperatures from 90 to 110°C in Icelandic geothermal systems; the relations shown in Figure 8 clearly indicate that geothermal fluids collected over this temperature range are supersaturated with respect to stilbite. Above these temperatures, the compositions of geothermal fluids tend to be at or below stilbite saturation.

Zones characterized by the occurrence of certain zeolites or zeolite assemblages may be controlled by changes in fluid chemistry. An example is a common diagenetic isograd reaction characterized by a distinct change in mineralogy from clinoptilolite-bearing assemblages to analcime-bearing assemblages (Fig. 4E). This reaction is often accompanied by a pronounced change in silica mineralogy from metastable cristobalite to quartz, which indicates that this zone boundary is not governed by univariant equilibrium, but rather results from a decrease in the activity of aqueous silica. A possible reaction governing this boundary is

$$\underset{\text{clinoptilolite}}{0.32\,Na_3Al_3Si_{15}O_{36}\cdot 12H_2O} \Leftrightarrow \underset{\text{analcime}}{Na_{0.96}Al_{0.96}Si_{2.04}O_6\cdot H_2O} + 2.76\,SiO_{2(aq)} + 2.2\,H_2O. \qquad (4)$$

The activitiy of aqueous silica in equilibrium with reaction (4) is shown as a function of temperature at 1.5 kbar in Figure 9 along with the solubility curves for quartz, low cristobalite, and amorphous silica calculated using the data for clinoptilolite-Na and analcime from Neuhoff (1999) and those of the silica polymorphs, $SiO_{2(aq)}$ and H_2O from SUPCRT92 (Johnson et al., 1992). Analcime is stable relative to clinoptilolite-Na at quartz saturation, whereas the opposite is true at cristobalite saturation. The trend in fluid composition accompanying the transition from clinoptilolite to analcime is depicted by the dark arrow in Figure 9. It thus appears that stability of clinoptilolite (and poten-

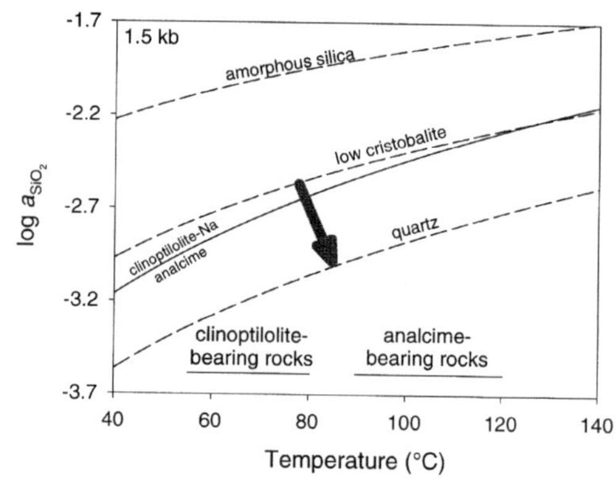

FIG. 9. Stabilities of Na-zeolites as a function of temperature and silica activity at 1.5 kbar. Shown for comparison are the activities of silica in equilibrium with quartz, low cristobalite, and amorphous silica. Arrow indicates trend in fluid composition responsible for the transition from clinoptilolite- to analcime-bearing rocks (temperatures associated with each from Iijima, 1978, 1988).

tially other very silicic zeolites) is promoted by the persistence of metastable silica polymorphs such as cristobalite.

Formation of zeolites in petroleum systems during diagenesis in petroleum reservoirs often leads to substantial destruction of porosity, thereby reducing reservoir quality (e.g., Crossey et al., 1984; Helmold and van de Kamp, 1984). This is largely a consequence of the large molar volumes of zeolites. One of the main controls on calcium zeolite stability in these systems is the activity of aqueous CO_2; at relatively high CO_2 activities, calcium zeolites are destabilized relative to assemblages containing calcite, quartz, and clay minerals (Zen, 1961; Senderov, 1973). Carbon dioxide is often in redox equilibrium with organic acids in petroleum field waters (Shock, 1988); thus, calcium zeolite stability with respect to carbonate-bearing assemblages can be alternatively represented in terms of the organic acid activity and a redox variable such as the fugacity of oxygen. For instance, equilibrium between laumontite and the assemblage calcite + quartz + kaolinite in fluid where acetic acid (CH_3COOH) is abundant can be represented by the reaction

$$2 \text{ laumontite} + CH_3COOH + 2\, O_2 \Leftrightarrow 2 \text{ calcite} + 2 \text{ kaolinite} + 4 \text{ quartz} + 6\, H_2O. \quad (5)$$

Figure 10 depicts equilibrium of reaction (5) as a function of oxygen fugacity and the activity of acetic acid. It can be seen that the stability of laumontite is favored in fluids with low oxygen fugacities (relatively reducing) and low organic acid activities (or low total carbon). The grey region in Figure 10 illustrates the range of oxygen fugacities and acetic acid activities commonly observed in petroleum fields (after Shock, 1988). In these fluids, the calcite + quartz + kaolinite assemblage is stable relative to laumontite, which is consistent with observations of laumontite (and other calcium zeolites) being replaced by carbonates after interaction with oilfield waters (Helmold and van de Kamp, 1984; Christiansen et al., 1999).

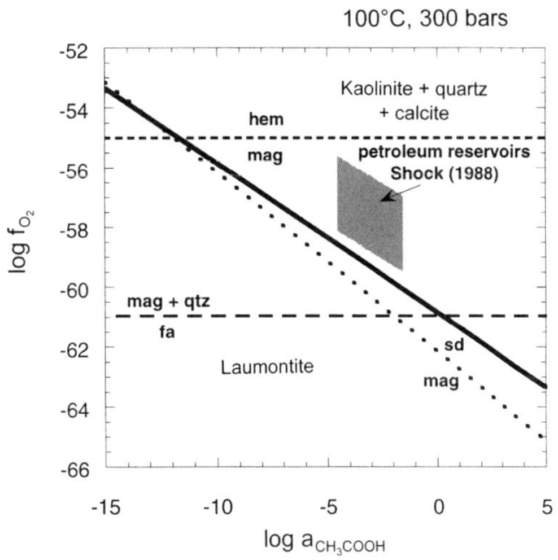

FIG. 10. Stability of laumontite relative to the assemblage calcite + kaolinite + quartz (separated by solid line) as a function of oxygen fugacity and acetic acid activity in aqueous solution at 100°C, 300 bars. Dashed lines represent the positions of several mineral buffers (mag = magnetite; qtz = quartz; fa = fayalite; hem = hematite; sid = siderite). Grey region shows the range of fluid compositions common in oilfield brines at similar temperatures and pressures (from Shock, 1988).

Dehydration of zeolites

Since their discovery by Cronstedt (1756), zeolites have attracted considerable attention for their remarkable ability to reversibly gain and lose molecular water while retaining their crystal structures. Reversible dehydration influences phase relations among naturally occurring zeolites and coexisting minerals, aqueous electrolyte solutions, hydrocarbon fluids, and gases in the Earth's crust. For example, Clapeyron slope constraints from experiments on reactions involving analcime require that the variable hydration state of this mineral as a function of temperature and pressure be accounted for in calculation of high-temperature univariant phase equilibria (Helgeson et al., 1978). Similarly, Carey and Bish (1996) demonstrated that phase equilibria involving clinoptilolite are strongly influenced by reversible hydration in this mineral. It is thus clear that rigorous thermodynamic consideration of the stability of natural zeolites necessitates careful assessment of water site occupancies in these minerals as functions of temperature, pressure, and water fugacity in geologic systems.

Water in zeolites (hereafter referred to as $H_2O_{(zeol)}$) is present as molecular H_2O within the intracrystalline channels characteristic of the crystal structures of zeolites. This can be seen in Figure 1, which depicts a cross-sectional view of the crystal structure of laumontite, showing the intracrystalline channel defined by Al- and Si-centered tetrahedra. The channel shown in Figure 1 contains both Ca^{+2} ions and H_2O molecules, the latter being distributed over four distinct sites (labeled W1, W2, W5, and

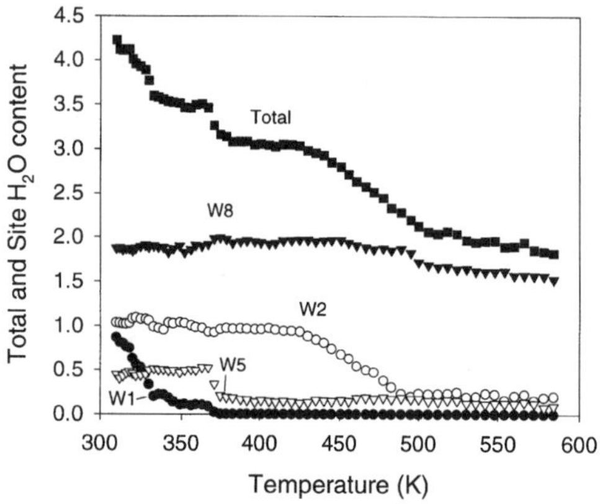

FIG. 11. Occupancies of water sites in laumontite as a function of temperature (Ståhl et al., 1996).

W8 in the figure). Water molecules on the W2 and W8 sites complete the sixfold coordination of the Ca^{+2} ion (the other three nearest Ca^{+2} neighbors being framework oxygens). In contrast, the W1 and W5 sites (the latter having half occupancy when laumontite is fully hydrated) occur near the center of the channel and are only loosely bound to the framework and/or other water molecules.

Both types of water molecules (those hydrating extraframework cations and those that do not) are common in many zeolite frameworks (see review in Gottardi and Galli, 1985). The different bonding environments of the distinct $H_2O_{(zeol)}$ sites in zeolites lead to substantial variation in the energetic properties of water on these sites. This is demonstrated in Figure 11 (after Ståhl et al., 1996), which depicts water contents of the four water sites in laumontite and total water content (values on vertical axis represent water contents per 12 framework oxygens) as a function of temperature. The decrease in both total and individual water site occupancies with increasing temperature during Ståhl et al.'s (1996) experiments illustrates the progressive dehydration exhibited by zeolites with increasing temperature. It can be seen in Figure 11 that the degree of water loss as a function of temperature varies among the individual sites, which indicates that these sites are not energetically equivalent. Although the water sites in some other zeolites display similar evidence for being energetically distinct (e.g., mesolite and scolecite; Ståhl and Hanson, 1994), data are generally not available to assess the thermodynamic properties of individual sites. Consequently, most thermodynamic calculations of $H_2O_{(zeol)}$ occupancy assume that water occurs on a single crystallographic site (or several energetically equivalent sites). This approximation, although not physically realistic, is convenient (e.g., Carey and Bish, 1996) to describe the thermodynamic consequences of partial dehydration in zeolites. Its success indicates that differences between the thermodynamic properties of water on different sites are relatively small.

Equilibrium between homologous (i.e., differing only in water content) hydrated ($Z \cdot nH_2O$) and dehydrated (Z) components of zeolites and stable liquid or vapor phase water ($H_2O_{(l\ or\ v)}$) may be represented by the general reaction:

$$Z - nH_2O \Leftrightarrow Z + n\ H_2O_{(l\ or\ v)}\ , \qquad (6)$$

where n is the number of moles of water in the fully hydrated zeolite. Standard states used for the species in reaction (6) vary between studies; in our recent work, we adopt a standard state of unit activity at all temperatures and pressure for Z and $Z \cdot nH_2O$, both having the framework structure of the fully hydrated form. The logarithmic mass action expressions for reaction (6) may be written as:

$$\log K_{(11)} \Leftrightarrow n \log a_{H_2O(l\ or\ v)\,)} + \log a_Z - \log a_{Z \cdot nH_2O}, \qquad (7)$$

where $K_{(7)}$ is the equilibrium constant for reaction (7) and a_i represents the activity of the subscripted species i. The activities of the mineral components can be evaluated by

$$a_i = X_i\, \gamma_i\ , \qquad (8)$$

where X and γ are the mol fraction and activity coefficient of the component i. Assuming that $\gamma_Z = \gamma_{Z \cdot n\, H_2O} = 1$ (corresponding to an ideal solution) appears to be a reasonable approximation for some zeolites (e.g., analcime; see below), recent thermogravimetric studies of equilibrium between clinoptilolite and water (Carey and Bish, 1996) nonetheless demonstrate that mixing between hydrated and dehydrated clinoptilolite is not ideal.

An example of the effect of temperature and pressure on the water content of zeolites is depicted in Figure 12, which shows isopleths of analcime composition (given as $X_{analcime}$, the mol fraction of hydrated analcime in the sample) calculated using data from Neuhoff (1999) for the reaction

$$\text{analcime} \Leftrightarrow \text{dehydrated analcime} + H_2O. \qquad (9)$$

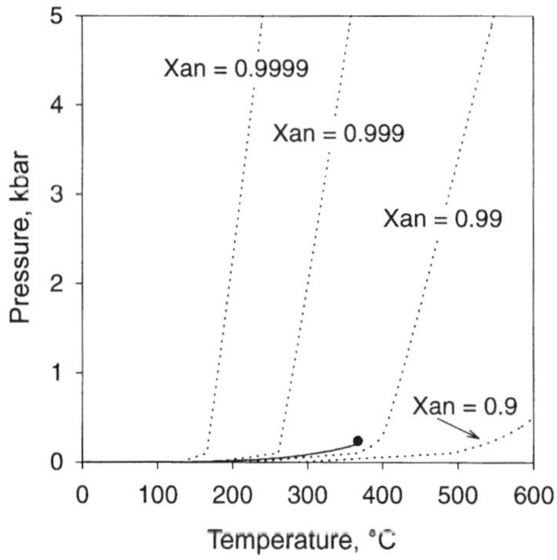

FIG. 12. Mol fraction of hydrated analcime (X_{an}) as a function of temperature and pressure calculated using data from Neuhoff (1999).

The isopleths in Figure 12, which represent constant values of K for reaction (9), have positive slopes consistent with the positive entropy and volume of reaction (9). This observation holds for all zeolites for which relevant data are available, and reflects the fact that the volume contraction and entropy change of the zeolite structure upon dehydration are both smaller (per mol H_2O) than the standard molar volume and entropy of liquid water. Note the relatively small departure from full water occupancy over the range of temperature and pressure in Figure 12. Thermodynamic calculations indicate that the mol fraction of the hydrated forms of most rock-forming zeolites is greater than 0.99 at the temperatures and pressures where liquid water is stable and under which most zeolites form in nature (Neuhoff and Bird, 1999). The consequence of this degree of partial dehydration on ΔG°_r for reactions involving zeolites is on the order of a few tens of calories per mol of water in the reaction, which is considerably smaller than the errors in ΔG°_f for zeolites determined from calorimetric measurements or retrieved from phase equilibria. However, partial dehydration can dramatically affect zeolite phase relations under some conditions where dehydration is more extensive than depicted in Figure 12. Conditions where extensive dehydration of zeolites in geologic systems occurs include equilibria involving analcime at elevated temperatures, where complete dehydration may occur (Helgeson et al., 1978), equilibria with water vapor (Carey and Bish, 1996), and dehydration of the W1 site in laumontite (Fig. 1), which reversibly dehydrates in air under room conditions (Neuhoff and Bird, 1999).

Ion exchange

The relatively weak ionic bonds between interchannel cations and the negatively charged framework in zeolites (for example, the Ca^{+2} ion in laumontite shown in Figure 1), along with the porous nature of zeolite frameworks, result in ready exchange of these ions between zeolites and aqueous electrolyte solutions. This process can be represented by the general reaction

$$(A_i^{+z(i)})_{\sigma i} + Z \cdot n_i\, H_2O + \sigma_j\, A_j^{+z(j)}{}_{(aq)} \Leftrightarrow (A_j^{+z(j)})_{\sigma j} + Z \cdot n_j\, H_2O + \sigma_i\, A_i^{+z(i)}{}_{(aq)} + (n_i - n_j)\, H_2O_{liquid}, \qquad (10)$$

where Z is the alumino-silicate framework of the zeolite, $A_i^{+z(i)}$ and $A_j^{+z(j)}$ represent cations i and j within the zeolite, $A_j^{+z(i)}{}_{(aq)}$ and $A_i^{+z(j)}{}_{(aq)}$ are the aqueous ions of i and j, z(i) and z(j) are the charges on ions of i and j, n_i and n_j are the number of water molecules in homoionic forms of Z containing i and j, and σ_i and σ_j represent stoichiometric coefficients for i and j in Z. The stoichiometric coefficients σ_i and σ_j are given by $\sigma_i = y/z(i)$ and $\sigma_j = y/z(j)$ where y is the number of mols of Al per formula unit in the zeolite. Note that the framework cations, i.e., Al^{+3} and Si^{+4}, are not involved in reaction (10). However, the water contents of zeolites are dependent on the identity of extraframework cations, which is accounted for explicitly in reaction (10). Rates of ion-exchange reactions between zeolites and aqueous solutions are widely variable, depending on size and charge of the exchanging cations, structure and Si/Al ratio of the zeolite, and temperature (Rao and Rees, 1966; Yang et al., 1997). Divalent cations have slower diffusion rates than monovalent ones in a given zeolite and cation diffusion rates are faster in Si-rich zeolites than Al-rich ones (Tsitsishvili et al., 1992). However, in most cases ion-exchange reaction rates in zeolites are fast enough to be readily reversed in the laboratory (equilibrium often established within days to weeks at 298 K). Ion-exchange reactions by zeolites in volcanic aquifers have been shown to deplete groundwaters in Ca^{+2} and Mg^{+2} relative to Na^+ and K^+ (White et al., 1980) and deplete geothermal fluids in Cs^+, Ra^+, and Ba^{+2} relative to K^+, Na^+, and Ca^{+2} (Keith et al., 1983; Sturchio et al., 1989).

TABLE 2. Selectivity Sequences of Some Natural Zeolites

Zeolite	Si/Al (ideal)	Alkali-metal cation-selectivity sequence	Alkaline-earth cation-selectivity sequence
Clinoptilolite	5	Cs>Rb>K>Na>Li	Ba>Sr>Ca>Mg
Heulandite	3.5	K>Rb>Na>Li	Sr>Ba>Ca>Mg
Chabazite	2	Cs>K>Rb>Na>Li	Ba>Sr>Ca>Mg
Eirionite	3	Rb>Cs ≥ K>Na>Li	Ba>Sr>Ca>Mg
Ferrierite	5	Cs>Rb>K>Na	Ba>Sr>Ca>Mg
Mordenite	5	Cs>Rb ≅ K>Na>Li	Ba>Sr>Ca>Mg
Phillipsite	2.66	Cs>K>Rb>Na>Li	Ba>Sr>Ca>Mg

Sources: Based on Colella, 1996 and Tsitsishvili et al., 1992 and references therein.

The extent of ion-exchange and propensity for exchanging one ion for another are described by the ion-exchange capacity and cation selectivity of zeolites. Ion-exchange capacity is defined as the amount of ion (usually normalized to charge, rather than molar, units) that can be exchanged in or out of the structure. Theoretically, the maximum ion-exchange capacity is controlled by the amount of Al^{+3} (or Fe^{+3}) in the framework of the zeolite; experimentally determined ion-exchange capacities are generally 5 to 15% lower than that suggested by framework composition (Pabalan, 1994). Cation selectivity refers to the relative affinity of a given mineral to incorporate different ions into its framework (Krauskopf and Bird, 1995). Cation selectivity sequences displayed by zeolites generally correlate with structural and chemical properties (Colella, 1996; Tsitsishvili et al., 1992). Table 2 summarizes experimentally determined selectivity sequences for several zeolites, and reveals that zeolites generally favor larger cations to smaller ones. This is particularly true for the alkaline earth metal cations, whereas the selectivity sequences for alkali-metal cations more often deviate from this pattern.

Sherry (1969) explained the observed preference of zeolites for larger ions, with low charge-to-radius ratios, as a result of relatively low anionic field strength of the zeolite framework. The apparent selectivity of zeolites for larger cations in exchange equilibrium with an aqueous solution is a result of the strong preference of the aqueous solution for the smaller cations. In other words, the difference in Gibbs energies of hydration between larger and smaller cations with equivalent charge is more important for the Gibbs energy of an ion-exchange reaction than corresponding Gibbs energy differences for the ions in the zeolite framework. The anionic field strength, or the charge density of the zeolite framework, is a function of the framework Si/Al ratio. Low Si/Al zeolites such as heulandite, chabazite, erionite, and phillipsite (see Table 1) are characterized by a higher anionic field strength than are high Si/Al zeolites such as clinoptilolite, mordenite, and ferrierite. Low Si/Al zeolites therefore should be more likely to deviate from the general selectivity pattern (large > small). This is indeed the case, as can be seen from Table 2. Clinoptilolite, mordenite, and ferrierite, having the highest Si/Al ratios of the zeolites in Table 2, are the only ones that display regular selectivity series (larger ions preferred over smaller) for the alkali-metal cations. This also applies to zeolites that have variable Si/Al ratios, as demonstrated by Shibue (1981; Fig. 13A) in the case of phillipsite.

Ion-exchange behavior in zeolites is frequently investigated via isothermal, isobaric experiments in which a zeolite is equilibrated with aqueous electrolyte solutions containing ions of a given binary exchange pair (for instance, Na^+ and Ca^{+2}) in order to determine the compositions of coexisting zeolites and solutions at equilibrium. The resulting data are typically represented in "isotherm" diagrams that depict the charge equivalent fraction of one ion in the aqueous phase as a function of the corresponding quantity in the zeolites. Examples of experimental ion-exchange isotherms of Ca^{+2} and Sr^{+2} for Na^+ in clinoptilolite and Ca^{+2} for Na^+ in phillipsite are given in Figure 13. This diagram illustrates: (1) that the preference of phillipsite for Na^+ compared to Ca^{+2} increases with increasing Si/Al ratio (Figure 13A); (2) the significant effect of temperature on Na^+-Ca^{+2} exchange in phillipsite corresponding to

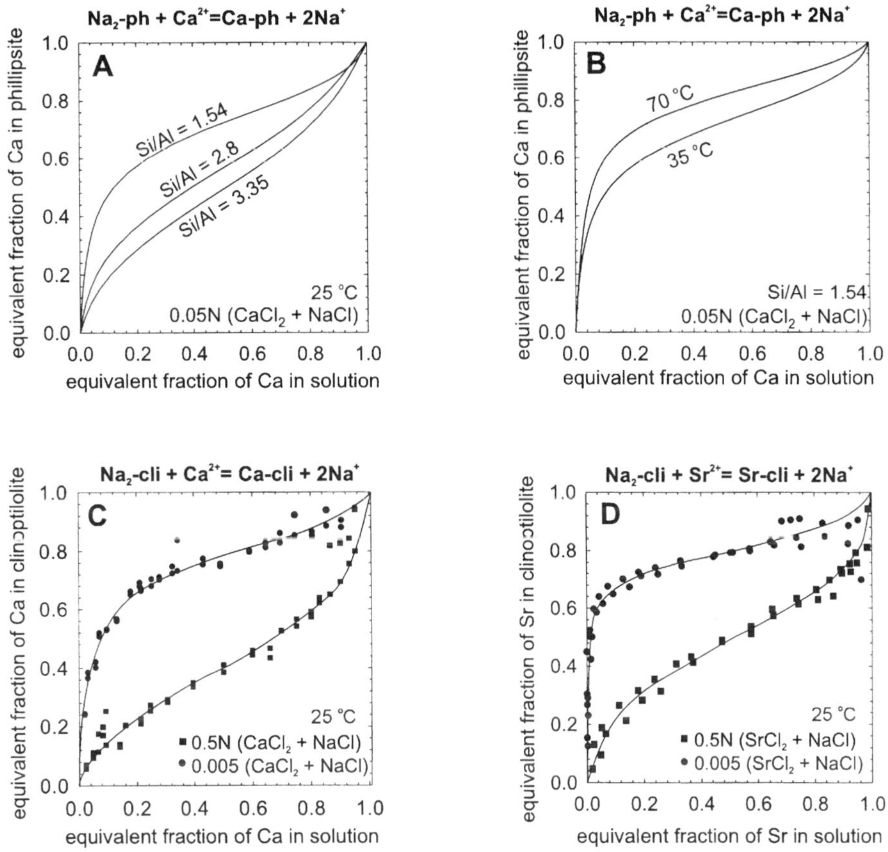

FIG. 13. Experimentally determined binary ion-exchange isotherms representing reactions between zeolites (phillipsite and clinoptilolite) and aqueous solutions, plotted as a function of the charge equivalent mol-fraction in solution versus the same quantity in the zeolite. A. Isotherms for Ca-Na exchange in phillipsites with three different Si/Al ratios. It can be seen that the preference of phillipsite (ph) for Na^+ increases with increasing Si/Al ratio of the mineral. Data are from Shibue (1981; Si/Al = 3.35 and Si/Al = 1.54), and Barrer and Munday (1971; Si/Al = 2.8). B. Experimental ion-exchange isotherms for Al-rich phillipsite (Shibue, 1981) at 70°C and 35°C, showing the effect of temperature on exchange equilibria. C. Ca-Na exchange in clinoptilolite (based on Pabalan, 1994). D. Sr-Na exchange in clinoptilolite (based on Pabalan and Bertetti, 1999). Figures C and D illustrate the "concentration-valence" effect of total concentration on heterovalent ion-exchange reactions, resulting in increased preference of the zeolite for divalent cations compared to monovalent at low ionic strengths.

ΔH^o_r of 7 kcal/mole between 35° and 70°C (Fig. 13B); (3) the dependence of isotherm shape on total concentration of the solution when heterovalent ions are involved in the exchange reaction (concentration-valence effect; Figs. 13C and D); and (4) the preference of these zeolites for large ions such as Sr^{+2} (Fig. 13D).

Aside from the studies by White et al. (1980), Sturchio et al. (1989), and Keith et al. (1983) mentioned above, little work has been done to assess the effects of ion-exchange equilibria between zeolites and aqueous solutions in natural systems. This in part stems from a lack of thermodynamic data for zeolite ion-exchange reactions at temperatures and pressures under which zeolites occur in nature (ion-exchange isotherms are generally determined at 298 K, whereas most zeolite occurrences are at higher temperatures). Typically, equilibrium constants for ion-exchange reactions in zeolites are thought to be relatively independent of temperature and pressure because ΔH^o_r and the standard volume of reaction (ΔV^o_r) are assumed to be relatively small (Helfferich, 1962). This presumption is, however, contradicted by experimental observations (see Fig. 13A;

Petrova et al., 1997; White et al., 1999) and theoretical estimates (Fridriksson et al., 1999a; Neuhoff, unpubl. calculations) of $\Delta H^o r$ and $\Delta V^o r$ for ion-exchange reactions. Theoretical predictions of the temperature dependence of K for zeolite ion-exchange reactions is complicated by poorly characterized water stoichiometry in homoionic end-member zeolites (see Reaction 10); the choice of water stoichiometry can have a profound effect on the predicted ΔH^{o}_{r} of these reactions (Fridriksson et al., 1999a).

Zeolite-Facies Metamorphism of Tertiary Lavas in the North Atlantic Region

After the formal recognition of the zeolite facies by Fyfe et al. (1958) and Coombs et al. (1959), most of the detailed studies of zeolites in geologic systems focused on their occurrence in sedimentary and pyroclastic rocks. This was largely a consequence of their importance as diagenetic minerals in petroleum systems, alteration phases in tuffs slated for nuclear waste disposal, and the occurrence of high-purity zeolite deposits that are actively mined for inexpensive ion-exchange substrates and desiccants. However, at nearly the same time that Fyfe et al. (1958) and Coombs et al. (1959) established that zeolites were widespread products of burial metamorphism in New Zealand, Walker (1951, 1959a, 1959b, 1960a, 1960b) made similar observations in his pioneering work on the Tertiary lavas in Ireland and eastern Iceland. Walker (1959a, 1960b) identified regionally extensive mineral assemblages that defined depth-controlled "zeolite zones" formed during burial metamorphism. The regional patterns of mineral distribution identified by Walker, which we review below, have subsequently been identified in Icelandic geothermal systems (Kristmannsdóttir and Tómasson, 1978), the Faeroe Islands (Jørgensen, 1984), East Greenland (Larsen et. al., 1989; Neuhoff et al., 1997), West Greenland (Christiansen et al., 1999), the Deccan Traps in India (Sukheswala et al., 1974), the Parana lavas in Brazil (Murata et al., 1987), and the North Shore Volcanic Group, Minnesota (Schmidt, 1993). We interpret the petrotectonic histories of these deposits below, drawing largely from our studies of zeolite distribution and paragenesis in eastern Iceland and East Greenland (Neuhoff et al., 1995, 1997, 1999; Neuhoff and Bird, 1997; Fridriksson et al., 1999b).

The geology of all the zeolite deposits in the North Atlantic region is characterized by thick sequences of basaltic lavas produced by magmatism associated with crustal rifting and the Iceland mantle plume. Glacial unroofing has exposed spectacular cross sections (~2 km deep in Iceland and >5 km deep in Greenland) through paleo-groundwater aquifers developed in the lavas. The basalts are relatively homogeneous, consisting of low magnesia/ferrobasalt or olivine tholeiite compositions (Wood, 1977; Larsen et al., 1989) with minor interbedded sediments and pyroclastic deposits (Wager, 1947; Walker, 1960b; Wood, 1977; Saemundsson, 1980; Larsen et al., 1989; Gústafsson, 1992). In Iceland, the basalts range in age from 13.6 Ma to recent (Watkins and Walker, 1977; Fridleifsson, 1982) and in Greenland from ~55 to 59 Ma (M. Storey, pers. commun., 1996). Both sequences initially were relatively flat lying, with little subsequent structural deformation other than localized intrusions, faults (Walker 1960b; Deer, 1976; Saemundsson, 1979; Brooks and Nielsen, 1982; Fridleifsson, 1982; Upton, 1988; Gústafsson 1992), and progressive tilting of the lavas towards the center of volcanism (particularly in eastern Iceland) due to loading of the crust during volcanism (Pálmason, 1981).

Low-temperature alteration and metamorphism of the lavas can be described in terms of three paragenetic stages: (I) near-surface alteration characterized by precipitation of celadonite and amorphous silica along vesicle walls, with limited celadonite formation and iron oxide/hydroxide precipitation in the matrix of the lavas; (II) regional burial metamorphism leading to the formation of zeolites and mafic phyllosilicates in the pore spaces and matrices of the lavas; and (III) local hydrothermal alteration around dikes and intrusions that formed complex assemblages of zeolites, silica minerals, mafic phyllosilicates, and carbonates. The relative timing of these stages is clearly defined based on crosscutting relationships. Stage I minerals (where they occur) are always the first minerals to form in primary pore spaces, and are followed by those formed during burial metamorphism (Stage II). Local hydrothermal alteration (Stage III) is developed in and around brittle deformation features that crosscut primary pores and Stage I and II mineral assemblages.

During Stage II, groundwater flowing through permeable zones irreversibly reacted with the basalts forming distinct parageneses of secondary minerals distributed in depth-controlled zones. Mineral parageneses developed in pore spaces during Stage II can be broadly characterized by the sequence

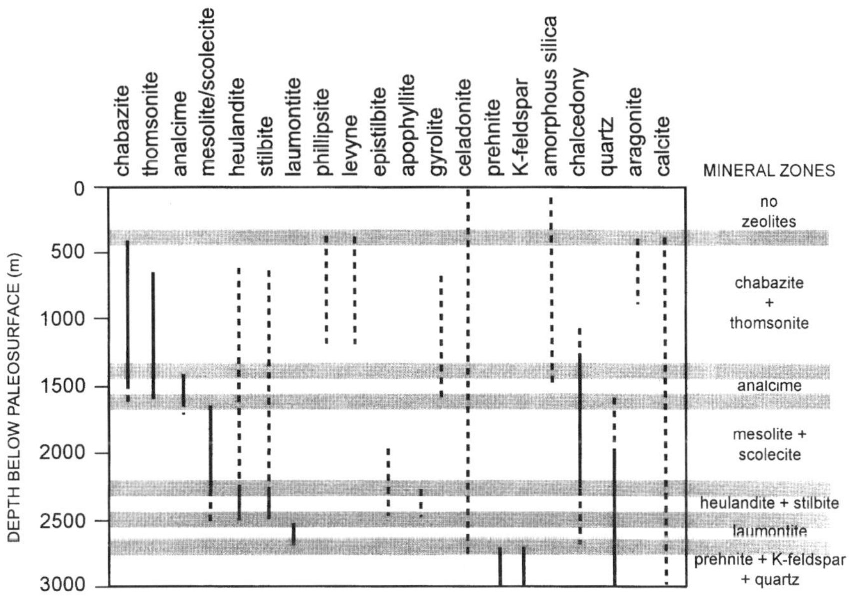

FIG. 14. Distribution of mesoscopic secondary minerals filling primary pore spaces in East Greenland flood basalts as a function of depth below paleosurface. Frequency of occurrence of given mineral with depth is denoted by solid (always present) and dashed (occasionally present) lines. Thick horizontal lines correspond to depth below paleosurface (assuming 400 m of zeolite-free lavas; Larsen et al., 1989) of the chabazite + thomsonite isograd (~400 m), analcime isograd (~1400 m), mesolite-scolecite isograd (~1600 m), heulandite + stilbite isograd (~2300 m), laumontite isograd (~2500 m), and prehnite isograd (~2700 m). After Neuhoff et al. (1997).

$$\underset{\text{early}}{\text{mafic phyllosilicates} \rightarrow \text{index zeolite(s)}} \rightarrow \underset{\text{late}}{\pm \text{ calcite} \pm \text{late-stage zeolites,}}$$

where the mafic phyllosilicates include smectites, interlayered chlorite/smectite minerals, and chlorites, and "index zeolites" are the minerals that define the "zeolite zones." The mineralogy of these zones in East Greenland is detailed in Figure 14 (Neuhoff et al., 1997). Index minerals for the zeolite zones (chabazite + thomsonite, analcime, mesolite-scolecite, heulandite + stilbite, and laumontite) are modally the most abundant secondary minerals filling primary porosity in the lavas, and are paragenetically the earliest zeolites. Near isograds, zeolites of the deeper zone can rarely be observed following lower-grade zeolites in paragenetic sequences; for instance, at the analcime to mesolite-scolecite zone boundary, analcime forms rims in vesicles that are filled with mesolite. Such instances are rare, however, and confined to regions immediately surrounding the isograds. *Zeolites from deeper zones are never observed forming at the expense of those characteristic of shallower zones.* These observations are consistent with the geologic observations summarized below concerning the relative timing of zeolite formation. Precipitation of zeolites in the matrix of the lavas and replacement of feldspars is only observed near primary macroscopic porosity, suggesting that fluid flow exerted a strong control on alteration. Later calcite, zeolites (e.g., stilbite, heulandite), prehnite, and other silicates listed in Figure 14 (indicative of late Stage II or Stage III alteration) are locally precipitated on top of, or pseudomorphically replace, index zeolite assemblages, but are rarely modally abundant.

Regional mapping of the distribution of the zeolite zones depicted in Figure 14 reveals several distinct geologic relationships between the mineral isograds bounding these zones and their host lavas. These relations can be seen in Figure 15, which depicts the regional distribution of zeolite zones in one cross section from the Berufjörður region of eastern Iceland (modified from Walker (1960b) based on our mapping). Inspection of Figure 15 reveals that the zeolite zones: (1) are of relatively uniform thickness throughout the lavas; (2) crosscut

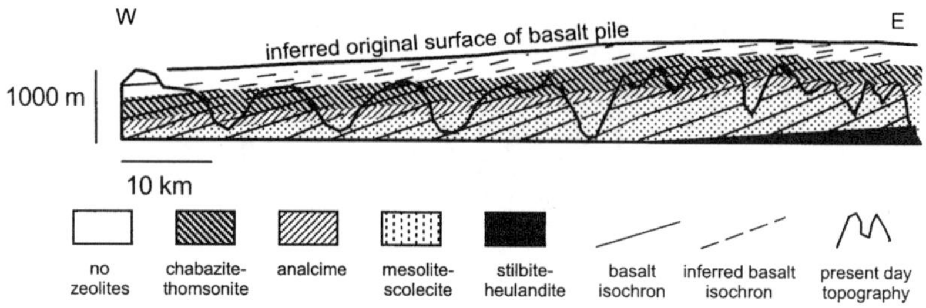

FIG. 15. Cross section showing the regional distribution of zeolite zones in eastern Iceland (after Walker, 1960b; Neuhoff et al., 1999).

the lava stratigraphy; and (3) are parallel to paleotopographic surfaces. The uniform thickness of the zeolite zones suggests that heat flow during metamorphism did not vary laterally in any significant manner. The orientation of the zone boundaries with respect to the paleotopographic surface indicates that basal heat flow controlled the thermal gradients during zeolite formation. The discordance between zeolite isograds and the volcanic stratigraphy indicates that the present isograds formed after extrusion and synvolcanic tilting of the lavas within the section.

Crosscutting relationships between the Stage I alteration, the lavas hosting Stage II alteration, and the intrusions responsible for Stage III alteration provide a means of constraining the timing and duration of zeolite formation. In the case of East Greenland, zeolite zones mineralize all but the uppermost flows of the 57 to 55 m.y. lavas (M. Storey, pers. commun., 1996). The zones appear to have formed rapidly during and just after eruption of the middle-series lavas but before prograde metamorphic replacement of regional zeolite zones by hydrothermal activity (Manning and Bird, 1991) associated with the Skaergaard intrusion (55 m.y.; Brooks and Gleadow, 1977; Hirschman et al., 1996) and the coastal dike swarm (most of which pre-dates the Skaergaard intrusion; Nielsen, 1978). The timing of the zeolite zones is thus indistinguishable, within error, from available radiometric dates for magmatic events in the province. The short duration of zeolitization can be explained by placing the mineral parageneses in the context of the petrotectonic and hydrologic evolution of the lavas.

Compositions of the first minerals to precipitate during Stage II, mafic phyllosilicates, exhibit temporal changes in composition from smectite-rich to chlorite-rich compositions, suggesting that temperature increased during precipitation of these phases by analogy to the distribution of mafic phyllosilicates in geothermal systems (Kristmannsdóttir, 1976; Schiffman and Fridleifsson, 1991). This suggests that mafic phyllosilicates form during burial of the lavas, in response to initial rapid release of iron and magnesium during hydrolysis of olivine and basaltic glass. The zeolites that define the zeolite zones did not precipitate until after mafic phyllosilicate formation and burial ceased, which explains the lack of textural evidence for progressive development of the zeolite-zone assemblages and the discordance between the zeolite isograd orientations and the volcanic stratigraphy. Precipitation of the zeolite-zone assemblages led to almost complete destruction of the primary porosity in the lavas, which prevented further water-rock interaction that might have overprinted the zeolite zones. Only when significant secondary porosity was created during brittle deformation associated with intrusions could later (Stage III) alteration occur.

Zeolite Zones as Indicators of Structural and Thermal Evolution in Large Igneous Provinces

The formation of zeolite zones shortly after the cessation of volcanism in basaltic terrains in the North Atlantic region makes them important indicators of the thermal and structural state of the crust during this stage in the evolution of large igneous provinces. For instance, the distribution of regional zeolite zones and isograds provides information about thermal gradients in the upper crust at the end of volcanism. In addition, the uniform thicknesses, planar geometry, and regional extent of zeolite zones provide distinct metamorphic marker horizons that can be used to distinguish synvolcanic structural deformation from later processes. When combined with detailed studies of regional structural and

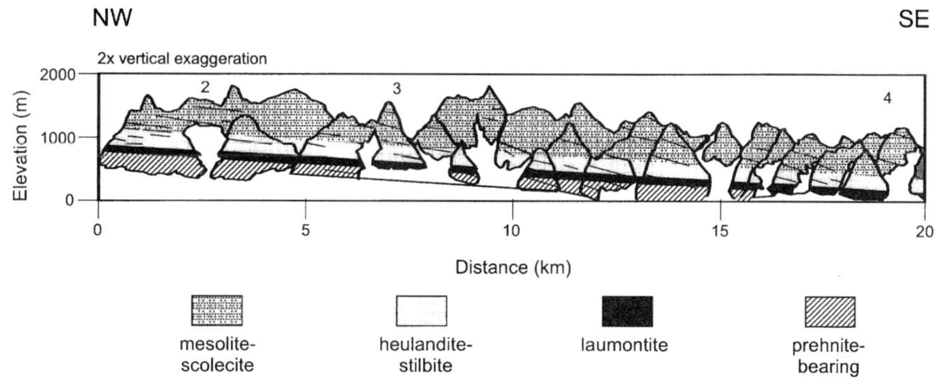

FIG. 16. Photogrammetrically measured cross section in Borggraven region in East Greenland (after Neuhoff et al., 1997). Strike of section is approximately perpendicular to strikes of basalt flows and zeolite isograds. Dips of selected lava flows are indicated by thin discontinuous solid lines; prominent stratigraphic horizon is shown in solid black. Locations of three sampling traverses (B2, B3, and B4; Fig. 1) are noted above section. Zeolite zone annotations are based on detailed sampling at these localities and selected localities between them. Apparent dip of zeolite isograds in cross section is ~ 2.5° SE. Note systematic increase in basalt-flow dips toward southeast.

stratigraphic studies (e.g., Walker, 1960b; Pedersen et al., 1997), regional mapping of zeolite zones can play an integral part in unraveling the tectonic and thermal history of large igneous provinces.

The strong correlation between the appearance of isograds delineating the zeolite zones and paleodepth indicates that the distribution of index zeolite assemblages is controlled by temperature and/or pressure. This is supported by observations of zeolite occurrence in Icelandic geothermal systems (Fig. 4D; Kristmannsdóttir and Tómasson, 1978), where zeolite isograds are clearly temperature dependent (Fig. 4D). Extrapolation from the temperatures shown in Figure 4D (where the corresponding pressures are at or near liquid-vapor saturation for water) to burial metamorphic systems is predicated on the assumption that the equilibria controlling the isograds have steep Clapeyron slopes. Thermodynamic calculations of Clapeyron slopes support this assumption (Neuhoff, 1999), as do the observations of Iijima (1978, 1988) that temperatures associated with zeolite isograds developed in tuffaceous sediments in Japan are independent of pressure between 1 and 2 kbar.

Independent estimates of regional basal heat flow during metamorphism provide a means of validating empirical temperature estimates based on the data in Figure 4. Comparison of the distribution of zeolite zones in Figures 4B–4D, 14, and 15 and the temperatures shown in Figure 4D indicates that the thermal gradients during alteration in eastern Iceland and East Greenland were ~60 ± 5°C/km and 40 ± 5°C/km, respectively. Combined with a thermal conductivity for water-saturated basalt of 1.5 to 2.0 W/mK (Oxburgh and Agrell, 1982), these values suggest regional heat flows of 1.9 to 3.2 and 1.3 to 2.2 heat flow units, respectively, during zeolite-facies metamorphism. For the eastern Iceland example, the calculated heat flow is within the range of present-day heat-flow values on the flanks of the volcanic rift (Flóvenz and Saemundsson, 1993), which is the setting of Stage II metamorphism (Neuhoff et al., 1999). The values for East Greenland are consistent with heat-flow estimates proposed by Manning et al. (1993) and Bird et al. (1985) for roughly coeval metamorphism of basalts east of the Skaergaard intrusion, and similar to heat flow in present-day continental rift systems (Lysak, 1992).

The concordance between the zeolite isograds and paleosurfaces of basalt plateaus and the common discordance between the isograds and the lava stratigraphy makes the orientations of the isograds important markers for assessing the timing of deformation in the lavas. This is illustrated in Figure 16, which depicts the distribution of zeolite zones through a portion of the coastal flexure zone (Wager and Deer, 1938) in East Greenland (from Neuhoff et al., 1997). Lava dips increase from 4° to 11° SE toward the coast in the section of Figure 16, whereas zeolite isograds dip monotonically ~2.5° SE. This pronounced dip in the zeolite isograds is in contrast to their horizontal attitudes in structurally undisturbed portions of the province. Discordance between zeolite isograds and the lavas indicates that tilting in the region of Figure 16 was initiated prior to regional metamorphism. This stage of tilting may well be synchronous with volcanism, considering the short period of time (<1 m.y.) after volcanism in which the isograds were established. Synvolcanic tilting of the lavas near the present-day coast is consistent with progressive focusing of volcanism along the rift (Pedersen et al., 1997), which led to crustal sagging in the coastal areas (e.g., Larsen and Marcussen, 1992). A similar relationship between zeolite isograds and synvolcanic structure is observed in eastern Iceland (Fig. 15; Walker, 1960b).

Potential causes of the pronounced dip of the isograds near the coast in Figure 16 include regional topographic slopes, heat-flow variations, or postmetamorphic structural deformation. The first two hypotheses appear unlikely. Regional topographic gradients of several degrees are uncommon in plateau basalts (e.g., Baker et al., 1987; see also Fig. 15 where the paleosurface and zeolite isograds dip ~ 0.5°), and the isograds are not telescoped as would be expected in regions of higher heat flow (e.g., Schiffman et al., 1984). The present-day orientation of the isograds is consistent, however, with deformation associated with the dramatic flexuring observed in East Greenland (e.g., near Kangerlussuaq; see Wager and Deer, 1938). Zeolite isograds thus serve as temporal marker horizons distinguishing synvolcanic deformation, which resulted in the discordance between isograd dips and lava dips, from postmetamorphic (and thus postvolcanic) deformation associated with regional flexuring that tilted the zeolite zones.

Zeolite Zones: Chemical Filters in Basaltic Crust

Basalts are the most abundant rock type in the Earth's crust (Wedepohl, 1969); the vast quantities of basalt erupted onto the Earth's surface bring with them crustally significant quantities of major and trace elements. These basalts form extensive aquifers for the storage and migration of groundwater, where water-rock reactions lead to dissolution and hydrolysis of primary phases releasing the chemical components of the basalts into the aquifers. The strong zonation of zeolites and other minerals formed during this interaction leads to substantial variations in physical and chemical properties of the crust with depth. One example of this can be seen in the variations in the ion-exchange selectivities of various zeolites (Table 2). Ion-exchange reactions between groundwater and zeolites in basaltic aquifers can lead to substantial variations in the mobility and concentrations of metals with depth in these systems, as the zeolites that characterize the zones have very different ion-exchange capacities and selectivities (Table 2). Below we focus on one example of this phenomenon, in which the petrogenetically and environmentally important element strontium appears to have been concentrated in the heulandite-stilbite zone in the Tertiary lavas in eastern Iceland through ion-exchange reactions.

Although strontium is one of the least abundant alkaline-earth metals, composing only 0.04% of the Earth's crust (Mason and Moore, 1982), it is an environmentally and petrologically important element. The ionic radius and bonding characteristics of strontium are similar to those of the significantly more abundant element calcium, and strontium commonly substitutes for calcium in feldspars, clays, carbonates, sulfates, and zeolites. The radiogenic isotope of strontium, ^{90}Sr (half-life of 28 years), is one of the principal components of radioactive waste derived from uranium fission. Input of ^{90}Sr into the environment through nuclear explosions and leakage from storage facilities provides earth scientists a means to evaluate strontium migration in the atmosphere-ocean (see Bowen and Roether, 1973) and ocean-river ecosystems (i.e., Columbia River Plume; Park et al., 1965). However, the physiological consequences of ^{90}Sr contamination, such as cancer, may be severe, since human and animal bones selectively concentrate strontium (Wyckoff and Doberenz, 1968). Significant quantities of ^{90}Sr are known to have leaked at U.S. Department of Energy sites such as the Idaho National Engineering Labs (INEL; McCurry et al., 1994) and the Hanford Nuclear Reservation (Illman, 1993). Both the Hanford Nuclear Site and INEL are situated atop basaltic aquifers, geologic formations similar to those in eastern Iceland and East Greenland. At the Hanford site in particular, the secondary mineralogy of the underlying aquifer is dominated by clays and heulandite-structure zeolites that are similar to the heulandite-stilbite–zone assemblages depicted in Figure 14. The analog relationship between the Hanford site and the basaltic aquifers developed in the North Atlantic region motivates our ongoing research of Sr mobility in the latter systems.

Strontium concentrations in meteoric groundwaters of Iceland (Elderfield and Greaves, 1981) are typically 1 to 3 orders of magnitude greater than in average North Atlantic precipitation (2 to 4 ppb in

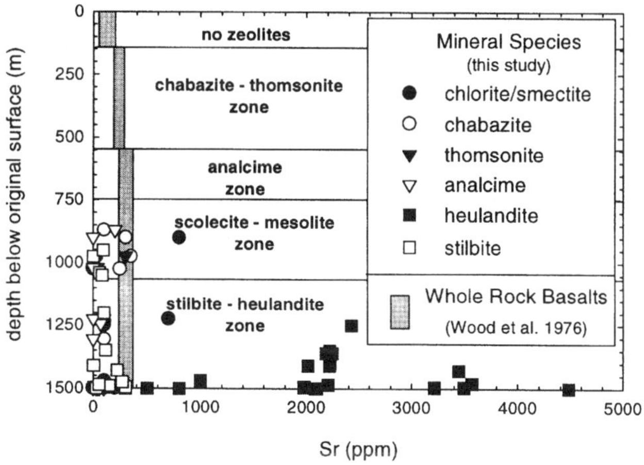

FIG. 17. Concentrations of Sr in secondary minerals (Neuhoff, unpubl. data) and lavas (Wood et al., 1976) from eastern Iceland as a function of depth below the original (pre-glaciation) surface of the lava pile.

Iceland; Daux et al., 1994). Thus, the dissolution of strontium from the basalts must be the major source of this element in aquifer fluids. Basalts in eastern Iceland (Fig. 17) and East Greenland typically contain 100 to 670 ppm strontium (Gerasimovskiy et al., 1976; Wood et al., 1976; Larsen et al., 1989). Strontium in unaltered basalt is concentrated in plagioclase feldspars (~300 to 1100 ppm), with lesser amounts in olivine (~100 to 175 ppm), and pyroxene (~150 to 300 ppm; Gerasimovskiy et al., 1976). Strontium concentrations in interstitial glass are not well documented; however, partition coefficients for strontium suggest that plagioclase is enriched two to four times relative to coexisting melts (Drake and Weill, 1975; Blundy and Wood, 1991; Bindeman et al., 1998).

Alteration of basalts often leads initially to rapid loss of strontium; release continues at a reduced rate thereafter (e.g., Yaalon et al., 1974). Interstitial glass dissolves more rapidly than all other phases during alteration (Gíslason and Eugster, 1987a; Hearn et al., 1989, 1990), suggesting that release from this phase is important in the rapid initial release of strontium (on the order of 10^{-1} $g_{Sr} m^{-3} a^{-1}$). The continued release of strontium during alteration reflects its high concentration in plagioclase, which hydrolyzes two to three orders of magnitude more slowly than the interstitial glass (Gíslason and Eugster, 1987a), resulting in Sr release rates on the order of 10^{-10} $g_{Sr} m^{-3} a^{-1}$ (Brantley et al., 1998).

Present-day groundwater in Iceland and predicted fluid compositions from the East Greenland basaltic paleo-aquifers are both dilute near-neutral to slightly basic solutions of meteoric origin (Taylor and Forester, 1979; Gíslason and Eugster, 1987b; Bird et al., 1988). Strontium speciation in these groundwaters is predominantly in the form of Sr^{+2}; concentrations of complexes such as $SrCO_3$(aq), $SrCl^+$, and $SrOH^+$ are all several orders of magnitude lower (aqueous speciation from EQ3NR; Wolery, 1991). Groundwater transport of strontium is largely controlled by the primary porosity structures of the lava flows, resulting in strong anisotropy of hydraulic conductivity (Freeze and Cherry, 1979; Manning and Bird, 1991). Tops and bottoms of flows are highly porous (>5 to 10%) because of gas vesicles and brecciation (see Davis, 1969), with hydraulic conductivities commonly between 10^{-9} and 10^{-5} m/s. Central portions of lava flows have low porosities (0.1 to 1%) and hydraulic conductivities (10^{-11} to 10^{-8} m/s; Davis, 1969). This heterogeneous distribution of porosity and permeability results in dominant vectors of fluid flow parallel to the basalt stratigraphy (Kononov, 1978). Nearly vertical fractures produced by faults and dikes locally increase the vertical component of permeability in the aquifer (Geptner and Sokolova, 1989), whereas precipitation of secondary minerals significantly reduces effective porosity and permeability during the geologic evolution of the aquifer (Neuhoff et al., 1999). This is particularly true at depth, where mineral precipitation nearly occludes all porosity, leading to significant reductions in permeability at the crustal scale (Manning and Ingebritsen, 1999).

Strontium appears to have been concentrated with depth in the basaltic paleoaquifers of eastern Iceland by dissolution of primary phases in the upper portions of the aquifer, transport by groundwater, and precipitation and ion exchange into zeolites in the heulandite-stilbite zone. This is illustrated in Figure 17, where Sr concentrations of the lavas (Wood et al., 1976) are compared to the Sr contents of secondary minerals (Neuhoff and Fridriksson, unpubl. data). Strontium concentrations in the lavas generally increase with depth (Fig. 17). Wood et al. (1976) argued that this geochemical pattern is not a result of temporal magmatic variation because: (1) Sr is not correlated with Zr in the altered Tertiary lavas, as observed in unaltered Quaternary lavas in Iceland; and (2) plagioclase, which generally has high Sr contents, is not more modally abundant in the deeper lavas than in the less Sr-rich shallower lavas. Rather, this trend appears to be evidence for mobilization of Sr during alteration from the upper levels of the lava pile, and transfer to greater depths where it is concentrated through mineralogic attentuation.

The conclusion of Wood et al. (1976) that Sr is concentrated at depth through incorporation in authigenic minerals is supported by the secondary mineral compositions shown in Figure 17. Inspection of the mineral analyses shows that Sr is most concentrated in heulandite, which is most modally abundant at depths below ~1 km from the paleosurface. Strontium concentrations in heulandite are typically an order of magnitude higher than in unaltered lavas (Fig. 17). Heulandite displays a strong ion-exchange selectivity for Sr (Table 2), particularly in dilute aqueous solutions (see Fig. 13D for the isostructural mineral clinoptilolite). The restricted depth range over which heulandite is a modally important alteration phase generally coincides with the region of Sr enrichment noted by Wood et al. (1976). Thus, the Sr enrichment of the lavas at depths corresponding to the heulandite-stilbite zone (Figs. 4 and 15) may at least in part result from its concentration in heulandite. The heulandite-stilbite zone in basaltic aquifers thus acts as a chemical filter for Sr, removing it from groundwater and limiting its further transport. Chemical filtration of Sr from groundwater by heulandite in Tertiary paleoaquifers in eastern Iceland serves as a potential analog for the chemical behavior of radioactive waste plumes at sites like Hanford, where ^{90}Sr contaminant plumes may likely come in contact with heulandite-group zeolites in the subsurface.

Concluding Remarks

In this communication, we review some major aspects of both the physical chemistry and geology of natural zeolites. The patterns of occurrence of zeolites in geologic systems provide essential information about the chemical, thermobarometric, and structural evolution of shallow portions of the Earth's crust. Because of this, the study of natural zeolite parageneses and compositions has proven a valuable aid in petroleum and geothermal resource development. As illustrated by the example of Sr mobility given above, the occurrence of zeolites can influence the mobility of contaminants, such as radioactive waste. Further and more informed use of zeolites as indicators of conditions within the Earth would be facilitated by more complete knowledge of their stability, crystal chemistry, and geologic occurrence. Future geologic, theoretical, and experimental insights into the controls on zeolite parageneses and compositions hopefully will provide this knowledge.

Acknowledgments

The authors wish to thank Prof. Konrad Krauskopf for his continuing encouragement of our investigations into the geochemistry and stability of natural zeolites. Through his own pioneering studies in geochemistry, and his profound public service in the area of radioactive waste disposal (in which zeolites play a prominent role), he has been a constant example of the goals to which we aspire as scientists.

This work was funded by the Danish Lithosphere Center, the Geological Survey of Denmark and Greenland, the U.S. National Science Foundation (grants EAR-84-18129, EAR-86-06256, and EAR-95-06469 to Bird), the Petroleum Research Fund of the American Chemical Society (grant ACS-PRF 31742-AC2 to DKB), the McGee and Shell Fund of Stanford University, and the Geological Society of America. Additional support was provided by graduate fellowships from the U.S. National Science Foundation and the U.S. Environmental Protection Agency (to Neuhoff). Careful reviews by C. Manning and R. Newton significantly improved the presentation of ideas in this communication.

REFERENCES

Abercrombie, H. J., Hutcheon, I. E., Bloch, J. D., and de Caritat, P., 1994, Silica activity and the smectite-illite reaction: Geology, v. 22, p. 539–542.

Armbruster, T., and Kohler, T., 1992, Rehydration and dehydration of laumontite: A single-crystal X-ray study at 100-K: Neues Jahrb. Mineral.-Monatshefte, v. 1992, p. 385–397.

Artioli, G., and Galli, E., 1999, Gonnardite: Re-examination of holotype material and discreditation of tetranatrolite: Amer. Mineral., v. 84, p. 1445–1450.

Baker, V. R., Greeley, R., Komar, P. D., Swanson, D. A., and Waitt, R. B., Jr., 1987, Columbia and Snake River plains, *in* Graf, W. L., ed., Geomorphic systems of North America, The Decade of North American Geology: Boulder, Geol. Soc. Amer., v. 2, p. 403–468.

Baldar, N. A., and Whittig, L. D., 1968, Occurrence and synthesis of soil zeolites, *in*, Proc., Soil Sci. Soc. Amer., v. 32, p. 235–238.

Bargar, K. E., and Beeson, M. H., 1981, Hydrothermal alteration in research drill hole Y-2, Lower Geyser Basin, Yellowstone National Park, Wyoming: Amer. Mineral., v. 66, p. 473–490.

Barrer, R. M., and Munday, B. M., 1971, Cation exchange reactions of a sedimentary phillipsite: Jour. Chem. Soc.-A, p. 2905–2909.

Berman, R.G., 1988, Internally-consistent thermodynamic data for minerals in the system Na_2O-K_2O-CaO-MgO-FeO-Fe_2O_3-Al_2O_3-SiO_2-TiO_2-H_2O-CO_2: Jour. Petrol., v. 29, p. 445–522.

Betz, V., 1981, Famous mineral localities; zeolites from Iceland and the Faeroes: Mineral. Record, v. 12, p. 5–26.

Bindeman, I. N., Davis, A. M., and Drake, M. J., 1998, Ion microprobe study of plagioclase-basalt partition experiments at natural concentration levels of trace elements: Geochim. et Cosmochim. Acta, v. 62, p. 1175–1193.

Bird, D. K., Schiffman, P., Elders, W. A., Williams, A. E., and McDowell, S. D., 1984, Calc-silicate mineralization in active geothermal systems: Econ. Geol., v. 79, p. 671–695.

Bird, D. K., Manning, C. E., and Rose, N. M., 1988, Hydrothermal alteration of Tertiary layered gabbros, East Greenland: Amer. Jour. Sci., v. 288, p. 405–457.

Bird, D. K., Rosing, M. T., Manning, C. E., and Rose, N. M., 1985, Geologic field studies of the Miki Fjord Area, East Greenland: Bull. Geol. Soc. Denmark, v. 34, p. 219–235.

Blundy, J. D., and Wood, B. J., 1991, Crystal-chemical controls on the partitioning of Sr and Ba between plagioclase feldspar, silicate melts, and hydrothermal solutions: Geochim. et Cosmochim. Acta, v. 55, p. 193–209.

Boles, J. R., 1977, Zeolites in deep-sea sediments, *in* Mumpton, F. A., ed., Mineralogy and geology of natural zeolites: Mineral. Soc. Amer., Rev. Mineral., v. 4, p. 137–163.

Bowen, V. T., and Roether, W., 1973, Vertical distribution of strontium-90, cesium-137, and tritium near 45° North in the Atlantic: Jour. Geophys. Res., v. 78, p. 6277–6285.

Bowers, T. S., and Burns, R. G., 1990, Activity diagrams for clinoptilolite; susceptibility of this zeolite to further diagenetic reactions: Amer. Mineral., v. 75, p. 601–619.

Brantley, S. L., Chesley, J. T., and Stillings, L. L., 1998, Isotopic ratios and release rates of strontium measured from weathering feldspars: Geochim. et Cosmochim. Acta, v. 62, p. 1493–1500.

Breck, D. W., 1974, Zeolite molecular sieves; structure, chemistry, and use: New York, John Wiley & Sons, 771 p.

Brooks, C. K., and Gleadow, A. J. W., 1977, A fission-track age for the Skaergaard Intrusion and the age of the East Greenland basalts: Geology, v. 5, p. 539–540.

Brooks, C. K., and Nielsen, T. F. D., 1982, The E Greenland continental margin; a transition between oceanic and continental magmatism: Jour. Geol. Soc. Lond., v. 139, p. 265–275.

Broxton, D. E., Bish, D. L., and Warren, R. G., 1987, Distribution and chemistry of diagenetic minerals at Yucca Mountain, Nye County, Nevada: Clays and Clay Minerals, v. 35, p. 89–110.

Capaldi, G., Civetta, L., and Gasparini, P., 1971, Fractionation of the ^{238}U decay series in the zeolitization of volcanic ashes: Geochim. et Cosmochim. Acta, v. 35, p. 1067–1072.

Carey, J. W., and Bish, D. L., 1996, Equilibrium in the clinoptilolite-H_2O system: Amer. Mineral., v. 81, p. 952–962.

Casey, W. H., 1988, Entropy production and the Ostwald step rule: Jour. Phys. Chem., v. 92, p. 226–227.

Cho, M., Maruyama, S., and Liou, J. G., 1987, An experimental investigation of heulandite-laumontite equilibrium at 1000 to 2000 bar P fluid: Contrib. Mineral. Petrol., v. 97, p. 43–50.

Christiansen, F. G., Boesen, A., Bojesen-Koefoed, J., Dallhoff, F., Dam, G., Neuhoff, P. S., Pedersen, A. K., Pedersen, G. K., Stannius, L. S., and Zinck-Joergensen, K., 1999, Petroleum geological activities onshore West Greenland in 1997: Geol. Greenland Surv. Bull., v. 180, p. 10–17.

Colella, C., 1996, Ion exchange equilibria in zeolite minerals: Mineralium Deposita, v. 31, p. 554–562.

Coombs, D. S., 1960, Lower grade mineral facies in New Zealand, *in*, Proc. 21st Int. Geol. Cong., Copenhagen, v. 13, p. 339–351.

Coombs, D. S., Alberti, A., Armbruster, T., Artioli, G., Colella, C., Galli, E., Grice, J. D., Liebau, F., Mandarino, J. A., Minato, H., Nickel, E. H., Passaglia, E., Peacor, D. R., Quartieri, S., Rinaldi, R., Ross, M., Sheppard, R. A., Tillmanns, E., and Vezzalini, G.,

1997, Recommended nomenclature for zeolite minerals; report of the Subcommittee on Zeolites of the International Mineralogical Association, Commission on New Minerals and Mineral Names: Can. Mineral., v. 35, p. 1571–1606.

Coombs, D. S., Ellis, A. J., Fyfe, W. S., and Taylor, A. M., 1959, The zeolite facies, with comments on the interpretation of hydrothermal syntheses: Geochim. et Cosmochim. Acta, v. 17, p. 53–107.

Cornu, F., 1908, Über die Paragenese der Minerale, namentlich die der Seolithe: Österreichische Zeitschrift für Berg- und Hüttenwesen, v. 56, p. 89–93.

Cronstedt, A.F., 1756, Rön och beskrifning om en obekänt bärg art, som kallas Zeolites: Svenska Vetenskaps Akademiens Handlingar, Stockholm, v. 17, p. 120–123.

Crossey, L. J., Frost, B. R., and Surdam, R. C., 1984, Secondary porosity in laumontite-bearing sandstones: Clastic diagenesis, v. 37, p. 225–237.

Daux, V., Crovisier, J. L., Hemond, C., and Petit, J. C., 1994, Geochemical evolution of basaltic rocks subjected to weathering; fate of the major elements, rare earth elements, and thorium: Geochim. et Cosmochim. Acta, v. 58, p. 4941–4954.

Davis, S. N., 1969, Porosity and permeability of natural materials, *in* Weist, D., ed., Flow through porous media: Washington, DC, Amer. Chem. Soc., p. 53–89.

Deer, W. A., 1976, Tertiary igneous rocks between Scoresby Sund and Kap Gustav Holm, East Greenland, *in* Escher, A., and Watt, W. S., eds., Geology of Greenland: Copenhagen, Geol. Surv. Greenland, p. 404–429.

Dibble, W. E., Jr., and Tiller, W. A., 1981, Kinetic model of zeolite paragenesis in tuffaceous sediments: Clays and Clay Minerals, v. 29, p. 323–330.

Donahoe, R. J., Guldman, S., and Liou, J. G., 1984, Synthesis and characterization of zeolites in the system Na_2O-K_2O-Al_2O_3-SiO_2-H_2O: Clays and Clay Minerals, v. 32, p. 433–443.

Drake, M. J., and Weill, D. F., 1975, Partition of Sr, Ba, Ca, Y, Eu^{2+}, Eu^{3+}, and other REE between plagioclase feldspar and magmatic liquid; an experimental study: Geochim. et Cosmochim. Acta, v. 39, p. 689–712.

Elderfield, H., and Greaves, M. J., 1981, Strontium isotope geochemistry of Icelandic geothermal systems and implications for sea water chemistry: Geochim. et Cosmochim. Acta, v. 45, p. 2201–2212.

Eskola, P., 1939, Die Metamorphen Gesteine, *in* Barth, T. F. W., Correns, C. W., and Eskola, P., eds., Die Entstehung der Gesteine: Berlin, Springer, p. 263–407.

Fairburn, H. W., 1943, Packing in ionic minerals: Bull. Geol. Soc. Amer., v. 54, p. 1305–1374.

Flóvenz, Ó. G., and Saemundsson, K., 1993, Heat flow and geothermal processes in Iceland: Tectonophys., v. 225, p. 123–138.

Freeze, R. A., and Cherry, J. A., 1979, Groundwater: Englewood Cliffs, NJ, Prentice-Hall, 604 p.

Frey, M., de Capitani, C., and Liou, J. G., 1991, A new petrogenetic grid for low-grade metabasites: Jour. Metamor. Geol., v. 9, p. 497–509.

Fridleifsson, I. B., 1982, The Iceland Research Drilling Project in relation to the geology of Iceland: Jour. Geophys. Res. B, v. 87, p. 6363–6369.

Fridriksson, Th., Neuhoff, P. S., and Bird, D. K., 1999a, Ion exchange equilibria between heulandite and geothermal fluids [abs.]: Geol. Soc. Amer. Abs. Prog., p. A27.

Fridriksson, Th., Neuhoff, P. S., Bird, D. K., and Arnórsson, S., 1999b, Clays and zeolites record alteration history at Teigarhorn, eastern Iceland, *in* Ármannsson, H., ed., Proc. Fifth Int. Symp. on Geochemistry of the Earth's Surface, p. 377–380.

Fyfe, W. S., Turner, F. J., and Verhoogen, J., 1958, Metamorphic reactions and metamorphic facies: Geol. Soc. Amer. Memoir, no. 73, 259 p.

Geptner, A. R., and Sokolova, G. V., 1989, Migration of petrogenic elements in zeolite formation in Icelandic basalts: Lithol. Min. Resources, v. 24, p. 1–10.

Gerasimovskiy, V. I., Laktionova, N. V., and Kovalenker, V. G., 1976, Strontium and barium in Iceland lavas: Geochem. Int., v. 13, p. 143–149.

Gíslason, S. R., and Arnórsson, S., 1990, Saturation state of natural waters in Iceland relative to primary and secondary minerals in basalts, *in* Spencer, R. J., and Chou, I. M., eds., Fluid-mineral interactions; a tribute to H. P. Eugster: Geochem. Soc., Spec. Publ., v. 2, p. 373–393.

Gíslason, S. R., and Eugster, H. P., 1987a, Meteoric water-basalt interactions; I, A laboratory study: Geochim. et Cosmochim. Acta, v. 51, p. 2827–2840.

______, 1987b, Meteoric water-basalt interactions; II, A field study in N.E. Iceland: Geochim. et Cosmochim. Acta, v. 51, p. 2841–2855.

Gottardi, G., 1989, The genesis of zeolites: Eur. Jour. Mineral., v. 1, p. 479–487.

Gottardi, G., and Galli, E., 1985, Natural zeolites: Berlin, Springer-Verlag, 409 p.

Gruner, J. W., 1950, An attempt to arrange silicates in the order of reaction energies at relatively low temperatures: Amer. Mineral., v. 35, p. 137–148.

Gústafsson, L. E., 1992, Geology and petrography of the Dyrfjöll central volcano, eastern Iceland: Berliner Geowissenschaftliche Abhandlungen, v. 138A, 101 p.

Hawkins, D. B., 1981, Kinetics of glass dissolution and zeolite formation under hydrothermal conditions: Clays and Clay Minerals, v. 29, p. 331–340.

Hay, R. L., 1963, Stratigraphy and zeolitic diagenesis of the John Day Formation of Oregon: Univ. Calif. Publ. Geol. Sci., v. 42, p. 199–262.

______, 1966, Zeolites and zeolitic reactions in sedimentary rocks: Geol. Soc. Amer. Spec. Pap. no. 85, 130 p.

______, 1977, Geology of zeolites in sedimentary rocks, *in* Mumpton, F. A., ed., Mineralogy and geology of natural zeolites: Mineral Soc. Amer. Rev. Mineral., v. 4, p. 53–64.

Hay, R. L., and Sheppard, R. A., 1977, Zeolites in open hydrologic systems, in Mumpton, F.A., ed., Mineralogy and geology of natural zeolites: Mineral Soc. Amer. Rev. Mineral., v. 4, p. 93–102.

Hearn, P. P., Jr., Steinkampf, W. C., Horton, D. G., Solomon, G. C., White, L. D., and Evans, J. R., 1989, Oxygen-isotope composition of ground water and secondary minerals in Columbia Plateau basalts; implications for the paleohydrology of the Pasco Basin: Geology, v. 17, p. 606–610.

Hearn, P. P., Jr., Steinkampf, W. C., White, L. D., and Evans, J. R., 1990, Geochemistry of rock-water reactions in basalt aquifers of the Columbia River plateau, *in* Doe, B. R., ed., U.S. Geol. Surv. Workshop on Environmental Geochemistry: Reston, VA, U. S. Geol. Surv., p. 63–68.

Helfferich, F., 1962, Ion exchange: New York, McGraw-Hill, 624 p.

Helgeson, H. C., Delany, J. M., Nesbitt, H. W., and Bird, D. K., 1978, Summary and critique of the thermodynamic properties of rock-forming minerals: Amer. Jour. Sci., v. 278A, 229 p.

Helmold, K. P., and van de Kamp, P. C., 1984, Diagenetic mineralogy and controls on albitization and laumontite formation in Paleogene arkoses, Santa Ynez Mountains, California: Amer. Assoc. Petrol. Geol. Memoir, v. 37, p. 239–276.

Hirschmann, M. M., Renne, P. R., and McBirney, A. R., 1996, ^{40}Ar-^{39}Ar age of the Skaergaard intrusion [abs.]: EOS (Trans. Amer. Geophys. Union), v. 77, p. F822.

Holland, T. J. B., and Powell, R., 1998, An internally consistent thermodynamic data set for phases of petrological interest: Jour. Metamor. Geol., v. 16, p. 309–343.

Horrebow, N., 1752, Tilforladelige Efterretninger om Island: Copenhagen.

Iijima, A., 1978, Geological occurrences of zeolite in marine enviroments, *in* Sand, L. B., and Mumpton, F. A., eds., Natural zeolites: Oxford, UK, Pergamon Press, p. 175–198.

______, 1988, Application of zeolites to petroleum exploration, *in* Kalló, D., and Sherry, H. S., eds., Occurrence, properties and utilization of natural zeolites: Budapest, Akadémiai Kiadó, p. 29–37.

Illman, D. L., 1993, Researchers take up environmental challange at Hanford, Stark pollution problems at former nuclear production site call forth unprecedented technological effort that may stretch out for decades: Chem. Engineer. News, June 21, p. 9–21.

Jakobsson, S. P., and Moore, J. G., 1986, Hydrothermal minerals and alteration rates at Surtsey Volcano, Iceland: Geol. Soc. Amer. Bull., v. 97, p. 648–659.

Johnson, J. W., Oelkers, E. H., and Helgeson, H. C., 1992, SUPCRT92; a software package for calculating the standard molal thermodynamic properties of minerals, gases, aqueous species, and reactions from 1 to 5000 bar and 0 to 1000 degrees C: Computers & Geosciences, v. 18, p. 899–947.

Jørgensen, O., 1984, Zeolite zones in the basaltic lavas of the Faeroe Islands: Ann. Soc. Scientar. Faroensis, Suppl., v. 9, p. 71–91.

Kasai, T., Maeda, H., Matsui, K., Kurnia, D. F., Nakayama, N., and Mizota, T., 1994, Hydration enthalpy for synthetic and cation exchanged A-type zeolites with special reference to zeolite heat pump media: Mineral. Jour., v. 17, p. 170–180.

Kastner, M., 1979, Zeolites, *in* Burns, R. G., ed., Marine minerals: Mineral. Soc. Amer. Rev. Mineral., v. 6, p. 111–122.

Keith, T. E. C., Thompson, J. M., and Mays, R. E., 1983, Selective concentration of cesium in analcime during hydrothermal alteration, Yellowstone National Park, Wyoming: Geochim. et Cosmochim. Acta, v. 47, p. 795–804.

Kisch, H. J., 1987, Correlation between indicators of very low grade metamorphism, *in* Frey, M., ed., Low temperature metamorphism: Glasgow, Blackie, 349 p.

Kononov, V. I., 1978, Subsurface chemical discharge in Iceland and its geologic significance: Transactions, v. 233, p. 212–214.

Kostov, I., 1969, Zoning in the development of the volcanogenic zeolites: Neues Jahrb. Mineral. Abhand., v. 111, p. 264–278.

Korzhinskii, D. S., 1959, Physicochemical basis of the analysis of the paragenesis of minerals: New York, Consultants Bureau, Inc., 142 p.

Krauskopf, K. B., and Bird, D. K., 1995, Introduction to geochemistry: New York, McGraw-Hill, 647 p.

Kristmannsdóttir, H., 1976, Types of clay minerals in hydrothermally altered basaltic rocks, Reykjanes, Iceland: Joekull, v. 26, p. 30–39.

Kristmannsdóttir, H., and Tómasson, J., 1978, Zeolite zones in geothermal areas in Iceland, *in* Sand, L. B., and Mumpton, F. A., eds., Natural zeolites: Oxford, UK, Pergamon Press, p. 277–284.

Lane, D. L., Rawson, S. A., Allen, C. C., and Burnell, J. R., 1989, Coupled transport and chemical reactions in basalt-groundwater flow-through experiments 1: Alteration phase distributions and theoretical considerations: Chem. Geol., v. 76, p. 327–340.

Larsen, H. C., and Marcussen, C., 1992, Sill-intrusion, flood basalt emplacement and deep crustal structure of the Scoresby Sund region, East Greenland, *in* Storey, B. C., et al., eds., Magmatism and the causes of continental break-up: Geol. Soc. Lond. Spec. Publ. 68, p. 365–386.

Larsen, L. M., Watt, W. S., and Watt, M., 1989, Geology and petrology of the lower Tertiary plateau basalts of the Scoresby Sund region, East Greenland: Grøn. Geol. Undersog. Bull. 157, 164 p.

Liou, J. G., 1970, Synthesis and stability relation of wairakite $CaAl_2Si_4O_{12}*2H_2O$: Contrib. Mineral. Petrol., v. 27, p. 259-282.

______, 1971a, P-T stabilities of laumontite, wairakite, lawsonite, and related minerals in the system $CaAl_2Si_2O_8$-SiO_2-H_2O: Jour. Petrol., v. 12, p. 379–411.

______, 1971b, Stilbite-laumontite equilibrium: Contrib. Mineral. Petrol., v. 31, p. 171–177.

Liou, J. G., de Capitani, C., and Frey, M., 1991, Zeolite equilibria in the system $CaAl_2Si_2O_8$-$NaAlSi_3O_8$-SiO_2-H_2O: New Zealand Jour. Geol. Geophys., v. 34, p. 293–301.

Liou, J. G., Maruyama, S., and Cho, M., 1985, Phase equilibria and mineral parageneses of metabasites in low-grade metamorphism: Mineral. Mag., v. 49, p. 321–333.

Lysak, S. V., 1992, Heat flow variations in continental rifts: Tectonophys., v. 208, p. 309–323.

Manning, C. E., and Bird, D. K., 1991, Porosity evolution and fluid flow in the basalts of the Skaergaard magma-hydrothermal system, East Greenland: Amer. Jour. Sci., v. 291, p. 201–257.

Manning, C. E., and Ingebritsen, S. E., 1999, Permeability of the continental crust: Implications of geothermal data and metamorphic systems: Rev. Geophys., v. 37, p. 127–150.

Manning, C. E., Ingebritsen, S. E., and Bird, D. K., 1993, Missing mineral zones in contact metamorphosed basalts: Amer. Jour. Sci., v. 293, p. 894–938.

Mason, B., and Moore, C. B., 1982, Principles of geochemistry: New York, John Wiley & Sons, 344 p.

McCurry, M., Estes, M., Fromm, J., Welhan, J. A., and Barrash, W., 1994, Three-dimensional chemical structure of the INEL aquifer system near the Idaho Chemical Processing Plant: Proc. Symp. on Engineering Geology and Geotechnical Engineering, v. 30, p. 207–219.

Meier, W. M., 1988, Atlas of zeolite structure types: Guilford, UK, Butterworth, 155 p.

Merino, E., 1975a, Diagenesis in Tertiary sandstones from Kettleman North Dome, California; I, Diagenetic mineralogy: Jour. Sediment. Petrol., v. 45, p. 320–336.

______, 1975b, Diagenesis in Tertiary sandstones from Kettleman North Dome, California; II, Interstitial solutions; distribution of aqueous species at 100 degrees C and chemical relation to the diagenetic mineralogy: Geochim. et Cosmochim. Acta, v. 39, p. 1629–1645.

Miyashiro, A., and Shido, F., 1970, Progressive metamorphism in zeolite assemblages: Lithos, v. 3, p. 251–260.

Moore, C. H., Ortoleva, P., and Chadam, J., 1986, Computer modeling of the formation of zeolite zones in altered basalts due to transport/reaction coupling under isothermal conditions, *in* Hitchon, B. C., ed., Fifth Int. Symp. on Water-Rock Interaction, Reykjavik: Rotterdam, Balkema, p. 501–502.

Morse, J. W., and Casey, W. H., 1988, Ostwald processes and mineral paragenesis in sediments: Amer. Jour. Sci., v. 288, p. 537–560.

Mortier, W. J., 1982, Compilation of extraframework sites in zeolites: Guilford, UK, Butterworth, 67 p.

Mumpton, F. A., 1977a, Utilization of natural zeolites, *in* Mumpton, F.A., ed., Mineralogy and geology of natural zeolites: Mineral. Soc. Amer. Rev. Mineral., v. 4, p. 177–204.

______, 1977b, Natural zeolites, *in* Mumpton, F. A., ed., Mineralogy and geology of natural zeolites: Mineral. Soc. Amer. Rev. Mineral., v. 4, p. 1–17.

______, 1988, Development of uses for natural zeolites: A critical commentary, *in* Kalló, D., and Sherry, H. S., eds., Occurrence, properties and utilization of natural zeolites: Budapest, Akadémiai Kiadó, p. 333–366.

Murata, K. J., Formoso, M. L. L., and Roisenberg, A., 1987, Distribution of zeolites in lavas of southeastern Parana Basin, state of Rio Grande do Sul, Brazil: Jour. Geol., v. 95, p. 455–467.

Murphy, W. M., Pabalan, R. T., Prikryl, J. D., and Goulet, C. J., 1996, Reaction kinetics and thermodynamics of aqueous dissolution and growth of analcime and Na-clinoptilolite at 25 degrees C: Amer. Jour. Sci., v. 296, p. 128–186.

Neuhoff, P. S., 1999, Thermodynamic properties and parageneses of rock-forming zeolites: Unpubl. Ph.D. thesis, Stanford Univ., Stanford, California.

Neuhoff, P. S., and Bird, D. K., 1994, Theoretical analysis of laumontite stability in organic acid-rich fluids [abs.]: Geol. Soc. Amer. Abs. Prog.. v. 26, p. A278.

______, 1997, Mineralogic constraints on aquifer geochemistry during regional zeolite facies metamorphism of Icelandic basalts [abs.]: Geol. Soc. Amer. Abs. Prog.. v. 29, p. V449.

______, 1999, Partial dehydration of laumontite: Thermodynamic constraints and petrogenetic implications [abs.]: Geol. Soc. Amer. Abs. Prog., v. 31, p. A221.

Neuhoff, P. S., Bird, D. K., and O'Day, P. A., 1995, Strontium and barium incorporation into natural heulandites during basalt-meteoric water interaction, eastern Iceland [abs.]: Geol. Soc. Amer. Abs. Prog.. v. 27, p. A253.

Neuhoff, P. S., Fridriksson, Th., Arnórsson, S., and Bird, D. K., 1999, Porosity changes and mineral paragenesis during low-grade metamorphism at Teigarhorn, eastern Iceland: Amer. Jour. Sci., v. 299, p. 467–501.

Neuhoff, P. S., Watt, W. S., Bird, D. K., and Pedersen, A. K., 1997, Timing and structural relations of regional zeolite zones in basalts of the East Greenland continental margin: Geology, v. 25, p. 803–806.

Nielsen, T. F. D., 1978, The Tertiary dike swarms of the Kangerdlugssuaq area, East Greenland; An example of magmatic development during continental break-up: Contrib. Mineral. Petrol., v. 67, p. 63–78.

Noh, J. H., and Boles, J. R., 1993, Origin of zeolite cements in the Miocene sandstones, North Tejon oil fields, California: Jour. Sediment. Petrol., v. 63, p. 248–260.

Ostwald, W. Z., 1897, Studien über die Bildung und Umwaldung fester Korper. 1. Abhandlung: Überstattigung und Überkaltung: Zeit. für Physikalische Chemie, v. 22, p. 289–330.

Ott, G., Karotke, E., and Althaus, E., 1986, Measurement of compressibility and high-pressure thermal expansion of minerals with a helium-pressurised heated X-ray camera: High Temperatures. High Pressures, v. 16, p. 549–551.

Oxburgh, E. R., and Agrell, S. O., 1982, Thermal conductivity and temperature structure of the Reydarfjordur borehole: Jour. Geophys. Res., B, v. 87, p. 6423–6428.

Pabalan, R. T., 1994, Thermodynamics of ion exchange between clinoptilolite and aqueous solutions of Na^+/K^+ and Na^+/Ca^{2+}: Geochim. et Cosmochim. Acta, v. 58, p. 4573–4590.

Pabalan, R. T., and Bertetti, F. P., 1999, Experimental and modeling study of ion exchange between aqueous solutions and the zeolite mineral clinoptilolite: Jour. Solution Chem., v. 28, p. 367–393.

Pálmason, G., 1981, Crustal rifting, and related thermomechanical processes in the lithosphere beneath Iceland: Geol. Rundschau, v. 70, p. 244–260.

Park, K., George, M. J., Miyake, Y., Saruhashi, K., Katsauragi, Y., and Kanazawa, T., 1965, Strontium-90 and cesium-137 in Columbia River plume, July 1964: Nature, v. 208, p. 1084–1085.

Pedersen, A. K., Watt, M., Watt, W. S., Larsen, L. M., and Kusznir, N. J., 1997, Structure and stratigraphy of the early Tertiary basalts of the Blosseville Kyst, East Greenland, *in* Roberts, A. M. P., ed., Symposium on geological processes at passive margins: Jour. Geol. Soc. Lond., v. 154, p. 565–570.

Petrova, N., Filizova, L., and Kirov, G., 1997, Calorimetric study of ion-exchange on clinoptilolite and mordenite at different temperatures, *in* Kirov, G., Filizova, L., and Petrov, O., eds., Natural zeolites—Sofia '95: Sofia, PENSOFT Publ., p. 173-180.

Phillips, B. L., and Kirkpatrick, R. J., 1994, Short-range Si-Al order in leucite and analcime; determination of the configurational entropy from ^{27}Al and variable-temperature ^{29}Si NMR spectroscopy of leucite, its Cs- and Rb-exchanged derivatives, and analcime: Amer. Mineral., v. 79, p. 1025–1031.

Ragnarsdóttir, K. V., 1993, Dissolution kinetics of heulandite at pH 2-12 and 25 degrees C: Geochim. et Cosmochim. Acta, v. 57, p. 2439–2449.

Rao, A., and Rees, L. V. C., 1966, Kinetics of ion exchange in mordenite: Trans. Faraday Soc., v. 62, p. 2505–2512.

Rengarten, N. V., 1950, Laumontite and analcime from Lower Jurassic deposits in the North Caucasus: Dokl. Akad. Nauk SSSR, v. 70, p. 485–488 (in Russian).

Robert, C., and Goffé, B., 1993, Zeolitization of basalts in subaqueous freshwater settings; field observations and experimental study: Geochim. et Cosmochim. Acta, v. 57, p. 3597–3612.

Ross, M., Flohr, M. J. K., and Ross, D. R., 1992, Crystalline solution series and order-disorder within the natrolite mineral group: Amer. Mineral., v. 77, p. 685–703.

Saemundsson, K., 1980, Outline of the geology of Iceland: Joekull, v. 29, p. 7–28.

Savage, D., Cave, M. R., Haigh, D., Milodowski, A. E., and Young, M. E., 1993, The reaction kinetics of laumontite under hydrothermal conditions: Eur Jour. Mineral., v. 5, p. 523–535.

Schiffman, P., Elders, W. A., Williams, A. E., McDowell, S. D., and Bird, D. K., 1984, Active metasomatism in the Cerro Prieto geothermal system, Baja California, Mexico; a telescoped low-pressure, low-temperature metamorphic facies series: Geology, v. 12, p. 12–15.

Schiffman, P., and Fridleifsson, G. Ó., 1991, The smectite-chlorite transition in drillhole NJ-15, Nesjavellir geothermal field, Iceland; XRD, BSE and electron micro probe investigations: Jour. Metamor. Geol., v. 9, p. 679–696.

Schmidt, S. T., 1993, Regional and local patterns of low-grade metamorphism in the North Shore Volcanic Group, Minnesota, USA: Jour. Metamor. Geol., v. 11, p. 401–414.

Seki, Y., 1961, Pumpellyite in low-grade regional metamorphism: Jour. Petrol., v. 2, p. 407–423.

Senderov, E. E., 1973, Effect of CO_2 on the stability of laumontite: Geochem. Int., v. 10, p. 139–147.

______, 1983, Metastable systems in zeolite formation: Geochem. Int., v. 20, p. 66–74.

Sherry, H. S., 1969, The ion-exchange properties of zeolites, *in* Marinski, J. A., and Dekker, M., eds., Ion exchange: New York, Marcel Dekker, p. 89–133.

Shibue, Y., 1981, Cation-exchange reactions of siliceous and aluminous phillipsites: Clays and Clay Minerals, v. 29, p. 397–402.

Shock, E. L., 1988, Organic acid metastability in sedimentary basins: Geology, v. 16, p. 886–890.

Smyth, J. R., 1982, Zeolite stability constraints on radioactive waste isolation in zeolite-bearing volcanic rocks: Jour. Geol., v. 90, p. 195–201.

Spear, F. S., 1993, Metamorphic phase equilibria and pressure-temperature-time paths: Washington, DC, Mineral. Soc. Amer., 799 p.

Ståhl, K., and Hanson, J., 1994, Real-time X-ray synchrotron powder diffraction studies of the dehydration processes in scolecite and mesolite: Jour. Appl. Crystallog., v. 27, p. 543–550.

Ståhl, K., Artioli, G., and Hanson, J. C., 1996, The dehydration process in the zeolite laumontite: A real-time synchrotron X-ray-powder diffraction study: Phys. Chem. Minerals, v. 23, p. 328–336.

Stoiber, R. E., and Davidson, E. S., 1959, Amygdule mineral zoning in the Portage Lake Lava series, Michigan copper district; Part 1: Econ. Geol., v. 54, p. 1250–1277.

Sturchio, N. C., Bohlke, J. K., and Binz, C. M., 1989, Radium-thorium disequilibrium and zeolite-water ion exchange in a Yellowstone hydrothermal environment: Geochim. et Cosmochim. Acta, v. 53, p. 1025–1034.

Sukheswala, R. N., Avasia, R. K., and Gangopadhyay, M., 1974, Zeolites and associated secondary minerals in the Deccan Traps of West India: Mineral. Mag., v. 39, p. 658–671.

Surdam, R. C., 1977, Zeolites in closed hydrologic systems, *in* Mumpton, F. A., ed., Mineralogy and geology of natural zeolites: Mineral. Soc. Amer. Rev. Mineral., v. 4, p. 65–91.

Surdam, R. C., and Boles, J. R., 1979, Diagenesis of volcanic sandstones: Soc. Econ. Paleontol. Mineral. Spec. Publ. no. 26, p. 277–342.

Surdam, R. C., Boese, S. W., and Crossey, L. J., 1984, The chemistry of secondary porosity: Clastic Diagenesis, v. 37, p. 127–149.

Taylor, H. P., Jr., and Forester, R. W., 1979, An oxygen and hydrogen isotope study of the Skaergaard Intrusion and its country rocks; a description of a 55-m.y. old fossil hydrothermal system: Jour. Petrol., v. 20, p. 355–419.

Thompson, A. B., 1970, Laumontite equilibria and the zeolite facies: Amer. Jour. Sci., v. 269, p. 267–275.

Tissot, B. P., and Welte, D. H., 1984, Petroleum formation and occurrence: Berlin, Springer Verlag, 699 p.

Tschernich, R. W., 1992, Zeolites of the world: Phoenix, AZ, Geoscience Press, 563 p.

Tsitsishvili, G. V., Andronikashvili, T. G., Kirov, G. N., and Filizova, L. D., 1992, Natural zeolites: Chichester, UK, Ellis Horwood, 295 p.

Upton, B. G. J., 1988, History of Tertiary igneous activity in the N Atlantic borderlands, *in* Morton, A. C., and Parson, L. M., eds., Early Tertiary volcanism and the opening of the NE Atlantic: Geol. Soc. Lond. Spec. Publ., no. 39, p. 429–453.

van Reeuwijk, L. P., 1974, The thermal dehydration of natural zeolites: Wageningen, The Netherlands, H. Veenman and Zonen B.V., 88 p.

van Santen, R. A., 1984, The Ostwald Step Rule: Jour. Phys. Chem., v. 88, p. 5768–5769.

Wager, L. R., 1947, The stratigraphy and tectonics of Knud Rasmussens Land and the Kangerdlugssuaq region, Part 4, *in* Geological investigations in East Greenland: Copenhagen, Nyt Nordisk Forlag, 64 p.

Wager, L. R., and Deer, W. A., 1938, A dike swarm and crustal flexure in east Greenland: Geol. Mag., v. 75, p. 39–46.

Walker, G. P. L., 1951, The amygdale minerals in the Tertiary lavas of Ireland; I, The distribution of chabazite habits and zeolites in the Garron plateau area, County Antrim: Mineral. Mag., v. 29, p. 773–791.

______, 1959a, Geology of the Reydarfjoerdur area, eastern Iceland: Quart. Jour. Geol. Soc. Lond., v. 114, p. 367–393.

______, 1959b, Some observations on the Antrim basalts and associated dolerite intrusions: Proc. Geol. Assoc., v. 70, p. 179–205.

______, 1960a, The amygdale minerals in the Tertiary lavas of Ireland; III, Regional distribution: Mineral. Mag., v. 32, p. 503–527.

______, 1960b, Zeolite zones and dike distribution in relation to the structure of the basalts of eastern Iceland: Jour. Geol., v. 68, p. 515-528.

Watkins, N. D., and Walker, G. P. L., 1977, Magnetostratigraphy of eastern Iceland: Amer. Jour. Sci., v. 277, p. 513–584.

Wedepohl, K. H., 1969, Composition and abundance of common igneous rocks: New York, Springer, p. 227–249.

White, A. F., Claasen, H. C., and Benson, L. V., 1980, The effect of glass dissolution on the water chemistry in a tuffaceous aquifer, Ranier Mesa, Nevada: U. S. Geol. Surv. Water Supply Paper 1535-Q, 30 p.

White, D. A., Nattkemper, A., and Rautiu, R., 1999, Evaluation of the effect of temperature on potassium clinoptilolite ion exchange: Nuclear Technol., v. 127, p. 212–217.

Wilkin, R. T., and Barnes, H. L., 1998, Solubility and stability of zeolites in aqueous solution; I, Analcime, Na-, and K-clinoptilolite: Amer. Mineral., v. 83, p. 746–761.

Wolery, T. J., 1991, EQ3NR. A computer program for geochemical aqueous speciation solubility calculations: Theoretical manual, user's guide and related documentation (Version 7.0). Livermore, CA, Lawrence Livermore National Laboratory.

Wood, D. A., 1977, A quantitative crystal fractionation model for silicic lavas in Iceland [abs.]: EOS (Trans. Amer. Geophys. Union), v. 58, p. 529.

Wood, D. A., Gibson, I. L., and Thompson, R. N., 1976, Elemental mobility during zeolite facies metamorphism of the Tertiary basalts of eastern Iceland: Contrib. Mineral. Petrol., v. 55, p. 241–254.

Wyckoff, R. W. G., and Doberenz, A. R., 1968, The strontium content of fossil teeth and bones: Geochim. et Cosmochim. Acta, v. 32, p. 109–115.

Yaalon, D. H., Brenner, I., and Koyumdjisky, H., 1974, Weathering and mobility sequence of minor elements on a basaltic pedomorphic surface, Galilee, Israel: Geoderma, v. 12, p. 233–244.

Yang, P., Stoltz, J., Armbruster, T., and Gunter, M., 1997, Na, K, Rb, and Cs exchange in heulandite single-crystals: Diffusion kinetics: Amer. Mineral., v. 82, p. 517–525.

Yoder, H. S., Jr., and Weir, C. E., 1960, High-pressure form of analcite and free energy change with pressure of analcite reactions: Amer. Jour. Sci., v. 258-A, p. 420–433.

Zen, E. A., 1961, The zeolite facies; an interpretation: Amer. Jour. Sci., v. 259, p. 401–409.

The Dolomite Problem Revisited[1]

HEINRICH D. HOLLAND AND HEIDE ZIMMERMANN

Department of Earth and Planetary Sciences, Harvard University, Cambridge, Massachusetts 02138

Abstract

A reassessment of the abundance of dolomite in carbonate sediments has confirmed that carbonates deposited during the past 150 Ma contain, on average, less dolomite than Proterozoic and Paleozoic carbonates. The lower dolomite content of the more recent carbonate sediments results from the increase in the deposition of $CaCO_3$ in deep-sea sediments, and to the difficulty of dolomitizing deep-sea $CaCO_3$ by reaction with cold, unevaporated seawater. The decrease in the rate of dolomite formation during the past 150 Ma has led to an increase in the output of oceanic Mg^{+2} by the reaction of seawater with clay minerals and with ocean-floor basalts. The increase in the output of marine Mg^{+2} into these reservoirs has been brought about by an increase in the Mg^{+2} concentration of seawater. During the past 40 Ma, the concentration of Mg^{+2} in seawater has probably increased by ~18 mmol/kg, and probably has been accompanied by an equimolar increase in the concentration of SO_4^{-2}.

The Dolomite Problem

IN HIS CLASSIC PAPER of 1909, Daly regretted the incompleteness of the 900 samples of carbonate rocks which he used to estimate the time variation of the ratio of calcium to magnesium in carbonate sediments, but he felt that the number was sufficient "to give a fairly accurate idea of the variation of the ratio through geologic time," and that the labor of searching for additional analyses of Tertiary and more recent limestones "did not seem necessary, since it is well known that these later limestones are usually very low in magnesium" (Daly, 1909, p. 164). The number of analyses of carbonate sediments has increased a great deal since 1909. A decrease in the dolomite/limestone ratio since the Triassic has been supported by the new data, but the fine structure of this ratio during the Phanerozoic, and even more so during the Precambrian, is still poorly defined. This poor definition is inherent in the problem of determining the percentage of dolomite in carbonate sediments on a worldwide basis. A very large number of analyses is required to define the percentage of dolomite in even a single carbonate province. The most extensive study of such a province is that of Vinogradov and Ronov (1956) for the distribution of dolomite and limestone on the Russian platform. Figure lA is a plot of their data for their "previously published data" and for their "documented samples." The agreement between the two sets of data is reasonable for most periods, but not for the Permian or the Jurassic. More than 8000 specimens are represented in each data set. The Jurassic is the period for which the sampling was least extensive.

Figure 2 combines the Vinogradov and Ronov (1956) data with the results of Sochava and Podkovyrov (1995) for the compositional evolution of Meso- and Neoproterozoic carbonate rocks in the Urals and in Siberia. Their study is based on about 1200 analyses of Riphean and Vendian sections. The results of the two studies are broadly consistent and complementary. They indicate that the dolomite content of this large carbonate province was high during much of the Proterozoic and Paleozoic, and that the dolomite content of Mesozoic and Cenozoic carbonates is minor. The interpretation of variations in the dolomite content of carbonates from period to period and between the lower, middle, and upper Riphean is still problematic.

Figure 1B summarizes the Phanerozoic data of Daly (1909) and those of Schmoker et al. (1985) for North America. The basic pattern of their data is similar to that of Vinogradov and Ronov (1956) in Figure 1A, but there are significant differences. These may be real, or perhaps result from uncertainties in the sampling. Daly's (1909) compilation contains less than 5% of the number of samples in Vinogradov and Ronov's (1956) data set. Schmoker et al. (1985) based their curve on the 993 limestone formations and 297 dolomite formations in the Geonames data file maintained by the U.S. Geological Survey. The volumetric ratio of limestones and dolomites was estimated in two ways: (1) the number of limestone and dolomite formations was assumed to be proportional to the relative volumes of the two

[1]This paper is dedicated with affection to Konnie Krauskopf on the occasion of a Symposium in his honor.

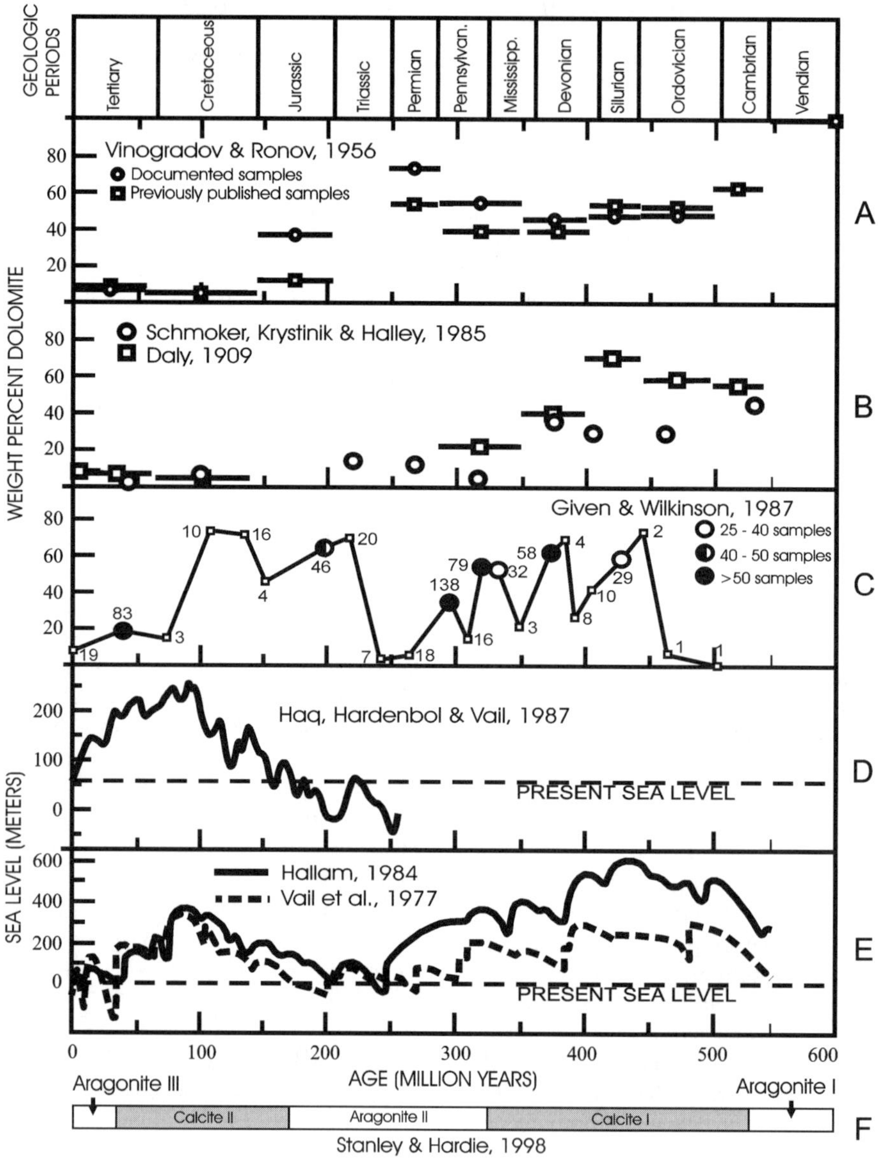

FIG. 1. A. The percentage of dolomite in Phanerozoic carbonate sediments of the Russian Platform (Vinogradov and Ronov, 1956), based on 8351 previously published analyses and 8847 specimens in their analyses of documented samples. B. The percentage of dolomite in Phanerozoic carbonate sediments of the North American Platform in the compilation of 900 analyses by Daly (1909) and the compilation of Schmoker et al. (1985) based on 1290 limestone and dolomite formations. C. Percentage of dolomite in Phanerozoic carbonate sediments (Given and Wilkinson, 1987). The numbers indicate the number of analyses for each data point. D. Mean sea level during the Mesozoic and Cenozoic (Haq et al., 1987). E. Mean sea level during the Phanerozoic (Vail et al., 1977; Hallam, 1984). F. Secular variation in the mineralogy of Phanerozoic nonskeletal marine carbonates (Stanley and Hardie, 1998).

rock types, and (2) type-section thicknesses of formally named formations with primary lithologies of limestone were summed, and the total thickness of limestone and dolomite type sections was assumed to be proportional to the relative volume of the two rock types. Although Schmoker et al. (1985) consid-

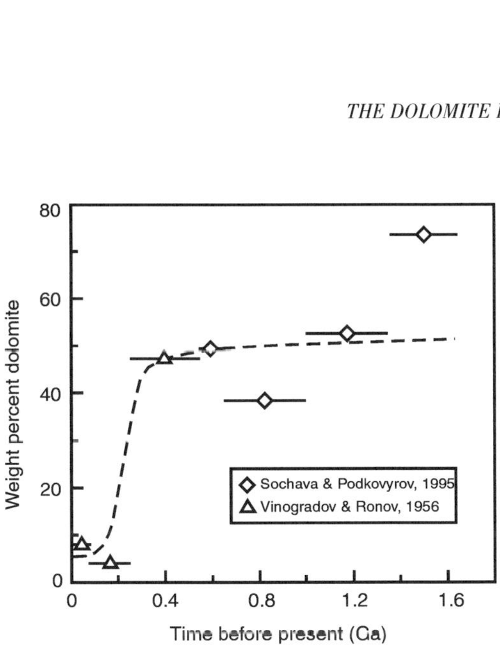

FIG. 2. The percentage of dolomite in carbonate sediments of the Russian platform during the last 1.6 Ga. The compilation of Sochava and Podkovyrov (1995) is based on ~1200 analyses. The three data points from the Vinogradov and Ronov (1956) compilation represent their averages for the Cenozoic, Mesozoic, and Paleozoic.

ered their results to be of sufficient accuracy to be useful, the dolomite abundances in Figure 1B can only be regarded as a rather rough measure of changes in the relative abundance of dolomite and limestone in North America during the Phanerozoic.

A very different variation of the percentage of dolomite in Phanerozoic carbonates was proposed by Given and Wilkinson (1987) (Fig. 1C), who suggested that the variation of dolomite abundance during the Phanerozoic is cyclical, roughly mimicking changes in the stand of sea level as shown in Figures 1D and IE (Vail et al., 1977; Hallam, 1984; Haq et al., 1987). Unfortunately, the database used by Given and Wilkinson (1987) is inadequate. Since their proposed dolomite abundance during the Cambrian and Ordovician is based on a single analysis for each period, their low-dolomite anchor at the beginning of the Phanerozoic is not meaningful (see Zenger, 1989). If one deletes all of the points in Figure 1C for which the total number of analyses is less than 25, the pattern of dolomite abundance during the Phanerozoic agrees reasonably well with that of Figures 1A and 1B except for the lower Jurassic. This final discrepancy disappears if one deletes all of the points in Figure 1C for which the total number of analyses is less than 50. The differences between Figures 1B and 1C are therefore largely, if not solely, caused by the small number of samples in Given and Wilkinson's (1987) database.

The rough agreement between the North American and Eurasian data sets is heartening, but an average of the two does not necessarily define the worldwide change in dolomite abundance during the Phanerozoic. An average of the two data sets suffers from at least two major omissions—data for the dolomite content of the extensive Mesozoic and Cenozoic carbonate platforms that border the margins of several continents, and data for deep-sea carbonates. Unfortunately, the available data for the dolomite content of carbonate platforms formed during the last 150 Ma are quite incomplete. Even if the analytical data in oil company records were released, it would be difficult to make more than a rough estimate of their dolomite/limestone ratio. The most intensively studied of the platforms is probably the Florida-Bahamas Platform Complex. This was part of one of the most extensive carbonate systems in the Earth's geologic history, extending nearly 7000 km along the eastern North American continental margin. The geology of the Cretaceous to Oligocene sediments of the platform has been summarized by Randazzo (1997), and the geology of the Miocene to Holocene sediments by Scott (1997). The Mesozoic and Paleogene sediments contain extensive evaporites. Many of the carbonate sediments are dolomites or partially to pervasively dolomitized limestones. In north-central Florida near the crest of the Ocala Platform, the Paleocene and Eocene sections consist of 50 to 75% dolostone and some of the limestones are dolomitic (T. M. Scott, pers. comm., 1999). The Miocene Hawthorn Group sediments consist largely of dolostone/dolomite in the northern two-thirds of the Florida Peninsula. There is very little dolomite in the Plio-Pleistocene section. No reliable estimates exist for the dolomite/limestone ratio of the bank as a whole. However, there is no reason to believe that it is significantly less than that of the Paleozoic carbonate sediments in the Vinogradov-Ronov (1956) compilation.

Dolomite also is a common component of the late Cenozoic sediments of the Bahamas carbonate platforms (Supko, 1977; Vahrenkamp et al., 1991), but there is at present little published information for the composition of the deeper parts of these platforms. In the Tertiary of Libya, some cores consist entirely of dolomite; an informal estimate places the dolomite content of these carbonates at 50 ± 30% (C. Pendexter, pers. commun., 1999). If the dolomite content of these carbonate sediments is representative of Mesozoic and Cenozoic carbonates as a whole, the estimates of the dolomite content of

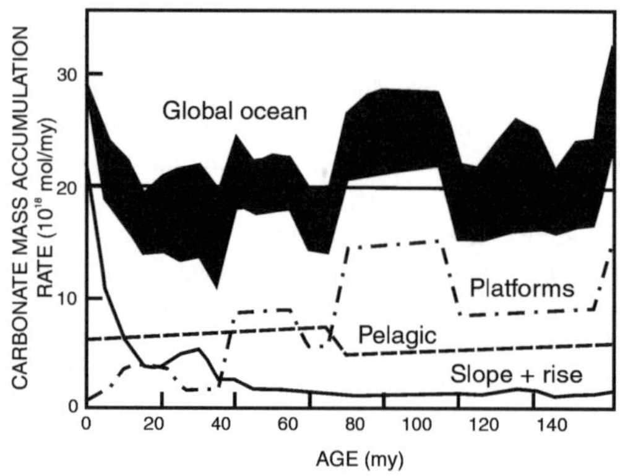

FIG. 3. The accumulation rate of carbonates on platforms, slopes and rises, in the deep ocean, and in the oceans as a whole (Opdyke and Wilkinson, 1998).

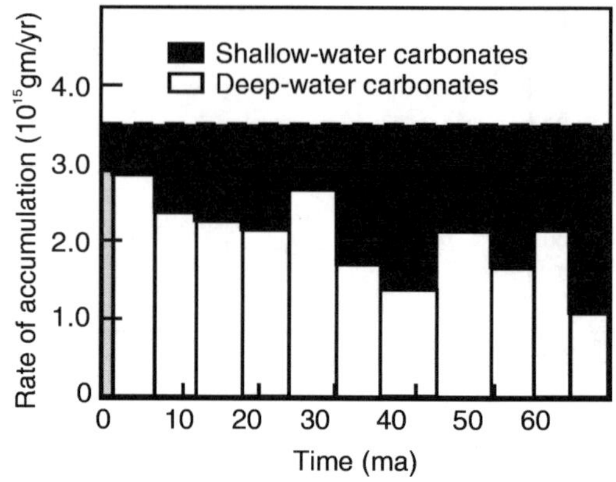

FIG. 4. The rate of accumulation of shallow- and deep-water carbonates during the past 60 Ma (Hay, 1994).

Mesozoic and Cenozoic carbonates by Daly (1909), Vinogradov and Ronov (1956), and Schmoker et al. (1985) are much too low. There are, however, compelling reasons to believe that the dolomite content of platform carbonates is not representative of the dolomite content of all carbonate sediments deposited during the past 150 Ma. Prior to the evolution of foraminifera and coccoliths, carbonate production in the open ocean almost certainly was minimal. Since then, a progressively larger percentage of carbonate sediments has been deposited on the ocean floor. Dolomitization of these carbonates has been minor. Lumsden (1985) found only 108 samples in a total of 50,000 DSDP deep-marine sediment samples that contained 5% or more dolomite. These samples were taken from 127 sites in the Atlantic, the Indo-Pacific, the Gulf-Caribbean, the Mediterranean, the Red Sea, and the Black Sea. Dolomitization of deep-sea sediments seems to be extensive only where sulfate in interstitial water is removed by bacterial reduction to sulfide in the presence of large amounts of organic matter (Garrison et al., 1984). The association of deep-sea carbonates with large amounts of organic matter is currently restricted to a few fairly near-shore settings, and the dolomite produced in these areas does not increase significantly the percentage of dolomite in deep-sea carbonates as a whole.

Opdyke and Wilkinson (1988) (see Fig. 3) and Hay (1994) (see Fig. 4) have published estimates of the mass accumulation rate of carbonate sediments in the global oceans, in pelagic settings, on platforms, and in slope-rise areas. Like all such compilations, those in Figures 3 and 4 are only semiquantitative. It is clear, however, that during the past 100 Ma the percentage of carbonates deposited in the deeper parts of the oceans has increased greatly at the expense of the percentage of shallow-water carbonates. Hay (1994) estimated the fraction of shallow-water carbonates today to be roughly 17% of the total. If these carbonates are half dolomite, half limestone, and if the dolomite content of deep-sea carbonates is ~2%, the overall dolomite content of modern carbonate sediments is ~10%. This figure, which is probably somewhat of an overestimate, is much less than the average dolomite content of Proterozoic and Paleozoic carbonates. The original proposition by Daly (1909) therefore seems to be correct: the dolomite content of carbonate sediments has decreased considerably since the Paleozoic; but his regrets regarding the incompleteness of his data set were justified. Mesozoic and Cenozoic carbonate platforms were very much underrepresented, and their dolomite content has turned out to be much greater than that of the carbonates in his collection. This source of error in Daly's (1909) compilation has, however, been more than offset by the very low dolomite content of deep-sea carbonates, which were virtually unknown in 1909.

Changes in Dolomite Abundance and the Composition of Seawater

The simplest explanation for the decline in dolomite abundance during the last ~150 Ma is the increase in the deposition of deep-sea carbonates, which are difficult to dolomitize by reaction with cold, unevaporated seawater. Other explanations are unlikely to be correct. It could be argued that the Mg^{+2}/Ca^{+2} ratio of average river water has dropped during the past 150 Ma, and that not enough Mg^{+2} is currently added to the oceans to generate more than a small percentage of dolomite in marine carbonates. However, Meybeck's (1988) average world river water contains 0.15 mmol/kg Mg^{+2} and 0.35 mmol/kg of Ca^{+2}. If all of this Mg^{+2} and Ca^{+2} were removed from the oceans as constituents of dolomite and calcite, dolomite would account for ~50 wt% of the total carbonate output of the oceans, a percentage comparable to that of Pre-Mesozoic carbonate sediments, and much greater than the apparent percentage of dolomite in Cenozoic carbonate sediments as a whole.

A second potential explanation for the low dolomite content of Cenozoic carbonates proposes that the CO_2 input to the atmosphere has declined, and that not enough CO_2 is available to precipitate a significant fraction of the river input of Mg^{+2} to the oceans as a constituent of carbonate minerals (see for instance Holland, 1969). If this were the case, much of the Mg^{+2} brought to the oceans today would be removed as a constituent of Mg-silicates generated by reverse weathering reactions. Although reverse weathering reactions do occur (Mackenzie et al., 1981; Michalopoulos and Aller, 1995; Hover et al., 1999), most of the Mg^{+2} that is removed from the oceans with clay minerals is apparently balanced by the release of Ca^{+2} (see, for instance, Martin and Sayles, 1994). Most of the released Ca^{+2} is then precipitated as a constituent of calcite and aragonite. Such cation exchange reactions do not involve a change in the CO_2 balance of the atmosphere-ocean system. They simply replace the deposition of the $MgCO_3$ component of dolomite by the deposition of $CaCO_3$.

A third potential explanation rests on the proposition that much of the dolomite in carbonate sediments is much younger than the age of the enclosing sediments. However, the abundance of dolomite in Cenozoic carbonate platforms indicates that dolomite in young shallow-water carbonates can be formed syndepositionally or within a few million years of sedimentation. Late-burial dolomites seem to be restricted geographically. Their textures are clearly distinguishable from those of early diagenetic dolomite (Montañez and Reid, 1992; Malone et al., 1996), and syndepositional dolomite has been recognized as the precursor of many late-burial dolomites (Land et al., 1975; Zenger, 1981; Holail et al., 1988; Kupecz, 1989; Montañez, 1989; Gregg and Shelton, 1990). It is therefore unlikely that the distribution of dolomite in the geologic record has been strongly affected by the dolomitization of limestones long after sedimentation.

The imbalance in the Mg^{+2} budget of the ocean as a result of the decrease in dolomite abundance during the past 150 Ma has probably been restored by a gradual increase in the Mg^{+2} output of the oceans into non-carbonate sinks. Two of these sinks are of major importance: ocean-floor basalts and marine clays. Seawater that cycles through ocean-floor basalts at significantly elevated temperatures loses essentially all of its Mg^{+2} (e.g., Mottl and Wheat, 1994; Von Damm, 1995; Wheat and Mottl, 2000). Today, the seawater flux through mid-ocean ridges is ~ 3×10^{16} gm/yr (Mottl, 1994). This accounts for a Mg^{+2} loss of ~ 4×10^{13} gm/yr—i.e., ~35% of the river flux of Mg to the oceans. The flux of seawater through basalts on ridge flanks is not yet well defined (Mottl and Wheat, 1994) but could be very significant. Seawater also reacts with ocean-floor basalts at the interface with overlying sediments, but the loss of Mg^{+2} in these settings appears to be minor (McDuff, 1981).

The loss of Mg^{+2} by cation exchange with river clays is initially rapid (see for instance Drever, 1971 and Hover et al., 1999) and continues for many years, as shown by Mg^{+2} profiles in interstitial seawater in marine sediments (see for instance Martin and Sayles, 1994). The loss of Mg^{+2} by these reactions probably amounts to a significant fraction of the oceanic Mg^{+2} budget. It therefore is easy to account for a rate of Mg^{+2} loss from the oceans that is sufficient to balance the riverine Mg^{+2} input rate.

Qualitatively, the effect of a decrease in the dolomite output of the oceans on the chemistry of seawater is easy to predict. The Mg^{+2} content of seawater will rise until the rate of Mg^{+2} removal into ocean basalts and clay minerals has increased sufficiently to compensate for the decreased rate of Mg^{+2} removal as a constituent of dolomite. The SO_4^{-2} content of seawater will also rise. Dolomitization leads to the precipitation of gypsum and/or anhydrite in evaporitic settings via the reaction

$$2CaCO_3 + Mg^{+2} + SO_4^{-2} \rightarrow CaMg(CO_3)_2 + CaSO_4. \quad (1)$$

A decrease in dolomite formation in evaporitic settings, where most dolomite has almost certainly formed, therefore decreases the SO_4^{-2} output from the oceans and leads to an increase in the SO_4^{-2}

concentration of seawater. SO_4^{-2} loss during the cycling of seawater through ocean basalts at elevated temperatures, especially at mid-ocean ridges, and SO_4^{-2} reduction in sediments rich in organic matter therefore will increase until the marine sulfur cycle is again in balance. It is still difficult to predict the magnitude of the effects of a decrease in dolomite formation quantitatively, and we still have no samples of ancient seawater against which to check our predictions. Nevertheless, it seems worthwhile to construct a simple, first-order model for the system and to compare its predictions with the available data for the composition of seawater during the Phanerozoic.

The loss of the majority of Mg^{+2} from seawater is the sum of three terms

$$J = A + B + C, \quad (2)$$

where A is the loss of Mg^{+2} to dolomite, B the loss to ocean-floor basalts, and C the loss to clay minerals. The loss to dolomite can be parameterized to the form

$$A \approx k_A f(G)\, R_{SWC}\, m_{Mg^{+2}}, \quad (3)$$

where m_{Mg+2} is the Mg^{+2} concentration in seawater, R_{SWC} is the rate of shallow-water carbonate deposition, $f(G)$ is a function that depends on a variety of geographic variables, as well as on temperature and relative humidity, and k_A is a constant.

The loss of Mg^{+2} to ocean-floor basalts can be parameterized to the form

$$B \approx k_B\, R_{SFS}\, m_{Mg^{+2}}, \quad (4)$$

where R_{SFS} is the rate of seafloor spreading and k_B is a constant.

The loss to clays can be parameterized to the form

$$C \approx k_C\, m_{Mg^{+2}}, \quad (5)$$

where k_C is a constant.

In this simplified formulation the output of Mg^{+2} from the oceans is therefore

$$J = [k_A f(G) R_{SWC} + k_B R_{SFS} + k_C]\, m_{Mg+2} = a m_{Mg+2}. \quad (6)$$

The rate of change of the Mg^{+2} content of seawater is given by the expression

$$W_{SW} \frac{dm_{Mg^{+2}}}{dt} = I - J, \quad (7)$$

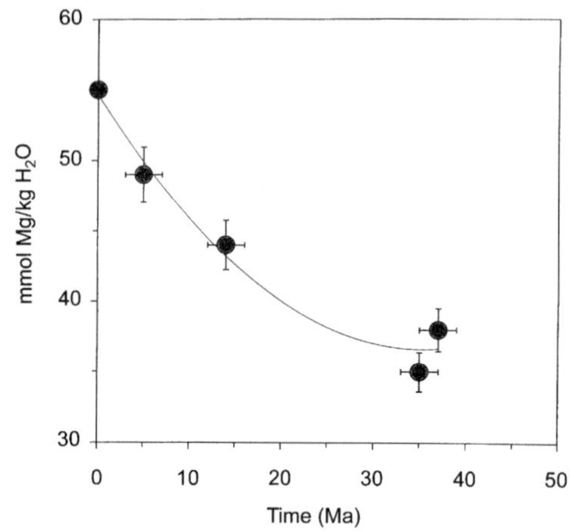

FIG. 5. The concentration of Mg^{+2} in seawater during the last 40 Ma as inferred from the composition of fluid inclusions in halite from marine evaporites (Zimmermann, 2000).

where W_{SW} is the mass of seawater and I is the rate of the river input of Mg^{+2} to the oceans. Inserting equation 6 into equation 7 and rearranging, one obtains

$$a = \frac{I - W_{SW} \frac{dm_{Mg^{+2}}}{dt}}{m_{Mg^{+2}}} \quad (8)$$

The value of the parameter "a" during the past 40 Ma can be calculated from the data in Figure 5, assuming that the river input of Mg^{+2} to the oceans has been equal to the present-day value of 6.1×10^{12} mol/yr. The data in Figure 5 are based on the composition of fluid inclusions in halite from four Tertiary marine evaporites (Zimmermann, 2000). These indicate that the Mg^{+2} concentration of seawater has increased by ~18 mmol/kg since the Eocene. The interpretation of these fluid-inclusion data is somewhat uncertain, because some of the observed Mg^{+2} depletion with respect to modern seawater may be the result of dolomitization within the evaporite basins. However, none of the four evaporites are associated with large carbonate platforms (Esteban, 1996; Garrett, 1996; Ortí, 1997; Sissingh, 1998), and it is unlikely that in-basin dolomitization accounts for a significant fraction of the observed depletion of Mg^{+2} in the inclusion brines with respect to modern seawater.

This observation and the simple, progressive increase in the Mg^{+2} concentration suggested by the

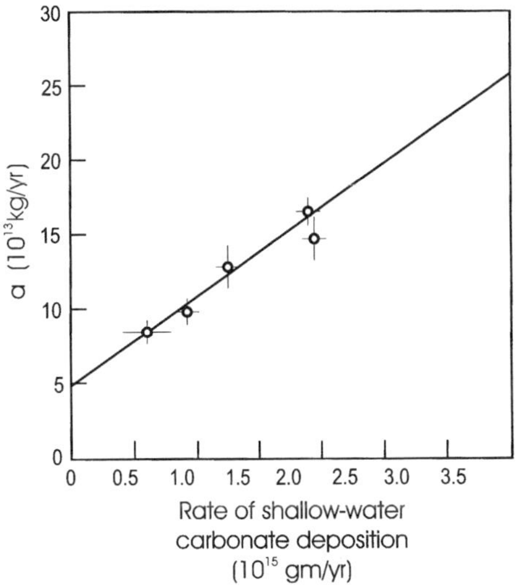

FIG. 6. The value of the parameter "a" in Equation 6 during the past 40 Ma, plotted against the rate of shallow-water carbonate accumulation.

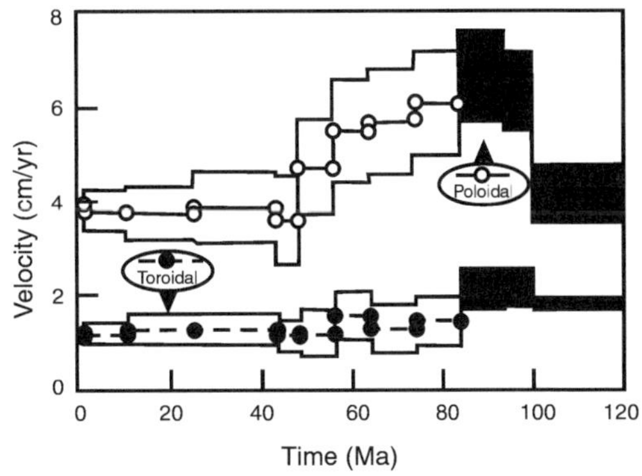

FIG. 7. Toroidal and poloidal rates of seafloor spreading during the past 120 Ma (Lithgow-Bertelloni et al. , 1993).

data in Figure 5 indicate that most of the differences between the composition of the inclusion brines reflect changes in the composition of seawater as a whole during the past 40 Ma. To test this notion, the value of the parameter "a" has been plotted in Figure 6 versus R_{SCW}, the rate of accumulation of marine carbonates deposited in shallow-water settings. The values of R_{SCW} were derived from Hay's (1994) data in Figure 4.

The linear relationship between the parameter "a" and the rate of shallow-water carbonate deposition is consistent with the simple model for the output of Mg^{+2} from the oceans. Since the rate of seafloor spreading seems to have been nearly constant during the past 40 Ma (see Fig. 7), a linear dependence of "a" on R_{SWC} is rather to be expected. The data in Figure 6 imply that if no shallow-water carbonate deposition were occurring—i.e., R_{SWC}= 0—the value of "a" would be 5.0×10^{13} yr^{-1}, and that if all carbonate deposition in the oceans were taking place in shallow-water settings, the value of "a" would be 23×10^{13} kg/yr. At steady state and at an Mg^{+2} input rate to the oceans equal to that of the present day, the Mg^{+2} concentration of seawater would then be 120 mmol/kg when $R_{SWC} = 0$, and 27 mmol/kg when $R_{SWC} = 3.5 \times 10^{15}$ gm/yr. This demonstrates the large effect that changes in R_{SWC} can exert on the Mg^{+2} concentration of seawater.

Before 40 Ma, the mineralogy and the composition of fluid inclusions in marine evaporites is well correlated with the stand of sea level (see Fig. 1). During the Cretaceous, Silurian, and Devonian, the sea-level stand was high and marine evaporites were greatly depleted in $MgSO_4$. During the Permian and Vendian, the sea-level stand was low, $MgSO_4$-rich evaporites were common, and the concentration of Mg^{+2} and SO_4^{-2} in fluid inclusions in halite was similar to those in evaporated modern seawater. The causes for this correlation are not completely clear. The stand of sea level is probably related to the rate of seafloor spreading. This is suggested by the data in Figure 7 but has been questioned by Heller et al. (1996). Inferences regarding the rate of seafloor spreading based solely on the stand of sea level prior to the Jurassic are therefore somewhat uncertain.

The stand of sea level also can affect the composition of seawater directly. High stands of sea level tend to be accompanied by extensive flooding of the continents. These shallow seas have often been sites of extensive carbonate deposition and frequently host large marine evaporites. Although the causes of the correlation between the apparent rate of seafloor spreading, the mineralogy of marine evaporites, the composition of fluid inclusions in halite from these evaporates, and changes in the mineralogy of marine carbonates during the Phanerozoic (Fig. 1F) (Sandberg, 1985; Stanley and Hardie, 1998, 1999) are not clear, it is unlikely that the correlation itself is fortuitous. We are finally approaching the time when we can quantify the compositional history of seawater during the Phanerozoic and relate that history to biological evolution and to the tectonic history of the Earth.

Conclusions

1. Previous compilations of the abundance of dolomite in Phanerozoic carbonates have suggested that dolomite became progressively less important as a constituent of these sediments during the Mesozoic and Cenozoic. This does not seem to be true for shallow-water carbonates. It does, however, seem to be true for carbonate sediments as a whole, because the dolomite content of deep-water carbonates, which have become progressively more important since the Jurassic, is very small.

2. The small dolomite content of deep-water carbonates results from the very slow rate of reaction of cold, unevaporated seawater with calcite and aragonite.

3. Much of the river input of Mg^{+2} to the oceans, which was formerly removed by the formation of dolomite, is almost certainly leaving the oceans by reaction with oceanic basalts and with marine clays.

4. This change has been brought about largely by an increase of ~18 mmol/kg in the Mg^{+2} concentration of seawater since the Eocene. The concentration of SO_4^{-2} also has increased during this period. The ratio $\Delta m_{Mg^{+2}}/\Delta m_{SO_4^{-2}}$ has probably been close to unity.

5. Between 40 and 150 Ma the Mg^{+2} and SO_4^{-2} concentration of seawater was affected both by changes in the worldwide rate of dolomite formation and by changes in the rate of seawater cycling through ocean-floor basalts.

6. The composition of seawater during the Phanerozoic has been related to the stand of sea level, because long-term changes in sea level appear to be a reflection, at least in part, of the rate of seafloor spreading, which has determined the rate of seawater cycling through ocean-floor basalts and the degree of flooding of the continents.

7. The chemistry of marine evaporites and the composition of fluid inclusions in halite of marine evaporites reflect both the composition of seawater as a whole and reactions of evaporating seawater with sediments in evaporite basins. In some instances, the effects of in-basin reactions are minor. The composition of fluid inclusions and the chemistry of the evaporites in such basins can be used to define the composition of the contemporary seawater.

Acknowledgments

This paper had its beginnings while one of us (HDH) was a Visiting Professor at the University of Hawaii. He benefitted greatly there from discussions with Ralph Moberly and Fred Mackenzie. Initially, the major problem of the paper was to define the time variation of the dolomite content of carbonate sediments. Disagreements with Fred Mackenzie and with Bruce Wilkinson were resolved, but only in part. More serious problems developed when copies of the first version of the paper were distributed to knowledgeable colleagues, including Alfred Fischer, who pointed out that the supposed time trend in dolomite abundance is "a huge error in sampling," because compilations of carbonate sediment compositions have not given sufficient weight to the dolomite content of the great carbonate platforms on continental margins. Correspondence and conversations with Bob Ginsburg, Tom Scott, Charles Pendexter, and Anthony Randazzo confirmed the validity of this criticism. Bill Hay, however, pointed out that the compilations that we had used for the dolomite content of Phanerozoic carbonate sediments were doubly in error, because they had also failed to include data for the large low-dolomite reservoir of deep-sea carbonates. When these are taken into account, the record indicates that the dolomite content of carbonate sediments has indeed decreased significantly since the Jurassic.

But why did the abundance of dolomite decrease during the past 150 Ma? In the first version of this paper HDH proposed that the decrease was due to a decrease in the global CO_2 degassing rate relative to the rate of silicate weathering. This was received with a good deal of skepticism, especially by Bob Berner, who pointed out that if most of the riverine Mg^{+2} input to the oceans is removed during the cycling of seawater through oceanic basalts, it is replaced in large part by Ca^{+2}. This trade does not affect the CO_2 balance of the ocean-atmosphere system, because the released Ca^{+2} is then precipitated as $CaCO_3$. Fred Mackenzie pointed out the potential importance of reverse weathering reactions, but conversations with Fred Sayles suggested that cation exchange reactions on clays are currently more important as a sink of marine Mg^{+2} than reverse weathering reactions.

For help with the problem of the rate of Mg^{+2} removal from seawater by reaction with ocean-floor basalts we turned to Mike Mottl. Rick O'Connell pointed out the likely course of the rate of seafloor spreading during the past 120 Ma, and Don Anderson emphasized some of the current uncertainties in these rates, particularly during the Cretaceous. The effects of changes in the rate of seafloor spreading on the composition of seawater has been somewhat of a contentious issue, but conversations with Laurie Hardie and Steve Stanley together with the discovery of the progressive change in the Mg^{+2} content of seawater during the Tertiary have begun to reduce our disagreements. The evolution of our thinking since the inception of this paper has been very significant, and the authors wish to thank the many people who have saved us from considerable embarrassment. These include Ray Siever, Alden Carpenter, Ken Deffeyes, John Grotzinger, and Andy Knoll. We also wish to thank Gary Ernst for organizing the symposium in honor of Konnie Krauskopf, where the paper was first presented in its present form, and NASA for support under Grant No. NAG5-4174 to Harvard University.

REFERENCES

Daly, R. A., 1909, First calcareous fossils and the evolution of the limestones: Geol. Soc. Amer. Bull., v. 20, p. 155–170.

Drever, J. I., 1971, Early diagenesis of clay minerals, Rio Ameca Basin, Mexico: Jour. Sediment. Petrol., v. 41, p. 982–994.

Esteban, M., 1996, An overview of Miocene reefs from Mediterranean areas: General trends and facies models, *in* Franseen, E. K., Esteban, M., et al., eds., Models for carbonate stratigraphy from Miocene reef complexes of Mediterranean regions: Concepts Sediment. Paleontol., v. 5, p. 3–53.

Garrett, D. E., 1996, Potash deposits: Processing, properties, and uses: London, Chapman and Hall, 734 p.

Garrison, R. E., Kastner, M., and Zenger, D. H., 1984, Dolomites of the Monterey Formation and other organic-rich units: Los Angeles, Soc. Econ. Paleontol. Mineral., 215 p.

Given, R. K., and Wilkinson, B. H., 1987, Dolomite abundance and stratigraphic age constraint on rates and mechanisms of Phanerozoic dolostone formation: Jour. Sediment. Petrol., v. 57, p. 1068–1078.

Gregg, J. M., and Shelton, K. L., 1990, Dolomitization and dolomite neomorphism in the back-reef facies of the Bonneterre and Davis Formations (Cambrian), southeastern Missouri: Jour. Sediment. Petrol., v. 60, p. 549–562.

Hallam, A., 1984, Pre-Quaternary sea level changes: Ann. Rev. Earth Planet. Sci., v. 12, p. 205–243.

Haq, B. U., Hardenbol, J., and Vail, P. R., 1987, Chronology of fluctuating sea levels since the Triassic: Science, v. 235, p. 1156–1167.

Hay, W. W., 1994, Pleistocene–Holocene fluxes are not the Earth's norm, *in* Material fluxes on the surface of the Earth: Washington, DC, National Academy Press/ Board on Earth Sciences and Resources of the National Research Council, 15–27.

Heller, P. L., Anderson, D. L., and Angevine, C. L., 1996, Is the middle Cretaceous pulse of rapid sea-floor spreading real or necessary?: Geology, v. 24, p. 491–494.

Holail, H., Lohmann, K. C., and Sanderson, I., 1988, Dolomitization and dedolomitization of Upper Cretaceous carbonates: Bahariya oasis, Egypt, *in* Skukla, V., and Baker, P. A., eds, Sedimentology and geochemistry of dolostones: SEPM Spec. Publ., No. 43, p. 191–207.

Holland, H. D., 1969, The abundance of CO_2 in the Earth's atmosphere through geologic time, *in* Ahrens, L. H., editor, Origin and distribution of the elements: Oxford, UK, Pergamon Press, p. 949–954.

Hover, V. C., Walter, L. M., Peacor, D. R., and Martini, A. M., 1999, Mg-smectite authigenesis in a marine evaporative environment, Salina Ometepec, Baja, California: Clays and Clay Minerals, v. 47, p. 252–268.

Kupecz, J. A., 1989, Petrographic and geochemical characterization of the lower Ordovician Ellenburger Group, West Texas: Unpubl. Ph.D. dissertation, Univ. of Texas, Austin, 158 p.

Land, L. S., Salem, M. R. I., and Morrow, D. W., 1975, Paleohydrology of ancient dolomites: Geochemical evidence: Amer. Assoc. Petrol. Geol. Bull., v. 59, p. 1602–1625.

Lithgow-Bertelloni, C., Richards, M. A., Ricard, Y., O'Connell, R. J., and Engebretson, D. C., 1993, Toroidal-poloidal partitioning of plate motions since 120 MA: Geophys. Res. Lett., v. 20, p. 375–378.

Lumsden, D. N., 1985, Secular variations in dolomite abundance in deep marine sediments: Geology, v. 13, p. 766–769.

Mackenzie, F. T., Ristvet, B. L., Thorstenson, D. C., Lerman, A., and Leeper, R. H., 1981, Reverse weathering and chemical mass balance in a coastal environment, *in* Marten, J. M., Bruton, J. D., Eisma, D., eds., River inputs to ocean systems: Geneva, UNEP and UNESCO.

Malone, M. J., Baker, P. A., and Burns, S. J., 1996, Recrystallization of dolomite: An experimental study from 50–200°C: Geochim. et Cosmochim. Acta, v. 60, p. 2189–2207.

Martin, W. R., and Sayles, F. L., 1994, Seafloor diagenetic fluxes, *in* Material fluxes on the surface of the Earth: Washington, DC, National Academy Press, Board on Earth Sciences and Resources of the National Research Council, p. 143–163.

McDuff, R. E., 1981, Major cation gradients in DSDP interstitial waters: The role of diffusive exchange between seawater and upper oceanic crust: Geochim. et Cosmochim. Acta, v. 45, p. 1705–1713.

Meybeck, M., 1988, How to establish and use world budgets of riverine materials, *in* Lerman, A., and Meybeck, M., eds., Physical and chemical weathering in geochemical cycles: Dordrecht, Kluwer Acad. Publ., p. 247–272.

Michalopoulos, P., and Aller, R. C., 1995, Rapid clay formation in Amazon Delta sediments: Reverse weathering and oceanic elemental cycles: Science, v. 270, p. 614–617.

Montañez, I. P., 1989, Regional dolomitization of Early Ordovician, Upper Knox Group, Appalachians: Unpubl. Ph.D. dissertation, Virginia Polytech. Inst. State Univ., 322 p.

Montañez, I. P., and Reid, J. F., 1992, Eustatic control on early dolomitization of cyclic peritidal carbonates: Evidence from the early Ordovician Upper Knox Group, Appalachians: Geol. Soc. Amer. Bull., v. 104, p. 872–886.

Mottl, M. J., 1994, *in* Kadko, D., et al., eds., Global impact of submarine hydrothermal processes: RIDGE Workshop Report, p. 1–15.

Mottl, M. J., and Wheat, C. G., 1994, Hydrothermal circulation through mid-ocean ridge flanks: Fluxes of heat and magnesium: Geochim. et Cosmochim. Acta, v. 58, p. 2225–2237.

Opdyke, B. N., and Wilkinson, B. H., 1988, Surface area control of shallow cratonic to deep marine carbonate accumulation: Paleoceanography, v. 3, p. 685–703.

Ortí, F., 1997, Evaporitic sedimentation in the South Pyrenean Foredeeps and the Ebro Basin during the Tertiary: A General View, *in* Busson, G., and Schreiber, B., eds., Sedimentary deposition in rift and foreland basins in France and Spain: New York, Columbia University Press.

Randazzo, A. F., 1997, The sedimentary platform of Florida: Mesozoic to Cenozoic, *in* Randazzo, A. F., and Jones, D. S., eds., The geology of Florida: Gainesville, FL, University Press of Florida, 39–56.

Sandberg, P. A., 1985, Nonskeletal aragonite and P_{CO_2} in the Phanerozoic and Proterozoic, *in* Sundquist, E. T., and Broecker, W. S., eds., The carbon cycle and atmospheric CO_2: Natural variations Archean to present: Amer. Geophysical Union, Geophys. Monog. 32, p. 585–594.

Schmoker, J. W., Krystinik, K. B., and Halley, R. B., 1985, Selected characteristics of limestone and dolomite reservoirs in the United States: Bull. Amer. Assoc. Petrol. Geol., v. 69, p. 733–741.

Scott, T. M., 1997, Miocene to Holocene history of Florida, *in* Randazzo, A. F., and Jones, D. S., eds., The geology of Florida: Gainesville, FL, Univ. Press of Florida, p. 57–67.

Sissingh, W., 1998, Comparative Tertiary stratigraphy of the Plune graben, Bresse graben, and Molasse basin: Correlation of Alpine foreland events: Tectonophys., v. 300, p. 249–284.

Sochava, A. V., and Podkovyrov, V. N., 1995, The compositional evolution of Meso- and Neoproterozoic carbonate rocks: Precamb. Res., v. 73, p. 283–289.

Stanley, S. M., and Hardie, L. A., 1998, Secular oscillations in the carbonate mineralogy of reef-building and sediment-producing organisms driven by tectonically forced shifts in seawater chemistry: Paleogeography, Paleoclimatology, Paleoecology, v. 144, p. 3–19.

Stanley, S. M., and Hardie, L. A., 1999, Hypercalcification: Paleontology links plate tectonics and geochemistry to sedimentology: GSA Today, v. 9, p. 2–7.

Supko, P. R., 1977, Subsurface dolomites, San Salvador, Bahamas: Jour. Sediment. Petrol., v. 47, p. 1063–1077.

Vahrenkamp, V. C., Swart, P. K., and Ruiz, J., 1991, Episodic dolomitization of Late Cenozoic carbonates in the Bahamas: Evidence from strontium isotopes: Jour. Sediment. Petrol., v. 61, p. 1002–1014.

Vail, P. R., Mitchum, R. W., and Thompson, S., 1977, Seismic stratigraphy and global changes of sea level 4, Global cycles of relative changes of sea level: AAPG Memoirs, v. 26, p. 82–97.

Vinogradov, A. P. and Ronov, A. B., 1956, Composition of the sedimentary rocks of the Russian platform in relation to the history of its tectonic movements: Geochemistry, v. 6, p. 533–559.

Von Damm, K. L., 1995, Controls on the chemistry and temporal variability of seafloor hydrothermal fluids, in seafloor hydrothermal systems: Physical, chemical, biological, and geological interactions: Amer. Geophys. Union, Geophys. Monog. 91, p. 222–247.

Wheat, C. G., and Mottl, M. J., 2000, Composition of pore and spring waters from Baby Bare: Global implications of geochemical fluxes from a ridge flank hydrothermal system: Geochim. et Cosmochim. Acta, v. 64, p. 629–642.

Zenger, D. H., 1981, Stratigraphy and petrology of the Little Falls Dolostone (Upper Cambrian), east-central New York: New York State Museum Map and Chart Series 34, 138 p.

______, 1989, Dolomite abundance and stratigraphic age: Constraints on rates and mechanisms of Phanerozoic dolostone formation—discussion: Jour. Sediment. Petrol., v. 59, p. 162–164.

Zimmermann, H., 2000, Tertiary seawater chemistry—implications from primary fluid inclusions in marine halite: Amer. Jour. Science, submitted.

Global Correlation of the 223 ka Pringle Falls Event

MICHAEL MCWILLIAMS

Department of Geological and Environmental Sciences, Stanford University, Stanford, California 94305-2115

Abstract

New $^{40}Ar/^{39}Ar$ ages link records of anomalous geomagnetic field behavior from western North America and New Zealand, and suggest that a reversal event occurred at 223 ± 4 ka (±1σ). Virtual geomagnetic poles from the Pringle Falls event recorded in both hemispheres lie on a common transition path, suggesting that the geomagnetic field during this interval was dominated by a dipolar component. The Pringle Falls event therefore may be a reverse polarity subchron within the Brunhes Normal Chron.

Introduction

REVERSE POLARITY MAGNETIZATIONS have been reported from within the 0–784 ka Brunhes Normal Chron (Baksi et al., 1992), but their interpretation is controversial. One opinion holds that the Brunhes contains as many as eight short reverse polarity subchrons (Champion et al., 1988; Spell and McDougall, 1992). A contrasting view is that there is no convincing evidence for even a single reverse polarity subchron during the Brunhes (Merrill and McFadden, 1994). In this latter interpretation, the only well documented Brunhes-age geomagnetic events are the Laschamps and Blake *reversal excursions*, defined as "apparent" reversals produced by relatively strong nondipole field fluctuations. Reversal excursions are distinct from *reversal events* that are "true" polarity subchrons related to reversals of the dipole field (Merrill and McFadden, 1994).

Some apparent reversals reported in the Brunhes Normal Chron are clearly artifacts that do not record actual geomagnetic field behavior (Verosub and Banerjee, 1977). A real event derived from an accurate geomagnetic field recording must therefore be either a reversal event (which should be essentially synchronous on a global scale) or a reversal excursion (which is unlikely to be either synchronous or global). A significant portion of the Brunhes Chron is characterized by low dipole field strength relative to the present value (Meynadier et al., 1992; Tric et al., 1992; Valet and Meynadier, 1993). A consequence is that an increase in the nondipole/dipole field strength ratio could lead to an increased frequency in the occurrence of reversal excursions but not necessarily reversal events, an argument invoked by those who classify the majority of Brunhes-age events as reversal excursions.

Merrill and McFadden (1994) offered three criteria to discriminate between reversal events and reversal excursions. First, a reversal event should be found at many locations in both hemispheres, with 180° opposite polarity observed at several sites in both hemispheres. Second, a reversal event should be found in different recording media—i.e., sediments and volcanic rocks. Third, evidence from the marine magnetic anomaly record is desirable.

Discrimination between synchronous reversal events and possibly asynchronous reversal excursions also depends critically on the ability to correlate candidate events on a global basis, but correlation is often made difficult in practice because of poor age control. The ~41–45 ka Laschamps event is often thought of as the best documented reversal event/excursion of the Brunhes Chron because anomalous directions of magnetization are observed at widely spaced localities and are indistinguishable in age. The Blake event is less accurately dated than the younger Laschamps event, but it has been observed at many localities. Most other candidate events within the Brunhes Chron are observed at only a single locality and/or have rather poor age control. This paper summarizes geochronological and paleomagnetic data for the Pringle Falls event, recognized here for the first time as a reversal event recorded by sediments and volcanic rocks at distant localities in both hemispheres for which precise age control confirms global synchroneity.

Records of the Pringle Falls Event

Western North America

The first detailed records of anomalous geomagnetic field behavior from a sedimentary sequence

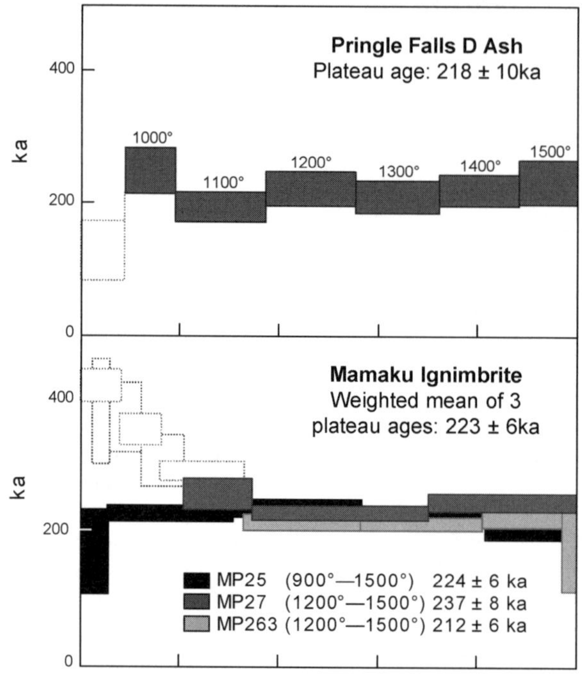

FIG. 1. $^{40}Ar/^{39}Ar$ ages from Pringle Falls D ash (top) and Mamaku Ignimbrite (bottom). Filled rectangles indicate steps used to calculate plateau age; uncertainties are $\pm 1\sigma$. Analytical details can be found in Herrero Bervera et al. (1994) and Tanaka et al. (1995). Neutron fluence monitor used is Taylor Creek sanidine with an assumed age of 29.92 Ma.

near Pringle Falls, Oregon were originally thought to represent the Blake event (Herrero-Bervera et al., 1989; Herrero-Bervera and Helsley, 1993). New $^{40}Ar/^{39}Ar$ ages (Fig. 1) from plagioclase in the "D" ash layer at Pringle Falls (Herrero-Bervero et al., 1994) indicate that the Pringle Falls event occurred at 218 ± 10 ka, significantly older than the commonly accepted 100–130? ka age for the Blake event. Moreover, the Pringle Falls event is found in sedimentary sequences at Summer Lake, Oregon (Negrini et al., 1988) and at Long Valley, California (Liddicoat and Bailey, 1989; Liddicoat 1990). Figure 2 illustrates the common virtual geomagnetic pole (VGP) paths for the Pringle Falls event recorded at these three localities.

North Island, New Zealand

The Mamaku Ignimbrite is the youngest of the large, sheet-forming ignimbrites of the Taupo Volcanic Zone. Anomalous field directions originally reported from a single site in the Mamaku ignimbrite (Cox 1971) have recently been confirmed (Shane et al., 1994; Tanaka et al., 1996).

Three new $^{40}Ar/^{39}Ar$ ages from plagioclase from the Mamaku Ignimbrite (Houghton et al., 1995) yield a weighted mean age of 223 ± 3 ka (Fig. 1), statistically indistinguishable from an isothermal fission-track age of 230 ± 12 ka for the same unit (Shane et al., 1994). Twenty-nine virtual geomagnetic poles (VGPs) for the Mamaku ignimbrite are plotted in Figure 2, along with VGPs from the Pringle Falls event from western North America.

Global Correlation of the Pringle Falls Event

Figures 1 and 2 show that the Pringle Falls event occurred at ~223 ± 4 ka, and was simultaneously recorded at four localities in both hemispheres. The agreement in time and space is remarkable. As discussed in detail in Herrero-Bervero et al. (1994), VGPs from the three sections in western North America are in good agreement. VGPs from volcanic rocks in New Zealand record only a fraction of the total event path and are not nearly as complete as the sedimentary records from North America. This

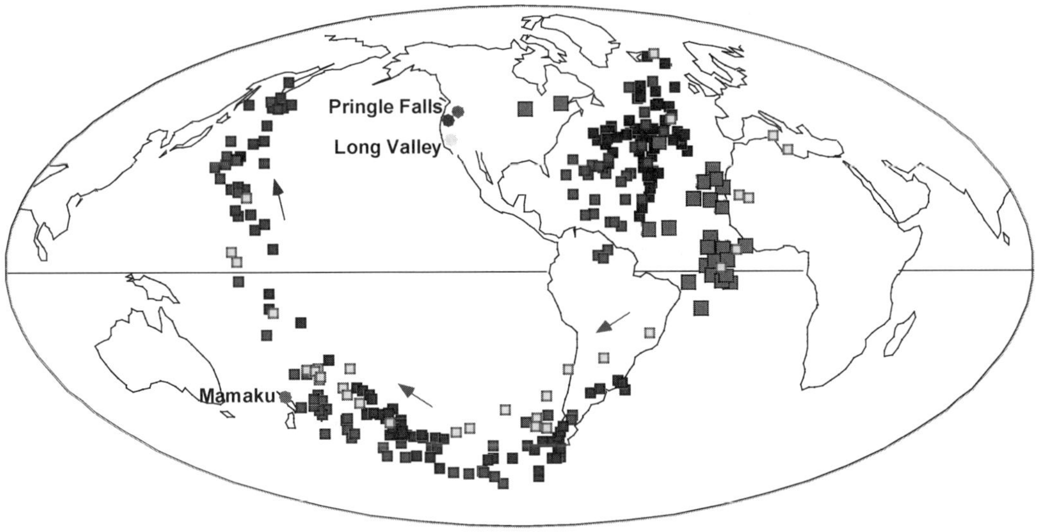

FIG. 2. Virtual geomagnetic pole paths for the Pringle Falls event. Squares are individual VGPs for the "A" segment of the event recorded at Pringle Falls, Oregon (2 records), Long Valley, California, and New Zealand. Circles are sampling sites, grey-tone coded to match respective points on the VGP path. Arrows show the order of the VGP path from older to younger. VGPs before and after the "A" segment lying above 60°N have been omitted for clarity.

is not unexpected, given the episodic nature of volcanism as opposed to more continuous sedimentation in a lacustrine environment. It is clear from Figure 2 that the VGP paths do not exactly coincide. This too is plausible, inasmuch as there was almost certainly a contribution from the nondipole field, which would produce directions differing slightly at localities around the globe.

The aforementioned criteria for discriminating between reversal events and reversal excursions (Merrill and McFadden, 1994) are exacting, and as a consequence it is not surprising that the authors find no convincing evidence for even a single reverse polarity subchron during the Brunhes. Were the entire reversal time scale judged with the same criteria, many extant polarity chrons would have to be rejected; few polarity transitions older than a few Ma have been observed in multiple rock types and reliably correlated on a global basis, primarily because of imprecise age control.

The available data from the Pringle Falls event meet some, but not all, of the standards established in Merrill and McFadden (1994). The first condition is partially met. Although four sites may not be considered as "many," they are at least distributed in both hemispheres. The Northern Hemisphere records from Pringle Falls and Long Valley do attain approximately opposite polarity, but the Southern Hemisphere record from New Zealand is punctuated by the volcanic recording process (Fig. 2). Discovery of the Pringle Falls event at more localities may completely satisfy this condition in the future. The second condition is clearly satisfied.

Meeting the third condition is perhaps the most difficult. If the Pringle Falls event lasted about 10^4 years (a reasonable approximation for the minimum time required for a N-R-N polarity transition without any intervening time latched in the R state), the anomaly width adjacent to a ridge spreading at 50 mm/year would be on the order of 0.5 km. Short-wavelength features of about this size have been inferred from the marine magnetic anomaly record, but the only Brunhes-age short event present in a widely accepted version of the polarity time scale (Cande and Kent, 1992) is the ~10^4 yr Emperor event centered at 490 ka. The Emperor event appears to be present in the marine magnetic anomaly record, but the principal land-based evidence for its existence (Champion et al., 1981) has been retracted (Champion et al., 1988). The original thermoremanent magnetization (TRM) acquired by oceanic basalts upon initial cooling is rapidly altered (Kent and Gee, 1994). If the time scale of the alteration is on the order of 1–2 × 10^4 yr (Gee and Kent,

1995) instead of 5×10^5 yr, as has long been assumed, it seems likely that evidence for short events such as that seen at Pringle Falls will be even more difficult to detect in the marine magnetic anomaly record.

Although not all the conditions set out in Merrill and McFadden (1994) have been fully met, three points suggest that the Pringle Falls event is at least as well documented as the Laschamps event. First, the detailed paleomagnetic records of the Pringle Falls event are much more complete and describe a long VGP path spanning the globe for Pringle Falls, as opposed to a few path segments for Laschamps. Second, radiometric age control is better, with a relative precision on the order of 2%, as opposed to a relative precision approaching 10% for Laschamps. Finally, the Pringle Falls event has been observed in at least as many localities as the Laschamps. Using the criteria outlined in Merrill and McFadden (1994), three well documented geomagnetic events are present in the most recent third of the Brunhes Normal Chron—the Laschamps and Blake events and now the Pringle Falls event.

Reversal Event or Reversal Excursion?

Reversal excursions may represent aborted reversals (Hoffman, 1992). If so, they are probably caused by the same physical process as reversal events and the term *reversal excursion* is superfluous. A contrasting view (Merrill and McFadden, 1994) is that excursions may represent a completely different process that is unrelated to the reversal mechanism and is more likely to occur when the nondipole/dipole field intensity ratio is relatively large. This view is based upon an analysis of the polarity time scale that suggests that the hypothesis of a common stability for the two polarity states cannot be rejected. The presence of reverse polarity subchrons within the Brunhes Normal Chron upsets this model, and as a consequence Merrill and McFadden (1994) have advocated that reversal excursions should not be grouped with reversal events.

The Pringle Falls event clearly qualifies as a well documented excursion, but is it a reversal event? Merrill and McFadden (1994) observed that the evidence for or against excursions as aborted reversals is equivocal. They noted that the Laschamps and Blake events occurred during a time when the nondipole/dipole field intensity ratio was relatively large and the dipole field intensity was relatively low. On these and other grounds, they suggested that these excursions and others should not be treated as aborted reversals.

Detailed studies of polarity transitions have shown that the Brunhes-Matuyama and other transitions in the 0–5 Ma range have two important distinguishing features. First, relative paleointensity records that span the Brunhes-Matuyama as well as older polarity transitions show large decreases just prior to the onset of polarity change (Valet and Meynadier, 1993; Meynadier et al., 1994; Verosub et al., 1996). Second, VGP paths for the Brunhes-Matuyama and other transitions appear to define a common longitudinal band suggestive of the influence of a strong dipole field component. The Pringle Falls event shares these same two characteristics—a leading paleointensity decrease (Meynadier et al., 1994) and a common VGP path as described here. It therefore seems reasonable to suggest that the Pringle Falls event is, in fact, a reversal event. Whether it should be labeled as a polarity subchron is arguable. One definition (Cande and Kent, 1992) holds that a subchron is defined by a pair of reversals, but the minimum length of the intervening polarity interval is unspecified. Accordingly, a pair of reversal events separated by even a very short time might qualify as a subchron. Regardless of the terminology, the Pringle Falls event may prove useful as a chronostratigraphic marker for the most recent third of the Brunhes Chron.

REFERENCES

Baksi, A. K., Hsu, V., McWilliams, M. O., and Farrar, E., 1992, $^{40}Ar/^{39}Ar$ dating of the Brunhes-Matuyama geomagnetic-field reversal: Science, v. 256, p. 356–357.

Cande, S. C., and Kent, D. V., 1992, A new geomagnetic polarity time scale for the late Cretaceous and Cenozoic: Journal of Geophysical Research, v. 97, p. 13,197–13,951.

Champion, D. E., Dalrymple, G. B., and Kuntz, M. A., 1981, Radiometric and paleomagnetic evidence for the Emperor reversed polarity event at 0.46 ± 0.05 m.y. in basalt lava flows from the eastern Snake River Plain, Idaho: Geophysical Research Letters, v. 8, p. 1055–1058.

Champion, D. E., Lanphere, M. A., and Kuntz, M. A., 1988, Evidence for a new geomagnetic reversal from lava flows in Idaho: Discussion of short polarity reversals in the Bruhnes and late Matuyama polarity chrons: Journal of Geophysical Research—Solid Earth and Planets, v. 93, p. 11667–11680.

Cox, A., 1971, Remanent magnetization and susceptibility of late Cenozoic rocks from New Zealand: New Zealand Journal of Geology and Geophysics, v. 22, p. 192–207.

Gee, J., and Kent, D. V., 1995, Magnetic hysteresis in young mid-ocean ridge basalts: Dominant cubic anisotropy: Geophysical Research Letters, v. 22, p. 551–554.

Herrero-Bervera, E. and Helsley, C. E., 1993, Global paleomagnetic correlation of the Blake geomagnetic polarity episode, *in* Applications of paleomagnetism to sedimentary geology: Society of Economic Paleontologists and Mineralogists, Special Publication 49, p. 71–82.

Herrero-Bervera, E., Helsley, C. E., Hammond, S. R., and Chitwood, L. A., 1989, A possible lacustrine paleomagnetic record of the Blake episode from Pringle Falls, Oregon, U.S.A.: Physics of the Earth and Planetary Interiors, v. 56, p. 112–123.

Herrero-Bervero, E., Helsley, C. E., Sarnawojcicki, A. M., Lajoie, K. R., Meyer, C. E., McWilliams, M. O., Negrini, R. M., Turrin, B. D., Donnelly-Nolan, J. M., and Liddicoat, J. C., 1994, Age and correlation of a paleomagnetic episode in the western U.S. by ^{40}Ar/^{39}Ar dating and tephrochronology: The Jamaica, Blake or a new polarity episode?: Journal of Geophysical Research—Solid Earth, v. 99, p. 24,091–24,103.

Hoffman, K. A., 1992, Dipolar reversal states of the geomagnetic field and core-mantle dynamics: Nature, v. 359, p. 789–794.

Houghton, B. F., Wilson, C. J. N., McWilliams, M. O., Lanphere, M. A., Weaver, S. D., Briggs, R. M., and Pringle, M. S., 1995, Chronology and dynamics of a large silicic magmatic system: Central Taupo Volcanic Zone, New Zealand: Geology, v. 23, p. 13–16.

Kent, D. V., and Gee, J., 1994, Grain size–dependent alteration and the magnetization of oceanic basalts: Science, v. 265, p. 1561–1563.

Liddicoat, J. C., 1990, Aborted reversal of the paleomagnetic field in the Brunhes Normal Chron in east-central California: Geophysical Journal International, v. 102, p. 747–752.

Liddicoat, J. C., and Bailey, R. A., 1989, Short reversal of the geomagnetic field about 280,000 years ago at Long Valley, California, *in* Lowes, F. J., ed., Geomagnetism and paleomagnetism: Norwell, MA, Kluwer, p. 137–153.

Meynadier, L., Valet, J. P., Bassinot, F. C., Shackleton, N. J., and Guyodo, Y., 1994, Asymmetrical saw-tooth pattern of the geomagnetic field intensity from equatorial sediments in the Pacific and Indian Oceans: Earth and Planetary Science Letters, v. 126, p. 109–127.

Meynadier, L., Valet, J. P., Weeks, R., Shackleton, N. J., and Hagee, V. L., 1992, Relative geomagnetic intensity of the field during the last 140 ka: Earth and Planetary Science Letters, v. 114, p. 39–57.

Merrill, R. T., and McFadden, P. L., 1994, Geomagnetic field stability; reversal events and excursions: Earth and Planetary Science Letters, v. 121, p. 57–69.

Negrini, R. M., Verosub, K. L., and Davis, J. O., 1988, The middle to late Pleistocene geomagnetic field recorded in fine-grained sediments from Summer Lake, Oregon and Double Hot Springs, Nevada, U.S.A.: Earth and Planetary Science Letters, p. 87, p. 173–192.

Shane, P., Black, T., and Westgate, J., 1994, Isothermal plateau fission-track age for a paleomagnetic excursion in the Mamaku Ignimbrite, New Zealand, and implications for late Quaternary stratigraphy: Geophysical Research Letters, v. 21, p. 1695–1698.

Spell, T., and McDougall, I., 1992, Revisions to the age of the Brunhes-Matuyama boundary and the Pleistocene geomagnetic polarity timescale: Geophysical Research Letters, v. 19, p. 1181–1184.

Tanaka, H., Turner, G. M., Houghton, B. F., Tachibana, T., Kono, M., and McWilliams, M. O., 1996, Palaeomagnetism and chronology of the central Taupo Volcanic Zone, New Zealand: Geophysical Journal International, v. 124, p. 919–934.

Tric, E., Valet, J. P., Tucholka, P., Paterne, M., Labeyrie, L., Guichard, F., Tauxe, L., and Fontugne, M., 1992, Paleointensity of the geomagnetic field during the last 80,000 years: Journal of Geophysical Research—Solid Earth, v. 97, p. 9337–9351.

Valet, J., and Meynadier, L. 1993, Geomagnetic field intensity and reversals during the past four million years: Nature, v. 366, p. 234–238.

Verosub, K. L., and Banerjee, S. K., 1977, Geomagnetic excursions and their paleomagnetic record: Reviews of Geophysics and Space Physics, v. 15, p. 145–155.

Verosub, K. L., Herrero-Bervera, E., and Roberts, A. P., 1996, Relative geomagnetic paleointensity across the Jaramillo Subchron and the Matuyama/Brunhes boundary: Geophysical Research Letters, v. 23, p. 467–470.

A Rapid In Situ Method for Determining the Ages of Uranium Oxide Minerals: Evolution of the Cigar Lake Deposit, Athabasca Basin

MOSTAFA FAYEK, T. MARK HARRISON, MARTY GROVE, AND CHRISTOPHER D. COATH

Department of Earth and Space Science and Institute of Geophysics and Planetary Physics, University of California, Los Angeles, California 90095-1567

Abstract

We present a rapid and accurate technique for making *in situ* U-Pb isotopic measurements of uranium oxide minerals that utilizes both electron and ion microprobes. U and Pb concentrations are determined using an electron microprobe, whereas the isotopic composition of Pb for the same area is measured using a high-resolution ion microprobe. The advantages of this approach are: mineral separation and chemical digestion are unnecessary; homogenous uranium oxide standards, which are difficult to obtain, are not required; and precise and accurate U-Pb ages on ~10 μm spots can be obtained in a matter of hours. We have applied our method to study the distribution of U-Pb ages in complexly intergrown uranium oxides from the unconformity-type Cigar Lake uranium deposit, Saskatchewan, Canada. *In situ* U-Pb results from early formed uraninite define a well-correlated array on concordia with upper and lower intercepts of 1467 ± 63 Ma and 443 ± 96 Ma (±1σ), respectively. The 1467 Ma age is interpreted as the minimum age of mineralization and is consistent with the age of clay-mineral alteration (~1477 Ma) and magnetization of diagenetic hematite (1650 to 1450 Ma) that is associated with these unconformity-type uranium deposits and early diagenesis of the Athabasca Basin sediments. *In situ* U-Pb isotopic analyses of uraninite and coffinite can document the Pb*/U heterogeneities that can occur on a scale of 15 to 30 μm, thus providing relatively accurate information regarding the timing of fluid interactions associated with the evolution of these deposits.

Introduction

THE URANIUM DEPOSITS of the Proterozoic Athabasca Basin (Fig. 1) are the highest-grade uranium deposits in the world (Nash et al., 1981). These deposits are termed "unconformity-type" because they generally form at or near the intersection between Hudsonian-age faults and the unconformity between Athabasca group sandstones and Aphebian metasediments. The fluids responsible for the deposition of these deposits are believed to be diagenetic-hydrothermal in origin and linked to regional-scale fluid-flow hydrodynamics, large-scale structures, time-temperature profiles, and fluid mixing (Hoeve and Quirt, 1984; Wilson and Kyser, 1987; Kyser et al., 1988; Kyser et al., 1990; Kotzer and Kyser, 1993, 1995; Fayek and Kyser, 1997; Fayek et al., 1997).

Although the detailed study of these deposits is important for reasons of their economic potential (they yield approximately one-third of the world's uranium supply) and use as natural analogues to understand the migration of uranium and other radionuclides from spent fuel repositories (e.g., Cramer, 1995), the extent and processes of alteration of uraninite (UO_2) in these deposits remain poorly understood. The fact that one of the essential chemical constituents of uraninite permits this phase to be used as a sensitive U-Pb chronometer has been exploited in attempts to assess the chemical evolution of uranium deposits.

Although three approaches have been investigated ($^{207}Pb/^{206}Pb$ ages, chemical Pb ages, and U-Pb ion microprobe ages), establishing the timing of initial uranium emplacement and subsequent alteration history have been challenging and incompletely realized goals. For example, it is very unusual to obtain concordant U-Pb data from uraninite and the degree of discordance is often much greater than five percent, leading to uncertainty in the interpretation of the U-Pb values (e.g., Cumming and Krstic, 1992). This reflects both the complexity of the U-Pb system in natural uraninites and the limitations of the analytical methods. For example, alteration of uraninite or radiation damage can

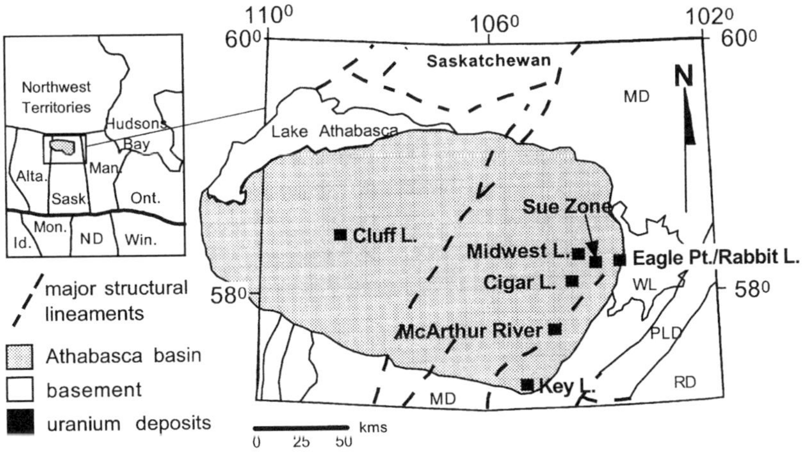

FIG. 1. Map showing the extent of the Athabasca Basin and location of the uranium deposits and major lithostructural domains in the crystalline basement of Saskatchewan (modified from Hoeve and Sibbald, 1978). Abbreviations: L = Lake; MD = Mudjatik Domain; PLD = Peter Lake Domain; R = River; RD = Rottenstone Domain; WD = Wollaston Domain; WL = Wollaston Lake.

result in loss of radiogenic Pb (Pb*), recrystallization, and changes in chemical composition (Kotzer and Kyser, 1995; Matzke and Wang, 1996; Fayek and Kyser, 1997). Previously applied dating methods are hindered by ambiguity produced by discordance (e.g., Cumming and Krstic 1992), the required assumption that all Pb in uraninite is radiogenic (i.e., chemical Pb ages; e.g., Fayek and Kyser, 1997), and the requirement of a uranium oxide standard homogenous in Pb*/U (i.e., U-Pb ion microprobe; e.g., Hollinger, 1988).

The wide range of ages (1514 Ma to 255 Ma) associated with multiple episodes of water-rock interactions and associated mineralization in the unconformity type uranium deposits of the Athabasca Basin, northern Saskatchewan, Canada, underscores both the mobility of U and Pb* during subsequent alteration events and the intrinsic limitations of these methods (Baadsgaard et al., 1984; Cumming and Krstic, 1992; Kotzer and Kyser, 1993; Fayek and Kyser, 1997). Although a considerable amount of effort has been directed at establishing the timing of initial uranium emplacement and subsequent alteration history, it seems unlikely that further chemical Pb or $^{207}Pb/^{206}Pb$ age measurements would substantially improve the present state of knowledge. However, the advantages of conventional U-Pb dating (i.e., use of concordia) and chemical Pb ages (i.e., *in situ* analysis and spatial resolution) are combined in ion microprobe U-Pb dating.

This paper discusses a methodology for obtaining high-precision *in situ* U-Pb isotopic analyses of uranium oxide minerals. We demonstrate how this method for measuring U and Pb concentrations by electron microprobe and Pb isotopes by ion microprobe precludes the need for a homogenous U-Pb uraninite standard, and show how the method can be applied to study chemically complex uranium oxides associated with unconformity-type uranium deposits.

Uranium oxides

Uraninite is the most common U^{4+}-bearing mineral. It is found in most uranium deposits, except where fluids have oxidized uranium to the uranyl (U^{6+}) state (Smith, 1984). Uraninite, which contains the heaviest naturally occurring element on Earth, preferentially incorporates the light isotope of oxygen (^{16}O). However, uraninite varies widely in composition from $UO_{2.07}$ to $UO_{2.25}$, is geochemically unstable in most surficial environments, and is susceptible to alteration and oxidation by surficial, oxidizing fluids (Grandstaff, 1976; Dyck, 1978, Smith, 1984; Burns et al., 1996), which make its isotopic and chemical properties potentially complex. Geochronological, petrographic, and isotopic studies of uranium minerals (Dyck, 1978, Baadsgaard et al., 1984, Cumming and Krstic, 1992, Kotzer and Kyser 1993, Fayek and Kyser, 1997) have documented the difficulties associated with establishing the age of initial emplacement of the uranium, the fluids responsible for uranium transport and deposition, and the source of the uranium. These problems largely reflect the extreme solubility of uraninite in an oxidizing environment. Therefore, precise, high-spatial-resolution U-Pb measurements of minerals with complex chemistry, such as uraninite, would greatly aid in unraveling the complex fluid histories associated with certain uranium deposits.

Previous ion microprobe studies of uranium oxide minerals

Previous ion microprobe investigations of the U-Pb isotope systematics of U-oxide minerals (Meddaugh, 1983; Holliger and Cathelineau, 1986, 1987, 1988; Fourel et al., 1988; Hollinger, 1988, 1991, 1994; Cathelineau et al., 1990) have revealed important information regarding the formation of certain types of uranium deposits, but are limited by possible matrix effects and the requirement of a concordant uraninite standard homogenous in Pb*/U.

Holliger (1988) developed a sample of uraninite from Vendée, France, for use as an ion microprobe standard. The U-Pb isotopic composition (i.e., $^{206}Pb/^{238}U$) of this material was characterized by microdrilling thin sections and analyzing the extracted material by isotope dilution methods. Five measurements yielded a concordant age of 540 Ma. Using an 8 keV Ar^+ primary ion beam focused on a 10 μm spot, Holliger (1988) found Pb^+ and UO_2^+ ions to be the most abundant secondary ions of Pb and U. A normalizing coefficient, α, of 0.72 ± 0.02 (2σ)—defined as the ratio of the true $^{206}Pb/^{238}U$ to the $^{206}Pb^+/^{270}UO_2^+$ value measured by the ion microprobe—was used to correct for instrumental mass fractionation between Pb^+ and UO_2^+ (Holliger, 1988). Holliger (1991) and Cathelineau et al. (1990) employed another sample of uraninite from Vendée that gave a concordant age of 285 Ma and exhibited remarkably homogenous U and Pb concentrations over large areas (>200 μm). The Pb/U ratio of the sample was determined by electron microprobe, and this ratio was used to calculate a revised α value of 0.63 ± 0.03. Cathelineau et al. (1990) reported uncertainties on $^{206}Pb/^{238}U$, $^{207}Pb/^{235}U$, and $^{206}Pb/^{207}Pb$ ratios of 0.5%, 1%, and 0.08% (2σ), respectively.

A limitation of the technique used by Holliger (1991) and Cathelineau et al. (1990) is that it requires a standard that is homogenous in Pb*/U and preferably concordant. Uranium oxide minerals having homogenous Pb*/U ratios over large areas are difficult to find, in part because the diffusion of Pb in uranium oxides is a rapid process (Brandt and Pershinov, 1968; Yershov, 1974; Janeczeck and Ewing, 1995). Thus uranium oxides easily lose Pb and have, typically, highly variable Pb*/U ratios (e.g., Kotzer and Kyser, 1993; Fayek and Kyser, 1997). A second disadvantage of their technique is that the α value calculated for uraninite cannot be used to correct for instrumental mass fractionation between Pb^+ and UO_2^+ ions in other phases such as coffinite ($USiO_4 \cdot nH_2O$), an alteration phase commonly found with uraninite, because the matrix of these phases is different from that of uraninite. Our own experience is that relatively small differences in uraninite composition can significantly alter the relative ionization efficiencies of these two species, essentially ruling out this approach for studying the complex compositions of the Athabasca Basin samples.

Analytical Techniques

Sample preparation

Sections of core from the Cigar Lake unconformity-type uranium deposit, northern Saskatchewan, Canada, were placed on double-sided adhesive tape within a 25 mm diameter brass mold, which was filled with epoxy resin. The hardened epoxy mounts were separated from the brass molds and polished using 600 to 2400 grit sandpaper and 3 μm to 0.1 μm diamond-polishing compounds. The mounts were then washed with a dilute soap solution, rinsed in deionized water, dried, and carbon coated. The samples then were examined for homogeneity and alteration by transmitted light microscopy, back-scattered electron imaging, and quantitative electron microprobe analysis. Chemical compositions of the uraninite (Table 1) were determined by wavelength dispersive spectroscopy using an automated CAMECA 50 X-ray microanalyzer operated at 15 keV, a beam diameter of 10 μm, and counting times of 40 s per element. Synthetic UO_2, grossular garnet, and galena were used as standards for U, Si and Ca, and Pb, respectively. Detection limits of the elements were on the order of 0.1 wt%. The program PAP was used to reduce the data for the various elements. Oxygen contents of uraninite were calculated by stochiometery assuming an ideal composition of UO_2. The mounts were then re-polished, cleaned to remove the carbon coat, and reflected light maps of the entire mount were made. A ~200 Å thick Au coat was sputter-deposited on the sample mount surface prior to ion microprobe analysis. The mounts

TABLE 1. Ion and Electron Microprobe U-Pb Data from Stage 1 Uraninite from the Cigar Lake Deposit

Spot No.	$^{207}Pb/^{206}Pb$	Error % (1σ)	$^{208}Pb/^{206}Pb$	Error % (1σ)	$^{207}Pb/^{206}Pb$ Age (Ma)	Electron-microprobe data[1] UO_2	PbO	SiO_2	CaO	U, %	Pb, %	^{206}Pb	^{207}Pb	^{208}Pb	^{235}U	^{238}U	$^{207}Pb/^{235}U$	Error, %	$^{206}Pb/^{238}U$	Error, %	ρ(x,y)
										W83C, 145-441.2											
1	0.08444	0.16	0.000239	3.4	1303±3	80.23	12.81	0.21	1.04	70.69	11.89	10.96	0.93	0.00262	0.51	70.2	1.818	1.8	0.156	1.8	0.997
2	0.08436	0.15	0.000228	2.2	1301±3	78.84	14.14	0.05	0.69	69.46	13.12	12.10	1.02	0.00275	0.5	69.0	2.041	1.7	0.175	1.7	0.996
3	0.08555	0.15	0.000167	3.6	1328±3	79.44	14.87	0.02	0.75	69.99	13.80	12.71	1.09	0.00212	0.5	69.5	2.157	1.6	0.183	1.6	0.996
4	0.08514	0.16	0.000090	3.8	1319±3	80.37	14.21	0.06	0.85	70.80	13.18	12.15	1.03	0.00109	0.51	70.3	2.029	1.7	0.173	1.7	0.996
5	0.08506	0.13	0.000076	3.7	1317±3	81.03	14.12	0.06	0.75	71.39	13.10	12.08	1.03	0.00092	0.514	70.9	1.998	1.7	0.170	1.7	0.997
6	0.08568	0.15	0.000079	4.1	1331±3	80.02	13.93	0.11	0.79	70.50	12.93	11.91	1.02	0.00093	0.508	70.0	2.010	1.7	0.170	1.7	0.997
7	0.08590	0.14	0.000064	3.7	1336±3	79.78	14.14	0.01	0.76	70.29	13.12	12.08	1.04	0.00077	0.506	69.8	2.051	1.7	0.173	1.7	0.997
8	0.08552	0.13	0.000088	3.6	1327±3	79.32	13.77	0.10	0.75	69.88	12.78	11.77	1.01	0.00104	0.503	69.4	2.000	1.7	0.170	1.7	0.998
9	0.08577	0.55	0.000305	1.7	1333±10	80.61	14.90	0.06	0.77	71.01	13.83	12.73	1.09	0.00389	0.511	70.5	2.136	1.7	0.181	1.6	0.954
10	0.08534	0.13	0.000075	5.0	1323±2	79.32	13.83	0.06	0.79	69.88	12.83	11.82	1.01	0.00089	0.503	69.4	2.005	1.7	0.170	1.7	0.998
11	0.08514	0.16	0.000087	4.8	1319±3	80.46	14.34	0.04	0.72	70.89	13.31	12.27	1.04	0.00107	0.510	70.4	2.046	1.7	0.174	1.7	0.996
12	0.08543	0.14	0.000242	2.1	1325±3	79.54	13.80	0.09	0.77	70.07	12.81	11.80	1.01	0.00285	0.505	69.6	1.997	1.7	0.170	1.7	0.997
13	0.08382	0.18	0.000142	3.5	1288±4	80.41	13.44	0.10	0.84	70.84	12.48	11.51	0.96	0.00164	0.510	70.3	1.891	1.8	0.164	1.8	0.995
14	0.08124	0.14	0.000152	3.4	1227±3	81.29	10.66	0.36	1.28	71.62	9.89	9.15	0.74	0.00139	0.516	71.1	1.441	2.1	0.129	2.1	0.998
15	0.08389	0.15	0.000120	3.4	1290±3	79.52	13.17	0.08	0.86	70.06	12.23	11.28	0.95	0.00136	0.504	69.6	1.876	1.8	0.162	1.8	0.997
16	0.08400	0.14	0.000092	4.4	1293±3	79.66	12.82	0.13	0.91	70.18	11.89	10.97	0.92	0.00101	0.505	69.7	1.824	1.8	0.157	1.8	0.997
										CS615A1[2]											
1	0.08220	0.24	0.000249	3.6	1250±9	78.27	12.38	0.22	1.09	68.95	11.49	10.62	0.87	0.00264	0.497	68.5	1.752	1.9	0.155	1.9	0.993
2	0.08169	0.26	0.000588	2.6	1238±13	80.09	12.45	0.29	1.05	70.55	11.55	10.67	0.87	0.00628	0.508	70.1	1.713	1.9	0.152	1.9	0.992
3	0.08397	0.20	0.000196	5.6	1292±8	77.34	14.11	0.04	0.85	68.13	13.10	12.08	1.01	0.00237	0.491	67.6	2.068	1.7	0.179	1.7	0.994
4	0.08245	0.22	0.000256	3.4	1256±9	78.93	13.01	0.15	0.90	69.54	12.07	11.15	0.92	0.00286	0.501	69.0	1.837	1.8	0.161	1.8	0.994
5	0.08124	0.22	0.000148	4.7	1227±8	81.80	11.78	0.07	0.89	72.07	10.93	10.11	0.82	0.00150	0.519	71.6	1.580	2.0	0.141	2.0	0.995
6	0.08246	0.26	0.000158	6.1	1256±16	78.76	11.81	0.23	1.13	69.39	10.96	10.12	0.83	0.00160	0.500	68.9	1.661	2.0	0.147	2.0	0.993
7	0.08302	0.21	0.000100	5.2	1270±8	78.65	13.95	0.03	1.08	69.29	12.95	11.95	0.99	0.00120	0.499	68.8	1.984	2.0	0.174	2.0	0.993

[1]All electron microprobe data are reported in wt%. Uncertainty in measuring the Pb and U concentrations is ±0.1 wt% (100 ppm).

[2]Spot locations for sample CS615A1 are shown in Figure 2.

were placed in steel sample holders where surface conductivity was approximately 5 to 10 ohms/cm. The entire assembly was then placed in the ion microprobe sample lock and held at high vacuum for a minimum of eight hours prior to the start of analysis.

Ion microprobe Pb isotopic analyses of uranium oxides

The analytical protocol for U-Pb measurements in uranium minerals using the CAMECA ims 1270 is similar to that used for U-Pb analyses of zircon (Quidelleur et al., 1997). A ~1 nA primary ion beam of O^- was focused to a 15 x 30 μm spot using a 100 μm aperture in the primary column. The secondary ion accelerating voltage was 10 kV with the extraction electrode at -10 kV relative to the sample surface. A mass resolving power of ~5500 with an energy window of 50 eV was used to obtain peaks with a ~0.020 amu flat top. Ions were detected by an ETP electron multiplier coupled with an ion counting system with an overall deadtime of 25 ns. The following species were detected sequentially by switching the magnetic field: 203.5 (background), $^{204}Pb^+$, $^{206}Pb^+$, $^{207}Pb^+$, and $^{208}Pb^+$. High-resolution mass scans indicate that isobaric interferences were fully resolved. A typical analysis lasted ~10 minutes, comprising 25 analysis cycles. Negligible common Pb ($^{204}Pb^+$) was detected.

The Pb isotope ratios determined by ion microprobe and the concentration of Pb measured by electron probe were combined to calculate the concentrations of ^{206}Pb and ^{207}Pb. Using the present-day $^{238}U/^{235}U$ ratio of 137.88 and the U concentration measured by electron probe, the amount of ^{238}U and ^{235}U were calculated. The concentrations of ^{238}U, ^{235}U, ^{206}Pb, and ^{207}Pb were then used to calculate $^{206}Pb/^{238}U$ and $^{207}Pb/^{235}U$ for each spot. The equations used are as follows:

$$x = {}^{207}Pb/{}^{235}U \quad (1)$$

$$= \{[Pb]\,\{{}^{207}Pb/({}^{206}Pb + {}^{207}Pb + {}^{208}Pb)\}\}/(g\,[U]) \quad (2)$$

$$= [Pb]C_7/(g\,[U]), \quad (3)$$

where [Pb] and [U] are the measured concentrations of Pb and U, g = 1/138.88 (i.e., $^{235}U/(^{238}U + ^{235}U)$), and $C_7 = {}^{207}Pb/({}^{206}Pb + {}^{207}Pb + {}^{208}Pb)$. Similarly,

$$y = {}^{206}Pb/{}^{238}U \quad (4)$$

$$= \{[Pb]\,\{{}^{206}Pb/({}^{206}Pb + {}^{207}Pb + {}^{208}Pb)\}\}/(f\,[U]) \quad (5)$$

$$= [Pb]C_6/(f\,[U]), \quad (6)$$

where f = 137.88/138.88 (i.e., $^{238}U/(^{238}U + ^{235}U)$), and $C_6 = {}^{206}Pb/({}^{206}Pb + {}^{207}Pb + {}^{208}Pb)$. This method thus avoids the need for a concordant uraninite or coffinite ion microprobe standard with homogenous U/Pb ratios because only the Pb isotopes are measured by ion microprobe, which exhibit negligible instrumental fractionation during sputtering (Meddaugh, 1983).

Error calculation

Uncertainties in the $^{206}Pb/^{238}U$ and $^{207}Pb/^{235}U$ ratios—σx and σy, respectively—and the covariance, σxy, were calculated by propagating the uncertainties for measuring Pb isotopic ratios by ion probe and the uncertainties in measuring the Pb and U concentrations by electron microprobe (Table 1) such that:

$$(\sigma x/x)^2 = [\sigma Pb/Pb]^2 + [\sigma U/U]^2 + (\sigma C_7/C_7)^2 \quad (7)$$

and

$$(\sigma y/y)^2 = [\sigma Pb/Pb]^2 + [\sigma U/U]^2 + (\sigma C_6/C_6)^2 \quad (8)$$

$$\sigma xy/(xy) = [\sigma Pb/Pb]^2 + [\sigma U/U]^2, \quad (9)$$

where σPb and σU are the uncertainties in measuring the Pb and U concentrations by electron probe (±0.1wt% or ± 100 ppm), respectively, and σC_6 and σC_7 are the errors in C_6 and C_7. The contribution to σxy arising from the correlation between the errors in C_6 and C_7 is negligible and has been omitted from [9]. The largest sources of error in this method is in the determination of U and Pb concentrations by electron microprobe. The correlation coefficient (ρ) reported in Table 1 is given by:

$$\rho = \sigma xy/\ \sigma x\ \sigma y. \quad (10)$$

Concordia intersections were calculated using the program ISOPLOT (Ludwig, 1993). Uncertainties in the ages are 1σ in every case.

Results and Discussion

The Cigar Lake unconformity-type uranium deposit is located along the eastern margin of the Athabasca Basin (Fig. 1). The deposit recently has been the focus of an international program to study uranium deposits as natural analogues for subsur-

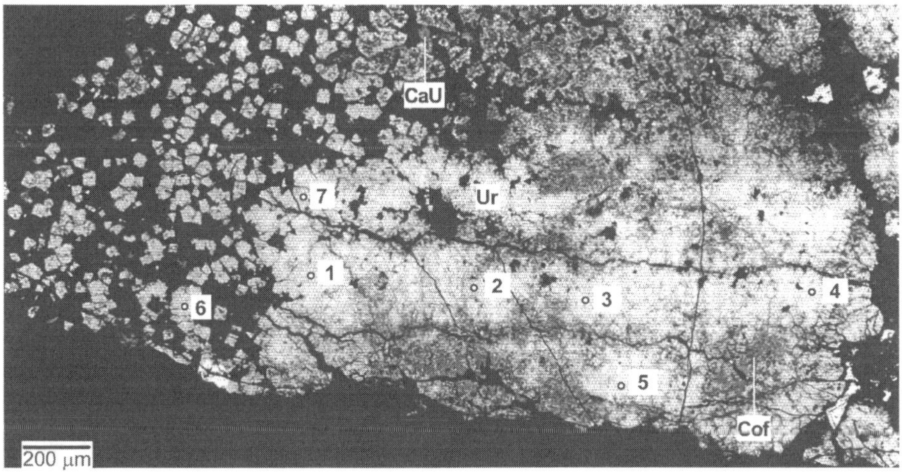

FIG. 2. Backscattered electron image of a sample from Cigar Lake (CS615 A1). Coffinite (Cof) and CaU-rich mineral (CaU) alteration appears as dark grey, whereas uraninite (Ur) is light grey. Labeled spots indicate analyses points for U-Pb measurements.

face spent nuclear fuel repositories (Cramer, 1995; Smellie and Karlsson, 1996).

Uraninite is the most common uranium ore mineral at Cigar Lake. It occurs as black to dark grey masses with a metallic to submetallic luster. Uraninite is highly fractured, with sulfides, arsenides, and clay and rust-colored carbonate minerals (mainly calcite) infilling fractures. X-ray diffraction revealed uraninite, coffinite, galena, chalcopyrite, and pyrite as the most abundant minerals in the clay-free fraction of the ore (Fayek et al., 1999).

As with other uranium deposits in the Athabasca Basin, three main stages of uraninite were observed in thin section and by backscattered imaging. All stages of uranium ore are variably altered to Ca-Si-rich hydrous uranium minerals and coffinite (Fayek et al., 1997). Uraninite occurs as masses, nodules, and euhedral cubes ranging in size from 0.1 cm to 0.5 cm. The nodules are composed of many botryoids that exhibit radial textures in thin section as a result of uniformly distributed radial shrinkage cracks (Fayek et al., 1997). Spaces between botryoids of uraninite are filled with clay minerals, Ni-Cu arsenides, and Ni-Co sulf-arsenides (Reyx and Ruhlmann, 1993; Fayek et al., 1997). However, Cu-Fe-sulfides are commonly associated with Stage 3 uraninite (Bruneton, 1987; Fayek and Kyser, 1997).

Figure 2 shows a backscattered electron image of Stage 1 uraninite (Ur), sample CS615 A1, from Cigar Lake. The uraninite occurs as euhedral cubes and masses, and is variably altered to coffinite (Cof) and small patches of Ca-Si–rich hydrous uranium minerals (CaU). Also shown are the spots that were analyzed by electron and ion microprobe.

Figure 3A shows a concordia plot of the U-Pb data from samples CS615 A1 and W 83C 145-441.2. When all 23 points are regressed together, an upper intercept of 1467 ± 63 Ma and lower intercept of 443 ± 96 Ma is obtained with an MSWD of 6.7. The age obtained from the regression of all the points is interpreted to represent the minimum age of mineralization, which corresponds with the age of clay mineral alteration (~1477 Ma) associated with the unconformity-type uranium deposits (Kotzer and Kyser, 1995) and the age of one of the magnetization events (B-type magnetization, 1600 to 1450 Ma) from the Athabasca basin (Kotzer et al., 1992).

In contrast, the concordia plot of data from Cigar Lake obtained by micro-drilling and conventional wet chemical techniques (Fig. 3B) (Cumming and Krstic, 1992) shows a great deal of scatter. In the latter study, the authors where unable to avoid μm size intergrowths of galena and coffinite. A four-point line (A) that they chose and interpreted as a possible limiting chord to most of the data has an MSWD of 6.4, which is rather large. However, they do get three data points that are collinear and give an upper intercept of 1287 ± 16 Ma. These points likely represent areas sampled that were devoid of coffinite, whereas the remaining scatter likely reflects a mixture of unaltered and altered uraninite and intergrowths of coffinite and galena. Therefore,

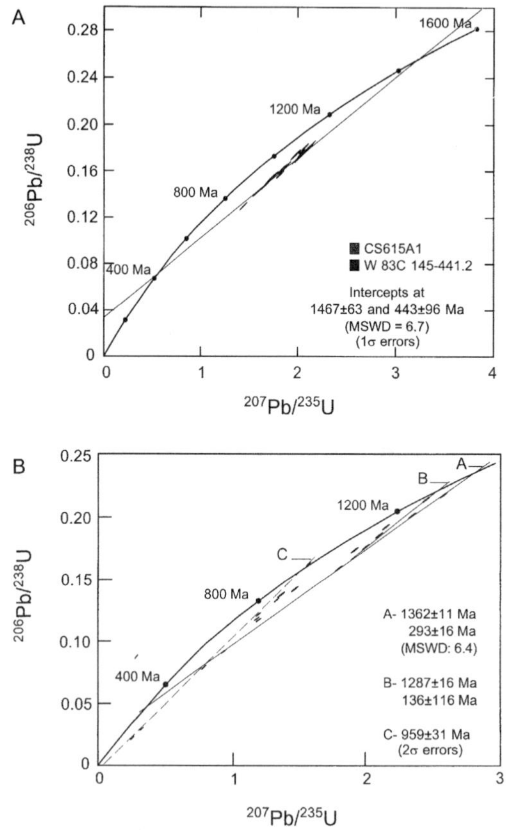

FIG. 3. A. U-Pb results from *in situ* isotopic analyses of Cigar Lake sample CS615 A1 and W 83C 145-441.2 (this study). Some of the analyses points are identified in Figure 2 and the data are reported in Table 1. B. Concordia diagram of U-Pb data obtained by conventional sampling and wet chemical techniques material (Cummings and Krstic, 1992). Line A is a proposed upper limit to the data field and lines B and C are based on fractions from spatially related material (Cummings and Krstic, 1992).

the *in situ* U-Pb data analyses may be indicating the timing and nature of the recrystallizing event, which prior to the *in situ* analyses was obscured by the wide variability in the data obtained by micro-drilling and conventional isotopic analyses.

Conclusions

We presented a technique that utilizes both the electron and ion microprobe to obtain rapid and precise *in situ* U-Pb isotopic analyses of uranium minerals. The advantages of this technique over conventional methods and previous ion microprobe techniques are: (1) mineral separation and chemical digestion are not required; (2) a uranium mineral standard with homogenous Pb*/U is not required; and (3) minor alteration phases, such as coffinite, and uranium minerals with a complex chemistry can be routinely analyzed for their U-Pb isotopic composition.

A concordia plot of U-Pb data, obtained by this technique from Stage 1 uraninite from the Cigar Lake uranium deposit, shows a single discordia with an upper intercept of 1467 ± 63 Ma and a lower intercept of 443 ± 96 Ma. The age obtained from the upper intercept corresponds with the age of the clay-mineral alteration associated with unconformity-type uranium deposits and age of magnetization from the Athabasca Basin.

In contrast, the concordia plot of data from Cigar Lake obtained by micro-drilling and conventional wet chemical techniques shows a great deal of scatter. Therefore, the *in situ* technique, when applied to other stages of mineralization and deposits, can potentially help constrain the chronological evolution of unconformity-type uranium deposits.

Acknowledgments

This study was supported by a grant from the U.S. Department of Energy. We thank Dr. R.C. Ewing for providing the Cigar Lake uraninite sample. We also thank Drs. Frank Kyte and Kevin McKeegan for their expert technical assistance. The UCLA ion microprobe laboratory is partially supported by a grant from the National Science Foundation Instrumentation and Facilities program. This manuscript has benefitted greatly from the review of Dr. Tom G. Kotzer.

REFERENCES

Allègre, C. J., Albarede, F., Grünenfelder, M., and Koppel, V., 1974, ^{238}U/^{206}Pb-^{235}U/^{207}Pb-^{232}Th/^{208}Pb zircon geochronology in alpine and non-alpine environments: Contrib. Mineral. Petrol., v. 43, p. 163–194.

Baadsgaard, H., Cumming, G. L., and Worden, J. M., 1984, U-Pb geochronology of minerals from the Midwest uranium deposit, northern Saskatchewan: Can. Jour. Earth Sci., v. 21, p. 642–648.

Brandt, S. B., and Pershinov, A. V., 1968, The PT-relationships for the migration capacity of lead under hydrothermal conditions: Geokhimiya, v. 6, p. 72–77 (in Russian).

Bruneton, P., 1987, Geology of the Cigar Lake uranium deposit (Saskatchewan, Canada), *in* Gilboy, C. F., and Vigrass, L. W., eds., Economic minerals of Saskatchewan: Sask. Geol. Soc. Spec. Publ., v. 8, p. 99–119.

Burns, P. C., Miller., M. L., and Ewing, R. C., 1996, U^{6+} minerals and inorganic phases: A comparison and hierarchy of crystal structures: Can. Mineral., v. 34, p. 845–880.

Cathelineau, M., Boiron, M. C., Holliger, P., and Poty, B., 1990, Metallogenesis of the French part of the Variscan orogen. Part II: Time-space relationships between U, Au, and Sn-W ore deposition and geodynamic events—mineralogical and U-Pb data: Tectonophys., v. 177, p. 59–79.

Cramer, J. J., 1995, The Cigar Lake uranium deposit: Analogue information for Canada's nuclear fuel waste disposal concept: AECL (Atomic Energy of Canada, Ltd.) Rep. 11204, 32 p.

Cumming, G. L., and Krstic, D., 1992, The age of unconformity-related uranium mineralization in the Athabasca Basin, northern Saskatchewan: Can. Jour. Earth Sci., v. 29, p. 1623–1639.

Dyck, W., 1978, The mobility and concentration of uranium and its decay products in temperate surficial environments, *in* Kimberley, M. M., ed., Short course in uranium deposits: Their mineralogy and origin, v. 3: Nepean, Ont., Mineral. Assoc. Canada, p. 57–92.

Fayek, M., Burns, P. C., Guo, Y-X, and Ewing, R. C., 1999, Micro-structures associated with uraninite alteration: Jour. Nucl. Mater. (in press).

Fayek, M., Janeczek, J., and Ewing, P. C., 1997, Mineral chemistry and oxygen isotopic analyses of uraninite, pitchblende, and uranium alteration minerals from the Cigar Lake deposit, Saskatchewan, Canada: Appl. Geochem., v. 12, p. 549–565.

Fayek, M., and Kyser, T. K., 1997, Characterization of multiple fluid events and rare-earth-element mobility associated with formation of unconformity-type uranium deposits in the Athabasca Basin, Saskatchewan: Can. Mineral., v. 35, p. 627–658.

Fourel, F., Lancelot, J. R., Allègre, C. J., and Dupré, B., 1988, Isotopic analyses of uraniferous minerals by both U-Pb and Sm-Nd methods: Chem. Geol., v. 70, p. 134.

Grandstaff, D. E., 1976, A kinetic study of the dissolution of uraninite: Econ. Geol., v. 71, p. 1493–1506.

Hoeve, J., and Quirt, D., 1984, Mineralization and host rock alteration in relation to clay mineral diagenesis and evolution of the middle-Proterozoic Athabasca Basin, northern Saskatchewan, Canada, *in* Sask. Res. Counc. Tech. Report, no. 187, 187 p.

Hoeve, J., and Sibbald, T. I. I., 1978, On the genesis of Rabbit Lake and other unconformity-type uranium deposits in northern Saskatchewan, Canada: Econ. Geol., v. 73, p. 1450–1473.

Holliger, P., 1988, Ages U-Pb définis in-situ sur pitchblende à l' analyseur ionique: Comptes Rendus Acad. Sci. Paris., ser. 2, v. 307, p. 367–373.

______, 1991, SIMS isotope analyses of U and Pb in uranium oxides: Geological and nuclear applications, *in* Benninghoven, A., Janssen, K. T. F., Tumpner, J., and Werner, H. W., eds., 8th Int. SIMS Conf. Proc., p. 719–722.

______, 1994, SIMS studies on the Oklo natural fission reactors, *in* Benninghoven, A., Nihei, Y., Shimizu, R., and Werner, H. W., eds., 9th Int. SIMS Conf. Proc., p. 906–910.

Holliger, P., and Cathelineau, M., 1986, Le Chronometre U-Pb en milieu uranifère: Application aux gisements hydrothermaux d'uranium liés spatialement au batholith de Mortagne (Vendée, France): Chron. Rech. Min., v. 485, p. 33–43.

______, 1987, Ion microprobe lead isotopic measurements on U-ore minerals: Application for in-situ determined Pb-Pb apparent ages: Terra Cognita, v. 7, p. 230.

______, 1988, In situ U-Pb age determination by secondary ion mass sepctrometry: Chem. Geol., v. 70, p. 173.

Janeczek, J., and Ewing, R. C., 1995, Mechanisms of lead release from uraninite in natural fission reactors in Gabon: Geochim. et Cosmochim. Acta., v. 59, p. 1917–1931.

Kotzer, T. G., and Kyser, T. K., 1993, O, U, and Pb isotopic and chemical variations in uraninite: Implications for determining the temporal and fluid history of ancient terrains: Amer. Mineral., v. 78, p. 1262–1274.

______, 1995, Petrogenesis of the Proterozoic Athabasca Basin, northern Saskatchewan, Canada, and its relation to diagenesis, hydrothermal uranium mineralization, and paleohydrogeology: Chem. Geol., v. 120, p. 45–89.

Kotzer, T. G., Kyser, T. K., and Irving, E., 1992, Paleomagnetism and the evolution of the fluids in the Proterozoic Athabasca Basin, northern Saskatchewan, Canada: Can. Jour. Earth Sci., v. 29, p. 1474–1491.

Kyser, T. K., Kotzer, T. G., and Sibbald, T. I. I., 1990, Oxygen, U-Pb, and Pb-Pb isotope systematics in uraninite from complex U-Au-PGE vein-type and unconformity-type U deposits in northern Saskatchewan, *in* Summary of investigations 1990: Saskatoon, Sask. Energy and Mines, Sask. Geol. Surv., Misc. Report, 90-4, p. 64–69.

Kyser, T. K., Wilson, M. R., and Ruhlmann, G., 1988, Stable isotopic constraints on the role of graphite in the genesis of unconformity-type uranium deposits: Can. Jour. Earth Sci., v. 26, p. 490–498.

Ludwig, K. R., 1993, ISOPLOT. A plotting and regression program for radiogenic-isotope data: U.S. Geol. Surv., Open File Report, 91-445, p. 1–42.

Matzke, H., and Wang, L.-M., 1996, High-resolution transmission electron microscopy of ion-irradiated uranium oxide: Jour. Nucl. Mater., v. 231, p. 155–158.

Meddaugh, W. S., 1983, Age and origin of uraninite in the Elliot Lake, Ontario uranium ores: Unpubl. Ph.D. dissertation, Harvard University.

Nash, J. T., Granger, H. C., and Adams, S. S., 1981, Geology and concepts of genesis of important types of uranium deposits: Econ. Geol., 75th Anniv. Vol., p. 63–117.

Quideller, X., Grove, M., Lovera, O. M., Harrison, T. M., and Yin, A., 1997, Thermal evolution and slip history of the Renbu Zedong Thrust, southeast Tibet: Jour. Geophys. Res., v. 102-B2, p. 2659–2679.

Reyx, J., and Ruhlmann, F., 1993, Etude metallographie des differentes associations minerales et caracterisation chimique des mineraux uraniferes du gisement de Cigar Lake (Saskatchewan, Canada): Can. Jour. Earth Sci., v. 30, p. 705–719.

Smellie, J., and Karlsson, F., 1996, A reappraisal of some Cigar Lake issues of importance to performance assessment: Stockholm, SKB Tech. Rep. (TR 96-08).

Smith, D. K., 1984, Uranium mineralogy, *in* Uranium geochemistry, mineralogy, geology, exploration, and resources: London, UK, Inst. Mining Metallurg., p. 43–88.

Wilson, M. R., and Kyser, T. K., 1987, Stable isotope geochemistry of alteration associated with the Key Lake uranium deposit: Econ. Geol., v. 82, p. 1450–1557.

Yershov, V. M., 1974, A method of examining the diffusion parameters of lead in uranium minerals: Geokhimiya, v. 10, p. 1565–1568.